QUASIHOMOGENEOUS DISTRIBUTIONS

NORTH-HOLLAND MATHEMATICS STUDIES 165

(Continuation of the Notas de Matemática)

NORTH-HOLLAND – AMSTERDAM • NEW YORK • OXFORD • TOKYO

QUASIHOMOGENEOUS DISTRIBUTIONS

Olaf von GRUDZINSKI

Mathematisches Seminar der Universität Kiel
Kiel, Federal Republic of Germany

1991

NORTH-HOLLAND – AMSTERDAM • NEW YORK • OXFORD • TOKYO

ELSEVIER SCIENCE PUBLISHERS B.V.
Sara Burgerhartstraat 25
P.O. Box 211, 1000 AE Amsterdam, The Netherlands

Distributors for the U.S.A. and Canada:

ELSEVIER SCIENCE PUBLISHING COMPANY, INC.
655 Avenue of the Americas
New York, N.Y. 10010, U.S.A.

Library of Congress Cataloging-in-Publication Data

Grudzinski, Olaf von, 1947-
Quasihomogeneous distributions / Olaf von Grudzinski.
p. cm. -- (North-Holland mathematics studies ; 165)
Includes bibliographical references and index.
ISBN 0-444-88670-2
1. Theory of distributions (Functional analysis) I. Title.
II. Series.
QA324.G78 1991
515'.782--dc20 90-23029
CIP

ISBN: 0 444 88670 2

Printed in The Netherlands

Contents

Introduction

A (generalized) function (or distribution) $f : X \longrightarrow \mathbb{C}$ on an open subset X of $\mathbb{R}^n$ is called *quasihomogeneous of degree* $m \in \mathbb{C}$ *and of type* $p \in \mathbb{R}^n \setminus \{0\}$ if

$$f(M_t x) = t^m f(x), \qquad x \in X, \; t \in]0,+\infty[,$$

where

$$(*) \qquad M_t x := (t^{p_1} x_1, \ldots, t^{p_n} x_n) .$$

Moreover, a partial differential operator $P(\partial)$ with constant coefficients is called *quasihomogeneous of type* p if its defining polynomial P is quasihomogeneous of type p.

If $p_j = 1$ for every $j \in \{1, \ldots, n\}$ then, of course, f (resp. $P(\partial)$) is called *homogeneous.*

Many classical differential operators are homogeneous or quasihomogeneous. For example, the Laplacian Δ or the wave operator $\partial_t^2 - \Delta_x$ are homogeneous while the heat operator $\partial_t - \Delta_x$ or the Schrödinger operator $i\partial_t - \Delta_x$ are quasihomogeneous of type $(2,1,\ldots,1)$.

Problem: Let $P(\partial)$ be a quasihomogeneous partial differential operator with constant coefficients; given a quasihomogeneous distribution $T \in \mathcal{D}'(\mathbb{R}^n)$, does there exist a quasihomogeneous solution $S \in \mathcal{D}'(\mathbb{R}^n)$ of the equation

$$(**) \qquad P(\partial) S = T \; ?$$

In particular, when is the answer in the affirmative for every quasihomogeneous T? (in this case - for the moment - we say that *$P(\partial)$ has the quasihomogeneous solvability property).*

Choosing T to be the Dirac distribution δ (which is quasihomogeneous for every p) one sees that the problem includes the question for the existence of quasihomogeneous fundamental solutions. As is well-known, for many of the classical differential operators the answer to this more special question is in the affirmative. For example, the wave operator has homogeneous fundamental solutions, and the Laplacian does so if $n \geq 3$. Moreover, the heat operator and the Schrödinger operator have quasihomogeneous fundamental solutions. However, as is equally well-known, sometimes the answer is in the negative. For example, in case $n = 2$ the Laplacian does not have homogeneous fundamental solutions, and the same is valid for the iterated Laplacian Δ^k if n is even and not larger than $2k$.

A major object of the text is to present a solution to the quasihomogeneous solvability problem posed above to which the answer was not previously known even for homogeneous distributions. In fact, under the special assumption "$p \in]0, +\infty[^n$" necessary and sufficient conditions for solvability are given, involving the defining polynomial P only. Using these conditions one can easily deduce that special $P(\partial)$ happen to have the quasihomogeneous solvability property while others - even if they admit quasihomogeneous fundamental solutions - do not. For example, it will be shown that the Schrödinger operator has the quasihomogeneous solvability property while the heat operator does not.

In order to provide transparent proofs and to obtain better insight into the nature of the problem I made the solvability theorems part of a systematic exposition of the basic theory of quasihomogeneous functions and distributions on X. In particular, the problems of existence and regularity of quasihomogeneous functions and distributions are studied and presentation formulas are given. Also included is the solvability theory of equations of the form

$$(***) \qquad qS = T$$

for quasihomogeneous distributions S and T on X where q is any quasihomogeneous real analytic function on X. From this the solution of the solvability problem for quasihomogeneous partial differential equations with constant coefficients is obtained via the Fourier transform.

I now give a short outline of the contents of the individual chapters. The basic material on quasihomogeneous functions is collected in *Chapter 1*. In particular, quasihomogeneous polynomial functions as well as positive quasihomogeneous differentiable functions are studied. Likewise, *Chapter 2* is devoted to the basic definitions and properties of quasihomogeneous distributions. In particular, it is concerned with meromorphic functions with values in the set of quasihomogeneous distributions.

The central tool for most parts of the rest of the text is the so-called *method of taking quasihomogeneous averages* which for homogeneous distributions was introduced by Gårding [6] and the elements of which - again for the homogeneous case - are also found in Volume 1 of Hörmander's monograph [11] on linear partial differential equations.

The method works well if X is *locally p-bounded*, i.e. if there exist positive continuous (resp. C^{∞}) functions on X which are quasihomogeneous of degree 1. The corresponding theory is expounded in *Chapters 3* (for functions) *and 4* (for distributions). In particular, if X is locally p-bounded then the equation (***) turns out to be always solvable in the set of quasihomogeneous distributions.

However, the situation is not so nice in case X is not locally p-bounded (for example if $X = \mathbb{R}^n$). Then even for functions it is difficult to develop a sufficiently general theory of quasihomogeneous averages. A convenient way to do this relies on meromorphic extension techniques. They, however, require the assumption "$p \in [0,+\infty[^n$". The corresponding theory is elaborated in *Chapter 5*. It turns out that for certain exceptional numbers m the averages obtained are not always quasihomogeneous but only *almost quasihomogeneous* in a sense made precise in Chapters 1 and 2.

For distributions on open subsets X of $\mathbb{R}^n$ which are not locally p-bounded the quasihomogeneous averages are introduced in *Chapter 6*. Since they depend on the quasihomogeneous averages of test functions one has to adhere to the assumption "$p \in [0,+\infty[^n$". Moreover, the flaws already found for functions in Chapter 5 entail that for certain values of m sometimes only so-called *almost quasihomogeneous distributions* (introduced in Chapter 2) are obtained as the result of the averaging procedure.

In *Chapter 7* the results of the preceding chapters are applied to the theory of quasihomogeneous multiplication equations on $\mathbb{R}^n$ and - via the Fourier transform - to the theory of quasihomogeneous linear partial differential equations on $\mathbb{R}^n$.

As a consequence of the results of Chapter 6, there do not always exist quasihomogeneous, but in general only almost quasihomogeneous solutions. In particular, in general only almost quasihomogeneous fundamental solutions exist. Necessary and sufficient conditions on q and T are given for the existence of quasihomogeneous solutions S of the equation (***). Several examples are treated in detail. In particular, for the heat and for the Schrödinger operator all the quasihomogeneous fundamental solutions which are invariant under the action of the orthogonal group on the space variables are determined.

Chapter 8 deals - still under the assumption "$p \in [0,+\infty[^n$" - with another aspect of the method of taking quasihomogeneous averages, namely that it allows to extend quasihomogeneous distributions on $\mathbb{R}^n \setminus \{0\}$ to the whole of $\mathbb{R}^n$.

Finally, in *Chapter 9* - under the assumption "$p \in]0,+\infty[^n$" - the singularities of quasihomogeneous distributions on $\mathbb{R}^n$ are studied. The most suitable tools for describing them are the quasihomogeneous wave front sets introduced by R. Lascar and Rodino. In Chapter 9 the necessary parts from the theory of quasihomogeneous wave front sets are presented in a way that keeps close to the presentation of (homogeneous) wave front sets in Hörmander's monograph. In a similar spirit quasihomogeneous wave front sets with respect to Gevrey classes are introduced and employed for the description of singularities of quasihomogeneous distributions. As an example, for the heat and for the Schrödinger operator quasihomogeneous wave front sets of the invariant fundamental solutions determined in Chapter 7 are computed.

One observes that the map $]0,+\infty[\longrightarrow GL(n;\mathbb{R})$, $t \mapsto M_t$, is a continuous (or rather real analytic) representation of multiplicative groups. Among all such representations the ones defined by (*) are distinguished in that they are in real diagonal form. It turns out that nearly everything of the theory sketched above can be extended to cover generalized functions which are quasihomogeneous with respect to an arbitrary continuous representations of $]0,+\infty[$ in $GL(n;\mathbb{R})$. In particular, representations which are in complex diagonal form are admitted. The only drawback is that the formulations of some of the results and many a proof become much more complicated than in the real diagonal case (this is particularly so if M is not semi-simple). So, from the outset we are going to build the theory for an arbitrary continuous representation $t \mapsto M_t$ of $]0,+\infty[$ in $GL(n;\mathbb{R})$, denoting by M the infinitesimal generator of the representation and speaking of "*quasihomogeneity of type M*".

For more detailed expository information see the introductions to each of the chapters and sections below.

The main body of the text consists of material not previously published. It requires only a basic knowledge of distribution theory as expounded, for example, in the basic chapters of Volume 1 of Hörmander's monograph. Of course, it is presumed that the reader has a basic knowledge of linear algebra, of holomorphic functions of one complex variable, of real variable theory, of Lebesgue integration theory, as well as of the rudiments of locally convex vector spaces. Otherwise the text is self-contained. Complete proofs are given. In this way it becomes accessible to graduates or even to advanced undergraduates.

Acknowledgements. I thank my colleagues and friends Dieter Blessenohl, Rudolf Schnabel, and foremostly Volker Wrobel for helpful discussions and for their encouragement.

Notation

Usually, I stick to the standard notation of Hörmander [11]. In particular, I use the following conventions:

$\mathbb{N}_0 := \mathbb{N} \cup \{0\}$; $\quad \mathbb{N}_k := \{n \in \mathbb{N};\ n \le k\}$, $k \in \mathbb{N}$.

The members of the standard basis of $\mathbb{R}^n$ are denoted by e_j, i.e. $(e_j)_k := \delta_{jk}$, $j,k \in \mathbb{N}_n$, where δ_{jk} denotes the Kronecker symbol.

Moreover, if V is a finite dimensional normed real vector space then

$K(x,r)$ (resp. $\overline{K}(x,r)$) := open (resp. closed) ball in V of radius r centered at x;

S_V := unit sphere in V (also denoted by S^{n-1} in case $V = \mathbb{R}^n$);

If X is any subset of V then χ_X := characteristic function of X and $\dot{X} := X \setminus \{0\}$.

If W is another finite dimensional $\mathbb{R}$-vector space W then

$L(V,W)$:= the $\mathbb{R}$-vector space of all linear maps $A: V \longrightarrow W$;

$GL(V,W)$:= set of all invertible elements of $L(V,W)$;

$L(V) := L(V,V)$ and $GL(V) := GL(V,V)$.

For $A \in L(V)$ and for any A-invariant subspace U of V the endomorphism of U induced by A is denoted by A_U.

V^* := dual space $L(V,\mathbb{R})$ of V;

The *dual* (or *transposed*) of a map $A \in L(V,W)$ is denoted by $A^* \in L(W^*,V^*)$.

If V and W are $\mathbb{C}$-vector spaces then for $\mathbb{K} \in \{\mathbb{R},\mathbb{C}\}$ the set of $\mathbb{K}$-linear maps $T: V \longrightarrow W$ is denoted by $L_{\mathbb{K}}(V,W)$.

Finally, for $x \in V$ and $\nu \in V^*$ $\nu(x)$ is also written as $\langle \nu, x \rangle$. Likewise, the duality bracket between distributions T and test functions φ is also denoted by $\langle T, \varphi \rangle$.

As for the special notation introduced in the text, a fairly comprehensive *list of symbols* is included below.

List of Symbols

Chapter I

(Almost) Quasihomogeneous Functions. Definitions and Basic Properties

In section (a) of the present chapter we collect some basic facts on continuous representations of the multiplicative group $]0,+\infty[$ in a finite dimensional vector space V. In particular, for the rest of the whole text we fix such a representation, denoting its infinitesimal generator by M. The basic material on quasihomogeneous functions (with respect to this representation) is introduced in section (b) of this chapter. Section (c) is devoted to the treatment of quasihomogeneous polynomial functions on V. If M is semi-simple they are easily described by means of mixed real-complex coordinates which are basic for much of the other material, as well. If M is not semi-simple then the natural candidates for quasihomogeneous polynomial functions fail to be so but possess a more general type of invariance property, called *almost quasihomogeneity*. Section (d) contains some special examples of quasihomogeneous polynomial functions in case M is not semi-simple.

In section (e) general almost quasihomogeneous functions are introduced, and their basic properties are exhibited. The full motivation for the notion of almost quasihomogeneity will come up in Chapter 2. Section (f) is concerned with smooth positive functions $\varkappa$ which are quasihomogeneous of degree 1 and with the hypersurfaces $S^{\varkappa}:=\varkappa^{-1}(1)$ associated with them. The set $\mathfrak{K}(M)$ of points where functions $\varkappa$ as above exist locally is described in section (g). Finally, in section (h), by means of such functions $\varkappa$ so-called quasihomogeneous polar coordinates are introduced and a few applications are given.

There are several results of a more technical character which are required in later chapters. They can be skipped on a first reading.

Once and for all we fix $n \in \mathbb{N}$ and an n-dimensional $\mathbb{R}$-vector space V, its norm usually being denoted by $|\cdot|$. Of course, sometimes we identify V with $\mathbb{R}^n$. Eventually we work with a scalar product $\langle \cdot, \cdot\cdot \rangle$ on V which also comes into the picture implicitly when we identify V with $\mathbb{R}^n$ and automatically associate the scalar product defined with respect to the standard basis $(e_1, \ldots, e_n)$ of $\mathbb{R}^n$ by

$$\langle x, y \rangle := \sum_{j=1}^{n} x_j y_j .$$

Note that since V is finite-dimensional we can work with any norm when dealing with convergence, continuity and differentiability.

(a) Representations of the Multiplicative Group $]0,+\infty[$

In this section we collect basic information on continuous representations of $]0,+\infty[$ in V, i.e. homomorphisms of the multiplicative group $]0,+\infty[$ into GL(V).

Example 1.1. *If $M \in L(V)$ then by*

$$\rho_M(t) := \exp((\log t)\, M)$$

a real analytic representation $\rho_M :]0,+\infty[\longrightarrow GL(V)$ is well-defined. ∎

An important special case is when the matrix $(M_{jk})_{j,k \in \mathbb{N}_n}$ of M with respect to some basis of V has diagonal form, i.e. when we can identify V with $\mathbb{R}^n$ in such a way that for some $p \in \mathbb{R}^n \setminus \{0\}$

(1.1) (a) $(Mx)_j = p_j x_j$, so that (b) $(\rho_M(t)x)_j = t^{p_j} x_j$, $x \in \mathbb{R}^n$, $j \in \mathbb{N}_n$.

Proposition 1.2. *If $\rho :]0,+\infty[\longrightarrow GL(V)$ is a continuous group representation then there exists a (unique) $M \in L(V)$ such that $\rho = \rho_M$. In particular, ρ is real analytic, and $M = \rho'(1)$.*

Proof. This is, of course, a special case of a general result from the theory of Lie groups. A direct proof is as follows. The first step is to show that ρ is differentiable. Since GL(V) is an open neighbourhood of $\rho(1) = \mathrm{Id}_V$ and since ρ is continuous we can find $a>1$ such that

$$A := \int_1^a \rho(s)\, ds$$

is invertible. Making use of the homomorphism property of ρ and substituting $s' = ts$ we obtain for every $t \in]0,+\infty[$ that

$$\rho(t) = \int_1^a \rho(t)\rho(s)\, ds\, A^{-1} = \int_1^a \rho(ts)\, ds\, A^{-1} = \frac{1}{t}\int_t^{at} \rho(s')\, ds'\, A^{-1}.$$

Since ρ is continuous the right-hand side is differentiable as a function of t, indeed.

Defining a differentiable representation $\tilde{\rho}$ of the additive group $\mathbb{R}$ in V by

$$\text{(1.2)} \qquad \tilde{\rho}(s) := \rho(e^s), \qquad s \in \mathbb{R},$$

differentiating the equation $\tilde{\rho}(s+t) = \tilde{\rho}(s)\tilde{\rho}(t)$ with respect to s, and evaluating the result at $s = 0$ one obtains

$$\tilde{\rho}'(t) = M\tilde{\rho}(t) \qquad \text{where} \qquad M := \tilde{\rho}'(0) = \rho'(1).$$

Since $\tilde{\rho}_M$ is a solution of the same differential equation satisfying the same initial condition it follows that $\tilde{\rho} = \tilde{\rho}_M$. ∎

Corollary 1.3. *If $\chi :]0,+\infty[\longrightarrow \dot{\mathbb{C}}$ is a continuous homomorphism of multiplicative groups then χ is real analytic, and with $m := \chi'(1)$ we have*

$$\chi(t) = t^m, \qquad t \in]0,+\infty[.$$

Proof. We apply Proposition 1.2 to $\rho := \Phi \circ \chi$ where $\Phi : \dot{\mathbb{C}} \longrightarrow GL(2;\mathbb{R})$ is the real analytic homomorphism of multiplicative groups defined by

$$\text{(1.3)} \qquad \Phi(a+ib) := \begin{pmatrix} a & -b \\ b & a \end{pmatrix}, \qquad a,b \in \mathbb{R}. \ ∎$$

For the whole text we now fix a non-trivial map $M \in L(V)$ and an n-tuple $p \in \mathbb{R}^n \setminus \{0\}$ and work with the representation ρ_M sometimes discussing the case where M is of the form (1.1.a).

Notation 1.4. Usually instead of $\rho_M(t)$ we shall write M_t.

The following properties of M_t are immediately verified.

Remark 1.5. *For every $t \in]0,+\infty[$ the following formulas hold:*

(1.4) $$M \circ M_t = M_t \circ M;$$

(1.5) $$\rho'_M(t) = \frac{1}{t} M \rho_M(t);$$

(1.6) $$\det M_t = t^{\mu} \quad \text{where} \quad \mu := \operatorname{tr} M;$$

(1.7) $$(M_t)^* = (M^*)_t. \quad \blacksquare$$

In order to analyze ρ_M more thoroughly we employ decompositions of V into M-invariant subspaces of V. First of all we decompose V according to

(1.8) $$V = V_{\mathbb{R}} \oplus V_{\mathbb{C}}$$

where

(1.8.a) $$V_{\mathbb{R}} := \bigoplus_{\lambda \in \sigma_{\mathbb{R}}(M)} G_M(\lambda)$$

denotes the direct sum of the *generalized eigenspaces* $G_M(\lambda) := \bigcup_{j \in \mathbb{N}} \ker(M - \lambda \operatorname{Id}_V)^j$ of M associated with the eigenvalues λ in the *real spectrum*

$$\sigma_{\mathbb{R}} := \sigma_{\mathbb{R}}(M) := \{\lambda \in \mathbb{R} ;\ M - \lambda \operatorname{Id}_V \text{ is not injective}\}$$

and where $V_{\mathbb{C}}$ is the unique M-invariant complement of $V_{\mathbb{R}}$. Since the real spectrum of the linear endomorphism $M_{V_{\mathbb{C}}}$ of $V_{\mathbb{C}}$ induced by M is empty one can provide $V_{\mathbb{C}}$ with the structure of a $\mathbb{C}$-vector space in such a way that $M_{V_{\mathbb{C}}}$ becomes $\mathbb{C}$-linear. This complex structure is unique if we require, in addition, that the endomorphism of $V_{\mathbb{C}}$ defined by $x \mapsto ix$ is a polynomial in $M_{V_{\mathbb{C}}}$ and that $\operatorname{Im} \lambda > 0$ for every eigenvalue λ of $M_{V_{\mathbb{C}}}$. This implies, in particular, that every M-invariant subspace of $V_{\mathbb{C}}$ becomes a complex subspace of $V_{\mathbb{C}}$. From now on we consider $V_{\mathbb{C}}$ as equipped with this $\mathbb{C}$-structure. Since we have to work with the mixed real-complex structure on V we introduce a unifying notation:

Notation 1.6.A. (i) For every $\lambda \in \mathbb{C}$ we set $K_\lambda := \mathbb{R}$ (resp. $\mathbb{C}$) if $\lambda \in \mathbb{R}$ (resp. $\mathbb{C} \setminus \mathbb{R}$) and $\tilde{M}_\lambda := M_{V_{K_\lambda}} - \lambda \operatorname{Id}_{V_{K_\lambda}}$.

(ii) By $\sigma = \sigma(M)$ we denote the set of all $\lambda \in \mathbb{C}$ such that $\tilde{M}_\lambda$ is not injective. Note that σ is the disjoint union of $\sigma_{\mathbb{R}}(M)$ and $\sigma_{\mathbb{C}}(M) := \sigma(M_{V_{\mathbb{C}}})$.

(iii) For every eigenvalue $\lambda\in\sigma$ of M we denote by $E_M(\lambda) := \ker \tilde{M}_\lambda$ the so-called *eigenspace of M with respect to λ*.

(iv) For every $\lambda\in\sigma$ we denote by

$$G_M(\lambda) := \bigcup_{j\in\mathbb{N}} \ker(\tilde{M}_\lambda)^j$$

the so-called *generalized eigenspace of M with respect to λ*. In particular, we have

$$\text{(1.8.b)}\qquad V_{\mathbb{C}} = \bigoplus_{\lambda\in\sigma_{\mathbb{C}}(M)} G_M(\lambda)\,.$$

(v) For arbitrary $x\in V$ and $\lambda\in\sigma$ we denote by x_λ the spectral projection of x onto $G_M(\lambda)$, i.e. each $x\in V$ is (uniquely) decomposed according to

$$\text{(1.9)}\qquad x = \sum_{\lambda\in\sigma} x_\lambda \quad \text{where } x_\lambda\in G_M(\lambda)\,.$$

(vi) $dim' V := \dim_{\mathbb{R}} V_{\mathbb{R}} + \dim_{\mathbb{C}} V_{\mathbb{C}}$.

We fix $\lambda\in\sigma$ and observe that

$$\text{(1.10.a)}\qquad \big((\rho_M\circ\exp)(s)\big)_{G_M(\lambda)} = \exp(s\lambda)\, P_\lambda(s)\,, \qquad s\in\mathbb{R}\,,$$

where

$$\text{(1.10.b)}\qquad P_\lambda(s) := \sum_{j=0}^{m-1} \frac{1}{j!} s^j \left(\tilde{M}_\lambda\right)^j_{G_M(\lambda)} \qquad \text{for some } m\le \dim' G_\lambda(M)\,.$$

Each of the generalized eigenspaces $G_M(\lambda)$ is the direct sum of M-cyclic subspaces. If U is such an M-cyclic subspace of $G_M(\lambda)$ of K_λ-dimension $d\in\mathbb{N}$ we let $\mathfrak{B} = (b_1,\dots,b_d)$ be a K_λ-basis of U such that

$$\text{(1.11)}\qquad N b_1 = 0 \quad\text{and}\quad N b_{j+1} = b_j \text{ for } j\in\mathbb{N}_{d-1} \quad\text{where}\quad N := (\tilde{M}_\lambda)_U\,.$$

This means that the matrix of N with respect to $\mathfrak{B}$ is of the form

$$\text{(1.11)'}\qquad N_{jk} = \begin{cases} 1 & \text{if } k=j+1 \\ 0 & \text{otherwise} \end{cases}, \qquad j,k\in\mathbb{N}_d\,.$$

It follows that the matrix of $P_\lambda(s)_U$ with respect to $\mathfrak{B}$ is of the form

$$\text{(1.10.b)'}\qquad \big(P_\lambda(s)_U\big)_{jk} = \begin{cases} s^\ell/\ell! & \text{if } \ell := k-j\ge 0 \\ 0 & \text{if } j>k \end{cases}, \qquad j,k\in\mathbb{N}_d\,.$$

Note that one obtains the matrix of $((\rho_M\circ\exp)(s))_U$ by multiplying (1.10.b)' by

$\exp(s\lambda)$. Recall that in case $\lambda \in \sigma \setminus \mathbb{R}$ one obtains the real matrix with respect to the real basis $\mathfrak{B}_{\mathbb{R}} := (b_1, ib_1, \dots, b_d, ib_d)$ out of the complex matrix by substituting the 2×2 matrix $\begin{pmatrix} a & -b \\ b & a \end{pmatrix}$ for every complex number $a+ib$.

Lemma 1.7. *Let $\lambda \in \sigma$, and let U be an M-invariant subspace of $G_M(\lambda)$ of K_λ-dimension $d \in \mathbb{N}$. Then there is a constant $C>0$ such that*

$$\frac{1}{C}(1+|\log t|)^{-d+1}\, t^{\mathrm{Re}\lambda}|x| \le |M_t x| \le C\,(1+|\log t|)^{d-1}\, t^{\mathrm{Re}\lambda}|x| \tag{1.12}$$

for arbitrary $x \in U$ and $t \in]0,+\infty[$.

Proof. It suffices to prove the assertion when U is M-cyclic. In this case the discussion following (1.10.b)' shows that there is a constant C such that the inequality on the right-hand side of (1.12) is valid. An application of this estimate to $M_{1/t}x$ instead of x leads to the inequality on the left-hand side of (1.12). ∎

A simple calculation in the case $d=2$ shows that the estimates (1.12) are sharp. We now make use of (1.12) to study the behaviour of $M_t x$ if t approaches the boundary of $]0,+\infty[$. Here it will be useful to work with

Notation 1.6.B. (i) For every subset τ of σ we define

(a) $G_M(\tau) := \bigoplus_{\lambda\in\tau} G_M(\lambda)$, and (b) $E_M(\tau) := \bigoplus_{\lambda\in\tau} E_M(\lambda)$;

(ii) (a) $\sigma_\pm := \{\lambda \in \sigma;\ \pm\mathrm{Re}\lambda > 0\}$; and (b) $\sigma_0 := \{\lambda\in\sigma;\ \mathrm{Re}\lambda = 0\}$.

(iii) For every $\nu \in \{+,-,0\}$ we denote by $P_\nu : V \longrightarrow V$ the spectral projection onto $G_M(\sigma_\nu)$.

(iv) The spectral projection onto $G_M(0)$ is denoted by $M_0 : V \longrightarrow V$.

Remark 1.8. *Let $x \in \dot{V} := V\setminus\{0\}$ and $s \in \{0,+\infty\}$. Set $a := \begin{cases} + & \text{if } s=0 \\ - & \text{if } s=+\infty\end{cases}$. Then*

(i) $\lim_{t\to s} M_t x$ exists in V if and only if $x \in G_M(\sigma_a) + \ker M$; if this is the case then

$$\lim_{t\to s} M_t x = M_0 x\,; \tag{1.13}$$

more precisely, if N_t denotes the endomorphism induced by M_t on $G_M(\sigma_a)+\ker M$ then $\lim_{t\to s} \|N_t - N_0\| = 0$.

(ii) If $x \in G_M(\sigma_a) + E_M(\sigma_o)$ *then* $\lim_{t\to s} \mathrm{dist}(M_t x, E_M(\sigma_o)) = 0$, *the limit being uniform if* x *stays in any compact subset of* $G_M(\sigma_a) + E_M(\sigma_o)$.

(iii) The following conditions are equivalent:

(a) $\limsup_{t\to s} |M_t x| = +\infty$;

(b) $\lim_{t\to s} |M_t x| = +\infty$;

(c) $x \notin G_M(\sigma_a) + E_M(\sigma_o)$;

the limit in (b) is uniform if x *varies in any compact subset of the complement of* $G_M(\sigma_a \cup \sigma_o)$ *in* V.

Corollary 1.9. *The map* $[0,+\infty[\times V \longrightarrow V$, $(t,x) \mapsto M_t x$, *is continuous provided that*

$$V = G_M(\sigma_+) + \ker M . \tag{1.14}$$ ■

Proof of Remark 1.8. It suffices to give the proof for the case $s = 0$, for an application of this case to $-M$ instead of M settles the case $s = +\infty$.

We fix $\lambda \in \sigma$ and define $Q_\lambda(s) := P_\lambda(s) x_\lambda$, $s \in \mathbb{R}$, where P_λ is given by (1.10.b). Then, since each component (with respect to a fixed basis of $G_M(\lambda)$) of Q_λ is a polynomial with respect to s of degree smaller than m, Q_λ does not vanish identically if and only if $x_\lambda \neq 0$. Moreover, Q_λ is constant if and only if x_λ belongs to the eigenspace $E_M(\lambda)$. It follows that in case $\mathrm{Re}\,\lambda > 0$ (see also (1.12)) we have

$$\lim_{s\to -\infty} \|\exp(\lambda s) P_\lambda(s)\| = 0$$

and in case $\mathrm{Re}\,\lambda < 0$ and $x_\lambda \neq 0$

$$\lim_{s\to -\infty} |\exp(\lambda s) Q_\lambda(s)| = +\infty$$

where in view of (1.12) the limit is uniform if x_λ stays in any compact subset of $G_M(\lambda)\setminus\{0\}$. Hence we suppose that $\lambda \in i\mathbb{R}$. If $x_\lambda \notin E_\lambda(M)$ then $\deg Q_\lambda$ is positive so that

$$\lim_{s\to -\infty} |\exp(\lambda s) Q_\lambda(s)| = \lim_{s\to -\infty} |Q_\lambda(s)| = +\infty .$$

Here, again the limits are uniform if x_λ stays in such a compact subset of $G_M(\lambda)$ that the degree of Q_λ does not change and the leading coefficient stays away from zero.

If $x_\lambda \in E_\lambda(M)$ then $Q_\lambda(s) = x_\lambda$. If, in addition, $\lambda \neq 0$ then one observes that $\exp(s\lambda)$ does not converge as $s \to \pm\infty$. From this all the assertions follow. ■

Next we describe the topological structure of the so-called "*quasihomogeneous rays*", the orbits under the action of $]0,+\infty[$ induced by ρ_M on V.

Proposition 1.10. *Let $x \in V$. Then by $i_x :]0,+\infty[\longrightarrow V$, $t \mapsto M_t x$, a real analytic function is defined, and*

(i) the following conditions are equivalent:

(a) i_x is immersive at some (resp. every) point of $]0,+\infty[$;

(b) i_x is non-constant;

(c) $x \notin \ker M$.

(ii) The following conditions are equivalent:

(d) i_x induces a homeomorphism onto its image $R_M(x) := i_x(]0,+\infty[)$;

(d)' there is a non-compact closed subinterval I of $]0,+\infty[$ such that i_x induces a homeomorphism of I onto $i_x(I)$;

(e) x does not belong to the set $E_M(\sigma_0)$;

(f) $R_M(x)$ is unbounded.

(iii) The following conditions are equivalent:

(g) i_x is not injective;

(g)' there exists $r>1$ such that for every $k \in \mathbb{Z}$ we have $i_x(r^k t) = i_x(t)$, $t \in]0,+\infty[$, and, in particular, $i_x([r^k, r^{k+1}[) = R_M(x)$;

(h) $x \in E_M(\sigma_0)$, and $\rho := \{\lambda \in \sigma_0;\ x_\lambda \neq 0\} \subset iw\mathbb{Z}$ for some $w \in \mathbb{R}$;

(j) $R_M(x)$ is compact.

(iv.A) $R_M(x)$ is a real analytic submanifold of V of dimension 1 if and only if it is non-trivial and compact or unbounded.

(iv.B) If $R_M(x)$ is bounded but non-compact then the dimension d of the $\mathbb{Q}$-vector space generated by $\{\lambda \in \sigma;\ x_\lambda \neq 0\}$ is not smaller than 2, and $\overline{R_M(x)}$ is a real-analytic submanifold of V of dimension d and is equal to the closure of $i_x(I)$ where I is any non-compact closed subinterval of $]0,+\infty[$.

The following lemma contains a part of the proof.

Lemma 1.11. *Let τ be a finite subset of $\dot{\mathbb{R}}$ and denote by d the dimension of the $\mathbb{Q}$-vector space generated by τ. Then the closure of the image of the map*

$$\Gamma: \mathbb{R} \longrightarrow \mathbb{C}^{\tau}, \quad t \mapsto (\exp(i\nu t))_{\nu\in\tau}, \tag{1.15}$$

is a real analytic (compact) submanifold of $\mathbb{C}^{\tau}$ of dimension d; in fact, it is diffeomorphic to a quotient of the d-dimensional torus by a finite subgroup.
Moreover, the image of Γ is closed if and only if $d=1$. Finally, if J is an unbounded subinterval of $\mathbb{R}$ then $\Gamma(J)$ is dense in $\overline{\Gamma(\mathbb{R})}$; more generally, for every non-empty open subset U of $\overline{\Gamma(\mathbb{R})}$ there is a finite subset R of J such that for every $w\in\overline{\Gamma(\mathbb{R})}$ there is $r\in R$ such that $(w_\nu \exp(i\nu r))_{\nu\in\tau}$ lies in U.

Proof. From τ we select a basis $(\nu_1,\dots,\nu_d)$ of $\sum_{\nu\in\tau}\mathbb{Q}\nu$. The remaining $\nu\in\tau$ are denoted by $\nu_{d+1},\dots,\nu_c$ where $c:=|\tau|$. Then we can fix numbers $q_{k,j}\in\mathbb{Z}$ and $\ell_j\in\dot{\mathbb{Z}}$ such that

$$\nu_{d+k} = \sum_{j=1}^{d} q_{k,j}\mu_j \quad \text{where } \mu_j := \nu_j/\ell_j, \qquad k\in\mathbb{N}_{c-d}. \tag{1.16}$$

Denoting the *d-dimensional torus* by $T^d := (S^1)^d$ we define $f: T^d \longrightarrow \mathbb{C}^c$ by

$$f(z) := \Big(z_1^{\ell_1},\dots,z_d^{\ell_d}, \prod_{j=1}^{d} z_j^{q_{1,j}},\dots,\prod_{j=1}^{d} z_j^{q_{c-d,j}}\Big).$$

Then f is a real analytic immersion onto a compact subset K of $\mathbb{C}^c$. Moreover, it is a homomorphism of multiplicative groups where, of course, the multiplication is component-wise. Its kernel is contained in the product of the subgroups consisting of the roots of unity of order ℓ_j, $j\in\mathbb{N}_d$. It follows that K is a real analytic submanifold of $\mathbb{C}^c$ and that f is a local diffeomorphism onto K.

Now let J be an unbounded subinterval of $\mathbb{R}$. Since $(\mu_1,\dots,\mu_d)$ is linearly independent over $\mathbb{Q}$, as well, the image of J under the map $\gamma:\mathbb{R}\longrightarrow T^d$ defined by

$$\gamma(t) := (\exp(i\mu_j t))_{j\in\mathbb{N}_d}$$

is dense in T^d. More precisely, it is not difficult to prove that

$$\lim_{R\to\infty} \frac{1}{R}\int_0^R f(\gamma(t))\,dt = \int_{T^d} f(z)\,dz, \qquad f\in C^0(T^d,\mathbb{C}), \tag{1.17}$$

where *dz* denotes the normalized Haar measure on T^d; in fact, if f is the restriction to T^d of a polynomial function then the condition (1.17) is verified by direct computation, and since by the Stone-Weierstraß theorem these functions are uniformly dense in $C^0(T^d,\mathbb{C})$ the general case follows from the special one by approximation (see Arnold [1], § 11.C, D, E).

Now we fix a non-empty open subset W of T^d. We first suppose that J contains $[b,+\infty[$ for some $b>0$. Let $v\in T^d$. Denoting by v^{-1} the inverse of v with respect to the multiplication of T^d and applying (1.17) to non-negative continuous functions f with support contained in $v^{-1}W$ one finds a number $t_v\in[b,+\infty[$ such that $\gamma(t_v)\in v^{-1}W$. This implies, in particular, that the set $\gamma([b,+\infty[)$ is dense in T^d, indeed. More generally, since the functions $T^d\longrightarrow T^d$, $z\mapsto z\gamma(t_v)$, are continuous one finds an open neighbourhood Z_v of v in T^d such that $z\gamma(t_v)\in W$ for every $z\in Z_v$. Since T^d is compact one can select a finite subset F of T^d such that the sets Z_v, $v\in F$, cover the whole of T^d. Setting $R_W:=\{t_v;\ v\in F\}$ we conclude that $(z\gamma(R_W))\cap W\neq\emptyset$ for every $z\in T^d$.

In case J is bounded from above we apply the case treated above to $-\mu_j$ instead of μ_j and obtain a finite subset R_W of J having the properties above.

It follows that the image of J under $f\circ\gamma$ is dense in K. Note that by (1.16) we have

$$\prod_{j=1}^{d}(\exp(i\mu_j t))^{q_{k,j}} = \exp(i\sum_{j=1}^{d} q_{k,j}\mu_j t) = \exp(i\nu_{d+k}t), \qquad k\in\mathbb{N}_{c-d}.$$

This implies that $(f\circ\gamma)(t)=(\exp(i\nu_j t))_{j\in\mathbb{N}_c}$. Now, if U is a non-empty open subset of K then $W:=f^{-1}(U)$ is a non-empty open subset of T^d, and the set $R:=R_W$ found above has the property

$$(f(z)\Gamma(R))\cap U = f((z\gamma(R))\cap W)\neq\emptyset \quad \text{for every } z\in T^d.$$

Since $f(T^d)=K=\overline{\Gamma(\mathbb{R})}$ the last part of the assertion is proved.

For the proof of the last but one part we observe that the image of $f\circ\gamma$ is equal to K if and only if $d=1$. ■

Proof of Proposition 1.10. By the chain rule and by (1.5) and (1.4) we have

$$\text{(1.18)} \qquad i_x'(t)=\frac{1}{t}M_t M x=\frac{1}{t}M_t i_x'(1).$$

This shows that i_x is immersive at some point if and only if it is immersive at every point of $]0,+\infty[$. Moreover, the equivalence *(a)⟺(b)* follows.

In addition, we see that i_x is immersive at 1 if and only if $Mx\neq 0$, i.e. the equivalence *(a)$\Longleftrightarrow$(c)* is valid.

(e)$\Longleftrightarrow$(f): The condition (f) means that there is $s\in\{0,+\infty\}$ such that

$$\limsup_{t\to s} |i_x(t)| = +\infty .$$

By Remark 1.8.(iii) this is equivalent to (e).

(iv.B) and (j)$\Rightarrow$(h): Suppose that $R_M(x)$ is bounded. Then the implication "(e)$\Rightarrow$(f)" already proven above shows that $x\in E_M(\sigma_0)$. We may assume that $x\notin \ker M$ so that $\tau := (-i\rho)\setminus\{0\}$ is a non-empty subset of $\dot{\mathbb{R}}$. We define an injective linear map $\alpha:\mathbb{C}^\tau \longrightarrow E_M(\sigma\cap i\dot{\mathbb{R}})$ by $\alpha(z) := \sum_{\nu\in\tau} z_\nu x_{i\nu}$. If $\Gamma:\mathbb{R}\longrightarrow\mathbb{C}^\tau$ denotes the map of Lemma 1.11 then $i_x = x_0 + \alpha\circ\Gamma\circ\log$, and by Lemma 1.11 the closure N of $R_M(x)$ is a compact d-dimensional real analytic submanifold of V where d is the dimension of the $\mathbb{Q}$-vector space $\sum_{\nu\in\tau}\mathbb{Q}\nu$. Moreover, by Lemma 1.11 $R_M(x)$ is compact if and only if $d=1$, i.e. there is $\nu_0\in\tau$ such that every $\nu\in\tau$ is of the form $q\nu_0$ for some $q\in\mathbb{Q}$, i.e. for some $k\in\dot{\mathbb{Z}}$ the number $w := \nu_0/k$ has the property $\rho\subset iw\mathbb{Z}$.

(h)$\Rightarrow$(g): if ρ and w are as in condition (h) we may suppose that $w\neq 0$. Hence, setting $s = 2\pi/w\in\dot{\mathbb{R}}$ we have $\lambda s\in 2\pi i\mathbb{Z}$ for every $\lambda\in\rho$ so that

$$\exp(sM)x = \sum_{\lambda\in\rho}\exp(\lambda s)x_\lambda = x = \exp(0M)x ,$$

i.e. i_x is not injective.

(g)$\Rightarrow$(g'): If i_x is not injective we find $t_1, t_2\in]0,+\infty[$ such that $t_1<t_2$ and $i_x(t_1) = i_x(t_2)$, i.e. $i_x(r) = x$ where $r := t_2/t_1 > 1$. It follows that $i_x(r^k t) = i_x(t)$ for arbitrary $t\in]0,+\infty[$ and $k\in\mathbb{Z}$.

(g')$\Rightarrow$(j): as the image of the compact interval $[1,r]$ under the continuous map i_x the set $R_M(x)$ is compact, itself. Hence the proof of *(iii)* is complete.

(d)'$\Rightarrow$(e): we prove the contraposition. Hence we assume that $x\in E_M(\sigma_0)$ so that by the implication "(f)$\Rightarrow$(e)" $R_M(x)$ is bounded. If $R_M(x)$ is compact then the implication "(j)$\Rightarrow$(g)' already proved above shows that the restriction of i_x to any non-compact closed subinterval of $]0,+\infty[$ is not even injective so that (d)' is violated for trivial reasons in this case. Hence we may assume that $R_M(x)$ is non-compact. Then by the assertion (iv.B) for arbitrary $t\in\mathbb{R}$ and $s\in\{0,+\infty\}$ we find a sequence $(t_n)_{n\in\mathbb{N}}$ in $]0,+\infty[$ tending to s as $n\to\infty$ and satisfying $i_x(t) = \lim_{n\to\infty} i_x(t_n)$. Since (t_n) does not converge to t this means that the condition (d)' is violated.

(e)⇒(d): The condition (e) means that the union of the sets

$$\tau_{\pm} := \{\lambda \in \sigma_{\pm} ;\ x_\lambda \neq 0\} \quad\text{and}\quad \tau_0 := \{\lambda \in \sigma_0 ;\ x_\lambda \notin E_M(\lambda)\}$$

is non-empty.

We fix $s \in \{0, +\infty\}$ and define a as in Remark 1.8. If $\tau_0 \neq \emptyset$ or $\tau_{-a} \neq \emptyset$ then we deduce from Remark 1.8.(iii) that

$$\lim_{t\to s} |M_t x| = +\infty .$$

On the other hand, if $\tau_0 = \emptyset = \tau_{-a}$ then it follows by Remark 1.8.(ii) that

$$\lim_{t\to s} \operatorname{dist}(M_t x, E_M(\sigma_0)) = 0 .$$

Putting everything together we conclude that (e) implies that the map

$$i :]0, +\infty[\longrightarrow V \setminus E_M(\sigma_0), \quad t \mapsto M_t x ,$$

is proper. Since the contraposition of the implication "(g)⇒(h)" already proven above tells us that i is injective if (e) is valid the condition (d) follows. Since the implication "*(d)⇒(d)'*" is trivial the proof of *(ii)* is complete.

(iv.A): If $x \notin \ker M$ and if $R_M(x)$ is compact then in the proof of (iv.B) we saw that $R_M(x)$ is a one-dimensional real-analytic submanifold of V. Suppose now that $R_M(x)$ is non-compact. Then by (iii) and (i) i_x is injective and immersive. Consequently, i_x is a homeomorphism onto its image $R_M(x)$ if and only if i_x is a real-analytic embedding. Since the latter is the case if and only if $R_M(x)$ is a real-analytic submanifold of V the assertion follows in view of the equivalence "(d)⟺(f)". ∎

Corollary 1.12. *(i) $\rho_M(]0,+\infty[)$ is bounded if and only if $\sigma \subset i\mathbb{R}$ and $E_M(\sigma) = V$.*

(ii) $\rho_M(]0,+\infty[)$ is compact if and only if $E_M(\sigma) = V$ and $\sigma \subset iw\mathbb{Z}$ for some $w \in \mathbb{R}$. If this is the case then there exists a real analytic representation $\hat{\rho}_M : S^1 \longrightarrow GL(V)$ of multiplicative groups and a positive number r such that

$$(1.19) \qquad \rho_M = \hat{\rho}_M \circ \psi_r \quad \text{where} \quad \psi_r(t) := t^{ir} .$$

Note that $\psi_r :]0,+\infty[\longrightarrow S^1$ is a real analytic homomorphism of multiplicative groups and a local diffeomorphism. ∎

(b) Quasihomogeneous Functions

The fact that $\rho_M :]0,+\infty[\longrightarrow GL(V)$, $t \mapsto M_t$, is a homomorphism of multiplicative groups implies that if $f : X \longrightarrow \mathbb{C}$ and $\chi :]0,+\infty[\longrightarrow \mathbb{C}$ are functions such that $f \circ M_t = \chi(t) f$, $t \in]0,+\infty[$, then χ is a homomorphism of multiplicative groups provided that f does not vanish identically. If, in addition, f is continuous then χ is continuous, as well, and hence by Corollary 1.3 of the form $\chi(t) = t^m$ for some $m \in \mathbb{C}$. So besides a function $f : X \longrightarrow \mathbb{C}$ we fix a complex number $m \in \mathbb{C}$ and define

Definition 1.13. f is said to be *quasihomogeneous of degree m (and of type M)* if and only if $f(M_t x) = t^m f(x)$ for arbitrary (x,t) belonging to the set

(1.20) $$X \times_M]0,+\infty[:= \{ (x,t) \in X \times]0,+\infty[;\ M_t x \in X \} .$$

If M is of the form (1.1.a) then here and in all the following similar definitions we say *"of type p"* instead of *"of type M"*.

First examples of quasihomogeneous functions are provided by polynomials (see section (c)). Whether or not positive functions on X exist which are quasihomogeneous of degree 1 is a fundamental question that will be dealt with in sections (f) and (g) and again in section 3.(c) below. For the present we are going to give some basic results showing how to obtain new quasihomogeneous functions out of given ones.

Remark 1.14. *Let W be an M-invariant subspace of V. Then the following assertions hold:*

(i) If f is quasihomogeneous of degree m and of type M then $f|_W$ is quasihomogeneous of degree m and of type M_W where by definition M_W is the element of L(W) induced by M.

(ii) Suppose that f is of the form $f = F \circ \pi_W$ where F is a function from V/W into $\mathbb{C}$ and where $\pi_W : V \longrightarrow V/W$ is the canonical projection. Then f is quasihomogeneous of degree m and of type M if and only if F is quasihomogeneous

of degree m and of type M^W *where* M^W *is by definition the element of* $L(V/W)$ *induced by M.*

Proof. *(i)*: this follows from $f|_W \circ (M_W)_t = f|_W \circ (M_t)_W = f \circ M_t|_W$.

(ii): this is a consequence of $f \circ M_t = F \circ (\pi_W \circ M_t) = F \circ (M^W)_t \circ \pi_W$. ■

The following assertion is a direct consequence of Definition 1.13 .

Remark 1.15. *If* $f: X \longrightarrow \mathbb{C}$ *is quasihomogeneous of degree m and* $q: X \longrightarrow \mathbb{C}$ *is quasihomogeneous of degree* $\ell \in \mathbb{C}$ *then* qf *is quasihomogeneous of degree* $m+\ell$. ■

Next we deal with linear differential operators preserving quasihomogeneity. It is convenient to describe them independently of the choice of special coordinates on V. To do this we have to look upon X as a "linear manifold" and upon $X \times V^*$ as its cotangent bundle. Namely, fixing any basis $\mathfrak{B} = (b_1, \ldots, b_n)$ of V we define linear charts $\mathfrak{k}_{\mathfrak{B}}: \mathbb{R}^n \longrightarrow V$ by $y \mapsto \sum_{j=1}^n y_j b_j$ and $\Phi_{\mathfrak{B}}: \mathbb{R}^n \times \mathbb{R}^n \longrightarrow V \times V^*$ by $\mathfrak{k}_{\mathfrak{B}} \times \mathfrak{k}_{\mathfrak{B}^*}$, i.e.

$$\Phi_{\mathfrak{B}}(y, \eta) := \Big(\sum_{j=1}^n y_j b_j , \sum_{j=1}^n \eta_j \beta^j \Big)$$

where by $\mathfrak{B}^* := (\beta^1, \ldots, \beta^n)$ we denote the basis of V^* dual to $\mathfrak{B}$. Note that if $\mathfrak{C}$ is another basis of V then

$$\Psi := \Phi_{\mathfrak{C}}^{-1} \circ \Phi_{\mathfrak{B}} \text{ equals } B^{-1} \times B^* \quad \text{where} \quad B := \mathfrak{k}_{\mathfrak{B}}^{-1} \circ \mathfrak{k}_{\mathfrak{C}} \tag{1.21}$$

and where here B^* denotes the adjoint of B with respect to the canonical scalar product on $\mathbb{R}^n$. To introduce the data defining differential operators we fix $r \in \mathbb{N}_0$ and define

Definition 1.16.A. A function $P: X \times V^* \longrightarrow \mathbb{C}$ is said to be a *copolynomial function on X of degree* $\leq r$ if and only if it is a polynomial of degree $\leq r$ with respect to the second variable, i.e. it is a C^∞ function with respect to the second variable such that $D_2^{r+1} P \equiv 0$.

Of course, in local coordinates this more explicitly means that $P_{\mathfrak{B}} := P \circ \Phi_{\mathfrak{B}}$ is a function $R: Y \times \mathbb{R}^n \longrightarrow \mathbb{C}$ of the form

$$R = \sum_{|\alpha| \leq r} R_\alpha \otimes G_\alpha \tag{1.22}$$

where $G_\alpha(\eta) := \eta^\alpha$, where $Y = X_{\mathfrak{B}} := (\mathfrak{k}_{\mathfrak{B}})^{-1}(X)$ and where $R_\alpha : Y \longrightarrow \mathbb{C}$ is given by $R_\alpha(y) = P_\alpha^{\mathfrak{B}}(y) := \frac{1}{\alpha!}(\partial_\eta^\alpha P_{\mathfrak{B}})(y,0)$. If $\mathfrak{C}$ is another basis of V then in view of (1.21) it follows by the chain rule that $P_\beta^{\mathfrak{C}}$ is a linear combination of the functions $P_\alpha^{\mathfrak{B}} \circ B$. This shows that the following definition is independent of the choice of $\mathfrak{B}$.

Definition 1.16.B. Let $\ell \in \mathbb{N}_0 \cup \{\infty\}$. P is called a *$C^\ell$-copolynomial function on X of degree $\leq r$* if and only if it is a copolynomial function on X of degree $\leq r$ such that its coefficients are C^ℓ functions, i.e.: the functions $R_\alpha = P_\alpha^{\mathfrak{B}}$ in (1.22) are C^ℓ functions.

Now we fix a copolynomial function P on X of degree $\leq r$ and for any $f \in C^r(X)$ define $P(x,\partial) f$ by

(1.23) $$(P(x,\partial) f) \circ \mathfrak{k}_{\mathfrak{B}} := P_{\mathfrak{B}}(y,\partial_y)(f \circ \mathfrak{k}_{\mathfrak{B}}) \qquad \text{where} \quad P_{\mathfrak{B}} := P \circ \Phi_{\mathfrak{B}}.$$

Remark 1.17. *Under the preceding assumptions, by (1.23) a linear differential operator $P(x,\partial)$ is well-defined independently of the choice of the basis $\mathfrak{B}$. For every C^r-function $f : X \longrightarrow \mathbb{C}$ and every $A \in GL(V)$ we have*

(1.24) $$P(x,\partial)(f \circ A) = [(P \circ (A^{-1} \times A^*))(x,\partial) f] \circ A .$$

Proof. First of all we are going to show that for any open subset Y of $\mathbb{R}^n$, any C^r function $g : Y \longrightarrow \mathbb{C}$, and any polynomial Q of degree $\leq r$ we have

(1.24)' $$Q(\partial)(g \circ B) = ((Q \circ B^*)(\partial) g) \circ B, \qquad B \in L(\mathbb{R}^n),$$

where here as in (1.21) we denote by $B^* \in L(\mathbb{R}^n)$ the adjoint of B with respect to the canonical scalar product on $\mathbb{R}^n$. Indeed, if Q and R are polynomials of degree $\leq r$ satisfying (1.24)' then the polynomials $Q + R$ and - if $\deg QR \leq r$ - QR satisfy (1.24)', as well. Hence, we may suppose that $Q = X_j$ for some $j \in \mathbb{N}_n$. But in this case (1.24)' is evident from the chain rule.

Next, for every function $R : Y \times \mathbb{R}^n \longrightarrow \mathbb{C}$ of the form (1.22) we deduce from (1.24)'

(1.24)'' $$R(y,\partial_y)(g \circ B) = \sum_{|\alpha| \leq r} R_\alpha \partial_y^\alpha (g \circ B) = \Big(\sum_{|\alpha| \leq r} R_\alpha \circ B^{-1} (G_\alpha \circ B^*)(\partial_y) g \Big) \circ B =$$
$$= \Big((R \circ (B^{-1} \times B^*))(y,\partial_y) g \Big) \circ B , \qquad g \in C^r(Y), \; B \in GL(\mathbb{R}^n).$$

Now, if $\mathfrak{C}$ is another basis of V then applying (1.24)" to $g=f\circ \mathfrak{k}_{\mathfrak{B}}$ and taking the equalities (1.21) and $P_{\mathfrak{C}}\circ\Psi = P_{\mathfrak{B}}$ into account we deduce that

$$P_{\mathfrak{C}}(z,\partial_z)(f\circ \mathfrak{k}_{\mathfrak{C}}) = [P_{\mathfrak{B}}(y,\partial_y)(f\circ \mathfrak{k}_{\mathfrak{B}})]\circ B,$$

i.e. the definition of $P(x,\partial)f$ does not depend on the special choice of $\mathfrak{B}$.

Finally, applying (1.24)" to $g:=f\circ \mathfrak{k}_{\mathfrak{B}}$ and $B:=(\mathfrak{k}_{\mathfrak{B}})^{-1}\circ A\circ \mathfrak{k}_{\mathfrak{B}}$ and taking the equalities $\Phi_{\mathfrak{B}}^{-1}\circ(A^{-1}\times A^*)\circ\Phi_{\mathfrak{B}} = B^{-1}\times B^*$ into account we conclude that

$$[P(x,\partial)(f\circ A)]\circ \mathfrak{k}_{\mathfrak{B}} = P_{\mathfrak{B}}(y,\partial_y)(g\circ B) =$$

$$= [(P_{\mathfrak{B}}\circ(B^{-1}\times B^*))(y,\partial_y)g]\circ B = [(P\circ(A^{-1}\times A^*))(x,\partial)f]\circ A\circ \mathfrak{k}_{\mathfrak{B}}. \quad \blacksquare$$

Note that the defining copolynomial function P can be reconstructed from the differential operator $P(x,\partial)$ via

(1.25) $$P(\cdot,\xi) = e^{-\langle\xi,\cdot\rangle}P(x,\partial)e^{\langle\xi,\cdot\rangle}, \qquad \xi\in V^*.$$

In Chapter 5 we shall also have to deal with separate differentiability properties with respect to a prescribed partition of the argument variables. In order to prepare this we fix two real subspaces V_1 and V_2 such that

(1.26) $$V = V_1\oplus V_2 .$$

Note that if for every $j\in\{1,2\}$ we denote by $\pi_j : V\longrightarrow V_j$ the canonical projection defined by (1.26) and identify V_j^* with the subspace $\{\nu\circ\pi_j ; \nu\in V^*\}$ then

(1.26)* $$V^* = V_1^*\oplus V_2^* .$$

Moreover, if $A\in L(V,V)$ is such that the subspaces V_1 and V_2 are invariant under A then A commutes with π_1 and π_2, and V_1^* and V_2^* are invariant under A^*. Finally, we fix $r,s\in\mathbb{N}_0\cup\{\infty\}$ and define

Definition 1.18. (i) By $C^{r,s}(X)$ we denote the space of all functions $f:X\longrightarrow\mathbb{C}$ such that $D_1^jD_2^kf$ exists and is continuous for arbitrary $j,k\in\mathbb{N}_0$ such that $j\le r$ and $k\le s$ (here the partial derivatives are taken with respect to the deomposition (1.26)).

(ii) A polynomial function $P:V^*\longrightarrow\mathbb{C}$ is said to be *of degree* $\le (r,s)$ if and only if the derivatives $D_1^{r+1}P$ and $D_2^{s+1}P$ vanish identically (here the partial derivatives are taken with respect to the decomposition (1.26)*).

(iii) A copolynomial function P on X is said to be *of degree* $\leq (r,s)$ if and only if $P(x,\cdot)$ is of degree $\leq (r,s)$ for every $x \in X$.

So a function $f: X \longrightarrow \mathbb{C}$ belongs to $C^{r,s}(X)$ if and only if $P(\partial)f$ is well-defined and continuous for every polynomial function $P: V^* \longrightarrow \mathbb{C}$ of degree $\leq (r,s)$. Of course, this can be expressed in coordinates as follows: if $\mathfrak{B} = (b_1, \ldots, b_n)$ is any basis of V such that

(1.27) $\quad (b_1, \ldots, b_d)$ is a basis of V_1 where $d := \dim V_1$, and $(b_{d+1}, \ldots, b_n)$ is a basis of V_2

then f belongs to $C^{r,s}(X)$ if and only if $\partial^\alpha f$ is well-defined and continuous for every α belonging to the set

(1.28) $$\mathbb{N}_o^n(r,s) := \{\alpha \in \mathbb{N}_o^n \,;\, |\alpha'| \leq r \text{ and } |\alpha''| \leq s\}$$

where we set $\alpha' := (\alpha_1, \ldots, \alpha_d)$ and $\alpha'' := (\alpha_{d+1}, \ldots, \alpha_n)$. Moreover, a copolynomial function P on X is of degree $\leq (r,s)$ if and only if the functions $R_\alpha = P_\alpha^{\mathfrak{B}}$ vanish identically if α does not belong to the set (1.28).

Remark 1.17'. *Let P be a copolynomial function on X of degree* $\leq (r,s)$. *Then for any* $f \in C^{r,s}(X)$ *by (1.23) a function* $P(x,\partial)f$ *on X is well-defined if* $\mathfrak{B} = (b_1, \ldots, b_n)$ *is chosen to be any basis of V satisfying (1.27).*

Moreover, if $A \in GL(V,V)$ *is such that* V_1 *and* V_2 *are A-invariant then* $P \circ (A^{-1} \times A^*)$ *is a copolynomial function on X of degree* $\leq (r,s)$, *as well, and the equation (1.24) is valid.*

Proof. In view of what was said in the text subsequent to Definition 1.18 one obtains the assertions either by following the proof of Remark 1.17 or by keeping the variables in V_1 (resp. V_2) fixed and applying Remark 1.17 with (V,A) replaced by (V_2, A_{V_2}) (resp. (V_1, A_{V_1})). ∎

We now describe the behaviour of quasihomogeneous functions under differentiation. For this we once and for all assume that the subspaces V_1 and V_2 in (1.26) are invariant under M and hence under M_t for every $t \in]0,+\infty[$.

Proposition 1.19. *Let $r \in \mathbb{N}_0$ and $\ell \in \mathbb{C}$, and let $P: X \times V^* \longrightarrow \mathbb{C}$ be a copolynomial function on X of degree $\le r$ which is quasihomogeneous of degree ℓ and of type $M \times (-M^*)$. Then for every C^r function $f: X \longrightarrow \mathbb{C}$ which is quasihomogeneous of degree m the function $P(x,\partial)f$ is quasihomogeneous of degree $m+\ell$.*

The assertion remains valid if P is a copolynomial function of degree $\le (r,s)$ with respect to the decomposition $(1.26)^*$ and if f belongs to $C^{r,s}(X)$.

Proof. Employing (1.24) and making use of the assumptions about f and P we deduce

$$(P(x,\partial)f) \circ M_t = (P(x,\partial)(f \circ M_t \circ M_{1/t})) \circ M_t =$$

$$= [P \circ (M_t \times M_{1/t}^*)](x,\partial)(f \circ M_t) = t^{\ell+m} P(x,\partial)f. \quad \blacksquare$$

Note that Remark 1.15 is a special case of Proposition 1.19. Another special case is

Corollary 1.20. *Let P be a polynomial function on V^* which is quasihomogeneous of degree $\ell \in \mathbb{C}$ and of type M^*, and suppose that f is a C^r function where $r := \deg P$. If f is quasihomogeneous of degree m then $P(\partial)f$ is quasihomogeneous of degree $m-\ell$.* ■

Next we introduce an important special operator, namely the directional derivative into the direction of the *quasihomogeneous rays* $R_M(x) := \{M_t x;\, t \in]0,+\infty[\}$, $x \in V$: the so-called *Euler operator with respect to M*. We denote the total derivative of a differentiable function f by Df.

Example 1.21. *The differential operator ∂_M defined by*

$$(1.29) \qquad (\partial_M f)(x) := Df(x) \cdot Mx, \qquad f \in C^1(X),\ x \in X,$$

is equal to $P_M(x,\partial)$ where the polynomial function $P_M: V \times V^* \longrightarrow \mathbb{R}$ is defined by $(x,\xi) \mapsto \langle \xi, Mx \rangle$. Note that P_M is quasihomogeneous of degree 0 and of type $M \times (-M^*)$. In coordinates the defining equation (1.29) reads as

$$(1.29)' \qquad \partial_M := \sum_{j=1}^{n} \Big(\sum_{k=1}^{n} M_{jk} x_k \Big) \partial_j . \quad \blacksquare$$

Proposition 1.19 shows that for every quasihomogeneous C^1 function f the function $\partial_M f$ is quasihomogeneous of the same degree. This, however, is also a trivial consequence of the following proposition which establishes *Euler's equation* for quasihomogeneous functions.

Proposition 1.22. *Suppose that $f: X \longrightarrow \mathbb{C}$ is differentiable.*

(i) If f is quasihomogeneous of degree m then $\partial_M f = mf$.

(ii) The converse implication is valid provided that for every $x \in X$ the set $\{t \in]0,+\infty[;\ M_t x \in X\}$ is an interval.

Proof. Setting $g(t,x) := t^{-m} f(M_t x)$ we obtain by (1.5) that

$$\partial_t g(t,x) = -m t^{-m-1} f(M_t x) + t^{-m} Df(M_t x)\cdot(\rho_M'(t)x) = t^{-m-1}(\partial_M f - mf)(M_t x).$$

From this the assertions are immediately deduced. ∎

We shall see in Proposition 1.58 below that under special assumptions on X and M the only differentiable quasihomogeneous functions on X are polynomial functions. These are studied in the following section.

(c) (Almost) Quasihomogeneous Polynomial Functions

In this section we describe the behaviour of arbitrary polynomial functions on V under composition with M_t. In particular, our aim is to determine the quasihomogeneous polynomial functions. If $V = \mathbb{R}^n$ and M is of the form (1.1.a) this is very easy as we shall first see.

Example 1.23. *Let $\alpha \in \mathbb{N}_o^n$. Then the monomial function $\mathbb{R}^n \ni x \mapsto x^\alpha$ is quasihomogeneous of type p and of degree*

$$\langle \alpha, p\rangle := \sum_{j=1}^{n} \alpha_j p_j . \quad ∎$$

If $P: \mathbb{R}^n \longrightarrow \mathbb{C}$ is a polynomial function we define by

$$Q_m P(x) := \sum_{\langle\alpha,p\rangle = m} \frac{1}{\alpha!} P^{(\alpha)}(0)\, x^\alpha , \qquad x \in \mathbb{R}^n,$$

the so-called *quasihomogeneous part* $Q_m P$ *of* P *of degree* m *(and of type* p*).* Of course, $Q_m P = 0$ if m does not belong to the set $Z(p) := \{\langle\alpha,p\rangle;\ \alpha \in Z^n\}$ for $Z = \mathbb{N}_0$. Taylor's formula shows that P is decomposed into its quasihomogeneous parts according to

$$P = \sum_{\ell \in \mathbb{N}_0(p)} Q_\ell P. \tag{1.30}$$

Moreover, it follows that P is quasihomogeneous of degree m if and only if P equals $Q_m P$.

If M has complex eigenvalues the above can be carried over if all the eigenvalues are *simple*, i.e. of algebraic multiplicity 1. For this it is appropriate to work with complex coordinates on $V_{\mathbb{C}}$ as fixed in

Convention 1.24.A. Set $d := \dim_{\mathbb{R}} V_{\mathbb{R}}$ and $c := \dim_{\mathbb{C}} V_{\mathbb{C}}$, and let $A = (a_1, \ldots, a_d)$ be an $\mathbb{R}$-basis of $V_{\mathbb{R}}$ and $B = (b_1, \ldots, b_c)$ be a $\mathbb{C}$-basis of $V_{\mathbb{C}}$. So via these bases one can identify $V_{\mathbb{R}}$ with $\mathbb{R}^d$ and $V_{\mathbb{C}}$ with $\mathbb{C}^c$. The (d+c)-tuple $(a_1, \ldots, a_d, b_1, \ldots, b_c)$ is called a *real-complex basis of V with respect to M*. By $(\eta_1, \ldots, \eta_{d+c})$ we denote the (d+c)-tuple of coefficient functionals associated with this basis. Note that $\eta_1, \ldots, \eta_c$ are real-valued and $\eta_{d+1}, \ldots, \eta_{d+c}$ are complex-valued. If $y = (\eta_1(x), \ldots, \eta_d(x))$ and $z = (\eta_{d+1}(x), \ldots, \eta_{d+c}(x))$ we also say that (y,z) are *the real-complex coordinates of* x.

Note that if we write $z = u + iv$ then $(y_1, \ldots, y_d, u_1, \ldots, u_c, v_1, \ldots, v_c)$ are the real coordinates of x with respect to the $\mathbb{R}$-basis $(a_1, \ldots, a_d, b_1, \ldots, b_c, ib_1, \ldots, ib_c)$ of V. Using real-complex coordinates we introduce special polynomial functions:

Notation 1.25.A. (i) $\mathfrak{A} := \mathbb{N}_0^d \times \mathbb{N}_0^c \times \mathbb{N}_0^c$; the $\alpha \in \mathfrak{A}$ are written as $\alpha = (\beta,\gamma,\delta)$ where $\beta \in \mathbb{N}_0^d$ and $\gamma, \delta \in \mathbb{N}_0^c$.

(ii) $x^\alpha := y^\beta z^\gamma \bar{z}^\delta$ for every $\alpha = (\beta,\gamma,\delta) \in \mathfrak{A}$ and every $x \in V$ with real-complex coordinates (y,z), i.e. more explicitly, $x^\alpha = \eta^\alpha(x)$ where

$$\eta^\alpha := \prod_{j=1}^{d} (\eta_j)^{\beta_j} \prod_{j=1}^{c} (\eta_{d+j})^{\gamma_j} (\overline{\eta_{d+j}})^{\delta_j};$$

mostly we shall denote the function η^α by x^α.

(iii) If $x \in V$ has real-complex coordinates (y,z) we write $x_j := y_j$ for $j \in \mathbb{N}_d$, $x_{d+j} := z_j$ and $x_{d+c+j} := \overline{z_j}$ for $j \in \mathbb{N}_c$, thereby looking upon $(x_1, \ldots, x_n)$ as some sort of *pseudo-real* coordinates of x.

Using this notation one can conveniently write down the action of real linear maps on V with the help of real-complex coordinates: Let H be a complex Banach space, let $T: V \longrightarrow H$ be an $\mathbb{R}$-linear map, and let $\xi \in V$; writing

$$\xi = \sum_{j=1}^{d} \xi_j a_j + \sum_{j=1}^{c} \mathrm{Re}(\xi_{j+d})\, b_j + \sum_{j=1}^{c} \mathrm{Im}(\xi_{j+d})\,(ib_j)$$

and noting that

$$\mathrm{Re}(\xi_{j+d}) = \frac{1}{2}(\xi_{j+d} + \xi_{j+d+c}) \quad \text{and} \quad \mathrm{Im}(\xi_{j+d}) = \frac{1}{2i}(\xi_{j+d} - \xi_{j+d+c})$$

we obtain

(1.31)
$$T\cdot\xi = \sum_{j=1}^{d} \xi_j T(a_j) + \sum_{j=1}^{c} \mathrm{Re}(\xi_{j+d})\, T(b_j) + \sum_{j=1}^{c} \mathrm{Im}(\xi_{j+d})\, T(ib_j) =$$
$$= \sum_{j=1}^{d} \xi_j T(a_j) + \sum_{j=1}^{c} \xi_{j+d}\left(\frac{1}{2}T(b_j) - \frac{i}{2}T(ib_j)\right) + \sum_{j=1}^{c} \xi_{j+d+c}\left(\frac{1}{2}T(b_j) + \frac{i}{2}T(ib_j)\right).$$

Now, if $g: X \longrightarrow H$ is a differentiable function one can apply (1.31) to $T = Dg(x)$ and obtains

(1.32)
$$Dg(x)\cdot\xi = \sum_{j=1}^{n} \xi_j (\boldsymbol{\partial}_j f)(x) = \sum_{\substack{\alpha\in\mathfrak{A} \\ |\alpha|=1}} \xi^{\alpha} (\boldsymbol{\partial}^{\alpha} f)(x), \qquad x\in X,\ \xi\in V,$$

where the derivatives $\boldsymbol{\partial}_j$ and $\boldsymbol{\partial}^{\alpha}$ are defined according to

Notation 1.25.B. (i) $\boldsymbol{\partial}^{\alpha} := \partial_y^{\beta}\partial_z^{\gamma}\partial_{\bar{z}}^{\delta}$, $\alpha = (\beta,\gamma,\delta) \in \mathfrak{A}$, where $\partial_{z_j} := \frac{1}{2}(\partial_{u_j} - i\partial_{v_j})$ and $\partial_{\overline{z_j}} := \frac{1}{2}(\partial_{u_j} + i\partial_{v_j})$ (here in $\boldsymbol{\partial}^{\alpha}$ we print the symbol ∂ in bold face in order to distinguish $\boldsymbol{\partial}^{\alpha}$ from the real derivative ∂^{α} for $\alpha \in \mathbb{N}_0^n$ - compare also the proof of Remark 1.43 below); instead of $\boldsymbol{\partial}^{\alpha} f$ we also write $f^{(\alpha)}$.

(ii) if $j\in\mathbb{N}_d$ we write $\boldsymbol{\partial}_j := \partial_{y_j}$; if $j \in \mathbb{N}_c$ we write $\boldsymbol{\partial}_{d+j} := \partial_{z_j}$ and $\boldsymbol{\partial}_{d+c+j} := \partial_{\overline{z_j}}$.

With the help of the notation introduced above Taylor's formula can be written in the usual form:

Lemma 1.26. *Let Ω be an open subset of V which is star-like with respect to the origin, let $r\in\mathbb{N}$, and let $f\in C^r(\Omega)$. Then for every $x\in\Omega$ we have*

$$f(x) = \sum_{\substack{\alpha\in\mathfrak{A} \\ |\alpha|\le r-1}} \frac{1}{\alpha!}(\boldsymbol{\partial}^{\alpha} f)(0)\, x^{\alpha} + \sum_{\substack{\alpha\in\mathfrak{A} \\ |\alpha|=r}} \frac{1}{\alpha!} x^{\alpha} \int_0^1 r(1-s)^{r-1}(\boldsymbol{\partial}^{\alpha} f)(sx)\, ds .$$

Proof. In view of (1.32) by the main theorem of calculus we have

$$f(x)-f(0) = \sum_{|\alpha|=1} x^{\alpha} \int_0^1 (\partial^{\alpha} f)(sx)\, ds \ .$$

This is the assertion for $r=1$. Following the proof of the standard formula one can now deduce the general case by induction on r: In order to derive the formula for $r+1$ from the one for r one fixes $\alpha \in \mathfrak{A}$ satisfying $|\alpha| = r$, applies the case $r=1$ to $(\partial^{\alpha} f)(sx)$ instead of $f(x)$, and substituting t for st, changing the order of integration and taking

$$\int_t^1 r(1-s)^{r-1} ds = (1-t)^r \qquad t \in [0,1],$$

into account obtains

$$\int_0^1 r(1-s)^{r-1} (\partial^{\alpha} f)(sx)\, ds - (\partial^{\alpha} f)(0) = \sum_{|\beta|=1} x^{\beta} \int_0^1 (1-t)^r (\partial^{\beta} \partial^{\alpha} f)(tx)\, dt \, .$$

Employing the convention $1/(-1)! := 0$ we deduce for every $\alpha \in \mathfrak{A}$ satisfying $|\alpha| = r+1$

$$\sum_{i=1}^{d} \frac{1}{(\beta-e_i)!} \frac{1}{\gamma!} \frac{1}{\delta!} + \sum_{j=1}^{c} \left(\frac{1}{\beta!} \frac{1}{(\gamma-e_j)!} \frac{1}{\delta!} + \frac{1}{\beta!} \frac{1}{\gamma!} \frac{1}{(\delta-e_j)!} \right) =$$

$$= \frac{1}{\alpha!} \left(\sum_{i=1}^{d} \beta_i + \sum_{j=1}^{c} (\gamma_j + \delta_j) \right) = \frac{1}{\alpha!} (r+1) \, .$$

Putting everything together one completes the induction step. ∎

Obviously, the product rule remains valid for the operators ∂_j. Consequently, one deduces the following version of the Leibniz formula by following the lines of the proof of the standard version.

Lemma 1.27. *For arbitrary $r \in \mathbb{N}$, $f, g \in C^r(X)$, and $\alpha \in \mathfrak{A}$ satisfying $|\alpha| \le r$ we have*

$$\partial^{\alpha}(fg) = \sum_{\beta \in \mathfrak{A}, \beta \le \alpha} \binom{\alpha}{\beta} \partial^{\beta} f \, \partial^{\beta-\alpha} g \, . \quad ∎$$

The Leibniz rule can be used to carry out the easy proof of

Proposition 1.28. *(i) For arbitrary $\alpha, \beta \in \mathfrak{A}$ we have*

$$\partial^{\beta} x^{\alpha} = \begin{cases} \dfrac{\alpha!}{(\alpha-\beta)!} x^{\alpha-\beta} & \text{if } \beta \le \alpha \\ 0 & \text{otherwise} \end{cases} .$$

(ii) Consequently, the restrictions of the functions x^{α}, $\alpha \in \mathfrak{A}$, to any non-empty open subset of V are linearly independent. ∎

For the study of the behaviour of a polynomial function $P:V \longrightarrow \mathbb{C}$ under the action of M_t it is most convenient to express P as a polynomial in the variables y, z, and $\overline{z}$ where (y,z) be real-complex coordinates. Here Lemma 1.26 yields

$$(1.33) \qquad P(x) = \sum_{\alpha \in \mathfrak{A}} \frac{1}{\alpha!} (\partial^{\alpha} P)(0)\, x^{\alpha}, \qquad x \in V.$$

In order to guarantee that the functions x^{α}, $\alpha \in \mathfrak{A}$, behave nicely under the action of M_t we postulate additional properties of the bases introduced in Convention 1.24.A.

Convention 1.24.B. (i) From now on we suppose that the bases A of $V_{\mathbb{R}}$ and B of $V_{\mathbb{C}}$ are chosen in such a way that they contain bases of the generalized eigenspaces $G_M(\lambda)$, $\lambda \in \sigma$. We define $\rho \in \mathbb{R}^d$ and $\zeta \in \mathbb{C}^c$ by the conditions

$$(1.34) \qquad a_j \in G_M(\rho_j),\ j \in \mathbb{N}_d, \quad \text{and} \quad b_j \in G_M(\zeta_j),\ j \in \mathbb{N}_c.$$

(ii) If necessary we even require that the real matrix $M_{\mathbb{R}}$ of $M_{V_{\mathbb{R}}}$ with respect to A and the complex matrix $M_{\mathbb{C}}$ of $M_{V_{\mathbb{C}}}$ with respect to B are in Jordan canonical form. This means, in particular, that the matrices $M_{\mathbb{R}} - \mathrm{diag}(\rho_1, \ldots, \rho_d)$ and $M_{\mathbb{C}} - \mathrm{diag}(\zeta_1, \ldots, \zeta_c)$ are nilpotent where by *diag(...)* we denote diagonal matrices. In this case we denote by $k_{\mathbb{R}}$ (resp. $k_{\mathbb{C}}$) the number of Jordan blocks in $M_{\mathbb{R}}$ (resp. $M_{\mathbb{C}}$), and for every $\ell \in I_{\mathbb{R}} := \mathbb{N}_{k_{\mathbb{R}}}$ (resp. $L \in I_{\mathbb{C}} := \mathbb{N}_{k_{\mathbb{C}}}$) denote by d_ℓ (resp. c_L) the dimension of the ℓth (resp. L^{th}) Jordan block of $M_{\mathbb{R}}$ (resp. $M_{\mathbb{C}}$). This means that for every $\ell \in I_{\mathbb{R}}$ (resp. $L \in I_{\mathbb{C}}$) the sequence $(a_{D_{\ell-1}+1}, \ldots, a_{D_\ell})$ (resp. $(b_{C_{\ell-1}+1}, \ldots, b_{C_\ell})$) is an M-cyclic $\mathbb{R}$-basis (resp. $\mathbb{C}$-basis) of its span over $\mathbb{R}$ (resp. $\mathbb{C}$) where

$$D_\ell := \sum_{j=1}^{\ell} d_j \quad \left(\text{resp. } C_L := \sum_{J=1}^{L} c_J\right).$$

Notation 1.29. $\alpha_M := \langle \beta, \rho \rangle + \langle \gamma, \zeta \rangle + \langle \delta, \overline{\zeta} \rangle$, $\alpha = (\beta, \gamma, \delta) \in \mathfrak{A}$.

Remark 1.30. *(i) For every $\alpha = (\beta, \gamma, \delta) \in \mathfrak{A}$ there are a number $N_\alpha \in \mathbb{N}_0$ and non-trivial polynomial functions $P_{\alpha,k}$, $1 \le k \le N_\alpha$, of degree $\le |\alpha|$ satisfying*

$$(1.35) \qquad \forall \beta \in \mathfrak{A}:\ \partial^{\beta} P_{\alpha,k}(0) \neq 0 \Rightarrow \beta_M = \alpha_M$$

such that

$$(1.36) \qquad x^{\alpha} \circ M_t = t^{\alpha_M}\Big(x^{\alpha} + \sum_{k=1}^{N_\alpha} \omega_k(t)\, P_{\alpha,k}\Big), \qquad t \in]0, +\infty[,$$

where the functions $\omega_k :]0,+\infty[\longrightarrow \mathbb{R}$, $k \in \mathbb{N}_0$, *are defined by*

$$\omega_k(t) := \frac{1}{k!} (\log t)^k . \tag{1.37}$$

(ii) Suppose that the assumption of Convention 1.24.B.(ii) is satisfied. Then the number N_α *is equal to*

$$\sum_{\ell=1}^{k_{\mathbb{R}}} \sum_{j=1}^{d_\ell} (d_\ell - j)\, \beta_{D_{\ell-1}+j} + \sum_{L=1}^{k_{\mathbb{C}}} \sum_{J=1}^{c_L} (c_L - J)\, (\gamma_{C_{L-1}+J} + \delta_{C_{L-1}+J})$$

where the data $k_{\mathbb{R}}$, d_ℓ, D_ℓ, *and* $k_{\mathbb{C}}$, c_L, C_L *are defined in Convention 1.24.B.(ii).*

In particular, x^α *is quasihomogeneous of degree* α_M *if and only if* $\beta_j = 0$ *for every* $j \in \mathbb{N}_d \setminus \{D_\ell \,;\, \ell \in I_{\mathbb{R}}\}$ *and* $\gamma_J = 0 = \delta_J$ *for every* $J \in \mathbb{N}_c \setminus \{C_L \,;\, L \in I_{\mathbb{C}}\}$.

Moreover, for every $\ell \in I_{\mathbb{R}}$ *(resp.* $L \in I_{\mathbb{C}}$*) the degree of* $P_{\alpha,k}$ *with respect to the variables* $y_{D_{\ell-1}+j}$, $1 \le j \le d_\ell$, *(resp.* $z_{C_{L-1}+J}$, *resp.* $\overline{z_{C_{L-1}+J}}$*) is not larger than*

$$\sum_{j=1}^{d_\ell} \beta_{D_{\ell-1}+j} \quad \Big(\text{resp. } \sum_{J=1}^{c_L} \gamma_{C_{L-1}+J} \,;\, \text{resp. } \sum_{J=1}^{c_L} \delta_{C_{L-1}+J}\Big).$$

<u>*Proof*</u>. Obviously, in view of

$$\omega_j\, \omega_k = \binom{j+k}{j} \omega_{j+k} , \qquad j,k \in \mathbb{N}_0 , \tag{1.38}$$

it suffices to prove the assertion for the case $|\alpha| = 1$, only. In this case the assertion <u>*(i)*</u> is an immediate consequence of (1.10). For the proof of <u>*(ii)*</u> we may assume that $k_{\mathbb{R}} + k_{\mathbb{C}} = 1$. In this case we denote by λ the single eigenvalue of M. If $k_{\mathbb{R}} = 1$ then λ is real, and the matrix of $t^{-\lambda}(M_{\mathbb{R}})_t$ is of the form (1.10.b)' where $s := \log t$. Hence we deduce that

$$t^{-\lambda|\alpha|}\, x^\alpha \circ M_t = \prod_{j=1}^{d} \Big(\sum_{i=0}^{d-j} \frac{s^i}{i!} y_{j+i} \Big)^{\beta_j} .$$

Since on the right-hand side the leading term (with respect to the powers of s) is equal to

$$\Big(\prod_{j=1}^{d} \frac{s^{d-j}}{(d-j)!} \Big)^{\beta_j} (y_d)^{|\beta|}$$

the assertion follows. If $k_{\mathbb{C}} = 1$ then in a similar way one obtains

$$t^{-\lambda|\gamma| - \overline{\lambda}|\delta|}\, x^\alpha \circ M_t = \prod_{J=1}^{c} \Big[\Big(\sum_{i=0}^{c-J} \frac{s^i}{i!} z_{J+i} \Big)^{\gamma_J} \Big(\sum_{i=0}^{c-J} \frac{s^i}{i!} \overline{z_{J+i}} \Big)^{\delta_J} \Big] .$$

Since here on the right-hand side the leading term is equal to

$$\Big(\prod_{J=1}^{c} [(c-J)!]^{-\gamma_J-\delta_J} s^{(c-J))(\gamma_J+\delta_J)}\Big) (z_c)^{|\gamma|} (\overline{z_c})^{|\delta|}$$

the assertion follows in this case, as well. ∎

Corollary 1.31. *Suppose that M is semi-simple, i.e. for the matrices $M_{\mathbb{R}}$ and $M_{\mathbb{C}}$ introduced in Convention 1.24.B above we have $M_{\mathbb{R}} = diag(\rho_1, \ldots, \rho_d)$ and $M_{\mathbb{C}} = diag(\zeta_1, \ldots, \zeta_c)$. Then for every $\alpha \in \mathfrak{A}$ the monomial function x^α is quasihomogeneous of type M and of degree α_M.* ∎

If M is not semi-simple then Remark 1.30 shows that x^α is not always quasihomogeneous but only *almost quasihomogeneous of order $\le N_\alpha$* in the sense of

Definition 1.32. Let $N \in \mathbb{N}_0$. A function $q_0 : X \longrightarrow \mathbb{C}$ is called *almost quasihomogeneous of degree m, of type M, and of order $\le N$* if and only if there is an N-tuple of functions $q_k : X \longrightarrow \mathbb{C}$, $k \in \mathbb{N}_N$, such that

$$t^{-m} q_0(M_t x) = \sum_{k=0}^{N} \omega_k(t)\, q_k(x), \qquad (x,t) \in X \times_M]0,+\infty[\quad \text{- see (1.20).} \tag{1.39}$$

Now we fix a polynomial function $P: V \longrightarrow \mathbb{C}$. In order to group the terms in (1.33) according to their behaviour under the action of M_t we introduce

Notation 1.33. Under the Convention 1.24 we set

(i) $\mathfrak{A}(M) := \{\alpha_M \,;\, \alpha \in \mathfrak{A}\}$;

(ii) $(Q_m P)(x) := \sum_{\alpha_M = m} \frac{1}{\alpha!} (\partial^\alpha P)(0)\, x^\alpha$, $\qquad x \in V$;

(iii) $\Lambda(P) := \{\ell \in \mathbb{C} \,;\, Q_\ell P \not\equiv 0\}$.

For reasons to become clear in a moment $Q_m P$ is called *the (almost) quasihomogeneous part $Q_m P$ of P of degree m (and of type M)*. Of course, $Q_m P \equiv 0$ if $m \notin \mathfrak{A}(M)$, i.e. $\Lambda(P) \subset \mathfrak{A}(M)$.

Proposition 1.34. *For every $m \in \mathfrak{A}(M)$ $Q_m P : V \longrightarrow \mathbb{C}$ is a polynomial function being independent of the choice of coordinates in Convention 1.24 and having the following properties:*

(i) $Q_m P$ *is almost quasihomogeneous of degree* m*; in case* M *is semi-simple* $Q_m P$ *is even quasihomogeneous of degree* m.

(ii.a) $$P = \sum_{\ell \in \Lambda(P)} Q_\ell P\,;$$

(ii.b) conversely, if $\mathcal{L}$ *is a finite subset of* $\mathbb{C}$ *and* $(P_\ell)_{\ell \in \mathcal{L}}$ *is a family of non-trivial polynomial functions on* V *such that for every* $\ell \in \mathcal{L}$ P_ℓ *is almost quasi-homogeneous of degree* ℓ *and such that*

$$P = \sum_{\ell \in \mathcal{L}} P_\ell \tag{1.40}$$

then $\mathcal{L} = \Lambda(P)$, *and* $P_\ell = Q_\ell P$ *for every* $\ell \in \mathcal{L}$.

(iii) P *is almost quasihomogeneous of degree* m *if and only if* $P = Q_m P$.

Proof. *(i)*: This is a consequence of Remark 1.30. *(ii.a)*: This is (1.33). For later purposes it is suitable to formulate a part of the proof of (ii.b) separately as

Remark 1.35. *For every* $\ell \in \mathfrak{A}(M)$ *the equations*

$$(Q_\ell P) \circ M_t = t^\ell \sum_{k=0}^{\infty} \omega_k(t)\, P_{\ell,k}\,, \qquad t \in]0,+\infty[, \tag{1.41}$$

define a sequence of polynomial functions $P_{\ell,k}$, $k \in \mathbb{N}_0$, *only finitely many of them not vanishing identically, having the following properties:*

(i) $P_{\ell,0} = Q_\ell P\,;$ *(ii)* $P_{\ell,k} = Q_\ell(P_{\ell,k})\,;$

(iii) $\deg P_{\ell,k} \le \deg P_{\ell,0} \le \deg P$; *and if* $Q_\ell P$ *is of degree* $\le (r,s)$ *with respect to an* M*-invariant decomposition (1.26) then the* $P_{\ell,k}$ *are so, as well.*

(iv) If M *is semi-simple then* $P_{\ell,k} = 0$ *for every* $k \in \mathbb{N}$.

Proof. The existence of the polynomial functions $P_{\ell,k}$ satisfying (1.41), (i), (ii), (iii), and (iv) is an immediate consequence of Remark 1.30. Their uniqueness is due to the fact that for every non-empty open subinterval I of $]0,+\infty[$

(1.42) the functions $I \longrightarrow \mathbb{C}$, $t \mapsto t^z \omega_j(t)$, $z \in \mathbb{C}$ and $j \in \mathbb{N}_0$, are linearly independent. ■

End of the proof of Proposition 1.34. *(ii.b)*: Let $(P_\ell)_{\ell \in \mathcal{L}}$ be as in assertion (ii.b), and for every $\ell \in \mathcal{L}$ let $(R_{\ell,k})_{k \in \mathbb{N}_0}$ be a sequence of functions such that (1.41) is valid with $(Q_\ell P, P_{\ell,k})$ replaced by $(P_\ell, R_{\ell,k})$. Inserting these equations into the equations for $P \circ M_t$ obtained from (i) and (1.40), likewise combining (ii.a) and (1.41), and making use of (1.42), again, we deduce that $\mathcal{L} = \{ \ell \in \mathfrak{A}(M);$ $P_{\ell,k} \neq 0$ for some $j \in \mathbb{N}_0\}$ and that $P_{\ell,k} = R_{\ell,k}$ for arbitrary $\ell \in \mathcal{L}$ and $k \in \mathbb{N}_0$. Since $P_{\ell,0} = Q_\ell P$ and $R_{\ell,0} = P_\ell$ and since by Remark 1.35 we have $P_{\ell,k} \neq 0$ for some $k \in \mathbb{N}_0$ if and only if $P_{\ell,0} \neq 0$ the assertion is proved.

(iii). "$\Rightarrow$" is the special case $\mathcal{L} = \{m\}$ of (ii.b). "$\Leftarrow$" follows from (i).

Finally, the fact that $Q_m P$ does not depend on the choice of coordinates is another consequence of (ii.b). ∎

As an immediate consequence of Corollary 1.31 and Proposition 1.34.(iii) one obtains

Corollary 1.36. *If M is semi-simple then every polynomial function which is almost quasihomogeneous is, in fact, quasihomogeneous.* ∎

Concerning products of polynomial functions one has

Corollary 1.37. *For any other polynomial function R on V we have*

(i)
$$Q_m(PR) = \sum_{\ell \in \mathfrak{A}(M)} Q_\ell P \; Q_{m-\ell} R$$

and

(ii)
$$(PR)_{m,j} = \sum_{z \in \mathfrak{A}(M)} \sum_{i=0}^{j} \binom{j}{i} P_{z,i} \, R_{m-z,j-i} \, .$$

Proof. *(i).* Since the right-hand side of the equation in (i) is almost quasihomogeneous of degree ℓ one obtains the assertion by computing PR with the help of Proposition 1.34.(ii.a) - applied to P and R - and by alluding to the uniqueness property (ii.b) of Proposition 1.34.

(ii). One has to compose both sides of formula (i) with M_t, and on the right-hand side one has to insert the formula (1.41) for P and the one for $(R, m-\ell)$ instead of (P,ℓ). Then the desired formula is obtained by the uniqueness of the functions $(PR)_{m,k}$ when one is taking (1.38) into account. ∎

For polynomial functions the converse of Remark 1.15 is valid:

Proposition 1.38. *Suppose that R is another polynomial function on V such that*

$PR \neq 0$. If PR is almost quasihomogeneous so are P and R; in particular, if PR is quasihomogeneous then so are P and R.

Proof. We set $D := \Lambda(P)$ and $E := \Lambda(R)$ - see Notation 1.33.(iii). In view of the assertions (i) and (ii.a) of Proposition 1.34 it suffices to show that

(1.43) $$|D| = 1 = |E|.$$

Note that D and E are non-empty finite subsets of $\mathbb{C}$. Moreover, if we fix $m \in \mathbb{C}$ such that $PR = Q_m(PR)$ then in view of Corollary 1.37.(i) for every $w \in \mathbb{C}$ the following implication holds:

(1.44) if the set $\mathfrak{I}_w := \{(a,b) \in D \times E \,;\, a+b = w\}$ consists of precisely one element then $w = m$.

In order to deduce (1.43) from this property we denote by $\langle \cdot, \cdot\cdot \rangle$ the real scalar product on $\mathbb{C}$ and observe that the union of the sets $\ker \langle k-k', \cdot \rangle$, $k, k' \in D$ such that $k \neq k'$, and $\ker \langle \ell - \ell', \cdot \rangle$, $\ell, \ell' \in E$ such that $\ell \neq \ell'$, is a subset of $\mathbb{C}$ of measure 0. Hence we can choose $z \in \mathbb{C}$ such that

(1.45) $$|D_z| = |D| \text{ and } |E_z| = |E|$$

where $C_z := \{\langle j,z \rangle \,;\, j \in C\}$. We choose $k, \ell \in D$ and $K, L \in E$ such that $\langle k,z \rangle = \min D_z$, $\langle K,z \rangle = \max D_z$, $\langle \ell,z \rangle = \min E_z$, and $\langle L,z \rangle = \max E_z$. Making use of the equation $\langle k'+\ell',z \rangle = \langle k',z \rangle + \langle \ell',z \rangle$ we derive from (1.45) that

(1.46) $$|\mathfrak{I}_{k+\ell}| = 1 = |\mathfrak{I}_{K+L}|.$$

By (1.44) it follows that $k+\ell = m = K+L$, i.e. $\mathfrak{I}_{k+\ell} = \mathfrak{I}_m = \mathfrak{I}_{K+L}$ so that by (1.46), again, we conclude that $k = K$ and $\ell = L$. This implies $|D_z| = 1 = |E_z|$ so that (1.43) follows from (1.45). ∎

Corollary 1.39. *Let R be another polynomial function such that $PR \neq 0$. If PR is a monomial function then so are P and R.*

Proof. It follows from Example 1.23 that as a monomial function PR is quasihomogeneous of type p for every $p \in \mathbb{N}^n$. By Proposition 1.38 the same remains valid for P and R. This, in turn, implies that P and R are monomial functions, as well. Indeed, let $\alpha, \beta \in \mathbb{N}_0^n$ be such that $P^{(\alpha)}(0) \neq 0 \neq P^{(\beta)}(0)$. Then $\langle \alpha,p \rangle = \langle \beta,p \rangle$ for every $p \in \mathbb{N}^n$, in particular, $(\alpha_j - \beta_j)k = \sum_{i \neq j}(\beta_i - \alpha_i)$ for arbitrary $j \in \mathbb{N}_n$ and $k \in \mathbb{N}$, i.e. $\alpha_j = \beta_j$ for every $j \in \mathbb{N}_n$, as claimed. ■

In view of Propositions 1.34.(iii) and 1.28.(ii) the dimension of the space $\mathfrak{P}_m(V)$ of polynomial functions on V which are almost quasihomogeneous of degree m is equal to the number of elements of the set

$$\{\alpha \in \mathfrak{A}\, ;\, \alpha_M = m\}\,. \tag{1.47}$$

This set is easily determined in case V is M-indecomposable: if $V = V_{\mathbb{R}}$ then it is obviously equal to $\{\beta \in \mathbb{N}_o^n\, ;\, |\beta| = m/\lambda\}$ where λ is the only eigenvalue of M; and if $V = V_{\mathbb{C}}$ it is described in

Remark 1.40. *Let $\lambda \in \mathbb{C} \setminus \mathbb{R}$, and suppose that $V = \mathbb{C}^n$ is M-cyclic and that λ is the only point in $\sigma(M)$.*

(i) If $Re\,\lambda = 0$ then the set (1.47) is equal to $\{(\gamma,\delta)\, ;\, \gamma,\delta \in \mathbb{N}_o^n,\ |\gamma| - |\delta| = m/\lambda\}$.

(ii) If $Re\,\lambda \neq 0$ then (1.47) equals $\{(\gamma,\delta)\, ;\, \gamma,\delta \in \mathbb{N}_o^n,\ |\gamma| = m_R,\ |\delta| = m_I\}$ where

$$m_R := \frac{1}{2}\left(\frac{Re\,m}{Re\,\lambda} + \frac{Im\,m}{Im\,\lambda}\right) \quad \text{and} \quad m_I := \frac{1}{2}\left(\frac{Re\,m}{Re\,\lambda} - \frac{Im\,m}{Im\,\lambda}\right);$$

in particular, in case $n = 1$ there is precisely one $\alpha \in \mathfrak{A}$ such that $\alpha_M = m$.

Proof. Let $\alpha = (\gamma,\delta) \in \mathfrak{A} = \mathbb{N}_o^n \times \mathbb{N}_o^n$. Then $\alpha_M = |\gamma|\lambda + |\delta|\bar{\lambda}$, and the assertion *(i)* is clear while *(ii)* is an immediate consequence of the following lemma. ■

Lemma 1.41. *If λ is as in Remark 1.40.(ii) then for every $z \in \mathbb{C}$ the equation*

$$r\lambda + s\bar{\lambda} = z \tag{1.48}$$

has a unique solution $(r,s) \in \mathbb{R}^2$ which is given by

$$r = \frac{1}{2}\left(\frac{Re\,z}{Re\,\lambda} + \frac{Im\,z}{Im\,\lambda}\right) \text{ and } s = \frac{1}{2}\left(\frac{Re\,z}{Re\,\lambda} - \frac{Im\,z}{Im\,\lambda}\right). \tag{1.48'}$$ ■

We close this section by considering polynomial functions P on the dual space V^* of V which are (almost) quasihomogeneous of type M^*. In order to emphasize this we are going to denote the functions $P_{\ell,k}$ associated to such a polynomial function P in Remark 1.35 by $P^*_{\ell,k}$. As a consequence of Remark 1.17 (resp. 1.17'), Proposition 1.34, and of (1.41) we note the following form of the chain rule.

Remark 1.42. *If $P: V^* \longrightarrow \mathbb{C}$ is a polynomial function of degree $\le r$ (resp. $\le (r,s)$) and if f belongs to $C^r(X)$ (resp. $C^{r,s}(X)$) then for every $t \in]0,+\infty[$ we have*

$$(1.49) \qquad P(\partial)(f \circ M_t) = \sum_{\ell \in \mathfrak{A}(M)} t^{\ell} \sum_{k=0}^{\infty} \omega_k(t) \, [P^*_{\ell,k}(\partial) f] \circ M_t \, ,$$

and

$$(1.49)' \qquad (P(\partial) f) \circ M_t = \sum_{\ell \in \mathfrak{A}(M)} t^{-\ell} \sum_{k=0}^{\infty} (-1)^k \omega_k(t) \, P^*_{\ell,k}(\partial) (f \circ M_t) \, . \quad \blacksquare$$

Finally, we show that the derivatives ∂^{α} introduced in Notation 1.25.B are induced by almost quasihomogeneous polynomial functions on V^*. To this end we have to introduce real-complex coordinates on V^* related to those for V introduced in Convention 1.24.A. As in Convention 1.24.A, by $(\eta_1, \ldots, \eta_d, \eta_{d+1}, \ldots, \eta_{d+c})$ we denote the (d+c)-tuple of coefficient functionals on V associated with the $\mathbb{R}$-basis A of $V_{\mathbb{R}}$ and the $\mathbb{C}$-basis B of $V_{\mathbb{C}}$ as introduced in Convention 1.24. Then

$$\mathfrak{B}^* := (\eta_1, \ldots, \eta_d, \operatorname{Re}\eta_{d+1}, \ldots, \operatorname{Re}\eta_{d+c}, \operatorname{Im}\eta_{d+1}, \ldots, \operatorname{Im}\eta_{d+c})$$

is the $\mathbb{R}$-basis of V^* dual to the $\mathbb{R}$-basis $\mathfrak{B} := (a_1, \ldots, a_d, b_1, \ldots, b_c, ib_1, \ldots, ib_c)$ of V. It follows that $(\eta_1, \ldots, \eta_d)$ is an $\mathbb{R}$-basis of $(V^*)_{\mathbb{R}}$ and that $(\operatorname{Re}\eta_{d+1}, \ldots, \operatorname{Re}\eta_{d+c}, \operatorname{Im}\eta_{d+1}, \ldots, \operatorname{Im}\eta_{d+c})$ is an $\mathbb{R}$-basis of $(V^*)_{\mathbb{C}}$. Moreover, if $J: V_{\mathbb{C}} \longrightarrow V_{\mathbb{C}}$, $v \mapsto iv$, denotes the complex structure introduced in the text preceding Notation 1.6.A then J^* is the corresponding complex structure on $(V^*)_{\mathbb{C}}$. It is easily seen that

$$(1.50) \qquad \operatorname{Im}\eta_j = -J^*(\operatorname{Re}\eta_j)$$

so that $(\operatorname{Re}\eta_1, \ldots, \operatorname{Re}\eta_c)$ is a $\mathbb{C}$-basis of V^*. In order to identify this as the $\mathbb{C}$-basis $(\eta_1, \ldots, \eta_c)$ of $(V^*)_{\mathbb{C}}$ which is dual to the $\mathbb{C}$-basis B of $V_{\mathbb{C}}$ we note that the map

$$(1.51) \qquad \mathfrak{J}: L_{\mathbb{R}}(V_{\mathbb{C}}, \mathbb{R}) \longrightarrow L_{\mathbb{C}}(V_{\mathbb{C}}, \mathbb{C}), \ \varkappa \mapsto \varkappa - iJ^*\varkappa \, ,$$

is a $\mathbb{C}$-isomorphism, its inverse being given by $\nu \mapsto \operatorname{Re}\nu$. We have

$$(1.51)' \qquad \mathfrak{J}\Big(\sum_{j=1}^{c} (\xi_j \operatorname{Re}\eta_j + \upsilon_j \operatorname{Im}\eta_j) \Big) = \sum_{j=1}^{c} (\xi_j - i\upsilon_j)\, \eta_j \, , \qquad \xi, \upsilon \in \mathbb{R}^c .$$

Convention 1.24*. Let $(\eta_1, \ldots, \eta_{d+c})$ be as above. We consider V^* as equipped with real-complex coordinates defined by the $\mathbb{R}$-basis $(\eta_1, \ldots, \eta_d)$ of $(V^*)_{\mathbb{R}}$ and the $\mathbb{C}$-basis $(2\eta_{d+1}, \ldots, 2\eta_{d+c})$ of $(V^*)_{\mathbb{C}}$ where the latter space is identified with $L_{\mathbb{C}}(V_{\mathbb{C}}, \mathbb{C})$ via the map $\mathfrak{J}$ defined in (1.51). This defines a real-complex basis

$$\mathfrak{C}^* := (\eta_1, \ldots, \eta_d, 2\eta_{d+1}, \ldots, 2\eta_{d+c}) \text{ of } V^*.$$

Note that under the identification of $(V^*)_{\mathbb{C}}$ and $L_{\mathbb{C}}(V_{\mathbb{C}},\mathbb{C})$ via $\mathfrak{J}$ the $\mathbb{R}$-basis $\mathfrak{B}^*$ of V^* becomes the $\mathbb{R}$-basis $(\eta_1,\ldots,\eta_d,\eta_{d+1},\ldots,\eta_{d+c},i\eta_{d+1},\ldots,i\eta_{d+c})$. Moreover, note that by Convention 1.24.B.(i) the real-complex basis $\mathfrak{E}^*$ contains a basis of $G_{M^*}(\lambda)$ for every $\lambda\in\sigma(M)=\sigma(M^*)$, and that (1.34) transforms into

$$(1.34)^* \qquad \eta_k\in G_{M^*}(\rho_k)\,,\ k\in\mathbb{N}_d\,, \quad \text{and} \quad \eta_{d+j}\in G_{M^*}(\zeta_j)\,,\ j\in\mathbb{N}_c\,.$$

Finally, we observe that (1.34) and $(1.34)^*$ imply that

$$(1.52) \qquad \alpha_{M^*}=\alpha_M\,, \qquad \alpha\in\mathfrak{A}\,.$$

The reason why the factor 2 has been introduced into the complex part of $\mathfrak{E}^*$ becomes clear in the proof of

Remark 1.43. *(i) For every $\alpha\in\mathfrak{A}$ there is a (unique) polynomial function $P_\alpha : V^*\longrightarrow\mathbb{C}$ such that*

$$(1.53) \qquad \boldsymbol{\partial}^\alpha=P_\alpha(\partial)\,.$$

P_α is almost quasihomogeneous of degree α_M and of type M^*. More precisely, P_α is the monomial function $x\mapsto x^\alpha$ which is associated with the real-complex basis $\mathfrak{E}^*$ of V^* (introduced in Convention 1.24) when in Notation 1.25.A V is replaced by V^*. In other words, for every $\xi\in V^*$ we have*

$$(1.53)' \qquad (a)\ \boldsymbol{\partial}^\alpha \exp\langle\xi,\cdot\rangle=\xi^\alpha\exp\langle\xi,\cdot\rangle\,, \quad \text{and} \quad (b)\ \boldsymbol{\partial}^\alpha\langle\xi,\cdot\rangle=\xi^\alpha \ \text{if}\ |\alpha|=1\,.$$

(ii) For every polynomial function $Q:V^*\longrightarrow\mathbb{C}$ of degree $r\in\mathbb{N}$ the Leibniz rule can be written as

$$(1.54) \qquad Q(\partial)(fg)=\sum_{\alpha\in\mathfrak{A}}\frac{1}{\alpha!}\,Q^{(\alpha)}(\partial)f\,\boldsymbol{\partial}^\alpha g\,, \qquad f,g\in C^r(X)\,,$$

where the derivatives $Q^{(\alpha)}$ of order α are formed with respect to the basis $\mathfrak{E}^*$.

Proof. *(i)*: By (ν,ξ,υ) we denote the real coordinates on V^* with respect to the $\mathbb{R}$-basis $\mathfrak{B}^*$ introduced in the text preceding Convention 1.24*. For every $k\in\mathbb{N}_d$ (resp. $j\in\mathbb{N}_c$) we define $P_k : V^*\longrightarrow\mathbb{C}$ (resp. $Q_j : V^*\longrightarrow\mathbb{C}$) by $(P_k\circ\mathfrak{k}_{\mathfrak{B}^*})(\nu,\xi,\upsilon):=\nu_k$ (resp. $(Q_j\circ\mathfrak{k}_{\mathfrak{B}^*})(\nu,\xi,\upsilon):=\frac{1}{2}(\xi_j-i\upsilon_j)$). Then

$$\partial_{y_k}=P_k(\partial)\,,\quad \partial_{z_j}=Q_j(\partial)\,,\ \text{and}\ \ \partial_{\overline{z}_j}=\overline{Q_j}(\partial)\,.$$

It follows that (1.53) is valid for

$$P_\alpha := \prod_{k=1}^{d} P_k^{\beta_k} \prod_{j=1}^{c} Q_j^{\gamma_j} \overline{Q}_j^{\delta_j} .$$

In view of (1.51)' P_α turns out to be nothing else than the monomial function $x \mapsto x^\alpha$ which comes about if in Notation 1.25.A V is replaced by V^* and if V^* is equipped with the real-complex basis $\mathfrak{E}^*$ introduced in Convention 1.24*. Hence Remark 1.30 - applied to (V^*, M^*) instead of (V, M) - shows that P_α is almost quasihomogeneous of degree α_{M^*} which equals α_M by (1.52). Finally, *(1.53.a)'* is a consequence of (1.25) and (1.53); and in view of the chain rule *(1.53.b)'* follows from (1.53.a)'.

(ii). Taking (i) into account and applying Proposition 1.28.(i) to (V^*, M^*) instead of (V, M) one obtains for arbitrary $\alpha, \beta \in \mathfrak{A}$ that

$$P_\alpha^{(\beta)} = \begin{cases} \beta!\binom{\alpha}{\beta} P_{\alpha-\beta} & \text{if } \alpha \geq \beta \\ 0 & \text{otherwise} \end{cases} .$$

Hence, choosing $c_\alpha \in \mathbb{C}$ such that $Q = \sum_{\alpha \in \mathfrak{A}} c_\alpha P_\alpha$ - which is possible in view of (i) and (1.33) - one deduces that

$$\sum_{\alpha \in \mathfrak{A}, \alpha \geq \beta} c_\alpha \binom{\alpha}{\beta} \partial^{\alpha-\beta} = \frac{1}{\beta!} Q^{(\beta)}(\partial), \qquad \beta \in \mathfrak{A}.$$

Consequently, the assertion follows by Lemma 1.27. ∎

(d) Non-trivial Examples of Quasihomogeneous Polynomial Functions in case M is Not Semi-simple

If M is semi-simple then according to Corollary 1.36 every almost quasihomogeneous polynomial function on V is quasihomogeneous. So in this case Proposition 1.34.(iii) describes all quasihomogeneous polynomial functions on V. In the present section we show how to obtain quasihomogeneous polynomial functions in the presence of multiple eigenvalues. First of all we give a coordinate-free reformulation of the content of the last but one part of Remark 1.30.(ii), denoting the canonical projection onto the real vector space

$$V' := V \Big/ \bigoplus_{\lambda \in \sigma} \widetilde{M}_\lambda(G_M(\lambda))$$

by $\pi' : V \longrightarrow V'$: *if $P: V' \longrightarrow \mathbb{C}$ is a polynomial function which is quasihomogeneous* *of degree m and of type M' where $M' \in L(V')$ is induced by M then $P \circ \pi'$ is a polynomial function on V which is quasihomogeneous of degree m and of type M.* Note that M' has the same eigenvalues as M but of algebraic multiplicity 1.

Are there any other quasihomogeneous polynomial functions on V? Under special assumptions the answer is "no". In any case, the number of quasihomogeneous polynomials is rather restricted under the presence of multiple eigenvalues. To see this more clearly we assume for the rest of the present section that V be M-cyclic. Denoting by λ the only eigenvalue of M we let $\mathfrak{B} = (b_1, \dots, b_n)$ be a K_λ-basis of V satisfying (1.11) so that the matrix of N with respect to $\mathfrak{B}$ is of the form (1.11)'. It follows that M_t is given by

$$(1.55) \qquad (M_t x)_j = t^\lambda \sum_{i=0}^{n-j} \frac{1}{i!} s^i x_{j+i} \quad \text{where } s := \log t , \qquad j \in \mathbb{N}_n ,$$

where the points x of V are denoted by the coordinates $(x_1, \dots, x_n) \in (K_\lambda)^n$ with respect to the basis $\mathfrak{B}$. We suppose that $n \geq 2$. The following result states only a very weak restrictive condition shared by all quasihomogeneous polynomial function on V.

Proposition 1.44. *Under the preceding assumptions every polynomial function $P: V \longrightarrow \mathbb{C}$ which is quasihomogeneous of degree $m \neq 0$ vanishes on the 1-dimensional subspace $(\tilde{M}_\lambda)^{n-1}(V)$.*

The proof is a by-product of more precise considerations. We begin with

The case "$\lambda \in \mathbb{R}$". Here it follows from Example 1.21 that

$$\partial_M = \sum_{j=1}^{n} (\lambda x_j + x_{j+1}) \partial_j \qquad (\text{here we set } x_{n+1} := 0).$$

We fix a polynomial function $P: V \longrightarrow \mathbb{C}$ which is almost quasihomogeneous of degree m, i.e.

$$P = \sum_{\alpha \in I_d} P_\alpha X^\alpha$$

where $d := m/\lambda$, $I_d := \{\alpha \in \mathbb{N}_0^n ;\ |\alpha| = d\}$, and $P_\alpha := P^{(\alpha)}(0)/\alpha!$. We are going to try to find necessary and sufficient conditions on the coefficients P_α which ensure that P is quasihomogeneous. To this end we employ Euler's equation. Since

$$\partial_M X^\alpha = \sum_{\substack{j=1 \\ \alpha_j \geq 1}}^{n} (\lambda x_j + x_{j+1}) \alpha_j X^{\alpha - e_j}$$

we conclude that

$$(\partial_M - m)\,P = \sum_{\alpha\in I_d} P_\alpha \sum_{\substack{j=1\\ \alpha_j\ge 1}}^{n-1} \alpha_j\, x^{\alpha-e_j+e_{j+1}} = \sum_{j=1}^{n-1} \sum_{\substack{\alpha\in I_d\\ \alpha_j\ge 1}} \alpha_j P_\alpha\, x^{\alpha-e_j+e_{j+1}} =$$

$$= \sum_{j=1}^{n-1} \sum_{\substack{\beta\in I_d\\ \beta_{j+1}\ge 1}} (\beta_j+1)\, P_{\beta+e_j-e_{j+1}}\, x^\beta = \sum_{\beta\in I_d} L_\beta(P)\, x^\beta$$

where

$$L_\beta(P) := \sum_{j\in J_\beta} (\beta_j+1)\, P_{\beta+e_j-e_{j+1}} \qquad \text{with} \qquad J_\beta := \{ j\in \mathbb{N}_{n-1};\ \beta_{j+1}\ge 1\}.$$

It follows that P is quasihomogeneous of degree m if and only if

(1.56) $\qquad L_\beta(P) = 0 \quad$ for every $\beta\in I_d$.

Note that J_β consists of one element only if and only if β is of the form

$$\beta = (d-\beta_{\ell+1},0,\dots,0,\beta_{\ell+1},0,\dots,0) \quad \text{with } \beta_{\ell+1}\in\mathbb{N}_d \text{ for some } \ell\in\mathbb{N}_{n-1}.$$

Hence, (1.56) implies that

(1.57.a) $\qquad P_{(r,d-r,0,\dots,0)} = 0, \qquad r\in\mathbb{N}_d,$

and - in case $n\ge 3$ -

(1.57.b) $\qquad P_{(r-1,0,\dots,0,1,d-r,0,\dots,0)} = 0, \qquad r\in\mathbb{N}_d.$

In case $n=2$ the equations (1.57.a) mean that

Remark 1.45. *Suppose that under the preceding assumptions we have $n=2$. Then P is quasihomogeneous of degree m if and only if P does not depend on the variable x_1.* ■

In case $n\ge 3$, however, there are choices of β such that the equation (1.56) relates at least two of the coefficients of P to each other. We restrict ourselves to the case $n=3$ when besides the equations (1.57.a) and (1.57.b) there appear equations involving two of the coefficients of P, only: for the triples $\beta\in I_d$ satisfying $\beta_2\ge 1$ and $\beta_3\ge 1$ the equation (1.56) leads to

(1.57.c) $\qquad P_{(k,\ell,d-k-\ell)} = -\frac{\ell+2}{k}\, P_{(k-1,\ell+2,d-k-\ell-1)}, \qquad k\in\mathbb{N}_{d-1},\ 0\le \ell\le d-k-1.$

Exploiting this we are going to derive

Proposition 1.46. *Suppose that under the preceding assumptions we have $n=3$. Then the polynomial functions*

$$\mathbb{R}^3 \ni (x,y,z) \mapsto (y^2-2xz)^s z^{d-2s}, \qquad s \in \mathbb{N}_0,\ s \le d/2, \tag{1.58}$$

constitute a basis of the space of polynomial functions on $\mathbb{R}^3$ which are quasihomogeneous of degree $m=d\lambda$.

Proof. Since $(y+sz)^2-2(x+sy+s^2z/2)z = y^2-2xz$ the functions are quasihomogeneous. The rest of the assertion is proved by induction on d. If $d=1$ then the assertion is obvious. Now suppose that $d \ge 2$, and let P be quasihomogeneous of degree $d\lambda$. Suppose first that z divides P, i.e. $P(x,y,z) = z\,Q(x,y,z)$. It then follows by Proposition 1.38 that Q is quasihomogeneous of degree $m-\lambda = (d-1)\lambda$ so that by applying the induction hypothesis to Q we deduce that P is a linear combination of the functions (1.58). Hence we suppose that z does not divide P, i.e. $P_{(r,d-r,0)} \ne 0$ for some $r \in \{0\} \cup \mathbb{N}_d$. In view of (1.57.a) this means that $c := P_{(0,d,0)} \ne 0$. From (1.57.c) we deduce by induction that

$$P_{(k,d-2k,k)} = (-1)^k \frac{d(d-2)\cdots(d-2k+2)}{k!}\, c, \qquad k \in \mathbb{N},\ k \le d/2.$$

If d were odd, i.e. $k := (d-1)/2 \in \mathbb{N}$ then, in particular, we would have $P_{(k,1,k)} \ne 0$ which contradicts (1.57.b) for $r=k+1$. Hence $j := d/2$ belongs to $\mathbb{N}$, and, consequently, by $R(x,y,z) := P(x,y,z) - c\,(y^2-2xz)^j$ a polynomial function R is defined which is quasihomogeneous of degree $d\lambda$ and which is divisible by z. Hence the proof of the induction step is completed by an application of the first part of the induction step already dealt with above. ∎

The case "$\lambda \in \mathbb{C}\setminus\mathbb{R}$". Since here $\mathfrak{B}$ is a $\mathbb{C}$-basis of V we denote the coordinates of a point z in V with respect to $\mathfrak{B}$ by $(z_1,\dots,z_n)$. It follows from Example 1.21 and from (1.32) and (1.11) that

$$\partial_M = \sum_{j=1}^{n} [(\lambda z_j + z_{j+1})\partial_j + (\overline{\lambda z_j} + \overline{z_{j+1}})\overline{\partial}_j]$$

where here we set $z_{n+1} := 0$, $\partial_j := \partial_{z_j}$ and $\overline{\partial}_j := \partial_{\overline{z}_j}$. Now we fix a polynomial function P on V which is almost quasihomogeneous of degree m. This means that

$$P = \sum_{(\alpha,\beta)\in\mathfrak{A}_m} P_{\alpha,\beta}\, z^\alpha \overline{z}^\beta$$

where $\mathfrak{A}_m := \{(\alpha,\beta) \in \mathbb{N}_o^n \times \mathbb{N}_o^n;\ |\alpha|\lambda + |\beta|\bar{\lambda} = m\}$ and $P_{\alpha,\beta} := (\partial_z^\alpha \partial_{\bar{z}}^\beta P)(0)/(\alpha!\beta!)$. Our aim is to find conditions on the coefficient $P_{\alpha,\beta}$ which are necessary and sufficient for P to be quasihomogeneous of degree m. Here, again, we make use of Euler's equations. Since

$$\partial_M z^\alpha \bar{z}^\beta = \sum_{\substack{j=1\\ \alpha_j \geq 1}}^n (\lambda z_j + z_{j+1})\,\alpha_j\, z^{\alpha-e_j}\,\bar{z}^\beta + \sum_{\substack{j=1\\ \beta_j \geq 1}}^n (\bar{\lambda}\bar{z}_j + \overline{z_{j+1}})\,\beta_j\, z^\alpha \bar{z}^{\beta-e_j}$$

we conclude that

$$(\partial_M - m)\,P = \sum_{(\alpha,\beta)\in\mathfrak{A}_m} P_{\alpha,\beta}\Big[\sum_{\substack{j=1\\ \alpha_j\geq 1}}^n \alpha_j\, z^{\alpha-e_j+e_{j+1}}\,\bar{z}^\beta + \sum_{\substack{j=1\\ \beta_j\geq 1}}^n \beta_j\, z^\alpha \bar{z}^{\beta-e_j+e_{j+1}}\Big] =$$

$$= \sum_{j=1}^{n-1} \sum_{\substack{(\alpha,\beta)\in\mathfrak{A}_m\\ \alpha_j\geq 1}} \alpha_j\, P_{\alpha,\beta}\, z^{\alpha-e_j+e_{j+1}}\,\bar{z}^\beta + \sum_{j=1}^{n-1} \sum_{\substack{(\alpha,\beta)\in\mathfrak{A}_m\\ \beta_j\geq 1}} \beta_j\, P_{\alpha,\beta}\, z^\alpha \bar{z}^{\beta-e_j+e_{j+1}} =$$

$$= \sum_{j=1}^{n-1} \sum_{\substack{(\gamma,\beta)\in\mathfrak{A}_m\\ \gamma_{j+1}\geq 1}} (\gamma_j+1)\, P_{\gamma+e_j-e_{j+1},\beta}\, z^\gamma \bar{z}^\beta +$$

$$+ \sum_{j=1}^{n-1} \sum_{\substack{(\alpha,\delta)\in\mathfrak{A}_m\\ \delta_{j+1}\geq 1}} (\delta_j+1)\, P_{\alpha,\delta+e_j-e_{j+1}}\, z^\alpha \bar{z}^\delta =$$

$$= \sum_{(\alpha,\beta)\in\mathfrak{A}_m} L_{\alpha,\beta}(P)\, z^\alpha \bar{z}^\beta$$

where

$$L_{\alpha,\beta}(P) := \sum_{j\in J_\alpha} (\alpha_j+1)\, P_{\alpha+e_j-e_{j+1},\beta} + \sum_{j\in J_\beta} (\beta_j+1)\, P_{\alpha,\beta+e_j-e_{j+1}}$$

with $J_\alpha := \{j \in \mathbb{N}_{n-1};\ \alpha_{j+1} \geq 1\}$. It follows that P is quasihomogeneous of degree m if and only if

(1.59) $$L_{\alpha,\beta}(P) = 0\,, \qquad (\alpha,\beta)\in\mathfrak{A}_m\,.$$

Applying this to $\alpha = (k-1,1,0,\dots,0)$ and $\beta = (\ell,0,\dots,0)$ and to $\alpha = (k,0,\dots,0)$ and $\beta = (\ell-1,1,0,\dots,0)$ we obtain

(1.60.a) $$P_{(k,0,\dots,0),(\ell,0,\dots,0)} = 0\,, \qquad (k,\ell)\in\mathbb{N}_o^2\setminus\{0\},\ k\lambda + \ell\bar{\lambda} = m\,.$$

In passing we can now give the

Proof of Proposition 1.44. In view of (1.11) we have $(\tilde{M}_\lambda)^{n-1}(V) = K_\lambda b_1$. Consequently, if $\lambda \in \mathbb{R}$ (resp. $\lambda \notin \mathbb{R}$) then the assertion follows from (1.57.a) for $r = d$ (resp. from (1.60.a)). ■

Next we observe that the analogue of Remark 1.45 for $\lambda \in \sigma \setminus \mathbb{R}$ is false: if $n=2$ then assuming that $V = \mathbb{C}\times\mathbb{C}$ and that M is given by the matrix $\left(\begin{smallmatrix}\lambda & 1\\ 0 & \lambda\end{smallmatrix}\right)$ we see that $M_t z = t^{\lambda}(z_1 + Lz_2, z_2)$ where $L := \log t$, and since

$$(z_1+Lz_2)\overline{z_2} - (\overline{z_1}+L\overline{z_2})\,z_2 = z_1\overline{z_2} - \overline{z_1}\,z_2$$

it follows for arbitrary $k,r,s \in \mathbb{N}_0$ that the function defined by

$$(1.61) \qquad z \mapsto (z_1\overline{z_2} - \overline{z_1}z_2)^k z_2^r\, \overline{z_2}^s$$

is quasihomogeneous of degree $k\lambda + k\overline{\lambda} + r\lambda + s\overline{\lambda}$. In fact, in case $n=2$ essentially these are all quasihomogeneous polynomial functions on V:

Proposition 1.47. *Suppose that under the assumptions stated at the beginning of the present section we have $n=2$ and $Im\,\lambda \neq 0$. Then a basis of the space of polynomial functions on V which are quasihomogeneous of degree m is given by the functions of the form (1.61) where (k,r,s) runs through the elements of the set*

$$\mathfrak{I}_\lambda := \begin{cases} \{(k,r,s)\in \mathbb{N}_o^3\,;\; r-s = d := m/\lambda\} & \text{if } Re\,\lambda = 0 \\ \{(k,r,s)\in \mathbb{N}_o^3;\; k \le \min\{m_R, m_I\},\; r = m_R - k,\; s = m_I - k\} & \text{if } Re\,\lambda \neq 0 \end{cases},$$

the numbers m_R and m_I being defined in Remark 1.40.(ii).

Proof. We fix a polynomial function on V which is quasihomogeneous of degree m. Besides (1.60.a) one obtains from (1.59) the following conditions:

$$(1.60.b) \qquad P_{(k,\ell),(r,0)} = 0\,, \qquad ((k,\ell),(r,0)) \in \mathfrak{A}_m,\; k \ge 1,$$

$$(1.60.c) \qquad P_{(k,0),(r,s)} = 0\,, \qquad ((k,0),(r,s)) \in \mathfrak{A}_m,\; r \ge 1,$$

and

$$(1.60.d) \qquad P_{(k+1,\ell-1),(r-1,s+1)} = -\frac{r}{k+1} P_{(k,\ell),(r,s)}\,, \quad ((k,\ell),(r,s)) \in \mathfrak{A}_m,\; \ell, r \ge 1.$$

Now, the proof that P is of the desired form is done by induction on deg P, the case $\deg P = 1$ being obvious. If P is divisible by z_2 or $\overline{z_2}$ then we apply the induction hypothesis to the quotient (which is quasihomogeneous, as well) and obtain the assertion for P. Hence we suppose that P is divided neither by z_2 nor by $\overline{z_2}$. This means that both the sets $\mathfrak{I} := \{(\alpha,\beta)\in\mathfrak{A}_m\,;\; \alpha_2 = 0,\; P_{\alpha,\beta}\neq 0\}$ and $\mathfrak{I}' := \{(\alpha,\beta)\in\mathfrak{A}_m\,;\; \beta_2 = 0,\; P_{\alpha,\beta}\neq 0\}$ are non-empty. Note that in view of the conditions (1.60.c) and (1.60.b) we have $\mathfrak{I} = \{(\alpha,\beta)\in\mathfrak{A}_m\,;\; \alpha_2 = 0 = \beta_1,\; P_{\alpha,\beta}\neq 0\}$ and $\mathfrak{I}' := \{(\alpha,\beta)\in\mathfrak{A}_m\,;\; \beta_2 = 0 = \alpha_1,\; P_{\alpha,\beta}\neq 0\}$. In fact, we are going to show that

(1.62) (a) $\mathfrak{I} \subset \{((k,0),(0,k));\ k \in \mathbb{N}_o\}$; (b) $\mathfrak{I}' \subset \{((0,k),(k,0));\ k \in \mathbb{N}_o\}$;

(c) $\mathfrak{K} := \{k \in \mathbb{N}_o;\ ((k,0),(0,k)) \in \mathfrak{I}\} = \{k \in \mathbb{N}_o;\ ((0,k),(k,0)) \in \mathfrak{I}'\}$;

(d) $P_{(k,0),(0,k)} = (-1)^k P_{(0,k),(k,0)}$, $k \in \mathfrak{K}$.

We can then deduce that the polynomial function R defined by

$$R(z) := P(z) - \sum_{k \in \mathfrak{K}} P_{(k,0),(0,k)} (z_1\overline{z_2} - \overline{z_1}z_2)^k$$

is divisible by z_2 (and, in fact, by $\overline{z_2}$). Applying the first part of the induction step to R we conclude that P is of the desired form which completes the proof of the induction step.

In order to prove (1.62) we fix $((k,0),(0,s)) \in \mathfrak{I}$ and $((0,\ell'),(r',0)) \in \mathfrak{I}'$ and deduce from (1.60.d) that

(1.63) $$P_{(k,0),(0,s)} = (-1)^j \frac{j!}{k(k-1)\cdots(k-j+1)} P_{(k-j,j),(j,s-j)}, \quad j \le \min\{k,s\},$$

and

(1.63)' $$P_{(0,\ell'),(r',0)} = (-1)^j \frac{r'(r'-1)\cdots(r'-j+1)}{j!} P_{(j,\ell'-j),(r'-j,j)}, \quad j \le \min\{\ell',r'\}.$$

In <u>*the case "Re $\lambda \neq 0$"*</u> we deduce from Remark 1.40.(ii) that $k = m_R = \ell'$ and $s = m_I = r'$. If $m_I \le m_R$ then (1.63) shows that $P_{(k-s,s),(s,0)} \neq 0$ which in view of (1.60.b) implies that $k=s$, i.e. $m_I = m_R$ and the equation in (1.62.d) is valid for $k = m_R$; in particular, we have $\mathfrak{K} = \{m_R\}$.

If $m_I \ge m_R$ then (1.63)' shows that $P_{(\ell',0),(r'-\ell',\ell')} \neq 0$ which in view of (1.60.c) implies that $\ell' = r'$, i.e. $m_I = m_R$ and the equation in (1.62.d) is valid for $k = m_R$; in particular, we have $\mathfrak{K} = \{m_R\}$, again.

Since in both subcases we have shown that $m_I = m_R$ the conclusions of both subcases are valid, i.e. (1.62) is proved.

Finally, we come to <u>*the case "Re $\lambda = 0$"*</u>. Here $\mathfrak{A}_m = \{(\alpha,\beta) \in \mathbb{N}_o^n \times \mathbb{N}_o^n;\ |\alpha| - |\beta| = d\}$ so that $k - s = d = \ell' - r'$. If $d \ge 0$ then (1.63) shows that $P_{(k-s,s),(s,0)} \neq 0$ which in view of (1.60.b) implies that $k=s$, i.e. $d=0$, and the equation in (1.62.d) holds.

On the other hand, if $d \le 0$ then (1.63)' shows that $P_{(\ell',0),(r'-\ell',\ell')} \neq 0$ which in view of (1.60.c) implies that $r' = \ell'$, i.e. $d = 0$, and the equation in (1.62.d) holds with k replaced by r'.

Since, again, in both subcases we have shown that $d = 0$ the conclusions of both subcases are valid, i.e. (1.62) is proved. ■

(e) Almost Quasihomogeneous Functions

In the present section we give a short discussion of general almost quasihomogeneous functions as defined in Definition 1.32 above. In section (c) this notion was motivated by the appearance of almost quasihomogeneous polynomial functions which are not quasihomogeneous. However, in view of Corollary 1.36 in case M is semi-simple this motivation becomes pointless. In sections (c) and (d) of Chapter 2 below we shall meet almost quasihomogeneous functions (and distributions) in a quite different context, thereby obtaining a more serious motivation which is also valid in case M is semi-simple.

Below we have to make use of two simple properties of the functions ω_j. The first one is an immediate consequence of the homomorphy property of the logarithm and of the binomial formula:

$$(1.64) \qquad \omega_j(st) = \sum_{k=0}^{j} \omega_k(s)\,\omega_{j-k}(t), \qquad j \in \mathbb{N}_0,\ s,t \in \,]0,+\infty[.$$

The second one is obvious:

$$(1.65) \qquad \omega_j(1/t) = (-1)^j\,\omega_j(t), \qquad j \in \mathbb{N}_0,\ t \in \,]0,+\infty[.$$

We fix $N \in \mathbb{N}$.

Lemma 1.48. *Suppose that $q_0 : X \longrightarrow \mathbb{C}$ is almost quasihomogeneous of degree m and of order $\leq N$. Then for every $k \in \mathbb{N}_N$ the function q_k appearing in (1.39) is uniquely determined and almost quasihomogeneous of degree m and of order $\leq N-k$; more precisely, we have*

$$(1.39)_k \qquad t^{-m} q_k(M_t x) = \sum_{\ell=0}^{N-k} \omega_\ell(t)\, q_{k+\ell}(x), \qquad (x,t) \in X \times_M]0,+\infty[.$$

In particular, the so-called almost quasihomogeneous order of q_0 *is well-defined by*

$$\operatorname{ord}_M(q_0) := \max\{k \in \mathbb{N}_N \cup \{0\}\,;\ q_k \neq 0\}.$$

Proof. Since $X \times_M]0,+\infty[$ is an open subset of $X \times]0,+\infty[$ the uniqueness of the functions q_k is a consequence of (1.42).

For the proof of $(1.39)_k$ we fix $(x,s) \in X \times_M]0,+\infty[$. Since X is open we find an open subinterval I of $]0,+\infty[$ such that $1 \in I$ and $M_{st}x \in X$ for arbitrary $t \in I$. Applying (1.39) twice and taking (1.64) into account we obtain

$$t^m \sum_{k=0}^{N} \omega_k(t)\, q_k(M_s x) = q_0(M_t M_s x) = q_0(M_{st} x) =$$

$$= (st)^m \sum_{j=0}^{N} \omega_j(st)\, q_j(x) = (st)^m \sum_{j=0}^{N} \sum_{k=0}^{j} \omega_k(t)\, \omega_{j-k}(s)\, q_j(x) =$$

$$= t^m \sum_{k=0}^{N} \omega_k(t)\, a_k$$

where

$$a_k := s^m \sum_{j=k}^{N} \omega_{j-k}(s)\, q_j(x) = s^m \sum_{\ell=0}^{N-k} \omega_\ell(s)\, q_{k+\ell}(x)\,.$$

By (1.42) it follows that $q_k(M_s x) = a_k$, as desired. ∎

The function q_k will also be called *the* k^{th} *order deficiency of* q_0. It is not only uniquely determined by q_0 but can also easily be computed out of q_0. To show this we make use of the following more precise version of (1.42):

Remark 1.49. *Let* $(t_j)_{j\in\mathbb{N}_N}$ *be a sequence of pairwise different numbers in the set* $]0,+\infty[\setminus\{1\}$. *Then the* $N\times N$*-matrix*

$$(1.66) \qquad \left(\omega_k(t_j)\right)_{(j,k)\in\mathbb{N}_N\times\mathbb{N}_N}$$

is invertible.

Proof. We set $s_j := \log t_j$. It then follows that

$$\det\left(\omega_k(t_j)\right)_{(j,k)\in\mathbb{N}_N\times\mathbb{N}_N} = \prod_{k=1}^{N} \frac{1}{k!} \prod_{j=1}^{N} s_j\; V(s_1,\ldots,s_N)$$

where $V(s_1,\ldots,s_N)$ denotes the Vandermonde determinant of $s_1,\ldots,s_N$. ∎

By subtracting $q_0(x)$ from both sides of (1.39), inserting $t=t_j$, and applying the inverse of the matrix (1.66) one obtains

Remark 1.50. *Let* $q_0 : X \longrightarrow \mathbb{C}$ *be almost quasihomogeneous of degree* m *and of order* $\leq N$. *Then for every* $k\in\mathbb{N}_N$ *its* k^{th} *order deficiency is given by*

$$q_k = \sum_{j=1}^{N} L_{kj}\,(t_j^{-m} q_0 \circ M_{t_j} - q_0) \quad \text{on the set } \{x\in X;\ \forall j\in\mathbb{N}_N:\ M_{t_j}x\in X\}$$

where $(L_{kj})_{(k,j)\in\mathbb{N}_N\times\mathbb{N}_N}$ *denotes the inverse of the matrix (1.66).* ∎

Since $X \times_M]0,+\infty[$ is an open subset of $X \times]0,+\infty[$ one obtains the following consequence of Remark 1.50.

Proposition 1.51. *Let $q_O : X \longrightarrow \mathbb{C}$ be a function which is almost quasihomogeneous of degree m and of order $\leq N$.*

(i) If q_O belongs to $\mathcal{L}^1_{loc}(X)$ so do its deficiencies.

(ii) Let $r \in \mathbb{N}_O \cup \{\infty, \omega\}$; if q_O is a C^r function so are its deficiencies. ■

Almost quasihomogeneous functions of degree 0 with constant deficiencies and quasihomogeneous functions of arbitrary degree are related in a simple way:

Remark 1.52. *(i) Let $f : X \longrightarrow \mathbb{C}$ be almost quasihomogeneous of degree 0 such that its 1st order deficiency is identically equal to a constant $\ell \in \mathbb{C}$; then $\exp \circ f$ is quasihomogeneous of degree ℓ.*

(ii) Conversely, if $q : X \longrightarrow]0,+\infty[$ is quasihomogeneous of degree $\ell \in \mathbb{C}$ then $\log q$ is almost quasihomogeneous of degree 0, its 1st order deficiency being identically equal to ℓ.

Proof. *(i)*: $\exp(f(M_t x)) = \exp(f(x) + \ell \log t) = t^\ell \exp(f(x))$.

(ii): $\log q(M_t x) = \log(t^\ell q(x)) = \ell \log t + \log q(x)$. ■

How to obtain almost quasihomogeneous functions of higher order is shown in

Remark 1.53. *Let $\nu : X \longrightarrow \mathbb{C}$ be almost quasihomogeneous of degree 0 and order 1 such that its 1st order deficiency is identically equal to 1. Then for every $j \in \mathbb{N}_O$ the function $\nu_j := \nu^j / j!$ is almost quasihomogeneous of degree 0 and order j such that in case $j \geq 1$ its 1st order deficiency equals ν_{j-1}.*

Proof. By the binomial formula we have for every $(x,t) \in X \times_M]0,+\infty[$

$$\nu_j(M_t x) = \frac{1}{j!} (\nu(x) + \log t)^j = \frac{1}{j!} \sum_{i=0}^{j} \binom{j}{i} \nu(x)^{j-i} (\log t)^i = \sum_{i=0}^{j} \omega_i(t)\, \nu_{j-i}(x).$$ ■

The preceding remark is also a special case of the following immediate consequence of Definition 1.32 and of (1.38).

Remark 1.54. *If f (resp. g) is almost quasihomogeneous of degree m (resp. ℓ) and of order $\leq N$ (resp. L) then fg is almost quasihomogeneous of degree $m+\ell$ and of order $\leq N+L$ such that for every $k \in \mathbb{N}_{N+L}$ its kth order deficiency is equal to*

$$\sum_{\substack{j=0 \\ j\geq k-L}}^{N} \binom{k}{j} f_j g_{k-j}$$

where h_j denotes the jth order deficiency of $h \in \{f,g\}$. ∎

In view of Proposition 1.51.(ii) one obtains the following proposition by combining Proposition 1.34.(iii) and the conditions (1.49)' and (1.38).

Proposition 1.55. *Let $r \in \mathbb{N}$, let $q_0 : X \longrightarrow \mathbb{C}$ be a C^r function which is almost quasihomogeneous of degree m and of order $\leq N$, and let $P: V^* \longrightarrow \mathbb{C}$ be a polynomial function of degree $\leq r$ which is almost quasihomogeneous of degree ℓ and of type M^*. Then $P(\partial)q_0$ is almost quasihomogeneous of degree $m-\ell$ and of type M.* ∎

The treatment of almost quasihomogeneous differential operators with variable coefficients applied to almost quasihomogeneous functions is postponed until almost quasihomogeneous distributions have been introduced (see Proposition 2.35 below). Likewise the infinitesimal characterization of differentiable almost quasihomogeneous functions will be carried out in the setting of almost quasihomogeneous distributions (see Proposition 2.31 below). Instead we show that (almost) quasihomogeneous functions on X can be extended to the smallest open subset of V containing X which is invariant under the action of M_t. This result is required for the proof of Proposition 2.37 below.

Definition 1.56. If Y is any subset of V then

(i) its *quasihomogeneous hull (of type M)* Y_M is defined by $Y_M := \bigcup_{t\in]0,+\infty[} M_t Y$;

(ii) Y is called *quasihomogeneous (of type M)* if and only if $Y = Y_M$.

Y_M is obviously quasihomogeneous. Note that Y_M is open if Y is so.

Proposition 1.57. *If q is almost quasihomogeneous of degree m and of order $\leq N$*

then there exists a unique almost quasihomogeneous extension $q_M : X_M \longrightarrow \mathbb{C}$ *of the same degree and order. It is defined by*

$$(1.67)\qquad q_M(M_t x) := t^m \sum_{\ell=0}^{N} \omega_\ell(t)\, q_\ell(x)\,, \qquad (x,t) \in X \times]0,+\infty[\,,$$

and has the following properties:

(i) For every $k \in \mathbb{N}_N$ *the* k^{th} *order deficiency of* q_M *is equal to* $(q_k)_M$ *where* q_k *denotes the* k^{th} *order deficiency of* q.

(ii) If $q \in \mathcal{L}^1_{loc}(X)$ *then* $q_M \in \mathcal{L}^1_{loc}(X_M)$.

(iii) If $r \in \mathbb{N} \cup \{\infty, \omega\}$ *is such that* $q \in C^r(X)$ *then* $q_M \in C^r(X_M)$.

Proof. If q_M is an almost quasihomogeneous extension of q it must satisfy (1.67). On the other hand, let $r, s, t \in]0,+\infty[$ and $x, y \in X$ such that $M_r(M_t x) = M_s y$. Then $y = M_{rt/s} x$, and it follows by $(1.39)_k$ that

$$s^m \sum_{k=0}^{N} \omega_k(s)\, q_k(y) = s^m \sum_{k=0}^{N} \omega_k(s)\, (rt/s)^m \sum_{\ell=0}^{N-k} \omega_\ell(rt/s)\, q_{k+\ell}(x) =$$

$$= r^m t^m \sum_{I=0}^{N} \Big(\sum_{k=0}^{I} \omega_{I-k}(rt/s)\, \omega_k(s) \Big)\, q_I(x)\,.$$

From (1.64) and (1.65) we deduce that

$$\sum_{k=0}^{I} \omega_{I-k}(rt/s)\, \omega_k(s) = \sum_{k=0}^{I} \sum_{j=0}^{I-k} \omega_j(rt)\, (-1)^{I-k-j}\, \omega_{I-k-j}(s)\, \omega_k(s) =$$

$$= \sum_{j=0}^{I} \omega_j(rt)\, E_j(s)$$

where

$$E_j(s) := \sum_{k=0}^{I-j} (-1)^{I-k-j} \binom{I-j}{k}\, \omega_{I-j}(s) = (1-1)^{I-j}\, \omega_{I-j}(s) = \delta_{Ij}\,.$$

Moreover, from (1.64) we observe that $\omega_I(rt) = \sum_{i=0}^{I} \omega_i(r)\, \omega_{I-i}(t)$. Combining this with the preceding equations and changing the order of summation we obtain

$$s^m \sum_{k=0}^{N} \omega_k(s)\, q_k(y) = r^m \sum_{i=0}^{N} \omega_i(r) \Big(t^m \sum_{I=i}^{N} \omega_{I-i}(t)\, q_I(x) \Big)\,.$$

The case "$r=1$" shows that q_M is well-defined by (1.67). Applying this to $(q_k, N-k)$ instead of (q,N) and taking Lemma 1.48 into account we conclude that for every $k \in \mathbb{N}_N$ a function $(q_k)_M : X_M \longrightarrow \mathbb{C}$ is well-defined by

$$(1.67)_k\qquad (q_k)_M(M_t x) := t^m \sum_{\ell=0}^{N-k} \omega_\ell(t)\, q_{k+\ell}(x)\,, \qquad (x,t) \in X \times]0,+\infty[$$

as well. The general case of the formula above can then be written as

$$q_M(M_r(M_t x)) = r^m \sum_{i=0}^{N} \omega_i(r)\,(q_i)_M(M_t x)\ ,$$

i.e. q_M is almost quasihomogeneous of degree m and of order $\leq N$; more precisely, $(1.39)_k$ holds with q_k replaced by $(q_k)_M$.

For the proof of the regularity properties *(ii)* and *(iii)* it suffices in view of Proposition 1.51 to observe that for every $z \in X_M$ there are $t \in]0,+\infty[$ and an open neighbourhood U of z in X_M such that $M_t(U) \subset X$ and

$$q_M|_U = t^{-m} \sum_{i=0}^{N} \omega_i(1/t)\,(q_i \circ M_t)|_U\ . \quad \blacksquare$$

Under special assumptions on X and M the only quasihomogeneous C^∞ functions on X are polynomial functions. This follows from

Proposition 1.58. *Suppose that $\sigma = \sigma_+$. Let $r \in \mathbb{N}$ be such that*

$$(1.68) \qquad Re\, m/r < \lambda_{min} := min\{ Re\,\lambda\ ;\ \lambda \in \sigma_+ \}\ .$$

And let $f \in C^r(V)$ be such that $f|_{\dot V}$ is almost quasihomogeneous of degree m . If f does not vanish identically then $m \in \mathfrak{A}(M)$, and f is a polynomial function of degree $\leq r-1$.

The proof relies on a lemma for which the following important notation is required.

Notation 1.59. (i) $X_\pm := X \setminus G_M(\sigma_0 \cup \sigma_\mp) = \{ x \in X\,;\ P_\pm(x) \neq 0 \}$. If M is of the form (1.1.a) then $X_\pm = \{ x \in X\,;\ x_j \neq 0$ for some $j \in J_\pm \}$ where $J_\pm := \{ j \in \mathbb{N}_n\,;\ \pm p_j > 0 \}$.

(ii) $X^o := X \cap M_o^{-1}(X) = X \cap M_o^{-1}(X \cap G_M(0))$.

Since X is an open subset of V so are $X_\pm$ and X^o. Moreover, for the same reason X^o equals $X \cap M_o^{-1}(M_o(X \setminus G_M(0)))$.

Lemma 1.60. *Suppose that (1.14) is valid. Let $f \in C^o(X)$. If the restriction of f to X_+ is almost quasihomogeneous of degree m then f itself is almost quasihomogeneous of degree m, and one of the following conditions holds:*

(α) f vanishes on X^o;

(β) $m = 0$, and f is quasihomogeneous on X^o;

(γ) $\operatorname{Re} m > 0$, and f vanishes on $X \setminus X_+$.

Proof. By Proposition 1.57 we may assume that X is quasihomogeneous. Let g be the restricition of f to X_+. We set $N := \operatorname{ord}_M(g)$, and suppose that $f \not\equiv 0$ on X^o. By Remark 1.50 for every $k \in \mathbb{N}_N$ the kth order deficiency of g extends to a continuous function $q_k : X \longrightarrow \mathbb{C}$. Let $x \in X_+$ such that $M_0 x \in X$. The equation (1.39) can be written in the form

$$(1.39)' \qquad f(M_t x) = t^m \left(f(x) + R_x(\log t) \right) \quad \text{where } R_x(s) := \sum_{k=1}^{N} \frac{1}{k!} q_k(x)\, s^k, \; s \in \mathbb{R}.$$

Since every point of $X \setminus X_+$ is in the closure of X_+ it follows from the continuity of f, q_k, and M_t that (1.39)' is valid for every $x \in X \setminus X_+$, as well. This shows that f is almost quasihomogeneous of degree m.

The continuity of f and $[0, +\infty[\ni t \mapsto M_t$ (see Corollary 1.9) implies that

$$(1.69) \qquad f(M_0 x) = \lim_{t \to 0} f(M_t x).$$

If $m = 0$ then it follows from (1.39)' and (1.69) that $R_x \equiv 0$, i.e. $q_k \equiv 0$ on X^o for every $k \in \mathbb{N}_N$. Hence, we suppose that $m \neq 0$ and $f(x) \neq 0$. Then (1.39)' and (1.69) imply that $\operatorname{Re} m > 0$ and $f(M_0 x) = 0$. ∎

Proof of Proposition 1.58. Let $\alpha \in \mathfrak{A}$ be such that $|\alpha| = r$. Since by Remark 1.43.(i) ∂^α is of the form $\partial^\alpha = P_\alpha(\partial)$ for some polynomial function $P_\alpha : V^* \longrightarrow \mathbb{C}$ which is almost quasihomogeneous of degree α_M and of type M^* Proposition 1.55 shows that the function $f^{(\alpha)}|_{\dot{V}}$ is almost quasihomogeneous of degree $m - \alpha_M$. Since $\operatorname{Re} \alpha_M \geq |\alpha| \lambda_{min}$ it follows from (1.68) that $\operatorname{Re}(m - \alpha_M) < 0$. Hence Lemma 1.60 implies that $f^{(\alpha)}$ vanishes identically, and in view of Taylor's formula as stated in Lemma 1.26 f is a polynomial function of degree $\leq r-1$, indeed. The assertion about m is then a consequence of Proposition 1.34.(iii). ∎

The converse of Lemma 1.60 is valid in the following form.

Proposition 1.61. *Suppose that (1.14) is valid. Let $q_0 : X_+ \longrightarrow \mathbb{C}$ be a continuous function which is almost quasihomogeneous of degree m and of order $\leq N$, its deficiencies being denoted by $q_1, \ldots, q_N$. If $\operatorname{Re} m > 0$ then for every $k \in \{0\} \cup \mathbb{N}_N$*

the extension $\dot{q}_k$ of q_k to X defined by $\dot{q}_k/_{X\setminus X_+} :\equiv 0$ is continuous.

Proof. By Proposition 1.57 we may assume that X is quasihomogeneous. By Proposition 1.51.(ii) we observe that the deficiencies of q_0 are continuous functions on X_+, as well.

Now we fix $x \in X\setminus X_+$, i.e. $P_+(x) = 0$, and choose $\varepsilon > 0$ so small that the compact neighbourhood $K := \{y \in V;\ |P_+ y| \leq \varepsilon,\ |M_0(y-x)| \leq \varepsilon\}$ of x is contained in X. Then $L := \{y \in V;\ |P_+ y| = \varepsilon,\ |M_0(y-x)| \leq \varepsilon\}$ is a compact subset of X_+.

Now let $(x_\ell)_{\ell \in \mathbb{N}}$ be a sequence in X_+ converging to x as $\ell \to \infty$. We choose $\ell_0 \in \mathbb{N}$ such that $x_\ell \in \mathring{K}$ for every $\ell \geq \ell_0$, and for every $\ell \geq \ell_0$ we fix $t_\ell \in]0, +\infty[$ such that $M_{t_\ell} x_\ell \in L$. Since $P_+(x_\ell)$ converges to $P_+(x) = 0$ as $\ell \to \infty$ it follows that t_ℓ tends to $+\infty$ as $\ell \to \infty$. By induction on N we are going to prove that

$$\lim_{\ell \to \infty} t_\ell^{\varepsilon}\, q_k(x_\ell) = 0\,, \qquad \varepsilon \in [0, \operatorname{Re} m[\,, \tag{1.70}$$

for every $k \in \mathbb{N}_N \cup \{0\}$. Note that in view of $\dot{q}_k(x) = 0$ the condition (1.70) for $\varepsilon = 0$ is the desired continuity property.

Suppose first that (1.70) is already proved for every $k \in \mathbb{N}_N$. By (1.39) it follows that

$$|q_0(x_\ell)| - \sum_{k=1}^{N} |\log t_\ell|^k\, |q_k(x_\ell)| \leq |t_\ell^{-m} q_0(M_{t_\ell} x_\ell)| \leq t_\ell^{-\operatorname{Re} m}\, c\,, \qquad \ell \geq \ell_0,$$

where $c := \sup\{|q_0(x)|;\ x \in L\}$. Since c is finite the induction hypothesis implies that (1.70) is valid for $k = 0$, as well.

Since for every $i \in \mathbb{N}_N$ the pair $(q_i, N-i)$ instead of (q_0, N) satisfies the assumptions, as well, an induction argument completes the proof of (1.70). ∎

(f) The Hypersurfaces $S^\varkappa$

In this section we discuss a simple but important method for constructing positive quasihomogeneous C^r functions on X out of C^r hypersurfaces in X and vice versa. We suppose that X is quasihomogeneous, feeling justified by Proposition 1.57. First of all we observe

Remark 1.62. *Let $f:X \longrightarrow \mathbb{C}$ be quasihomogeneous of degree m. If f has no zeros and if $\mathrm{Re}\, m \neq 0$ then by $\varkappa(x) := |f(x)|^{1/\mathrm{Re}\, m}$ a function $\varkappa: X \longrightarrow]0,+\infty[$ is defined which is quasihomogeneous of degree 1.* ∎

So in the following we deal with such functions $\varkappa$.

Proposition 1.63. *Let $r \in \mathbb{N} \cup \{\infty, \omega\}$. Then for every subset N of X the following conditions are equivalent:*

(a) There exists a (unique) C^r function $\varkappa: X \longrightarrow]0,+\infty[$ which is quasihomogeneous of degree 1 such that $N = S^{\varkappa} := \varkappa^{-1}(1)$;

(b) N is a C^r submanifold of X of codimension 1 such that every quasihomogeneous ray in X intersects N in exactly one point, the intersection being transversal, i.e. for every $x \in X$ the map i_x (defined in Proposition 1.10) is transversal to N such that $|i_x^{-1}(N)| = 1$.

Proof. (a) ⇒ (b): Let $x \in \varkappa^{-1}(1)$. Then $\varkappa \circ i_x = \mathrm{Id}_{]0,+\infty[}$ so that i_x induces a homeomorphism onto its image. Moreover, differentiating the relation $\varkappa(M_t x) = t$ with respect to t yields

$$D\varkappa(M_t x) \cdot i_x'(t) = 1\,, \qquad t \in]0,+\infty[\,. \tag{1.71}$$

Consequently, i_x is immersive so that its image, the quasihomogeneous ray $R_M(x)$ through x, is a real analytic submanifold of X, the tangent space $T_x R_M(x)$ of $R_M(x)$ at x being generated by $i_x'(1)$. Moreover, (1.71) shows that $D\varkappa(x) \neq 0$ so that $S^{\varkappa}$ is a C^r hypersurface at x. Since $\ker D\varkappa(x)$ is equal to $T_x S^{\varkappa}$ one deduces from (1.71) that V is the direct sum of $T_x S^{\varkappa}$ and $T_x R_M(x)$. Finally, if $y \in X$ and $t \in]0,+\infty[$ then $1 = \varkappa(M_t y) = t\,\varkappa(y)$ if and only if $t = 1/\varkappa(y)$.

(b) ⇒ (a): We define $\varkappa: X \longrightarrow]0,+\infty[$ by the condition

$$\{M_t x\,;\; t \in]0,+\infty[\} \cap N = \{M_{1/\varkappa(x)} x\}\,, \qquad x \in X.$$

Then for any $s \in]0,+\infty[$ we have

$$\{M_{s/\varkappa(M_s x)} x\} = \{M_t M_s x\,;\; t \in]0,+\infty[\} \cap N = \{M_{1/\varkappa(x)} x\}$$

so that $\varkappa(M_s x) = s\,\varkappa(x)$, i.e. $\varkappa$ is quasihomogeneous of degree 1 with $\varkappa^{-1}(1) = N$. To see that $\varkappa$ is C^r we fix $x \in X$, set $y := M_{1/\varkappa(x)} x \in N$, and choose an open neighbourhood U of y in X and a C^r submersion $f: U \longrightarrow \mathbb{R}$ such that $f^{-1}(0) = U \cap N$.

We set $W := \{(z,t) \in X \times]0,+\infty[;\ M_{1/t}z \in U\}$ and define a C^r function $g: W \longrightarrow \mathbb{R}$ by $(z,t) \mapsto f(M_{1/t}z)$. Then for any $(z,t) \in W$ we have: $g(z,t)=0 \iff M_{1/t}z \in N \iff t=\varkappa(z)$. Hence the implicit function theorem implies that $\varkappa$ is C^r near x provided that $\partial_2 g(x,\varkappa(x)) \neq 0$. Now, from the equation $g(x,t) = f(i_y(\varkappa(x)/t))$ we deduce

$$\partial_2 g(x,\varkappa(x)) = Df(y)\cdot(-\frac{1}{\varkappa(x)}\, i_y'(1)).$$

Since in view of the equality $\ker Df(y) = T_yN$ the transversality assumption means that $i_y'(1) \notin \ker Df(y)$ it follows that $\partial_2 g(x,\varkappa(x))$ does not vanish, as required. ■

Recall from the beginning of the proof of Proposition 1.63 that the condition (a) implies that i_x induces a homeomorphism onto its image for every $x \in X$. Hence, in view of Proposition 1.10.(ii) a necessary condition for the existence of a function $\varkappa$ as in condition (a) of Proposition 1.63 is that $X \cap E_M(\sigma_0)$ is empty. As we shall see later, in general this condition is by no means sufficient. Next we turn to the question when positive quasihomogeneous functions $\varkappa$ exist locally.

Lemma 1.64. *For every $x \in V$ the following conditions are equivalent:*

(a) there is a quasihomogeneous open neighbourhood U of x in V and a continuous (resp. real analytic) function $\varkappa: U \longrightarrow]0,+\infty[$ which is quasihomogeneous of degree 1;

(b) for some neighbourhood Y of x in V the set $\{t \in]0,+\infty[;\ Y \cap M_tY \neq \emptyset\}$ is relatively compact in $]0,+\infty[$;

(c) $x \notin E_M(\sigma_0)$, and the map

$$\{(y,M_ty);\ y \in V\setminus E_M(\sigma_0),\ t \in]0,+\infty[\} \longrightarrow]0,+\infty[,\ (y,M_ty) \mapsto t, \tag{1.72}$$

(which is well-defined by Proposition 1.10.(ii)) is continuous at (x,x).

Proof. *(a)⇒(b).* Let $\varkappa$ be a continuous function having the properties of condition (a), and let Y be any compact neighbourhood of x in U. Then $I := \inf \varkappa(Y) > 0$ and $S := \sup \varkappa(Y) < +\infty$. Hence for arbitrary $t \in]0,+\infty[$ and $y \in Y$ satisfying $M_ty \in Y$ we conclude that $t = \varkappa(M_ty)/\varkappa(y) \leq S/I$ and $\geq I/S$.

(b)⇒(c): Let Y be a neighbourhood of x having the properties postulated in (b). First of all we observe that by Proposition 1.10.(iii) the map $i_x:]0,+\infty[\longrightarrow V$, $x \mapsto M_tx$, is injective and $R_M(x)$ is non-compact. If x would belong to $E_M(\sigma_0$

then the assertions (ii) and (iv.B) of Proposition 1.10 would imply the existence of an unbounded sequence $(t_j)_{j\in\mathbb{N}}$ such that $(M_{t_j}x)_{j\in\mathbb{N}}$ converges to x as $j\to\infty$. Since this contradicts (b) we conclude that $x\notin E_M(\sigma_0)$. Now, to deduce the second part of (c) we let $(x_j)_{j\in\mathbb{N}}$ be a sequence in $X\setminus E_M(\sigma_0)$ and $(t_j)_{j\in\mathbb{N}}$ a sequence in $]0,+\infty[$ such that $\lim_{j\to\infty}(x_j, M_{t_j}x_j) = (x,x)$. This means, in particular, that for some $N\in\mathbb{N}$ we have $(x_j, M_{t_j}x_j)\in Y\times Y$ for $j\geq N$. Hence, by condition (b) (t_j) stays in a compact subset of $]0,+\infty[$. Hence, there is a subsequence of (t_j) converging to some $t\in]0,+\infty[$. By the continuity of the map $(y,s)\mapsto M_s y$ this implies that $x = M_t x$. Since the map i_x is injective we conclude that $t=1$. By the same argument we see that every subsequence of (t_j) has a subsequence converging to 1. This means that the sequence (t_j) itself converges to 1, as is to be shown.

(c)⇒(a). Let W be real subspace of V of codimension 1 not containing Mx, and let Y be an open neighbourhood of x in $V\setminus E_M(\sigma_0)$ such that

(1.73) $\qquad My\notin W$ for every $y\in Y$.

Moreover, we claim that by making Y still smaller if necessary we can achieve that for every $y\in N:=Y\cap(W+x)$ the quasihomogeneous ray $R_M(y)$ intersects N in precisely one point. Indeed, if not then we could find a sequence $(y_j)_{j\in\mathbb{N}}$ in $W+x$ converging to x and a sequence $(t_j)_{j\in\mathbb{N}}$ in $]0,+\infty[\setminus\{1\}$ such that $x_j := M_{t_j}y_j$ belongs to $W+x$ for every $j\in\mathbb{N}$ and such that $(x_j)_{j\in\mathbb{N}}$ converges to x, as well, as $j\to\infty$. It follows from (c) that $\lim_{j\to\infty} t_j = 1$. The main theorem of calculus and (1.18) show that for every $j\in\mathbb{N}$ we have

$$w_j := \frac{1}{t_j-1}\left(M_{t_j}y_j - y_j\right) = \int_0^1 M_{\tau_j(s)} M y_j \frac{ds}{\tau_j(s)}$$

where $\tau_j(s) := st_j + (1-s) = 1 + s(t_j-1)$. Consequently, $\lim_{j\to\infty} w_j = Mx$. Since $w_j\in W$ for every $j\in\mathbb{N}$ and since W is closed this contradicts (1.73), and the claim is proved.

Next we observe that in view of (1.18) and (1.73) the inverse function theorem shows that by making Y and hence N smaller we achieve that there exist an open neighbourhood Z of x and some $\varepsilon>0$ such that the map $N\times]1-\varepsilon,1+\varepsilon[\longrightarrow Z$, $(y,t)\mapsto M_t y$, is a real analytic diffeomorphism. It follows that $U:=N_M$ is an open subset of V. In view of the claim proven before and in view of (1.18) and (1.73) all the quasihomogeneous rays through the points of N intersect N transversally. Hence we may apply Proposition 1.63 to U instead of X and obtain the desired $\varkappa$. ∎

Notation 1.65. By $\mathfrak{R}(M)$ we denote the set of all points $x \in V$ satisfying one (and hence each) of the conditions of Lemma 1.64.

In fact, the points in $\mathfrak{R}(M)$ satisfy the condition (c) in a much stronger form:

Lemma 1.66. *The set $\mathfrak{M} := \{(x, M_t x);\ x \in \mathfrak{R}(M),\ t \in]0,+\infty[\}$ is a real-analytic submanifold of $V \times V$, and the map $\mathfrak{M} \longrightarrow]0,+\infty[$, $(x, M_t x) \mapsto t$, is well-defined and real-analytic.*

Proof. We set $Y := \mathfrak{R}(M)$. By $\Psi : Y \times]0,+\infty[\longrightarrow Y \times Y$ we denote the real analytic immersion defined by $(x,t) \mapsto (x, M_t x)$. Let $x \in Y$, and choose U and $\varkappa$ as in condition (a) of Lemma 1.64. Then by $(y,z) \mapsto \varkappa(z)/\varkappa(y)$ a map $\Phi : U \times U \longrightarrow]0,+\infty[$ is well-defined which is real-analytic. Since $\varkappa$ is quasihomogeneous of degree 1 it follows from Euler's equation that $D\varkappa$ has no zeros. Hence Φ is a submersion. Denoting by $\Pi_1 : U \times U \longrightarrow U$ the projection onto the first factor, we conclude that

$$(\Pi_1, \Phi) \circ \Psi|_{U \times]0,+\infty[} = \mathrm{Id}_{U \times]0,+\infty[}\,.$$

From this the assertions follow. ■

Via the condition (b) of Lemma 1.64 we deduce from Remark 1.8 the first inclusion of the following remark (employing Notation 1.59.(i) for $X = V$), the second inclusion being valid in view of Lemma 1.64.(c):

Remark 1.67. $V_+ \cup V_- \subset \mathfrak{R}(M) \subset V \setminus E_M(\sigma_0)$. ■

Note that the complement of $V_+ \cup V_-$ in V equals $G_M(\sigma_0)$. At first one might suspect that for the inclusion on the right-hand side actually equality holds. However, in case there exists an eigenvalue $\lambda \in \sigma_0 \setminus \{0\}$ of algebraic multiplicity ≥ 3 this is not so. This can be seen from the explicit description of the set $\mathfrak{R}(M)$ which is postponed to section (g) below. Now we are going to construct globally defined positive quasihomogeneous function $\varkappa$ on $V_\pm$. In case M is of the form (1.1.a) this is very easy:

Example 1.68. *Suppose that M is of the form (1.1.a).*

(i) If $p \in]0,+\infty[^n$ then $(X,N) = (\dot{\mathbb{R}}^n, S^{n-1})$ satisfies the condition (b) of Propo-

sition 1.63; the corresponding $\varkappa$ that satisfies (a) is denoted by $\varkappa_p$.

(ii) In general, if one sets

$$S_{\pm}^{p} := \{x \in X;\ \sum_{j \in J_{\pm}} x_j^2 = 1\}$$

then the condition (b) of Proposition 1.63 is satisfied for $(X,N) = (X_{\pm}, S_{\pm}^{p})$; the corresponding real analytic quasihomogeneous function $\varkappa$ is denoted by $\varkappa_p^{\pm}$.

<u>*Proof*</u>. <u>*(i)*</u>: Since the assumption on p means that $J_+ = \mathbb{N}_n$ and $S_+^p = S^{n-1}$ this is a special case of (ii).

<u>*(ii)*</u>: Let $x \in S_{\pm}^{p}$. We define $\xi \in V = \mathbb{R}^n$ by

$$\xi_j := \begin{cases} x_j & \text{if } j \in J_{\pm} \\ 0 & \text{otherwise} \end{cases} .$$

Then ξ is a normal unit vector to $S_{\pm}^{p}$ at x. Since

$$\pm\langle \xi, i_x'(1) \rangle = \sum_{j \in J_{\pm}} (\pm p_j)\, x_j^2 > 0$$

it follows that $i_x'(1)$, the generator of $T_x R_M(x)$, does not belong to $T_x S_{\pm}^{p}$. Since the map

$$]0,+\infty[\, \ni t \mapsto \sum_{j \in J_{\pm}} t^{2p_j} x_j^2$$

is strictly increasing (resp. decreasing), converging to $+\infty$ (resp. 0) as $t \to +\infty$ and to 0 (resp. $+\infty$) as $t \to 0$ the assertion follows. ■

For general M a similar construction works. For the proof we require

Lemma 1.69. *For every $\varepsilon > 0$ there exists a scalar product $\langle\langle \cdot, \cdot\cdot \rangle\rangle$ on V having the following properties - here $\|\cdot\|$ denotes the norm defined by $\langle\langle \cdot, \cdot\cdot \rangle\rangle$:*

(i) the generalized eigenspaces $G_M(\lambda)$, $\lambda \in \sigma$, are pairwise orthogonal;

(ii) for every $\lambda \in \sigma$ we have

$$(1.74) \qquad (Re\,\lambda - \varepsilon)\|x\|^2 \le \langle\langle x, Mx \rangle\rangle \le (Re\,\lambda + \varepsilon)\|x\|^2, \qquad x \in G_M(\lambda);$$

(iii) for every $x \in V$ we have

$$(1.75) \qquad \sum_{\lambda \in \sigma} t^{2Re\lambda - 2\varepsilon}\|x_\lambda\|^2 \le \|M_t x\|^2 \le \sum_{\lambda \in \sigma} t^{2Re\lambda + 2\varepsilon}\|x_\lambda\|^2, \qquad t \in [1,+\infty[,$$

and

$$(1.75)' \qquad \sum_{\lambda\in\sigma} t^{2Re\lambda+2\varepsilon} \|x_\lambda\|^2 \le \|M_t x\|^2 \le \sum_{\lambda\in\sigma} t^{2Re\lambda-2\varepsilon} \|x_\lambda\|^2, \qquad t\in]0,1].$$

Proof. Let $\lambda\in\sigma$, let U be an M-cyclic subspace of $G_M(\lambda)$, and let $(b_1,\dots,b_d)$ be a K_λ-basis of U which brings the matrix of $\tilde{M}_\lambda$ into Jordan canonical form, i.e. (1.11) holds. Setting $c_j := \varepsilon^{j-1} b_j$ we obtain another basis of U such that $Nc_1 = 0$ and $Nc_{j+1} = \varepsilon c_j$, $j\in\mathbb{N}_{d-1}$. We now define

$$\langle\langle \sum_{j=1}^d z_j c_j, \sum_{j=1}^d w_j c_j \rangle\rangle := \operatorname{Re} \sum_{j=1}^d z_j \overline{w_j}$$

and observe that

$$\langle\langle \sum_{j=1}^d z_j c_j, M\Big(\sum_{j=1}^d z_j c_j\Big) \rangle\rangle = \operatorname{Re}\lambda \,\Big\| \sum_{j=1}^d z_j c_j \Big\|^2 + \varepsilon \operatorname{Re} \sum_{j=1}^{d-1} z_j \overline{z_{j+1}}$$

so that by the Cauchy-Schwarz inequality we obtain

$$\Big| \langle\langle x, Mx \rangle\rangle - \operatorname{Re}\lambda \,\|x\|^2 \Big| \le \varepsilon \|x\|^2, \qquad x\in U.$$

Writing V as a direct sum of such spaces U and defining $\langle\langle \cdot,\cdot\cdot \rangle\rangle$ in such a way that these spaces are orthogonal to each other we arrive at the assertions *(i)* and *(ii)*.

For the proof of *(iii)* let $x\in G_M(\lambda)\setminus\{0\}$, define $\gamma:\mathbb{R}\longrightarrow V$ by $\gamma(s) := \exp(sM)x$, observe that $\gamma'(s) = M\gamma(s)$, and compute

$$\frac{d}{ds}\|\gamma(s)\| = \|\gamma(s)\|^{-1} \langle\langle \gamma(s), M\gamma(s) \rangle\rangle$$

so that by (1.74) one concludes that

$$\operatorname{Re}\lambda - \varepsilon \le \frac{d}{ds} \log \|\gamma(s)\| \le \operatorname{Re}\lambda + \varepsilon.$$

Integrating this one obtains

$$(\operatorname{Re}\lambda - \varepsilon)s \le \log\|\gamma(s)\| - \log\|\gamma(0)\| \le (\operatorname{Re}\lambda+\varepsilon)s$$

so that

$$\exp((\operatorname{Re}\lambda-\varepsilon)s) \le \|\gamma(s)\|/\|x\| \le \exp((\operatorname{Re}\lambda+\varepsilon)s), \qquad s\in]0,+\infty[$$

Applying the same reasoning to $s\mapsto\gamma(-s)$ instead of γ one deduces

$$\exp((\operatorname{Re}\lambda+\varepsilon)s) \le \|\gamma(s)\|/\|x\| \le \exp((\operatorname{Re}\lambda-\varepsilon)s), \qquad s\in]-\infty,0[$$

Inserting $s = \log t$ one arrives at (iii). ∎

Proposition 1.70. *Let $c,d\in]0,+\infty[$ be such that $1/c < Re\lambda < 1/d$ for every $\lambda\in\sigma_+$. Then there is a norm $\|\cdot\|$ on $G_M(\sigma_+)$ induced by a scalar product such that th* *condition (b) of Proposition 1.63 is satisfied for* $(X,N) = (X_+, S_+)$ *where*

$$S_+ := \{x \in X;\ \|P_+(x)\| = 1\} \tag{1.76}$$

and such that the corresponding real analytic function $\varkappa_+$ *defined by condition (a) of Proposition 1.63 satisfies the following estimates:*

$$\min\{\|P_+(x)\|^c, \|P_+(x)\|^d\} \le \varkappa_+(x) \le \max\{\|P_+(x)\|^c, \|P_+(x)\|^d\}, \quad x \in X_+. \tag{1.77}$$

Moreover, for every $x \in G_M(\sigma_+)\setminus\{0\}$ *the map* $]0,+\infty[\longrightarrow]0,+\infty[,\ t \mapsto \|M_t x\|$, *is strictly increasing.*

<u>*Proof*</u>. Setting $\alpha := 1/c$ and $\beta := 1/d$ and applying Lemma 1.69 we choose a scalar product $\langle\langle \cdot, \cdot\cdot \rangle\rangle$ on $G_M(\sigma_+)$ such that (1.74) is valid with $(\mathrm{Re}\lambda-\varepsilon, \mathrm{Re}\varepsilon+\varepsilon)$ replaced by (α,β) and such that

$$t^{\alpha}\|x\| \le \|M_t x\| \le t^{\beta}\|x\|, \qquad x \in G_M(\sigma_+),\ t \in [1,+\infty[. \tag{1.78}$$

Applying this to $M_{1/t}x$ instead of x we deduce

$$t^{\beta}\|x\| \le \|M_t x\| \le t^{\alpha}\|x\|, \qquad x \in G_M(\sigma_+),\ t \in]0,1]. \tag{1.78)'$$

We fix $x \in X_+$. Since by (1.18) and (1.74)

$$\frac{d}{dt}\|P_+ M_t x\|^2 = \frac{2}{t}\langle\langle P_+ M_t x, M P_+ M_t x\rangle\rangle \ge \frac{2}{t}\alpha\|M_t P_+ x\|^2 > 0$$

the function $]0,+\infty[\ni t \mapsto \|P_+ M_t x\|$ is strictly increasing so that $i_x^{-1}(S_+)$ consists of precisely one element. Moreover, i_x is transversal to S_+ at the point of intersection.

In order to prove the estimates (1.77) we observe that by the definition of $\varkappa := \varkappa_+$ we have $1 = \|M_{1/\varkappa(x)}P_+ x\|$ so that in case $\varkappa(x) < 1$ (resp. >1) we deduce from (1.78) (resp. (1.78)') that

$$\varkappa(x)^{-\alpha}\|P_+ x\| \le 1 \le \varkappa(x)^{-\beta}\|P_+ x\|, \quad \text{i.e.} \quad \|P_+ x\|^c \le \varkappa(x) \le \|P_+ x\|^d$$

$$(\text{resp.} \quad \varkappa(x)^{-\beta}\|P_+ x\| \le 1 \le \varkappa(x)^{-\alpha}\|P_+ x\|, \quad \text{i.e.} \quad \|P_+ x\|^d \le \varkappa(x) \le \|P_+ x\|^c).$$

Since $\varkappa(x) < 1$ (resp. >1) if and only if $\|P_+ x\| < 1$ (resp. >1) the proof is complete. ■

Note that for X_- one similarly obtains a real analytic quasihomogeneous function $\varkappa_- : X_- \longrightarrow]0,+\infty[$ by applying Proposition 1.70 to $-M$ instead of M and setting $\varkappa_- := 1/\varkappa_+$.

For the sake of easy reference we formulate a special form of the situation of Proposition 1.70 as a separate condition:

(1.79) the scalar product $\langle\cdot,\cdot\cdot\rangle$ on V is such that the generalized eigenspaces $G_M(\lambda)$, $\lambda\in\sigma$, are orthogonal to each other and such that the restriction of $\langle\cdot,\cdot\cdot\rangle$ to $G_M(\sigma_+)$ has the properties of Proposition 1.70 where c and d are given numbers satisfying the requirements of Proposition 1.70 ; and let $\varkappa_+$ be defined as in Proposition 1.70 .

In section 3.(c) below, in the context of quasihomogeneous averages, one finds necessary and sufficient conditions on X for the existence of functions $\varkappa$ satisfying the condition (a) of Proposition 1.63 (see Theorem 3.39 and Proposition 3.34). At present we suppose that such a function $\varkappa: X \longrightarrow]0,+\infty[$ is given and introduce a map which implicitly appeared in the proof of Proposition 1.63, already.

Notation 1.71. We define (i) $\rho_\varkappa: X \longrightarrow S^\varkappa$ by $x \mapsto M_{1/\varkappa(x)}x$ and

(ii) $\Phi_\varkappa: X \longrightarrow S^\varkappa \times]0,+\infty[$ by $x \mapsto (\rho_\varkappa(x), \varkappa(x))$.

Proposition 1.72. *(i) If $\varkappa$ is continuous then $\rho_\varkappa$ is a continuous retraction of X onto $S^\varkappa$ satisfying*

(1.80) $$\rho_\varkappa \circ M_t = \rho_\varkappa, \qquad t\in]0,+\infty[,$$

and $\Phi_\varkappa$ is a homeomorphism, its inverse being given by $(\vartheta,t)\mapsto M_t\vartheta$.

(ii) Let $r\in\mathbb{N}\cup\{\infty,\omega\}$; if $\varkappa$ is a C^r function then $\rho_\varkappa$ is a C^r submersion, and $\Phi_\varkappa$ is a C^r diffeomorphism.

Proof. Since the map $(t,x)\mapsto M_{1/t}x$ is real analytic the regularity properties of $\rho_\varkappa$ follow by the chain rule. The condition (1.80) and the retraction property are immediate. That $\rho_\varkappa$ is a submersion follows from the facts that $\rho_\varkappa|_{S^\varkappa}$ is equal to the identity map of $S^\varkappa$ and that by (1.80) one has $D\rho_\varkappa(M_t\vartheta) = D\rho_\varkappa(\vartheta)\circ M_{1/t}$ for arbitrary $\vartheta\in S^\varkappa$ and $t\in]0,+\infty[$. Finally, the assertions about $\Phi_\varkappa$ are consequences of the fact that the restriction to $S^\varkappa\times]0,+\infty[$ of the real analytic map $X\times]0,+\infty[\longrightarrow X$, $(\xi,t)\mapsto M_t\xi$, is inverse to $\Phi_\varkappa$. ■

In view of Proposition 1.72 it is easy to describe all quasihomogeneous functions on X with the help of $\varkappa$:

Remark 1.73. *(i) If* $g: S^{\varkappa} \longrightarrow \mathbb{C}$ *is any function then* $q := \varkappa^m g \circ \rho_{\varkappa}$ *is quasihomogeneous of degree m, and* $q_{/S^{\varkappa}} = g$ *; if* $\varkappa$ *and* g *are* C^r *so is* q.

(ii) A function $f: X \longrightarrow \mathbb{C}$ *is quasihomogeneous of degree m if and only if*

$$f = \varkappa^m (f_{/S^{\varkappa}}) \circ \rho_{\varkappa} . \quad \blacksquare$$

It is not difficult to obtain a similar description of almost quasihomogeneous functions on X. To this end we introduce certain basic almost quasihomogeneous functions associated with $\varkappa$:

Remark 1.74. *For every* $k \in \mathbb{N}_0$ *the function* $\eta_k := \omega_k \circ \varkappa$ *is almost quasihomogeneous of degree 0 and of order* $\leq k$ *such that*

$$\eta_k \circ M_t = \sum_{j=0}^{k} \omega_j(t)\, \eta_{k-j} . \tag{1.81}$$

Proof. By the binomial formula we have

$$\eta_k(M_t x) = \frac{1}{k!}(\log t + \log \varkappa(x))^k = \sum_{j=0}^{k} \omega_j(t)\, \omega_{k-j}(\varkappa(x)) . \quad \blacksquare$$

Remark 1.73'. *(i) If* $g_0, \ldots, g_N : S^{\varkappa} \longrightarrow \mathbb{C}$ *are arbitrary functions then by*

$$q_k := \sum_{j=0}^{N-k} \eta_j\, \varkappa^m\, g_{k+j} \circ \rho_{\varkappa} , \qquad 0 \leq k \leq N, \tag{1.82}$$

a sequence of functions is defined satisfying $q_{k/S^{\varkappa}} = g_k$ *such that* q_0 *is almost quasihomogeneous of degree* $\leq m$ *and of order* $\leq N$ *with* k^{th} *order deficiency* q_k*; moreover, if* $\varkappa$ *and the* g_j *are* C^r *so are the* q_k.

(ii) Suppose that q_0 *is almost quasihomogeneous of degree m and of order* $\leq N$*; then*

$$q_0 = \varkappa^m \sum_{k=0}^{N} \eta_k\, (q_{k/S^{\varkappa}}) \circ \rho_{\varkappa} \tag{1.83}$$

where (q_k) *is the sequence of deficiencies of* q_0 *defined by (1.39).*

Proof. (i): Taking (1.81) into account and changing the order of summation we deduce for every $t \in]0,+\infty[$ that

$$q_0 \circ M_t = \sum_{j=0}^{N} \sum_{i=0}^{j} \omega_i(t)\,\eta_{j-i}\, t^m \varkappa^m g_j \circ \rho_\varkappa = t^m \sum_{i=0}^{N} \omega_i(t) \sum_{j=i}^{N} \eta_{j-i}\, \varkappa^m g_j \circ \rho_\varkappa .$$

Since by definition the very last sum is equal to q_i the assertion follows.

(ii): Let $x \in X$ and choose $t \in]0,+\infty[$ such that $\vartheta := M_{1/t}x \in S^\varkappa$. Then $t = \varkappa(x)$, $\omega_k(t) = \eta_k(x)$, and $\vartheta = \rho_\varkappa(x)$ so that (1.83) follows from (1.39). ∎

(g) Determining the Set $\mathfrak{L}(M)$

In this section we give a fairly explicit description of the set $\mathfrak{L}(M)$ defined in Notation 1.65. This will be made use of in the proof of Remark 2.30 and in § 3.(c), only. The following lemma contains the basic part of the proof.

Lemma 1.75. *Let $d \in \mathbb{N}$ and $\zeta \in \mathbb{C}^{1+d}$, and let P be a polynomial of degree $\le d$.*

(i) If there are a sequence of polynomials P_n, $n \in \mathbb{N}$, of degree $\le d$ converging to P and an unbounded sequence $(s_n)_{n\in\mathbb{N}}$ in $\mathbb{C}$ such that $\lim_{n\to\infty} (P_n^{(j)}(s_n)/j!)_{0\le j\le d} = \zeta$ then

(α) $\deg P \le D := d/2$; (β) $\zeta_j = 0$ for $j > D$; (γ) if d is even then

$$\zeta_D = (-1)^D P^{(D)}(0)/D! . \tag{1.84}$$

(ii) Conversely, if the conditions (α), (β), and (γ) are valid then there is an analytic family of polynomials P_s, $s \in \dot{\mathbb{C}}$, of degree $\le d$ such that

$$\text{(a)} \lim_{|s|\to\infty} P_s^{(\ell)}(0) = P^{(\ell)}(0), \text{ and } \text{(b)} \lim_{|s|\to\infty} P_s^{(\ell)}(s) = \ell!\,\zeta_\ell , \quad 0 \le \ell \le d, \tag{1.85}$$

and such that the coefficients of P_s are real for every $s \in \dot{\mathbb{R}}$ provided that the coefficients of P and ζ are real.

For the proof of this and other assertions a simple formula is required:

Lemma 1.76. *Let $j,k,\ell \in \mathbb{N}_0$ such that $k \le j+\ell$. Then*

$$\sum_{i=0}^{\min\{\ell,\, j+\ell-k\}} (-1)^i \binom{\ell}{i} \binom{j+\ell-i}{k} = \begin{cases} 0 & \text{if } k < \ell \\ \binom{j}{k-\ell} & \text{if } k \ge \ell \end{cases} . \tag{1.86}$$

Proof. By the binomial formula we obtain

$$\sum_{i=0}^{\ell} \binom{\ell}{i} (-1)^i t^{j+\ell-i} = t^j (t-1)^\ell, \qquad t \in \mathbb{R}.$$

Differentiating both sides k-times with respect to t (with the right-hand side making use of Leibniz' formula) and evaluating the result at $t=1$ we deduce

$$\sum_{i=0}^{\min\{\ell,j+\ell-k\}} \binom{\ell}{i} (-1)^i \frac{(j+\ell-i)!}{(j+\ell-i-k)!} = \begin{cases} 0 & \text{if } k<\ell \\ \binom{k}{\ell} \frac{j!}{(j-k+\ell)!} \ell! & \text{if } k \geq \ell \end{cases}.$$

Dividing by k! we arrive at the desired result. ■

As a consequence we are going to derive

Lemma 1.77. *Let $L,n \in \mathbb{N}$ and $r \in \mathbb{N}_0$ be such that $r+n \leq L$. Then the matrix*

(1.87.a) $\qquad B := (b_{ij})_{r \leq i \leq r+n, L \leq j \leq L+n}$

where

(1.87.b) $\qquad b_{ij} := \binom{j}{i}$

is invertible, the first row of its inverse being given by

(1.88) $\qquad (-1)^{i+r} \binom{L+n-i}{L-r} \binom{i}{r} \Big/ \binom{L}{r}, \qquad r \leq i \leq r+n.$

Proof. We first deal with the case "$r=0$". That B is invertible is readily seen by computing its determinant: for every $L+1 \leq j \leq L+n$ subtracting the $(j-1)^{\text{th}}$ column from the jth one, making use of the relation $\binom{j}{i} - \binom{j-1}{i} = \binom{j-1}{i-1}$ and expanding with respect to the first row we see that $\det B$ is equal to the determinant of the $n \times n$-matrix with entries b_{ij}, $0 \leq i \leq n-1$, $L \leq j \leq L+n-1$. Hence we deduce by induction on n that $\det B = 1$.

Applying (1.86) to $(j, L+n-j)$ instead of (ℓ, j) and noting that the inequalities $j \geq L$ and $k \geq L$ imply $(L+n-j)+j-k = L+n-k \leq n \leq L \leq j$ and $L+n-k \leq n$ we see that the matrix C with entries

$$c_{ki} := \begin{cases} (-1)^i \binom{L+n-i}{k} & \text{if } 0 \leq i \leq L+n-k \\ 0 & \text{if } L+n-k < i \leq n \end{cases}, \qquad L \leq k \leq L+n,$$

has the property that $\Delta := CB$ is a lower triangular matrix, its diagonal entries Δ_{kk} being equal to 1. This shows that Δ (and - by the way - hence B) is invertible and that Δ^{-1} is a lower triangular matrix satisfying $(\Delta^{-1})_{kk} = 1$ for $L \le k \le L+n$. Since $B^{-1} = \Delta^{-1}C$ this implies that the first row of B^{-1} coincides with the first row of C. This proves (1.88) in case $r = 0$.

For the proof of *the case "r>0"* one observes that

$$\binom{j}{i} = \binom{j-r}{i-r}\binom{j}{r}\Big/\binom{i}{r}$$

so that the assertion follows from the case "r = 0" by the following standard fact. ∎

Remark 1.78. *Let (a_{ij}) be an invertible $n\times n$-matrix with complex entries, and let $b, c \in (\mathbb{C}\setminus\{0\})^n$. Then the $n\times n$-matrix $(b_i a_{ij} c_j)$ is invertible, its inverse being given by $\left(\frac{1}{c_i} d_{ij} \frac{1}{b_j}\right)$ where (d_{ij}) denotes the inverse of (a_{ij}).* ∎

Proof of Lemma 1.75. (i). We may assume that $\lim_{n\to\infty} |s_n| = +\infty$. Setting $a_{n,j} := P_n^{(j)}(0)/j!$ and $a_j := P^{(j)}(0)/j!$ we deduce by induction on k that

$$\text{(1.89)} \qquad \lim_{n\to\infty} \sum_{j=k}^{d} a_{n,j}\, s_n^{j-k+1} = \delta_{k1}\zeta_0 - a_{k-1}, \qquad k \in \mathbb{N}_d .$$

Indeed, for $k=1$ this is due to the fact that by the assumptions $P_n(s_n)$ converges to ζ_0 and $a_{n,0}$ to a_0 as $n\to\infty$; and if the equation in (1.89) is valid for some $k \in \mathbb{N}_{d-1}$ then writing the sum in the form $s_n c_{n,k}$ where

$$c_{n,k} := \left(\sum_{j=k}^{d} a_{n,j}\, s_n^{j-k}\right)$$

and using the fact that $|s_n|$ tends to $+\infty$ as $n\to\infty$ we conclude that $\lim_{n\to\infty} c_{n,k} = 0$ which implies the equality in (1.89) for $k+1$ instead of k because $(a_{n,k})_{n\in\mathbb{N}}$ tends to a_k as $n\to\infty$. Applying the same reasoning to the polynomials $\frac{1}{i!}P_n^{(i)}$, $0 \le i \le d$, we arrive at the system of equations

$$\lim_{n\to\infty} \sum_{j=i+k}^{d} a_{n,j}\, b_{ij}\, s_n^{j-i-k+1} = \delta_{k1}\zeta_i - a_{i+k-1}\, b_{i,i+k-1}, \qquad k \in \mathbb{N}_{d-i},$$

where b_{ij} is defined in (1.87.b). Substituting $L=i+k$ and setting $a_{n,j,L} := a_{n,j}\, s_n^{j-L+1}$ we rewrite this as

$$\text{(1.89)}_i \qquad \lim_{n\to\infty} \sum_{j=L}^{d} a_{n,j,L}\, b_{ij} = \delta_{i,L-1}\zeta_i - a_{L-1}\, b_{i,L-1} \qquad i+1 \le L \le d .$$

From this we are going to deduce that the sequences $(a_{n,j,L})_{n\in\mathbb{N}}$ converge for certain indices j and L. To this end we have to invert certain matrices with entries b_{ij}. We fix $L \in \mathbb{N}_d$ and observe that with $(1.89)_i$, $0 \le i \le L-1$, we have L equations for the d-(L-1) quantities $a_{n,j,L}$. So in order to be able to form square matrices of size d-L+1 we have to postulate $L \geq d-L+1$, i.e. $2L \geq d+1$ or $L > D$. Assuming this we set $n := d-L$ and fix a number $r \in \mathbb{N}_0$ such that $r \leq L-n-1 = 2L-d-1$ and denote by B_r the matrix defined in Lemma 1.77. Since by this lemma the matrix B_r is invertible we deduce from $(1.89)_i$ that

$$(1.90) \qquad \lim_{n\to\infty} (a_{n,j,L})_{L\leq j\leq d} = B_r^{-1} v_r \qquad \text{where} \quad v_r := \zeta_{L-1}\eta_r - a_{L-1}\beta_r$$

with

$$\eta_r := (\delta_{i,L-1})_{r\leq i\leq n+r} \quad \text{and} \quad \beta_r := (b_{i,L-1})_{r\leq i\leq n+r}.$$

Since $|s_n|$ tends to $+\infty$ as $n\to\infty$ this implies that

$$a_j = \lim_{n\to\infty} a_{n,j} = 0, \qquad L \leq j \leq d.$$

Since this is so, in particular, if L is equal to the smallest natural number L_0 strictly larger than D we have proved <u>*(α)*</u>. Moreover, since the condition (1.90) for $L = L_0$ shows that for every $L_0 \leq j \leq d$ the sequence of numbers a_{n,j,L_0}, $n \in \mathbb{N}$, is bounded we deduce that in case $L \geq L_0+1$ we even have $\lim_{n\to\infty} a_{n,j,L} = 0$ for $L \leq j \leq d$ so that $(1.89)_{L-1}$ in combination with (α) shows that <u>*(β)*</u> is valid.

In order to prove <u>*(γ)*</u> we assume that $D \in \mathbb{N}$ and set $L = D+1$. Then observing that r may be chosen between 0 and 1 we conclude from the fact that the left-hand side of (1.90) does not depend on r that

$$B_0^{-1} v_0 = B_1^{-1} v_1.$$

Since by our present choice of L we have $n = d-L = L-2$ we see that $\eta_0 = 0$ and $\eta_1 \neq 0$ so that

$$(1.91) \qquad a_D (B_1^{-1}\beta_1 - B_0^{-1}\beta_0) = \zeta_D B_1^{-1}\eta_1.$$

This already shows that ζ_D and a_D are proportional by a constant only depending on D. To compute the constant we first observe that by Lemma 1.77 we have

$$(B_r^{-1}\beta_r)_L = \sum_{i=0}^{n} (-1)^i \binom{L-r+n-i}{L-r} \binom{r+i}{r} b_{i+r,D} \Big/ \binom{L}{r}.$$

Since

$$\binom{r+i}{r} b_{i+r,D} = \binom{D-r}{i} \binom{D}{r}$$

we can apply Lemma 1.76 to (n+1, L-r, D-r) instead of (j, k, ℓ) - observing that $n+1+D-r \geq L-r$ and $\min\{D-r, n+1+D-r-(L-r)\} = n$ - and obtain that

$$(B_r^{-1}\beta_r)_L = \binom{n+1}{L-D} \binom{D}{r} \Big/ \binom{L}{r} = (n+1) \binom{D}{r} \Big/ \binom{L}{r}.$$

Moreover, by (1.88) again and in view of $n = d - L = D - 1$ we see that

$$(B_1^{-1}\eta_1)_L = (-1)^n \binom{1+n}{1} \Big/ \binom{L}{1} = (-1)^{D-1}\frac{D}{L}.$$

Since $L = D+1$ it follows that

$$(B_1^{-1}\beta_1 - B_0^{-1}\beta_0)_L = D\frac{D}{L} - D = -\frac{D}{L}.$$

Inserting this into (1.91) and dividing by $\frac{D}{L}$ we conclude that $a_D = (-1)^D \zeta_D$, as desired.

(ii). Let n be the largest natural number strictly smaller than D. Note that in case d is odd (resp. even) we have $n = D - 1/2$, i.e. $d - n = D + 1/2$ (resp. $n = D - 1$, i.e. $d - n = D+1$). For every $j \in \mathbb{N}_0$ such that $j \le D$ we set $a_j := P^{(j)}(0)/j!$. We are going to construct P_s in the following form

$$P_s := \sum_{0 \le j \le D} a_j T^j + \sum_{j=d-n}^{d} a_j(s) T^j.$$

where the coefficients $a_j(s)$, $d-n \le j \le d$, are to be defined in such a way that for every $i \in \{0\} \cup \mathbb{N}_n$ in $(1.89)_i$ equality holds for $L = i+1$ (and $a_{n,j} = a_j(s_n)$) even before the limit is taken. To carry this out we denote by $B(s)$ the matrix with entries

$$s^{j-i} b_{ij}, \qquad 0 \le i \le n, \; d-n \le j \le d.$$

Then by Lemma 1.77 and Remark 1.78 $B(s)$ is invertible, the entries of its inverse being given by

$$s^{I-j} d_{jI}, \qquad d-n \le j \le d, \; 0 \le I \le n,$$

where by d_{jI} we denote the entries of $B(1)^{-1}$. Now we define $a_{d-n}(s), \ldots, a_d(s) \in \mathbb{C}$ by

$$(a_j(s))_{d-n \le j \le d} := B(s)^{-1} \cdot (\zeta_I - a_I - \sum_{I+1 \le J \le D} b_{IJ}\, a_J\, s^{J-I})_{0 \le I \le n}, \quad s \in \mathbb{C}.$$

In view of what was observed about $B(s)^{-1}$ this means that

$$a_j(s) = \sum_{I=0}^{n} s^{I-j} d_{jI} (\zeta_I - a_I - \sum_{I+1 \le J \le D} b_{IJ}\, a_J\, s^{J-I}), \qquad d-n \le j \le d. \tag{1.92}$$

Since $n < D < d-n$ so that $I - j \le J - j \le -1$ for $I \le J \le D$ and $j \ge d-n$ we observe that $a_j(s) = O(1/|s|)$ as $|s| \to \infty$, i.e.

$$\lim_{|s| \to \infty} P_s = P.$$

Moreover, for every $0 \le i \le n$ we have

$$\frac{1}{i!} P_s^{(i)}(s) = \sum_{i \le j \le D} a_j b_{ij} s^{j-i} + R_i(s)$$

where

$$(1.93)_i \qquad R_i(s) := \sum_{j=d-n}^{d} b_{ij} s^{j-i} \Big(\sum_{I=0}^{n} s^{I-j} d_{jI} (\zeta_I - a_I - \sum_{I+1 \le J \le D} b_{IJ} a_J s^{J-I}) \Big) =$$

$$= \sum_{I=0}^{n} s^{I-i} (\zeta_I - a_I) \Big[\sum_{j=d-n}^{d} b_{ij} d_{jI} \Big] - \sum_{1 \le J \le D} a_J s^{J-i} \sum_{I=0}^{J-1} b_{IJ} \Big[\sum_{j=d-n}^{d} b_{ij} d_{jI} \Big].$$

Since by the definition of (d_{jI}) the sums in square brackets are equal to δ_{iI} the left-hand side coincides with

$$\zeta_i - a_i - \sum_{i+1 \le J \le D} b_{iJ} a_J s^{J-i}.$$

Inserting this into the formula above we obtain $\frac{1}{i!} P_s^{(i)}(s) = \zeta_i$, $0 \le i \le n$.

Now, in case d is even we obtain that $\frac{1}{D!} P_s^{(D)}(s) = a_D + R_D(s)$ where R_D is defined by $(1.93)_D$. Since in this case we have $n = D-1$ we observe that

$$R_D(s) = a_D \, (C(D) - 1) + O(1/|s|) \text{ as } |s| \to \infty$$

where

$$C(D) := 1 - \sum_{I=0}^{n} b_{ID} \Big[\sum_{j=d-n}^{d} b_{Dj} d_{jI} \Big].$$

It follows that

$$\lim_{|s| \to \infty} \frac{1}{D!} P_s^{(D)}(s) = C(D) \, a_D.$$

Finally, for $d-n \le i \le d$ we have $n-i < D-i < 0$ so that

$$\frac{1}{i!} P_s^{(i)}(s) = \sum_{j=i}^{n} b_{ij} s^{j-i} \Big(\sum_{I=0}^{n} s^{I-j} d_{jI} (\zeta_I - a_I - \sum_{I+1 \le J \le D} b_{IJ} a_J s^{J-I}) \Big) = O(1/|s|)$$

as $|s| \to \infty$. Hence in view of the condition (β) we have proved that for every $j \in \mathbb{N}_0$ satisfying $0 \le j \le d$ we have

$$\lim_{|s| \to \infty} P_s^{(j)}(s)/j! = (\tilde{\zeta})_j := \begin{cases} \zeta_j & \text{if } j \ne D \\ C(D) \, a_D & \text{if } j = D \end{cases}.$$

So in case d is odd the proof is complete. Hence we suppose that d is even. To complete the proof in this case it suffices, of course, to show that $\tilde{\zeta} = \zeta$. In view of (1.84) this is done once we have shown that

$$(1.94) \qquad C(D) = (-1)^D.$$

Note that it does not seem to be easy to verify (1.94) by direct computation. However, applying the assertion (i) to $\tilde{\zeta} \in \mathbb{C}^{1+d}$ instead of ζ, thereby noting that the sequence of polynomials $P_n := P_{s_n}$, $n \in \mathbb{N}$, satisfies the requirements of (i) if

$(s_n)_{n\in\mathbb{N}}$ is any sequence in $\mathbb{C}$ such that $|s_n|$ tends to $+\infty$ as $n\to\infty$, we obtain $C(D)\,a_D = (-1)^D a_D$. Since for a moment we may assume that $a_D \neq 0$ this implies (1.94), indeed. Hence the proof of Lemma 1.75 is complete. ∎

Now we come to the description of the set $V\setminus\mathcal{L}(M)$. For simplicity we give it with respect to a fixed M-cyclic decomposition $\mathfrak{Z}$ of $G_M(\sigma_0)$ although, certainly, $V\setminus\mathcal{L}(M)$ does not depend on the choice of such a decomposition. We assume that the decomposition $\mathfrak{Z}$ consists of M-invariant subspaces which are indecomposable with respect to M; in particular, for every $Z\in\mathfrak{Z}$ there is a (unique) $\lambda\in\sigma_0$ - denoted by λ_Z - such that $Z\subset G_M(\lambda)$.

Definition 1.79. Let $Z\in\mathfrak{Z}$ and $x\in G_M(\sigma_0)$. Then we denote

(i) by d_Z the K_{λ_Z}-dimension of Z;

(ii) by $i_Z(x)$ the smallest non-negative integer m such that the projection x_Z of x onto Z belongs to $\ker(\tilde{M}_{\lambda_Z})^m$;

(iii) $\mathfrak{Z}_o(x) := \{Z\in\mathfrak{Z};\ i_Z(x) = (d_Z+1)/2\}$, and $\sigma(x) := \{\lambda_Z;\ Z\in\mathfrak{Z}_o(x)\}$;

(iv) by $T(\rho)$ the closure of $\{(\exp(\lambda s))_{\lambda\in\rho};\ s\in\mathbb{R}\}$ in $\mathbb{C}^\rho$ for any non-empty subset ρ of $i\mathbb{R}$.

Proposition 1.80. *The set $V\setminus\mathcal{L}(M)$ consists of all $x\in G_M(\sigma_0)$ satisfying*

$$i_Z(x) \le (d_Z+1)/2, \qquad Z\in\mathfrak{Z}, \tag{1.95}$$

such that in case $\mathfrak{Z}_o(x)\neq\emptyset$ there exists $w\in T(\sigma(x))$ satisfying

$$w_{\lambda_Z} = (-1)^{i_Z(x)-1}, \qquad Z\in\mathfrak{Z}_o(x). \tag{1.96}$$

For later purposes we are going to prove a slightly more general result concerning a notion to be introduced in

Definition 1.81. Let $x,y\in G_M(\sigma_0)$.

(i) The point x is said to be *M-connected to y* if and only if there are sequences $(t_m)_{m\in\mathbb{N}}$ in $]0,+\infty[$ and (y_m) in $G_M(\sigma_0)$ such that

$$\text{(a)}\ \lim_{m\to\infty} t_m = +\infty\,; \quad \text{(b)}\ y = \lim_{m\to\infty} y_m\,; \quad \text{and} \quad \text{(c)}\ x = \lim_{m\to\infty} M_{t_m} y_m\,. \tag{1.97}$$

(ii) $\mathscr{C}_M$ denotes the set of points $x \in G_M(\sigma_o)$ which are M-connected to some $y \in G_M(\sigma_o)$.

Proposition 1.82. *(i) $\mathscr{C}_M$ consists of all points $x \in G_M(\sigma_o)$ satisfying (1.95).*

(ii) Let $x \in \mathscr{C}_M$ and $y \in G_M(\sigma_o)$; then x is M-connected to y if and only if $y \in \mathscr{C}_M$, $\mathfrak{Z}_o(y) = \mathfrak{Z}_o(x)$, and in case $\mathfrak{Z}_o(x) \neq \emptyset$ there exists $w \in T(\sigma(x))$ such that

$$(1.98) \qquad (y_Z)_j = (-1)^{j-1} w_{\lambda_Z} (x_Z)_j \quad \text{for } j = i_Z(x), \qquad Z \in \mathfrak{Z}_o(x).$$

Proof. We fix $Z \in \mathfrak{Z}$ and choose $b_Z \in Z$ such the vectors $b_j := (\tilde{M}_{\lambda_Z})^{d_Z - j}(b_Z)$, $j \in \mathbb{N}_{d_Z}$, form a K_{λ_Z}-basis $\mathfrak{B}_Z$ of Z. For every $z \in Z$ we denote the coordinates of z with respect to $\mathfrak{B}_Z$ by $(z_1, \dots, z_{d_Z})$. The definition of $i_Z(x)$ means that

$$(1.99) \qquad \text{(a)} \ (x_Z)_j = 0, \ j > i_Z(x); \quad \text{and} \quad \text{(b)} \ (x_Z)_j \neq 0 \ \text{if} \ j = i_Z(x) \ \text{and} \ x \neq 0.$$

Moreover, the matrix of $N := \tilde{M}_{\lambda_Z}$ with respect to the basis $\mathfrak{B}_Z$ is of the form (1.11)' so that

$$(M_t)_Z = \exp(s\lambda_Z)\, P_Z(s), \qquad s = \log t, \ t \in]0, +\infty[,$$

where the matrix of $P_Z(s)$ with respect to $\mathfrak{B}_Z$ is given by (1.10.b)'. This means that for every $z \in Z$ the coordinates of $\exp(-s\lambda_Z) M_t z$ with respect to $\mathfrak{B}_Z$ are given by

$$(\exp(-s\lambda_Z)\, M_t z)_i = \sum_{j=0}^{d_Z - i} \frac{1}{j!} z_{i+j}\, s^j = P_{Z,z}^{(i-1)}(s), \qquad 1 \leq i \leq d_Z,$$

where

$$P_{Z,z}(s) := \sum_{j=0}^{d_Z - 1} \frac{1}{j!} z_{j+1}\, s^j.$$

Now we suppose that x is M-connected to y and choose a sequence $(t_m)_{m \in \mathbb{N}}$ in $]0, +\infty[$ and a sequence $(y_m)_{m \in \mathbb{N}}$ in $G_M(\sigma_o)$ satisfying (1.97). In particular, the condition (1.97.a) means that

$$\lim_{n \to \infty} s_n = +\infty$$

where $s_n := \log t_n$. By choosing subsequences we achieve that

$$w_\lambda := \lim_{n \to \infty} \exp(-\lambda s_n) \ \text{ exists for every } \lambda \in \sigma_o.$$

In view of what was said above we conclude for every $i \in \mathbb{N}_{d_Z}$ we have

$$\lim_{n \to \infty} P_{Z,u_n}^{(i-1)}(0) = (y_Z)_i \qquad \text{and} \qquad \lim_{n \to \infty} P_{Z,u_n}^{(i-1)}(s_n) = w_{\lambda_Z} (x_Z)_i$$

where u_n denotes the projection of y_n onto Z. In view of (1.99) Lemma 1.75.(i) implies that 1° $i_Z(v)-1 \le (d_Z-1)/2$ for $v \in \{x,y\}$, i.e. both x and y satisfy (1.95), and 2° if d_Z is odd then $(y_Z)_j = (-1)^{j-1} w_{\lambda_Z}(x_Z)_j$ for $j=(d_Z+1)/2$. This yields the equation $\mathfrak{Z}_o(y) = \mathfrak{Z}_o(x)$ and the condition (1.98). Since in case $\mathfrak{Z}_o(x) \neq \emptyset$ the $\sigma(x)$-tuple $w := (w_\lambda)_{\lambda \in \sigma(x)}$ belongs to $T(\sigma(x))$ we see that the assertions *(i), "⊂"* and *(ii), "⇒"* are proved.

Conversely, we suppose that x and y satisfy (1.95) and that in case $\mathfrak{Z}_o(x) \neq \emptyset$ we can choose $w \in T(\sigma(x))$ satisfying (1.98). By Lemma 1.11 we find a sequence $(s_m)_{m\in\mathbb{N}}$ in $]0,+\infty[$ converging to $+\infty$ such that $w_\lambda = \lim_{j\to\infty} \exp(-\lambda s_j)$ for every $\lambda \in \sigma(x)$. By choosing a subsequence we achieve that

$$w_\lambda := \lim_{j\to\infty} \exp(-\lambda s_j)$$

exists for every $\lambda \in \sigma_o \setminus \sigma(x)$, as well. In view of (1.98) the assumption on $i_Z(x)$ and $i_Z(y)$ allows us for every $Z \in \mathfrak{Z}$ to apply Lemma 1.75.(ii) (its assertion being trivial for $d=0$) to the data $d := d_Z - 1$, $\zeta_j := w_{\lambda_Z}(x_Z)_{j-1}/(j-1)!$, and $P = P_{Z,y_Z}$. In this way we obtain a real analytic function $\Gamma_Z : \dot{\mathbb{R}} \longrightarrow (K_{\lambda_Z})^{d_Z}$ such that with the abbreviation $\gamma_Z := \mathfrak{k}_Z \circ \Gamma_Z$ (here $\mathfrak{k}_Z : (K_{\lambda_Z})^{d_Z} \longrightarrow Z$ denotes the canonical K_{λ_Z}-isomorphism associated with the basis $\mathfrak{B}_Z$) we have

$$\lim_{|s|\to\infty} \Gamma_Z(s)_i = \lim_{|s|\to\infty} (P_{Z,\gamma_Z(s)})^{(i-1)}(0) = (y_Z)_i$$

and

$$\lim_{|s|\to\infty} (P_{Z,\gamma_Z(s)})^{(i-1)}(s) = w_{\lambda_Z}(x_Z)_i, \qquad 1 \le i \le d_Z.$$

By our choice of $(s_m)_{m\in\mathbb{N}}$ and in view of the discussion at the beginning of the proof this implies that the condition (1.97) is satisfied for

$$y_m := \sum_{Z\in\mathfrak{Z}} \gamma_Z(s_m) \quad \text{and} \quad t_m := \log s_m. \quad ■$$

Proof of Proposition 1.80. First of all we are going to see that

(1.100) $\qquad x \notin \mathfrak{K}(M) \iff x$ is M-connected to itself.

In fact, if (1.97) is valid for $y=x$ then the condition (c) in Lemma 1.64 is violated. Conversely, if $x \notin \mathfrak{K}(M)$ then by the same lemma we choose sequences $(t_m)_{m\in\mathbb{N}}$ in $]0,+\infty[$ and $(y_m)_{m\in\mathbb{N}}$ in $G_M(\sigma_o)$ such that the sequence of points $M_{t_m}y_m$ tends to x, as well, and such that no subsequence of (t_m) stays in a compact subinterval of $]0,+\infty[$. By choosing subsequences we may assume that (t_m) converges to $+\infty$ or to 0. Since in the latter case the condition (1.97.c) remains valid with t_m replaced by $1/t_m$ this means that x is M-connected to itself, indeed. Now, in view of (1.100) one obtains the assertion as an immediate consequence of Proposition 1.82, in the proof of "$\subset$", when deducing (1.96) from (1.98), taking (1.99.b) into account. ■

Note that the condition (1.96) is not always satisfied: for example, if there exists $Z \in \mathfrak{Z}$ such that $\lambda_Z = 0$ and such that d_Z and $(d_Z-1)/2$ are odd then (1.96) does not hold for every $x \in \mathfrak{C}_M$ such that $i_Z(x) = (d_Z+1)/2$. Thus, in particular, in view of (1.100) Proposition 1.80 shows that x is not always M-connected to itself. This means that, in general, M-connectedness does not define an equivalence relation as it is not necessarily reflexive. That this, in fact, is the only obstruction is shown by

Corollary 1.83. *(i) Let $x,y \in G_M(\sigma_O)$. Then x is M-connected to y if and only if y is M-connected to x.*

(ii) The following conditions are equivalent:

(a) M-connectedness defines an equivalence relation on $\mathfrak{C}_M$;

(b) every $x \in \mathfrak{C}_M$ is M-connected to itself;

(c) if the set $\mathfrak{Z}_o := \{Z \in \mathfrak{Z};\ d_Z \text{ is odd}\}$ is non-empty there exists $w \in T(\sigma_{odd})$ - where $\sigma_{odd} := \{\lambda_Z;\ Z \in \mathfrak{Z}_o\}$ - such that

$$(1.96)_o \qquad w_{\lambda_Z} = (-1)^{(d_Z-1)/2}, \qquad Z \in \mathfrak{Z}_o.$$

(iii) The condition (c) in (ii) is satisfied if for every $Z \in \mathfrak{Z}$ either the number d_Z is even or d_Z-1 is divisible by 4.

Proof. *(i)*: this is an immediate consequence of the fact that $T(\sigma(x))$ carries the structure of a group when equipped with componentwise multiplication.

(ii). *(a)⇒(b)*: this is trivial.

(b)⇒(c): obviously, one can find an element $x \in \mathfrak{C}_M$ such that $\mathfrak{Z}_o(x) = \mathfrak{Z}_o$ so that the condition (1.96) becomes $(1.96)_o$.

(c)⇒(a): in view of (i) it suffices to show that M-connectedness is transitive. In view of $(1.96)_o$ this follows with the help of the group structure of $T(\sigma_{odd})$

by an application of Proposition 1.82.

(iii): This is obvious since the element $w \in \mathbb{C}^{\sigma_{odd}}$ defined by $w_\lambda := 1$ belongs to $T(\sigma_{odd})$. ∎

For the sake of easy reference we reformulate a part of the proof of Proposition 1.82 as

Lemma 1.84. *Let $x, y \in \mathcal{C}_M$ such that x is M-connected to y. Then there exist $w \in T(\sigma_O)$ and a real analytic function $\delta :]0, +\infty[\setminus \{1\} \longrightarrow G_M(\sigma_O)$ such that for every $s \in \{0, +\infty\}$*

$$\text{(1.101)} \qquad (a) \ \lim_{t \to s} \delta(t) = y, \quad \text{and} \quad (b) \ \lim_{t \to s} N_t \delta(t) = H_w^-(x)$$

where $H_w^{\pm} : G_M(\sigma_O) \longrightarrow G_M(\sigma_O)$ is defined by $x \mapsto ((w_\lambda)^{\pm 1} x_\lambda)_{\lambda \in \sigma_O}$ and where

$$\text{(1.102)} \qquad N_t := H_{\gamma(t)}^- \circ M_t, \quad \text{i.e.:} \quad M_t = H_{\gamma(t)}^+ \circ N_t,$$

with $\gamma :]0, +\infty[\longrightarrow T(\sigma_O)$ being defined by $t \mapsto (t^\lambda)_{\lambda \in \sigma_O}$.

Proof. By the equations following (1.99) in the present notation we have

$$N_t = \bigoplus_{Z \in \mathfrak{Z}} P_Z(\log t),$$

and the second equation in (1.102) holds. Hence, setting

$$\delta(t) := \sum_{Z \in \mathfrak{Z}} \gamma_Z(\log t), \qquad t \in]0, +\infty[\setminus \{1\},$$

we see that the last part of the proof of Proposition 1.82 implies the assertion of the lemma. ∎

(h) Quasihomogeneous Polar Coordinates

In this section we continue to suppose that X is quasihomogeneous (Definition 1.56). Let $\varkappa : X \longrightarrow]0, +\infty[$ be a C^1 function which is quasihomogeneous of degree 1. Then by Proposition 1.63 $S^\varkappa := \varkappa^{-1}(1)$ is a C^1 submanifold of X of codimension 1.

Notation 1.85. By $\tilde{\varkappa}$ we denote the Radon measure on $S^{\varkappa}$ induced by the differential form ω_M on V which is defined as follows:

for every $x \in V$ $\omega_M(x)$ is the contraction $Vol^o \lrcorner (Mx)$ of the canonical oriented volume form Vol^o (associated to the given scalar product on V) by the tangent vector Mx of the quasihomogeneous ray through x at the point x. More explicitly, in coordinates this means

$$\omega_M := \sum_{j=1}^{n} (-1)^{j-1} \Big(\sum_{k=1}^{n} M_{jk} x_k \Big) dx_1 \wedge \ldots \wedge dx_{j-1} \wedge dx_{j+1} \wedge \ldots \wedge dx_n .$$

Below the Lebesgue integral will be formed with respect to orthonormal coordinates (with repect to the given scalar product on V); in other words, integration will be carried out with respect to the density on X defined by Vol^o.

Proposition 1.86. *Let $f: X \longrightarrow \mathbb{C}$ be a function.*

(i) f is (Lebesgue-) integrable if and only if the function defined by

$$(t,\vartheta) \mapsto f(M_t\vartheta)\, t^{\mu-1}$$

belongs to $\mathscr{L}^1(]0,+\infty[\times S^{\varkappa};\, dt \otimes \tilde{\varkappa})$ where $\mu := tr M$ - see (1.6);

(ii) if this is the case then

$$\int_X f(x)\, dx = \int_{]0,+\infty[\times S^{\varkappa}} f(M_t\vartheta)\, t^{\mu-1}\, dt \otimes d\tilde{\varkappa}(\vartheta) .$$

(iii) f is Lebesgue-measurable if and only if the function $(t,\vartheta) \mapsto f(M_t\vartheta)$ is measurable with respect to $dt \otimes \tilde{\varkappa}$.

Proof. For arbitrary $x \in X$ and $\xi_1, \ldots, \xi_{n-1} \in V$ we have

$$\Big(\frac{1}{\varkappa} d\varkappa \wedge (\omega_M|_X)\Big)(x) \cdot (Mx, \xi_1, \ldots, \xi_{n-1}) =$$
$$= \frac{1}{\varkappa(x)} D\varkappa(x) \cdot Mx \; Vol^o(Mx, \xi_1, \ldots, \xi_{n-1}) .$$

Applying Euler's equation for $\varkappa$ (see Proposition 1.22) and denoting by Ω the n-form on V defined by $\Omega :\equiv Vol^o$ one can write this as

$$\frac{1}{\varkappa} d\varkappa \wedge (\omega_M|_X) = \Omega|_X . \tag{1.103}$$

We define a C^∞ map $\varphi :]0,+\infty[\times X \longrightarrow X$ by $(t,x) \mapsto M_t x$. Since

$$d(\varkappa\circ\varphi) = \varkappa\circ\pi_2\,dt + t\pi_2^*(d\varkappa)$$

where $\pi_2 :]0,+\infty[\times X \longrightarrow X$ denotes the canonical projection onto the second factor we deduce

$$\varphi^*\Big(\frac{1}{\varkappa}d\varkappa\wedge(\omega_M|_X)\Big) = \frac{1}{\varkappa\circ\varphi}d(\varkappa\circ\varphi)\wedge\alpha = \frac{1}{t}dt\wedge\alpha + \frac{1}{\varkappa\circ\pi_2}\pi_2^*(d\varkappa)\wedge\alpha \tag{1.104}$$

where

$$\alpha := \varphi^*(\omega_M|_X) .$$

Now, by the definition of φ^* we have for arbitrary $z=(t,x)\in]0,+\infty[\times X$ and $\zeta_1,\ldots$ $\ldots,\zeta_{n-1}\in\mathbb{R}\times V$ that

$$\alpha(z)\cdot(\zeta_1,\ldots,\zeta_{n-1}) = \omega_M(M_t x)\cdot(D\varphi(z)\cdot\zeta_1,\ldots,D\varphi(z)\cdot\zeta_{n-1}) .$$

Writing $\zeta_j=(\tau_j,\xi_j)$ where $\tau_j\in\mathbb{R}$ and $\xi_j\in V$, inserting the equality

$$D\varphi(z)\cdot\zeta_j = \tau_j\partial_1\varphi(z) + D\varphi(t,\cdot)(x)\cdot\xi_j ,$$

and making use of the multilinearity of $\omega_M(M_t x)$ we find an $(n-2)$-form β on $]0,+\infty[\times X$ such that

$$\alpha(z) = [\pi_2^*\varphi(t,\cdot)^*(\omega_M|_X) + dt\wedge\beta](z) .$$

From the definition of ω_M and from (1.4) and (1.6) we derive for arbitrary $t\in]0,+\infty[$, $x\in X$, and $\xi_1,\ldots,\xi_{n-1}\in V$ that

$$\begin{aligned}
&\varphi(t,\cdot)^*(\omega_M|_X)(x)\cdot(\xi_1,\ldots,\xi_{n-1}) = \\
&= \mathrm{Vol}^o\big(M(M_t x), M_t\xi_1,\ldots,M_t\xi_{n-1}\big) = \\
&= \det M_t\ \mathrm{Vol}^o(Mx,\xi_1,\ldots,\xi_{n-1}) = t^\mu\,\omega_M(x)\cdot(\xi_1,\ldots,\xi_{n-1}) .
\end{aligned}$$

It follows that the right-hand side of (1.104) is equal to

$$t^{\mu-1}dt\wedge\pi_2^*(\omega_M|_X) + \frac{1}{\varkappa\circ\pi_2}\pi_2^*(d\varkappa)\wedge\alpha .$$

Combining this with (1.103) and (1.104), introducing the C^1 map $\Psi := \varphi|_{]0,+\infty[\times S^\varkappa}$, and taking the equation $d\varkappa|_{S^\varkappa}\equiv 0$ into account we obtain

$$\Psi^*(\Omega|_X) = t^{\mu-1}dt\wedge(\omega_M|_{S^\varkappa}) .$$

Since by Proposition 1.72 Ψ is a C^1 diffeomorphism the assertions *(i)* and *(ii)* follow by the theory of Radon measures induced by differential forms. Finally, the assertion *(iii)* is a consequence of (i) and (ii). ■

For the following assertion on how to compute $d\tilde{\varkappa}$ we denote by $Vol^o_\varkappa$ the oriented volume form on $S^\varkappa$ induced by the given scalar product on V.

Lemma 1.87. *(i)* $\omega_{M}|_{S^\varkappa} = w_\varkappa Vol^o_\varkappa$ *where* $w_\varkappa(\vartheta) := 1/|grad\, \varkappa(\vartheta)|$, $\vartheta \in S^\varkappa$; *and consequently:*

$$\int_{S^\varkappa} g\, d\tilde{\varkappa} = \int_{S^\varkappa} g\, w_\varkappa\, d\widetilde{Vol}^o_\varkappa, \qquad g \in \mathcal{L}^1(S^\varkappa; \tilde{\varkappa}).$$

(ii) If U (resp. W) is an open subset of X (resp. $\mathbb{R}^{n-1}$*) and if* $\varphi: W \longrightarrow \mathbb{R}$ *is a* C^1 *function satisfying* $U \cap S^\varkappa = graph(\varphi)$ *then*

$$\int_{U\cap S^\varkappa} g\, d\tilde{\varkappa} = \int_W \frac{g}{|\partial_n \varkappa|}(y, \varphi(y))\, dy, \qquad g \in \mathcal{L}^1(S^\varkappa; \tilde{\varkappa}).$$

Proof. (i): Let $x \in S^\varkappa$. First of all one observes that by Euler's equation $grad\, \varkappa(x)$ is non-zero, i.e. $w_\varkappa(x)$ is well-defined and positive. Moreover, another way of writing Euler's equation for $\varkappa$ at x is

(1.105) $\qquad \langle \nu(x), Mx \rangle = w_\varkappa(x)$

where $\nu(x) := w_\varkappa(x)\, grad\, \varkappa(x)$. Consequently, since $\nu(x)$ is the outward normal unit vector to $S^\varkappa$ at x and hence is orthogonal to $T_x S^\varkappa$ one finds $\xi \in T_x S^\varkappa$ satisfying $Mx = w_\varkappa(x)\nu(x) + \xi$. It follows that

$$(Vol^o \lrcorner (Mx))\big|_{T_x S^\varkappa} = w_\varkappa(x)\, (Vol^o \lrcorner\, \nu(x))\big|_{T_x S^\varkappa}.$$

Since the factor of $w_\varkappa(x)$ on the right-hand side is equal to $Vol^o_\varkappa(x)$ the assertion follows.

(ii): Since $\varkappa(y, \varphi(y)) = 1$ we have $\partial_j \varkappa(y, \varphi(y)) + \partial_n \varkappa(y, \varphi(y))\, \partial_j \varphi(y) = 0$ for arbitrary $y \in W$ and $j \in \mathbb{N}_{n-1}$. Consequently, in view of Euler's equation for $\varkappa$ the derivative $\partial_n \varkappa(y, \varphi(y))$ does not vanish, and it follows that

$$1 + |grad\, \varphi(y)|^2 = |grad\, \varkappa(y, \varphi(y))|^2 / |\partial_n \varkappa(y, \varphi(y))|^2, \qquad y \in W.$$

Combining this with the assertion (i) one obtains (ii). ■

If the outward normal unit vectors $\nu(x)$, $x \in S^\varkappa$, are explicitly known then $w_\varkappa(x)$ can be computed from (1.105):

Example 1.88. *Suppose that the condition (1.14) is satisfied, and that* $\varkappa = \varkappa_+$ *is* *defined as in Proposition 1.70. Then*

$$1/|\operatorname{grad} \varkappa(\vartheta)| = \langle P_+(\vartheta), M\vartheta \rangle, \qquad \vartheta \in S^\varkappa.$$

Proof. Since by the choice of the scalar product in Proposition 1.70 we have $\nu(\vartheta) = P_+(\vartheta)$ the assertion follows from (1.105). ■

Combining Proposition 1.86 with the Fubini and the Fubini-Tonelli theorem one obtains a convenient integrability criterion. Here the integrability of the functions $\vartheta \mapsto f(M_t\vartheta)$ with respect to $\tilde{\varkappa}$ often easily follows if the support of these functions is compact. The last condition can sometimes be verified with the help of

Lemma 1.89. *(i) $K_M \cap S^\varkappa$ is compact for any compact subset K of X.*

(ii) Suppose that $X = V_+$ and that (1.14) is valid, and assume that (1.79) holds. In particular, let $\varkappa = \varkappa_+ : V_+ \longrightarrow]0,+\infty[$ be given as in Proposition 1.70. Then $\overline{K_M} \cap S^\varkappa$ is compact for every compact subset K of V.

Proof. *(i)*. This is a consequence of Lemma 3.29 and Proposition 3.22 below. A direct proof is as follows: Let $(x_\ell)_{\ell\in\mathbb{N}}$ be any sequence in K and $(t_\ell)_{\ell\in\mathbb{N}}$ be any sequence in $]0,+\infty[$ such that $y_\ell := M_{t_\ell}x_\ell$ belongs to $S^\varkappa$ for every $\ell \in \mathbb{N}$. By choosing subsequences we achieve that (x_ℓ) converges to some $x \in K$. In view of $t_\ell \varkappa(x_\ell) = \varkappa(y_\ell) = 1$ it follows that

$$\lim_{\ell\to\infty} t_\ell = \lim_{\ell\to\infty} 1/\varkappa(x_\ell) = 1/\varkappa(x) =: t$$

so that (y_ℓ) converges to $M_t x \in K_M \cap S^\varkappa$.

(ii). Let $(y_\ell)_{\ell\in\mathbb{N}}$ be a sequence in $\overline{K_M} \cap S^\varkappa$. We choose sequences (t_ℓ) in $]0,+\infty[$ and (x_ℓ) in K such that

$$\lim_{\ell\to\infty} (y_\ell - M_{t_\ell}x_\ell) = 0 .$$

By choosing subsequences one achieves that (x_ℓ) converges to some $x \in K$. It follows that

$$\lim_{\ell\to\infty} M_0 x_\ell = M_0 x \qquad \text{so that} \qquad \lim_{\ell\to\infty} M_0 y_\ell = M_0 x .$$

That y_ℓ belongs to $S^\varkappa$ means that $|P_+(y_\ell)| = 1$ - see (1.76). Hence, by choosing subsequences for a second time one achieves that $(P_+(y_\ell))_{\ell\in\mathbb{N}}$ converges to $P_+(z$

for some $z \in V$ as $\ell \to +\infty$. Consequently, (y_ℓ) converges to $y := M_0(x) + P_+(z)$. Since $(M_{t_\ell} x_\ell)_{\ell \in \mathbb{N}}$ converges to y, as well, one sees that y belongs to $\overline{K_M} \cap S^\varkappa$. ■

Combining Proposition 1.86 with the Fubini theorem one sees that for any locally integrable function $f: X \longrightarrow \mathbb{C}$ the function $S^\varkappa \longrightarrow \mathbb{C}$, $\vartheta \mapsto f(M_t \vartheta)$, is locally integrable with respect to $\tilde{\varkappa}$ for almost every $t \in]0,+\infty[$. But, of course, generally one does not know whether this is valid for $t=1$, i.e. whether $f|_{S^\varkappa}$ is locally integrable. With almost quasihomogeneous functions this is different:

Lemma 1.90. *Let $N \in \mathbb{N}$, and let $q_0: X \longrightarrow \mathbb{C}$ be almost quasihomogeneous of order $\leq N$. Then the following conditions are equivalent:*

(a) q_0 as well as its deficiencies $q_1, \ldots, q_N$ belong to $\mathcal{L}^1_{loc}(X)$;

(a)' $q_0 \in \mathcal{L}^1_{loc}(X)$;

(b) for arbitrary $k \in \mathbb{N}_N \cup \{0\}$ and $t \in]0,+\infty[$ the function $q_k \circ M_t|_{S^\varkappa}$ is locally integrable with respect to $\tilde{\varkappa}$;

(b)' for every $k \in \mathbb{N}_N \cup \{0\}$ the function $q_k|_{S^\varkappa}$ is locally integrable with respect to $\tilde{\varkappa}$.

Proof. The equivalence (a) $\Longleftrightarrow$ (a)' and the fact that the measurability of q_0 implies that of its deficiencies $q_1, \ldots, q_N$ are consequences of Remark 1.50. Moreover, if $q_k|_{S^\varkappa}$ is $\tilde{\varkappa}$-measurable then the function $(t,\vartheta) \mapsto (q_k \circ \rho_\varkappa)(M_t \vartheta)$ is $(dt \otimes \tilde{\varkappa})$-measurable so that $q_k \circ \rho_\varkappa$ is Lebesgue-measurable by Proposition 1.86.(iii). Hence, Remark 1.73'.(ii) shows that the functions $q_0, \ldots, q_N$ are Lebesgue-measurable if there restrictions to $S^\varkappa$ are $\tilde{\varkappa}$-measurable. Consequently, for the rest of the proof we may assume that the functions $q_0, \ldots, q_N$ are Lebesgue-measurable.

(a) $\Rightarrow$ (b): Let K be a compact subset of $S^\varkappa$. Then the set

$$L := \{ M_t x \,;\, t \in I,\ x \in K \} \tag{1.106}$$

is compact for every compact subinterval I of $]0,+\infty[$, say $I = [1/2, 2]$. In view of condition (a) it follows from Proposition 1.86 and Fubini's theorem that for every $k \in \{0\} \cup \mathbb{N}_N$ there exists $t_k \in I$ such that the function $K \longrightarrow \mathbb{C}$, $\vartheta \mapsto q_k(M_{t_k} \vartheta)$, is $\tilde{\varkappa}$-integrable. Since by Lemma 1.48 q_N is quasihomogeneous this shows that $q_N|_K$ is $\tilde{\varkappa}$-integrable. By induction one can now deduce that the same is valid for $q_{N-1}, q_{k-2}, \ldots, q_1$, and q_0. Indeed, let $k \in \{0\} \cup \mathbb{N}_{N-1}$ be such that the restriction of q_j to K is $\tilde{\varkappa}$-integrable for every $j \in \mathbb{N}_N$ satisfying $j > k$. Since the equation $(1.39)_k$ in Lemma 1.48 means that

$$(1.39)'_k \qquad q_k = t_k^{-m} q_k \circ M_{t_k} - \sum_{\ell=1}^{N-k} \omega_\ell(t_k)\, q_{k+\ell}$$

it follows that the restriction of q_k to K is $\tilde{\varkappa}$-integrable, as well. Another application of $(1.39)_k$ now shows that the restriction of $q_k \circ M_t$ to K is $\tilde{\varkappa}$-integrable for arbitrary $k \in \{0\} \cup \mathbb{N}_N$ and $t \in]0, +\infty[$.

<u>*(b)'⇒(a)'*</u>. Let H be a compact subset of X. Then by Lemma 1.89.(i) $K := \overline{H_M} \cap S^\varkappa$ is a compact subset of $S^\varkappa$, and $\varkappa(H)$ is contained in a compact subinterval I of $]0, +\infty[$. Since H is contained in the set (1.106) the equation (1.39), the Fubini-Tonelli theorem, and Proposition 1.86 immediately imply that q_0 is integrable on H. ■

Sometimes with the help of Proposition 1.86 one obtains local integrability even at points not belonging to $\mathfrak{K}(M)$. For the following proposition we drop the assumption concerning the existence of $\varkappa$.

Proposition 1.91. *Suppose that (1.14) is valid. Let $N \in \mathbb{N}_0$, and let $q_0 \in \mathcal{L}^1_{loc}(X_+)$ be almost quasihomogeneous of degree m and of order $\leq N$, its deficiencies being denoted by $q_1, \ldots, q_N$. Let $\dot{q}_k$ be the extension of q_k to X defined by $\dot{q}_k/_{X \setminus X_+} :\equiv 0$. Then the following conditions are equivalent:*

(a) $\dot{q}_k \in \mathcal{L}^1_{loc}(X)$ for every $k \in \{0\} \cup \mathbb{N}_N$;

(a)' $\dot{q}_0 \in \mathcal{L}^1_{loc}(X)$;

(b) $\operatorname{Re} m > -\mu$, or $\operatorname{supp} \dot{q}_k \subset X_+$ for every $k \in \{0\} \cup \mathbb{N}_N$.

<u>*Proof.*</u> <u>*(a)'⇒(a)*</u>: this is an immediate consequence of Remark 1.50.

For the proof of the other implications we suppose that (1.79) is valid and choose $\varkappa = \varkappa_+$ as in Lemma 1.89.(ii). If $\operatorname{supp} \dot{q}_k \subset X_+$ for every $k \in \{0\} \cup \mathbb{N}_N$ then there is nothing to be proved. So we fix $\ell \in \{0\} \cup \mathbb{N}_N$ and $x \in \operatorname{supp} \dot{q}_\ell \setminus X_+$ and choose $\varepsilon > 0$ so small that the compact neighbourhood $K := \{y \in V;\ |P_+ y| \leq \varepsilon,\ |M_0(y-x)| \leq \varepsilon\}$ of x is contained in X. Then

$$(1.107) \qquad \overline{K_M} = \{y \in V;\ |M_0(y-x)| \leq \varepsilon\} = K_M;$$

Consequently, by Lemma 1.89.(ii) the set $L := K_M \cap S^\varkappa$ is a compact subset of X_+.

Since K is equal to $\overline{K\setminus\ker M}$ and since K is compact one finds $R>0$ such that K is contained in the set $J:=\{M_t x\,;\, x\in L,\ t\in [0,R]\}$. Moreover, by (1.107) and by compactness, again, there is $\delta\in]0,R]$ such that $J\setminus K$ is contained in the compact subset $\{M_t x\,;\, x\in L,\ t\in [\delta,R]\}$ of X_+. Hence for every $k\in\{0\}\cup\mathbb{N}_N$ the assumption on q_k implies that $\dot{q}_k|_K$ is integrable if and only if $\dot{q}_k|_J$ is integrable, and we have to prove that the latter is the case for every $k\in\{0\}\cup\mathbb{N}_N$ if and only if $\operatorname{Re} m > -\mu$.

(a)⇒(b): we may assume that $\ell=\max\{k\in\{0\}\cup\mathbb{N}_N\,;\, x\in\operatorname{supp}\dot{q}_k\}$ and that ε is so small that $K\cap\operatorname{supp}\dot{q}_k=\emptyset$ for every $k\in\mathbb{N}_N$ such that $k>\ell$. It then follows by the equation $(1.39)^{\cdot}_{\ell}$ - see the proof of Lemma 1.90 - that

$$\dot{q}_\ell(M_t y) = t^m\,\dot{q}_\ell(y) \quad \text{for arbitrary } y\in K \text{ and } t\in]0,+\infty[\,. \tag{1.108}$$

Note that by Lemma 1.90 $q_\ell|_L$ is integrable with respect to $\tilde{\varkappa}$. If

$$\int_L |q_\ell(\vartheta)|\,d\tilde{\varkappa}(\vartheta)$$

were equal to zero then in view of (1.108) and Proposition 1.86 q_ℓ would vanish almost everywhere on the set L_M (which equals $K_M\setminus\ker M$) so that $\operatorname{supp}\dot{q}_\ell\cap\overset{\circ}{K}$ would be empty in contradiction to the choice of x and ℓ. Consequently, since by (a) $\dot{q}_\ell$ is locally integrable so that $\dot{q}_\ell|_J$ is integrable we deduce from Proposition 1.86 and (1.108) that the function $t\mapsto t^{m+\mu-1}$ is integrable on $]0,R]$. Since the latter is the case if and only if $\operatorname{Re} m+\mu>0$ the proof of the implication "(a)⇒(b)" is complete.

(b)⇒(a)': It follows by Proposition 1.86, by Fubini's theorem, and by (1.39) that

$$\int_K |\dot{q}_0(y)|\,dy \le \int_J |\dot{q}_0(y)|\,dy = \int_0^R \int_L |q_0(M_t\vartheta)|\,d\tilde{\varkappa}(\vartheta)\; t^{\mu}\,\frac{dt}{t} \le$$

$$\le \sum_{k=0}^{N} \int_L |q_k(\vartheta)|\,d\tilde{\varkappa}(\vartheta) \int_0^R |\log t|^k\, t^{\operatorname{Re} m+\mu-1}\,dt\,.$$

Since by Lemma 1.90 the q_k are $\tilde{\varkappa}$-integrable on the compact subset L of $X_+\cap S^{\varkappa}$ the assumption on m implies that the right-hand side of the preceding estimate is finite. Consequently, $\dot{q}_0$ is locally integrable on X. ■

We close this section by another application of Proposition 1.86:

Lemm 1.92. *Suppose that $\sigma=\sigma_+$. Let $q_0:\dot{V}\longrightarrow\mathbb{C}$ be continuous and almost quasi-*

homogeneous of degree m. Suppose that q_0 does not vanish identically. Then the restriction of q_0 to $V \setminus K(0,1)$ is integrable if and only if $\operatorname{Re} m < -\mu$.

Proof. First of all we observe that by Proposition 1.51.(ii) the kth order deficiency q_k of q_0 is continuous, as well, for every $k \in \mathbb{N}_N$ where $N := \operatorname{ord}_M(q_0)$. Let $\varkappa$ be the function $\varkappa_+$ of Proposition 1.70. Then $S^\varkappa = S^{n-1}$ is compact. It follows from Proposition 1.86 that q_0 is integrable on $V \setminus K(0,1)$ if and only if the function $g: [1,+\infty[\times S^\varkappa \longrightarrow \mathbb{C}$, $(t,\vartheta) \mapsto t^{\mu-1} q_0(M_t \vartheta)$, is integrable on $[1,+\infty[\times S^\varkappa$ with respect to $dt \otimes \tilde{\varkappa}$.

"$\Rightarrow$". Since q_0 is locally integrable it follows from Remark 1.50 that the restriction to $V \setminus K(0,1)$ of every deficiency of q_0 is integrable. Hence, in view of Lemma 1.48 we may assume that q_0 is quasihomogeneous of degree m. But in this case we have $g(t,\vartheta) = t^{m+\mu-1} q_0(\vartheta)$ so that in view of the compactness of $S^\varkappa$ and since $\int_{S^\varkappa} |q_0(\vartheta)| \, d\vartheta \neq 0$ the Fubini theorem shows that the function $t \mapsto t^{m+\mu-1}$ is integrable on $[1,+\infty[$, i.e. $\operatorname{Re} m + \mu < 0$.

"$\Leftarrow$". By the equation (1.39) we have

$$g(t,\vartheta) = \sum_{k=0}^{N} t^{m+\mu-1} \omega_k(t)\, q_k(\vartheta), \qquad (t,\vartheta) \in]0,+\infty[\times S^\varkappa.$$

Since the assumption on m implies that the functions $t \mapsto t^{m+\mu-1} \omega_k(t)$, $0 \le k \le N$, are integrable on $[1,+\infty[$ and since the continuous functions q_k are integrable on the compact set $S^\varkappa$ with respect to $\tilde{\varkappa}$ the Fubini-Tonelli shows that g is integrable on $[1,+\infty[\times S^\varkappa$ with respect to $dt \otimes \tilde{\varkappa}$, as desired. ∎

Chapter II

(Almost) Quasihomogeneous Distributions. Definitions and Basic Properties

Section (a) of this chapter contains the basic material on quasihomogeneous distributions. Among other things, distributions T which are quasihomogeneous of degree m are characterized as the solutions of Euler's equation $(\partial_M - m)T = 0$ where ∂_M denotes the directional derivative into the direction of the quasihomogeneous rays (see Example 1.21). Moreover, we shall see that the restrictions of quasihomogeneous distributions to the hypersurfaces $S^{\varkappa}$ are well-defined for any C^{∞} function $\varkappa : X \longrightarrow]0,+\infty[$ which is quasihomogeneous of degree 1. Section (b) contains some basic facts on the behaviour of quasihomogeneous distributions under the Fourier transform.

Section (c) is devoted to meromorphic functions of quasihomogeneous distributions. It turns out that their Laurent coefficients possess new invariance properties under composition with M_t: they are *almost quasihomogeneous* in the sense of Definition 2.28 below. In section (d) the basic facts on almost quasihomogeneous distributions are presented. Many properties carry over from sections (a) and (b). In particular, we shall see that a distribution T is almost quasihomogeneous of degree m if and only if for some $N \in \mathbb{N}_0$ it solves the generalized Euler equation $(\partial_M - m)^{N+1} T = 0$. Section (e) is about meromorphic functions of almost quasihomogeneous distributions. It turns out that the Laurent coefficients arising here do not have new invariance properties: they are almost quasihomogeneous themselves.

Finally, an appendix - section (f) - contains the basic facts on distributions invariant under the action of a compact subgroup of GL(V).

As in Chapter 1, there are quite a few results of a more technical nature required in later chapters.

As before, X is an open subset of V which at first is not required to be quasi-homogeneous. Considering X as an n-dimensional linear manifold we recall that the space $\mathcal{D}'(X)$ of distributions on X can be defined as the dual of the space of C_0^∞-densities on V (see for example § 6.3 in Hörmander [11]). In the following we shall identify $\mathcal{D}'(X)$ with the dual of $C_0^\infty(X)$ via the strictly positive density $\omega_X := Vol|_X$ where *Vol* denotes the volume form associated with the given scalar product on V:

$$\langle T, \varphi \rangle := \langle T, \varphi\, \omega_X \rangle, \qquad T \in \mathcal{D}'(X),\ \varphi \in C_0^\infty(X).$$

In particular, as usual every locally integrable function $f: X \longrightarrow \mathbb{C}$ (with respect to the Lebesgue measure dx induced by the given scalar product) induces a distribution T_f on X defined by

$$\langle T_f, \varphi \rangle := \int f \varphi \, dx, \qquad \varphi \in C_0^\infty(X).$$

Multiplication by functions and the action of differential operators are defined as in the standard way. For any $T \in \mathcal{D}'(X)$ by $\check{T}$ we denote the distribution on $(-X)$ defined by $\langle \check{T}, \varphi \rangle := \langle T, \check{\varphi} \rangle$ where $(\check{\varphi})(x) := \varphi(-x)$.

(a) Quasihomogeneous Distributions

We first observe that if $\varphi: V \longrightarrow \mathbb{C}$ is any function then

$$(2.1) \qquad \operatorname{supp} \varphi \circ M_{1/t} = M_t(\operatorname{supp} \varphi), \qquad t \in]0, +\infty[.$$

For any compact subset K of X we introduce the open subset

$$(2.2) \qquad J_M(K; X) := \{ t \in]0, +\infty[\, ; \ M_t(K) \subset X \}$$

of $]0, +\infty[$. It follows from (2.1) that $\varphi \circ M_{1/t}$ belongs to $C_0^\infty(X)$ for arbitrary $\varphi \in C_0^\infty(X)$ and $t \in J_M(\operatorname{supp} \varphi; X)$. Consequently, for any $f \in \mathcal{L}^1_{loc}(X)$ a change of variables shows that

$$\int_X f(M_t x)\, \varphi(x)\, dx = t^{-\mu} \int_X f(y)\, \varphi(M_{1/t} y)\, dy, \qquad t \in J_M(K; X)$$

where $\mu = \operatorname{tr} M$ - see (1.6). If f is continuous it follows that f is quasihomogeneous of degree m if and only if

$$t^{-\mu} \int_X f(y)\, \varphi(M_{1/t}y)\, dy = t^m \int_X f(x)\, \varphi(x)\, dx$$

for arbitrary $\varphi \in C_0^\infty(X)$ and $t \in J_M(\operatorname{supp}\varphi; X)$.

Definition 2.1. A distribution $T \in \mathcal{D}'(X)$ is said to be *quasihomogeneous of degree m (and of type M)* if and only if

$$t^{-\mu} \langle T, \varphi \circ M_{1/t} \rangle = t^m \langle T, \varphi \rangle \tag{2.3}$$

for arbitrary $\varphi \in C_0^\infty(X)$ and $t \in J_M(\operatorname{supp}\varphi; X)$.

Of course, if X is quasihomogeneous then the left-hand side of (2.3) is by definition equal to $\langle T \circ M_t, \varphi \rangle$, and T is quasihomogeneous of degree m if and only if

$$T \circ M_t = t^m T, \qquad t \in]0, +\infty[. \tag{2.3)'}$$

Example 2.2. *The Dirac distribution δ_0 (at 0) is quasihomogeneous of degree $-\mu$.* ∎

The analogue of Proposition 1.19 is valid:

Proposition 2.3. *Let $T \in \mathcal{D}'(X)$ be quasihomogeneous of degree m. Let $\ell \in \mathbb{C}$, and let $P: X \times V^* \longrightarrow \mathbb{C}$ be a C^∞-copolynomial function (in the sense of Definition 1.16) which is quasihomogeneous of degree ℓ and of type $M \times (-M)^*$. Then $P(x,\partial)T$ is quasihomogeneous of degree $m+\ell$.*

Proof. We fix $\varphi \in C_0^\infty(X)$ and $t \in J_M(\operatorname{supp}\varphi; X)$. Then by Remark 1.17 we have

$$P(x,-\partial)(\varphi \circ M_{1/t}) = t^\ell (P(x,-\partial)\varphi) \circ M_{1/t}$$

so that by (2.3)

$$t^{-\mu} \langle P(x,\partial)T, \varphi \circ M_{1/t} \rangle = t^{\ell-\mu} \langle T, (P(x,-\partial)\varphi) \circ M_{1/t} \rangle =$$

$$= t^{m+\ell} \langle T, P(x,-\partial)\varphi \rangle = t^{m+\ell} \langle P(x,\partial)T, \varphi \rangle. \quad ∎$$

Corollary 2.4. *Let $T \in \mathcal{D}'(X)$ be quasihomogeneous of degree m. Then*

(i) $P(\partial)T$ is quasihomogeneous of degree $m-\ell$ for every polynomial function P on V^* which is quasihomogeneous of degree $\ell \in \mathbb{C}$ and of type M^*;

(ii) qT is quasihomogeneous of degree $m+\ell$ for every $q \in C^{\infty}(X)$ which is quasihomogeneous of degree $\ell \in \mathbb{C}$. ■

The restrictions of quasihomogeneous distributions on X to open subsets of X are, of course, quasihomogeneous, as well. On the other hand we have

Proposition 2.5. *(i) If $T \in \mathcal{D}'(X)$ is quasihomogeneous of degree m then there exists a unique extension $T_M \in \mathcal{D}'(X_M)$ of T which is quasihomogeneous of degree m, as well.*

(ii) The map $T \mapsto T_M$ is linear and (weakly) continuous.

Proof. Let K be a compact subset of X_M. Since the sets $M_r X$, $r \in]0,+\infty[$, are open one finds a finite subset R of $]0,+\infty[$ such that $K \subset \bigcup_{r \in R} M_r X$. Let $(\varphi_r)_{r \in R}$ be a partition of unity on K subordinated to the covering $(M_r X)_{r \in R}$. For every $\varphi \in C_0^{\infty}(K)$ we set

$$(2.4) \qquad \langle T_M, \varphi \rangle := \sum_{r \in R} r^{m+\mu} \langle T, (\varphi_r \varphi) \circ M_r \rangle .$$

Now we fix $t \in]0,+\infty[$ and a finite subset S of $]0,+\infty[$ such that $M_t K \subset \bigcup_{s \in S} M_s X$, choose a partition of unity $(\psi_s)_{s \in S}$ on $M_t K$ subordinated to the covering $(M_s X)_{s \in S}$, and set $K_{r,s} := \operatorname{supp} \psi_s \cap \operatorname{supp}(\varphi_r \circ M_{1/t}) \cap M_t K$. If $x \in M_{1/s} K_{r,s}$ then $M_s x \in \operatorname{supp} \varphi_r \circ M_{1/t} = M_t \operatorname{supp} \varphi_r$ so that $M_{s/t} x \in \operatorname{supp} \varphi_r \subset M_r X$. This shows that $t_{r,s} := \frac{s}{rt} \in J_M(M_{1/s} K_{r,s}; X)$. Note that the last set is contained in $J_M(\operatorname{supp}(\psi_s(\varphi_r \varphi) \circ M_{1/t}) \circ M_s; X)$ because $\operatorname{supp}(\psi_s(\varphi_r \varphi) \circ M_{1/t})$ is a subset of $K_{r,s}$. It follows that

$$\begin{aligned} & t^{-\mu} \sum_{s \in S} s^{m+\mu} \langle T, (\psi_s \varphi \circ M_{1/t}) \circ M_s \rangle = \\ &= t^m \sum_{s \in S} \sum_{r \in R} r^{m+\mu}\, t_{r,s}^{m+\mu} \langle T, (\psi_s (\varphi_r \varphi) \circ M_{1/t}) \circ M_s \rangle = \\ &= t^m \sum_{r \in R} r^{m+\mu} \sum_{s \in S} \langle T, (\psi_s (\varphi_r \varphi) \circ M_{1/t}) \circ M_s \circ M_{1/t_{r,s}} \rangle = \\ &= t^m \sum_{r \in R} r^{m+\mu} \sum_{s \in S} \langle T, ((\psi_s \circ M_t)\, \varphi_r \varphi) \circ M_r \rangle = \\ &= t^m \sum_{r \in R} r^{m+\mu} \langle T, (\varphi_r \varphi) \circ M_r \rangle \end{aligned}$$

where the last equality comes about since $\sum_{s \in S} \psi_s \circ M_t$ is equal to 1 on a neighbourhood of $\operatorname{supp} \varphi$.

Now, the case "$t=1$" shows that the right-hand side of (2.4) does not depend on the choice of R and $(\varphi_r)_{r\in R}$, i.e. T_M is well-defined. Moreover, the general case shows that (2.3) is valid for T_M instead of T. From (2.4) it is obvious that the restriction of T_M to $C_0^\infty(K)$ is linear and continuous. If K is contained in X we may take $R=\{1\}$ so that (2.4) shows that $T_M|_X = T$. The assertion (ii) also follows from (2.4). ∎

Support and singular support of quasihomogeneous distributions are quasihomogeneous:

Proposition 2.6. *Let $T\in\mathcal{D}'(X)$ be quasihomogeneous. Then*

(i) $\operatorname{supp} T = (\operatorname{supp} T)_M \cap X$;

(ii) $\operatorname{sing\,supp} T = (\operatorname{sing\,supp} T)_M \cap X$.

Proof. (i). Let Y be the open subset $X\setminus \operatorname{supp} T$ of X. Then for every $x\in Y_M\cap X$ one finds an open neighbourhood U of x in $Y_M\cap X$ and a number $t\in\,]0,+\infty[$ such that $M_{1/t}U\subset Y$. Hence for every $\varphi\in C_0^\infty(U)$ we deduce from (2.1) that $\operatorname{supp}(\varphi\circ M_t)\subset Y$ and $t\in J_M(\operatorname{supp}(\varphi\circ M_t);X)$ so that by applying (2.3) to $\varphi\circ M_t$ instead of φ we obtain: $0 = t^{m+\mu}\langle T, \varphi\circ M_t\rangle = \langle T,\varphi\rangle$.

(ii). This time we set $Y := X\setminus \operatorname{sing\,supp} T$ and let $f\in C^\infty(Y)$ be such that $T|_Y = T_f$. Since by the argument preceding Definition 2.1 f is quasihomogeneous it extends to a quasihomogeneous C^∞ function $f_M: Y_M \longrightarrow \mathbb{C}$ by Proposition 1.57. Condition (i) implies that $T_{(f_M)}$ and T coincide on $Y_M\cap X$. Hence $Y_M\cap X\subset Y$, i.e. $Y_M\cap X = Y$, and the assertion follows. ∎

Corollary 2.7. *Suppose that (1.14) holds. Let $T\in\mathcal{D}'(V_+)$ be quasihomogeneous. If $\operatorname{sing\,supp} T\cap U$ is empty for some neighbourhood U of $\ker M$ then T is induced by a C^∞ function.*

Proof. By Remark 1.8.(i) the assumption (1.14) implies that $\lim_{t\to 0} M_t x = M_0 x\in U$ for every $x\in V_+$. Hence $(U\cap V_+)_M = V_+$, and the assertion follows by Proposition 2.6.(ii). ∎

We now come to *Euler's equation* for quasihomogeneous distributions. Let $T\in\mathcal{D}'(X)$

be fixed. We first have to determine the transpose (with respect to the distributional duality bracket) of the Euler operator ∂_M defined by (1.29).

Remark 2.8. *The transposed operator* $^t(\partial_M)$ *of* ∂_M *with respect to the duality bracket between* $\mathcal{D}'(X)$ *and* $C_0^\infty(X)$ *is equal to* $-\partial_M - \mu$. *Hence*

$$(2.5) \qquad {}^t(\partial_M - m) = -(\partial_M - m^*) \qquad \text{where} \qquad m^* := -m-\mu .$$

Proof. Let $T \in \mathcal{D}'(X)$ and $\varphi \in C_0^\infty(X)$. Recall that the transposed operator in question is defined by the equation $\langle \partial_M T, \varphi \rangle = \langle T, {}^t(\partial_M)\varphi \rangle$. Expressing ∂_M in coordinates via the formula (1.29)' we see that

$${}^t(\partial_M)\varphi = \sum_{j=1}^n (-\partial_j)\Big(\sum_{k=1}^n M_{jk}\, x_k\, \varphi\Big) = -\sum_{j=1}^n M_{jj}\,\varphi - \sum_{j=1}^n \Big(\sum_{k=1}^n M_{jk}\, x_k\Big)\partial_j \varphi = -\mu\varphi - \partial_M \varphi . \quad \blacksquare$$

Lemma 2.9. *The following conditions are equivalent:*

(a) T satisfies (2.3) for arbitrary $\varphi \in C_0^\infty(X)$ *and* $t \in]0,+\infty[$ *such that* $[1,t]$ *is contained in* $J_M(\operatorname{supp}\varphi; X)$;

(b) for every $\varphi \in C_0^\infty(X)$ *there is* $\varepsilon > 0$ *such that T satisfies (2.3) for every* $t \in]1-\varepsilon, 1+\varepsilon[$;

(c) $(\partial_M - m)\,T = 0$.

This is an immediate consequence of

Lemma 2.10. *Let* $\varphi \in C_0^\infty(X)$. *Then by* $g(t) := t^{-m-\mu}\langle T, \varphi\circ M_{1/t}\rangle$ *a* C^1 *function* $g: J_M(\operatorname{supp}\varphi; X) \longrightarrow \mathbb{C}$ *is well-defined satisfying*

$$g'(t) = t^{-m-\mu-1}\langle(\partial_M - m)\,T, \varphi\circ M_{1/t}\rangle .$$

Proof. That g is a C^1 (in fact, a C^∞) function is clear. Applying the chain rule twice and making use of (1.5) and (1.4) we obtain

$$\partial_t(\varphi\circ M_{1/t})(x) = D\varphi(M_{1/t}x)\cdot\big((-1/t^2)\tfrac{1}{1/t}\, M\, M_{1/t}x\big) =$$
$$= -\tfrac{1}{t} D(\varphi\circ M_{1/t})(x)\cdot Mx = -\tfrac{1}{t}\partial_M(\varphi\circ M_{1/t})(x) .$$

for arbitrary $x \in X$ and $t \in J_M(\operatorname{supp}\varphi; X)$. It follows that $g'(t)$ is equal to

$$-(m+\mu)\,t^{-m-\mu-1}\langle T, \varphi\circ M_{1/t}\rangle - t^{-m-\mu-1}\langle T, \partial_M(\varphi\circ M_{1/t})\rangle .$$

In view of (2.5) this equals the right-hand side of the desired equation. ■

An obvious consequence of Lemma 2.9 is

Proposition 2.11. *(i) If T is quasihomogeneous of degree m then $\partial_M T = mT$; (ii) the converse is valid provided that $J_M(K;X)$ is an interval for every compact subset K of X.* ■

Of course, the last condition is satisfied if X is quasihomogeneous.

In passing we are going to recall the explicit description of all homogeneous distributions on $\mathbb{R}$. To this purpose we require

Lemma 2.12. *Let $T \in \mathcal{D}'(]0,+\infty[)$ be homogeneous of degree m. Then there is $c \in \mathbb{C}$ such that*

$$\langle T, \varphi \rangle = c \int_0^{+\infty} \varphi(x)\, x^m \, dx, \qquad \varphi \in C_0^\infty(]0,+\infty[).$$

Proof. Since the distribution $S := x^{-m} T$ is homogeneous of degree 0 it follows by Euler's equation that $0 = 0 \cdot S = xS'$, i.e. $S' = 0$. Consequently, S is induced by a constant function, and since $T = x^m S$ the assertion follows. ■

Proposition 2.13. *The space $\mathfrak{H}'_m(\mathbb{R})$ of distributions on $\mathbb{R}$ which are homogeneous of degree m is a two-dimensional vector space spanned by (i) x_+^m, x_-^m if $-m \notin \mathbb{N}$, and by (ii) $\underline{x}^m$, $\delta_0^{(-m-1)}$ if $-m \in \mathbb{N}$ (here we adopt the notation of Hörmander [11], pp. 68 and 72).*

Proof. If $m \notin -\mathbb{N}$ then x_+^m and x_-^m are homogeneous by (3.2.7) on p. 71 in Hörmander [11]. If $m \in -\mathbb{N}$ then it follows by Example 2.2 and Corollary 2.4.(i) that $\delta_0^{(-m-1)}$ is homogeneous of degree m. Moreover, since $x_-^m = (x_+^m)^\vee$, $m \in \mathbb{C} \setminus (-\mathbb{N})$, we conclude from the formulas (3.2.10)' and (3.2.8) in [11] that $\underline{x}^m$ is homogeneous of degree m in case "$m \in -\mathbb{N}$".

On the other hand, fixing any non-trivial $T \in \mathfrak{H}'_m(\mathbb{R})$ we conclude from Lemma 2.12 that there are constants $c_\pm \in \mathbb{C}$ such that

$$T|_{\pm]0,+\infty[} = c_\pm \, x_\pm^m |_{\pm]0,+\infty[}.$$

Since $x_{\mp}^m$ vanishes on $\mp]0,+\infty[$ this means that the restrictions of the distributions T and $c_+ x_+^m + c_- x_-^m$ to $\mathbb{R}\setminus\{0\}$ coincide. This implies that for some complex polynomial P of one variable we have

$$T = c_+ x_+^m + c_- x_-^m + P(\partial)\,\delta_0 .$$

If $m \notin -\mathbb{N}$ then P vanishes identically since $\delta_0^{(j)}$ is homogeneous of degree $-1-j$, $j \in \mathbb{N}_0$. Moreover, if $-m \in \mathbb{N}$ then we conclude that $P = cx^{-m-1}$ for some $c \in \mathbb{C}$, and the left-hand side of the equation

$$T - (-1)^{-m} c_- \underline{x}^m - c\,\delta_0^{(-m-1)} = (c_+ - (-1)^{-m} c_-)\, x_+^m$$

(which is valid by the formula (3.2.10)' in Hörmander [11]) is homogeneous of degree m. Since in view of the formula (3.2.8) in [11] the distribution x_+^m is not homogeneous the right-hand side of the preceding equation must vanish so that T is a linear combination of $\underline{x}^m$ and $\delta_0^{(-m-1)}$, as was to be shown. ∎

Next we note an important consequence of Euler's equation:

Proposition 2.14. *If $T \in \mathcal{D}'(X)$ is quasihomogeneous then its analytic wave front set $WF_A(T)$ is contained in the set*

$$(2.6) \qquad \Gamma_M(X) := \{ (x,\xi) \in X \times \dot{V}^* \,;\ \langle \xi, Mx \rangle = 0 \} .$$

Proof. Since $\Gamma_M(X)$ is the characteristic set of the differential operator $\partial_M - m$ the assertion follows from Proposition 2.11.(i) and from Theorem 8.6.1 in Hörmander [11]. ∎

For the rest of the present section we suppose that we can fix a C^∞ function $\varkappa : X \longrightarrow]0,+\infty[$ which is quasihomogeneous of degree 1. By Proposition 1.63 the set $S^\varkappa = \varkappa^{-1}(1)$ is then a C^∞ hypersurface in X, and by Proposition 1.72 the map $\rho_\varkappa$ (see Notation 1.71) is a C^∞ submersion of X onto $S^\varkappa$. This leads to the following description of $\Gamma_M(X)$. Recall that if $\rho : X \longrightarrow Y$ is a C^∞ map of X into a C^∞ manifold Y then for any subset Γ of the cotangent bundle T^*Y of Y the set $\{(x, T_x^*\rho \cdot \eta) \,;\ x \in X,\ \eta \in T^*_{\rho(x)}Y,\ (\rho(x),\eta) \in \Gamma\}$ is denoted by $\rho^*(\Gamma)$.

Lemma 2.15. *(i) $\Gamma_M(X)$ equals $\rho_\varkappa^*(\dot{T}^*S^\varkappa)$ where $\dot{T}^*S^\varkappa$ denotes the complement of the zero section in $T^*S^\varkappa$.*

(ii) The intersection of $\rho_x^*(\dot{T}^*S^x)$ and the conormal bundle of S^x in X is empty.

Proof. (i): Let $x \in X$. Since ρ_x is a submersion the set $R := \rho_x^{-1}(\rho_x(x))$ is a submanifold of X such that its tangent space $T_x R$ is equal to $\ker T_x \rho_x$ and such that the annihilator $(T_x R)^o$ of $T_x R$ in $T_x^* X$ equals the image of $T_x^* \rho_x$. On the other hand, R is the image of the C^∞ map $i_x :]0,+\infty[\longrightarrow X$, $t \mapsto M_t x$, which in view of Lemma 1.64.(c) and Proposition 1.10 induces a diffeomorphism of $]0,+\infty[$ onto R. Consequently, the tangent space $T_x R$ is generated by $i_x'(1) = Mx$.

(ii): By $J : S^x \longrightarrow X$ we denote the inclusion map. Let $x \in S^x$ and $\eta \in T_x^* S^x$. Then $T_x^* \rho_x \cdot \eta$ belongs to $(T_x S^x)^o$ if and only if

$$0 = \langle T_x^* \rho_x \cdot \eta, T_x J \cdot v \rangle = \langle \eta, T_x(\rho_x \circ J) \cdot v \rangle = \langle \eta, v \rangle, \qquad v \in T_x S^x,$$

i.e. if and only if $\eta = 0$. ■

Combining Proposition 2.14 and Lemma 2.15 with Corollary 8.2.7 in Hörmander [11] one obtains

Theorem 2.16. *If T is quasihomogeneous then the restriction $T|_{S^x}$ to S^x is well-defined as the pullback by the inclusion map $J : S^x \longrightarrow X$.* ■

Conversely, since ρ_x is a C^∞ submersion it follows that every distribution on S^x can be pulled back by ρ_x to a distribution on X (see e.g. Chapter VI in Hörmander [11]). From the results of Chapter VIII in [11] one easily deduces

Theorem 2.17. *The map $C^o(S^x) \longrightarrow C^o(X)$, $\psi \mapsto \psi \circ \rho_x$, extends to a continuous linear map $\rho_x^* : \mathcal{D}'(S^x) \longrightarrow \mathcal{D}'(X)$ such that for every $v \in \mathcal{D}'(S^x)$ the distribution $T := \rho_x^*(v)$ has the following properties:*

(i) T is quasihomogeneous of degree 0;

(ii) $WF(T) = \rho_x^* WF(v) \subset \Gamma_M(X)$;

(iii) if x is real analytic then (ii) remains valid with WF replaced by WF_L (see Definition 8.4.3 in Hörmander [11]);

(iv) the restriction of T to S^x - which is well-defined by Theorem 2.16 - is equal to v. ■

Later we shall see that every $T \in \mathcal{D}'(X)$ which is quasihomogeneous of degree m is of the form

(2.7) $$T = \varkappa^m \rho_\varkappa^*(\nu)$$

for some $\nu \in \mathcal{D}'(S^\varkappa)$; in fact, one can take $\nu = T|_{S^\varkappa}$ (see Theorem 4.25 below).

(b) The Fourier Transform of Quasihomogeneous Distributions

We have to fix a few notational conventions by way of which the standard $\mathbb{R}^n$-theory of the Fourier transform is reformulated in a coordinate-free manner. First of all, by $\mathcal{S}(V)$ we denote the Fréchet space of all rapidly decreasing complex-valued C^∞ functions on V. By selecting any basis of V it can be defined as the space $\mathcal{S}(\mathbb{R}^n)$. For any $\varphi \in \mathcal{S}(V)$ the Fourier transform $\hat{\varphi}: V^* \longrightarrow \mathbb{C}$ is defined by

$$\hat{\varphi}(\xi) := \int_V \exp(-i\langle \xi, x\rangle)\, \varphi(x)\, dx\,, \qquad \xi \in V^*,$$

where $\langle \cdot, \cdot\cdot \rangle$ denotes the canonical duality bracket between V^* and V and where - as before - the integral is taken with respect to any orthonormal basis of V. The Fourier transform for functions $\varphi \in \mathcal{S}(V^*)$ is defined in the same way once it is settled in what sense the integration on V^* is to be understood. In accordance with the convention for V it suffices to fix a scalar product on V^*. To this end the given scalar product $\langle \cdot, \cdot\cdot \rangle$ on V is transported to V^* by way of the Riesz isomorphism $V \longrightarrow V^*$, $x \mapsto \langle x, \cdot \rangle$. A priori, for any $\varphi \in \mathcal{S}(V^*)$ the domain of definition of $\hat{\varphi}$ is V^{**}. However, via the canonical isomorphism $\iota: V \longrightarrow V^{**}$ defined by $\iota(x)(\xi) = \langle \xi, x \rangle$ we shall identify V^{**} with V. So $\varphi \mapsto \hat{\varphi}$ defines a continuous linear map $\mathcal{S}(V^*) \longrightarrow \mathcal{S}(V)$. Consequently, for any $T \in \mathcal{S}'(V)$ it makes sense to define $\hat{T} \in \mathcal{S}'(V^*)$ by the standard formula:

$$\langle \hat{T}, \varphi \rangle = \langle T, \hat{\varphi} \rangle\,, \qquad \varphi \in \mathcal{S}(V^*)$$

where $\mathcal{S}'(V)$ denotes the space of *temperate distributions* on V. In fact, by

$$\mathcal{F}_V : \mathcal{S}'(V) \longrightarrow \mathcal{S}'(V^*),\ \ T \mapsto \hat{T},$$

an isomorphism is defined extending the Fourier transform on $\mathcal{S}(V)$. The inverse of $\mathcal{F}_V$ is given according to the Fourier inversion formula:

$$(\mathcal{F}_V)^{-1} T = (2\pi)^{-n} \mathcal{F}_{V^*}(\check{T}), \qquad T \in \mathcal{S}'(V^*),$$

where $\langle \check{T}, \varphi \rangle := \langle T, \check{\varphi} \rangle$ and $\check{\varphi}(x) := \varphi(-x)$. Below, mostly we shall write $\mathcal{F}$ instead of $\mathcal{F}_V$ or $\mathcal{F}_{V^*}$.

In order to be able to formulate the rule for Fourier transforms of derivatives we observe that for every polynomial function $P: V \longrightarrow \mathbb{C}$ and for every $\zeta \in \mathbb{C}$ a polynomial function on V denoted by $x \mapsto P(\zeta x)$ is well-defined via coordinates since a linear change of coordinates leaves the homogeneous parts of polynomials invariant. In passing we note the following not completely trivial

Remark 2.18. *Let $P: V \longrightarrow \mathbb{C}$ be a polynomial function. If P is quasihomogeneous (resp. almost quasihomogeneous) of degree m then so is $P(\zeta(\cdot))$ for every $\zeta \in \mathbb{C}$.*

Proof. Working with coordinates and making use of (1.29)' one sees that

$$(\partial_M - m)\, P(\zeta(\cdot)) = ((\partial_M - m) P)(\zeta(\cdot)).$$

Consequently, the assertion follows from Proposition 1.22 in the quasihomogeneous case and from Proposition 2.31 and Remark 2.30 below in the almost quasihomogeneous case. ■

Now let $P: V^* \longrightarrow \mathbb{C}$ be a polynomial function. Then the rule for Fourier transforms of derivatives reads as usual

$$\mathcal{F}(P(D)T) = P\hat{T}$$

where $D := -i\partial$ and where the polynomial function $P(-i(\cdot))$ on V^* is defined as above with (V,ζ) replaced by $(V^*, -i)$ (compare (1.25)). Moreover, if $Q: V \longrightarrow \mathbb{C}$ is a polynomial function then

$$\mathcal{F}(QT) = Q(-D)\hat{T}$$

where for the differential operator $Q(-D)$ on V^* to be well-defined we have to consider Q as a polynomial function on V^{**} via the canonical isomorphism ι described above.

Finally, we recall that by a transformation of variables one obtains

$$(2.8) \qquad \mathcal{F}(T \circ A) = |\det A|^{-1}\, \hat{T} \circ (A^{-1})^*, \qquad A \in GL(V,V),$$

for every $T \in \mathcal{S}'(V)$. For $A = M_{1/t}$ this becomes

(2.8)' $\hat{T}\circ M_t^* = t^{-\mu}\,\mathcal{F}(T\circ M_{1/t}),$ $t\in]0,+\infty[.$

If $T\in\mathcal{S}'(V)$ then we observe that - since $C_0^\infty(V)$ is dense in $\mathcal{S}(V)$ and since the map $\mathcal{S}(V)\longrightarrow\mathcal{S}(V)$, $\varphi\mapsto\varphi\circ M_{1/t}$, is continuous - T is quasihomogeneous of degree m if and only if (2.3) holds for every $\varphi\in\mathcal{S}(V)$. We shall see below (Theorem 6.45) that in case $\sigma=\sigma_+$ every quasihomogeneous distribution on V is automatically temperate. Here we note as an immediate consequence of (2.8)':

Proposition 2.19. *Let $T\in\mathcal{S}'(V)$. Then T is quasihomogeneous of degree m and of type M if and only if $\hat{T}$ is quasihomogeneous of degree $m^* := -m-\mu$ and of type M^*.* ■

Corollary 2.20. *Let Q be a polynomial function on V^*, and denote by δ_o the Dirac-distribution at 0. Then $Q(D)\delta_o$ is quasihomogeneous of degree m^* and of type M if and only if Q is quasihomogeneous of degree m and of type M^*.*

Proof. Since $\mathcal{F}(Q(D)\delta_o)=T_Q$ this follows from Proposition 2.19 and from what was said in the text preceding Definition 2.1. ■

The assertion of Proposition 2.19 is also a consequence of the following lemma telling us what becomes of the Euler operator ∂_M under the Fourier transform.

Lemma 2.21. *Let $T\in\mathcal{S}'(V)$. Then $(\partial_{M^*}-m^*)\hat{T} = -\hat{d}$ where $d:=(\partial_M-m)T$.*

Proof. Computing with respect to real orthonormal coordinates we obtain

$$\xi_j\,\partial_k\hat{T} = -\xi_j\, i\,\mathcal{F}(x_kT) = (-i)^2\,\mathcal{F}(\partial_j(x_kT)) = -\delta_{jk}\hat{T} - \mathcal{F}(x_k\,\partial_jT).$$

From this we deduce by making use of (1.29)' that

$$\partial_{M^*}\hat{T} = -\mu\,\hat{T} - \mathcal{F}(\partial_MT) = -\mu\hat{T} - \hat{d} - m\hat{T}\,. \quad ■$$

The assertion of Proposition 2.6.(ii) carries over to wave front sets:

Proposition 2.22. *Let $T\in\mathcal{D}'(X)$ be quasihomogeneous. Then*

$$WF(T) = \{(M_tx, M_{1/t}^*\xi);\ (x,\xi)\in WF(T),\ t\in J_M(\{x\};X)\}.$$

Proof. It suffices to observe that by (2.8)' we have

$$\mathcal{F}(\varphi\circ M_{1/t}T)(M^*_{1/t}\xi) = t^m\,\mathcal{F}((\varphi T)\circ M_{1/t})(M^*_{1/t}\xi) = t^{m+\mu}\,\mathcal{F}(\varphi T)(\xi) \tag{2.9}$$

for arbitrary $\xi\in V^*$, $\varphi\in C_0^\infty(X)$, and $t\in J_M(\operatorname{supp}\varphi;X)$. ∎

We close this section with

Proposition 2.23. *Suppose that $\sigma=\sigma_+$ and that M is semi-simple. Let $T\in\mathcal{S}'(V)$ be such that the restriction of T to $\dot{V}$ is induced by an almost quasihomogeneous C^∞ function. Then sing supp $\hat{T}\subset\{0\}$.*

Proof. Let $g\in C^\infty(\dot{V})$ be such that $T|_{\dot{V}}=T_g$, and let $m\in\mathbb{C}$ be such that g is almost quasihomogeneous of degree m. We fix $\alpha\in\mathfrak{A}$ and choose $\ell\in\mathfrak{A}(M)$ such that $\operatorname{Re}\ell>\operatorname{Re}(m+\alpha_M)+\mu$. Applying Remark 1.43 to V^* instead of V and identifying V^{**} with V, by P_α we denote the polynomal function on V satisfying $P_\alpha(i\partial)=\partial^\alpha$ where here ∂^α is acting on distributions defined on V^*. Note that by Corollary 1.36 and Remarks 1.43 and 2.18 P_α is quasihomogeneous of degree α_M. Let $Q:V^*\longrightarrow\mathbb{C}$ be any polynomial function which is quasihomogeneous of degree ℓ and of type M^*. Note that by Remark 2.18 and Proposition 1.55 $Q(D)(P_\alpha|_{\dot{V}}g)$ is almost quasihomogeneous of degree $m+\alpha_M-\ell$ the real part of which is strictly smaller than $-\mu$. Hence, by Lemma 1.92 the restriction of $Q(D)(P_\alpha|_{\dot{V}}g)$ to $V\setminus K(0,1)$ is integrable. Choosing $\varphi\in C_0^\infty(X)$ equal to 1 on $K(0,1)$ one concludes that $(1-\varphi)\,Q(D)(P_\alpha T)$ belongs to $\mathcal{L}^1(V)$. Hence its Fourier transform is continuous. Since the support of $\varphi\,Q(D)(P_\alpha T)$ is compact its Fourier transform is analytic. Hence $Q\,\partial^\alpha\hat{T}=\mathcal{F}(Q(D)(P_\alpha T))$ is continuous. Since M is semi-simple, for every $\xi\in V^*\setminus\{0\}$ the polynomial function Q can be chosen such that $Q(\xi)\neq0$. Hence, it follows that $\partial^\alpha\hat{T}$ is continuous on $V^*\setminus\{0\}$. This implies the assertion. ∎

Actually, the assumption that M be semi-simple is superfluous. In fact, a more general and more precise version of Proposition 2.23 will be proved in Chapter 9 (see Theorem 9.34) showing, in particular, that the converse of Proposition 2.23 is valid, as well.

(c) Meromorphic Functions of Quasihomogeneous Distributions

Let Ω be a connected open subset of $\mathbb{C}$, and let $h:\Omega\longrightarrow\mathcal{D}'(X)$ be a *meromorphic function*, i.e. there is a discrete subset $\mathfrak{D}$ of Ω such that for every $\varphi\in C_0^\infty(X)$ by $\Omega\setminus\mathfrak{D}\ni z\mapsto\langle h(z),\varphi\rangle$ a meromorphic function $h_\varphi:\Omega\longrightarrow\mathbb{C}$ is defined, for every $z\in\mathfrak{D}$ the pole order of h_φ at z being bounded by a constant independent from φ. We fix $z_o\in\Omega$. Approximating the integral by Riemann sums and employing the Banach-Steinhaus theorem (note that the space $C_0^\infty(X)$ is barrelled) one deduces that for every $j\in\mathbb{Z}$ a distribution $a_j(z_o;h)\in\mathcal{D}'(X)$ is defined by

$$(2.10)\qquad \langle a_j(z_o;h),\varphi\rangle := \frac{1}{2\pi i}\int_{\gamma_\varepsilon}\langle h(z_o+\lambda),\varphi\rangle\,\lambda^{-j-1}\,d\lambda,\qquad \varphi\in C_0^\infty(X),$$

where $\gamma_\varepsilon:[0,2\pi]\longrightarrow\mathbb{C}$ is defined by $\gamma_\varepsilon(t):=\varepsilon e^{it}$ and where $\varepsilon\in]0,+\infty[$ has to be so small that the closed disc $\overline{K}(z_o,\varepsilon)$ is contained in Ω and contains no point of $\mathfrak{D}$ with the possible exception of z_o. For obvious reasons one calls $a_j(z_o;h)$ *the* j^{th} *Laurent coefficient of h at* z_o. Similarly, the number

$$ord(z_o;h) := \inf\{j\in\mathbb{Z};\ a_j(z_o;h)\neq 0\}$$

(which is finite by the definition of meromorphy) is called the *order of h at* z_o. The set $\mathfrak{Pol}(h)$ of poles of h is, of course, by definition equal to $\{z\in\Omega;\ ord(z;h)<0\}$. Note that by the Banach-Steinhaus theorem the Laurent series

$$\sum_{j=ord(z_o;h)}^{\infty}(z-z_o)^j\,a_j(z_o;h)$$

converges to h(z) uniformly on every bounded subset of $C_0^\infty(X)$ and uniformly for z in any compact subset of $K(z_o,r_o)\setminus\mathfrak{Pol}(h)$ where $r_o:=\mathrm{dist}(z_o,\mathfrak{Pol}(h)\setminus\{z_o\}\cup\mathbb{C}\setminus\Omega)$.

Moreover, let $g:\Omega\longrightarrow\mathbb{C}$ be a holomorphic function such that $g(z_o)=m$. In the present section we deal with the situation that h(z) is quasihomogeneous of degree g(z). In view of Proposition 2.5 it is no loss of generality that from now on we assume that X is quasihomogeneous.

The principle of analytic continuation shows:

Remark 2.24. *If there is a non-empty open subset Z of* $\Omega\setminus\mathfrak{Pol}(h)$ *such that for every* $z\in Z$ *the distribution h(z) is quasihomogeneous of degree g(z) then the same is valid for every* $z\in\Omega\setminus\mathfrak{Pol}(h)$. ∎

How the quasihomogeneity of h(z) is reflected in the Laurent coefficients of h is shown by

Proposition 2.25. *The following conditions are equivalent:*

(a) h(z) is quasihomogeneous of degree g(z) for every $z \in \Omega \setminus (\{z_o\} \cup \mathfrak{Pol}(h))$;

(b) for every $j \in \mathbb{Z}$ *we have*

$$(\partial_M - m)\, a_j(z_o;h) = \sum_{k=1}^{\infty} \frac{1}{k!} g^{(k)}(z_o)\, a_{j-k}(z_o;h)$$

(note that the sum is finite since $ord(z_o;h) > -\infty$*).*

Proof. By $\partial_M h$ we denote the meromorphic function on Ω mapping $z \in \Omega \setminus \mathfrak{Pol}(h)$ to $\partial_M(h(z))$. From (2.10) one deduces for every $j \in \mathbb{Z}$ that $a_j(z_o;\partial_M h) = \partial_M a_j(z_o;h)$ and

$$a_j(z_o;gh) = \sum_{k=0}^{\infty} \frac{1}{k!} g^{(k)}(z_o)\, a_{j-k}(z_o;h)\,.$$

As two meromorphic functions on the connected open set Ω coincide if and only if their Laurent expansions at z_o do so the condition (b) is equivalent to the equality $\partial_M h = gh$ which in view of Proposition 2.11 is equivalent to (a). ∎

If $g \equiv m$ then the condition (b) means that $a_j(z_o;h)$ is quasihomogeneous of degree m for every $j \in \mathbb{Z}$. However, in general the latter is valid for $j = ord(z_o;h)$, only. For example, if $g(z)=z$, $z \in \Omega$, then (b) reads as

(2.11) $$(\partial_M - m)\, a_j(z_o;h) = a_{j-1}(z_o;h)$$

so that $a_j(z_o;h)$ is definitely not quasihomogeneous of degree m if $j > ord(z_o;h)$ and if h does not vanish identically.

In order to obtain information on the behaviour of $a_j(z_o;h) \circ M_t$ for $j > ord(z_o;h)$ one could compute $a_j(z_o;h \circ M_t)$ in terms of the Laurent coefficients of $t^g h$. However, for fixed j only finitely many of them are involved. Therefore we prefer to examine the condition (b) for arbitrary finite sequences of distributions not regarding whether or not they appear as coefficients in a Laurent series.

So we fix $N \in \mathbb{N}$, distributions $T_o, \ldots, T_N$ on X, and a sequence of complex numbers c_k, $k \in \mathbb{N}_N$. The analogue of condition (b) becomes

(2.12) $$(\partial_M - m)\, T_j = \sum_{k=0}^{j-1} c_{j-k} T_k$$

for every $0 \le j \le N$. To formulate the equation for $T_j \circ M_t$ which is implied by (2.12) we set

$$c(z) := \sum_{\ell=1}^{N} c_\ell (z-m)^\ell , \qquad z \in \mathbb{C},$$

and for arbitrary $t \in]0,+\infty[$ and $k \in \mathbb{N}_0$ denote by $b_k(t)$ the kth Taylor coefficient of the function $z \mapsto t^{c(z)}$ at $z = m$, i.e.

$$(2.13) \qquad \sum_{k=0}^{\infty} b_k(t)\,(z-m)^k = t^{c(z)}, \qquad z \in \mathbb{C}.$$

Applying the binomial formula one obtains more explicitly that

$$(2.13)' \qquad b_k(t) = \sum_{j=0}^{k} b_{kj}\,(\log t)^j$$

where $b_{00} := 1$ and

$$b_{kj} := \sum_{\alpha \in A(k,j)} \prod_{\ell=1}^{N} c_\ell^{\alpha_\ell} / \alpha_\ell !$$

with

$$A(k,j) := \left\{ \alpha \in \mathbb{N}_0^N ;\ \sum_{\ell=1}^{N} \alpha_\ell = j,\ \sum_{\ell=1}^{N} \ell \alpha_\ell = k \right\} \qquad \text{if } k \in \mathbb{N}.$$

Proposition 2.26. *Under the preceding hypotheses the relations*

$$(2.14) \qquad T_j \circ M_t = t^m \Big(T_j + \sum_{k=0}^{j-1} b_{j-k}(t)\, T_k \Big), \qquad t \in]0,+\infty[,$$

hold for every $j \in \mathbb{N}_N \cup \{0\}$ if and only if (2.12) is valid for every $j \in \mathbb{N}_N \cup \{0\}$.

Proof. Let $\varphi \in C_0^\infty(X)$. For every $j \in \mathbb{N}_N \cup \{0\}$ we define funtions $g_j :]0,+\infty[\longrightarrow \mathbb{C}$ and $h_j :]0,+\infty[\longrightarrow \mathbb{C}$ by

$$g_j(t) := t^{-m} \langle T_j \circ M_t , \varphi \rangle \qquad \text{and} \qquad h_j(t) := \sum_{k=0}^{j} b_{j-k}(t) \langle T_k , \varphi \rangle .$$

By Lemma 2.10 g_j is differentiable satisfying

$$g_j'(t) = t^{-m-1} \langle ((\partial_M - m)\, T_j) \circ M_t , \varphi \rangle , \qquad t \in]0,+\infty[.$$

To compute h_j' we first differentiate both sides of (2.13) with respect to t and obtain

$$\sum_{k=0}^{\infty} b_k'(t)\,(z-m)^k = \tfrac{1}{t}\, c(z)\, t^{c(z)} .$$

Inserting (2.13) into the right-hand side, substituting the definition of c(z) and comparing coefficients we obtain that

$$b_k'(t) = \frac{1}{t}\sum_{\ell=0}^{k-1} c_{k-\ell}\, b_\ell(t)\,, \qquad t\in\,]0,+\infty[\,,\ k\in\mathbb{N}_N\cup\{0\}.$$

Inserting this with k replaced by j−k into the equation obtained by differentiating the defining equality for h_j, writing $L=k+\ell$, and changing the order of summation we arrive at

$$(2.15)\qquad h_j'(t) = \frac{1}{t}\sum_{L=0}^{j-1} c_{j-L}\,\langle \sum_{k=0}^{L} b_{L-k}(t)\,T_k\,, \varphi\rangle, \qquad t\in\,]0,+\infty[\,.$$

Now we suppose that (2.14) holds. Since this means that $g_j = h_j$ it follows in view of $b_{L-k}(1) = \delta_{Lk}$ that

$$\langle(\partial_M - m)T_j\,, \varphi\rangle = g_j'(1) = h_j'(1) = \sum_{L=0}^{j-1} c_{j-L}\,\langle T_L, \varphi\rangle$$

so that (2.12) is valid.

Conversely, we suppose that (2.12) holds and that (2.14) is already proved for every $L\in\mathbb{N}_{j-1}\cup\{0\}$ instead of j. Inserting this into (2.15) we see that

$$h_j'(t) = t^{-m-1}\sum_{L=0}^{j-1} c_{j-L}\,\langle T_L\circ M_t, \varphi\rangle, \qquad t\in\,]0,+\infty[\,.$$

By (2.12) this implies that $h_j' = g_j'$. Since in view of $b_k(1) = \delta_{k0}$ we have $g_j(1) = \langle T_j, \varphi\rangle = h_j(1)$ it follows that $h_j = g_j$, i.e. (2.14) holds. Since by Proposition 2.11 the condition (2.14) for $j=0$ is equivalent to (2.12) for $j=0$ the assertion follows by induction. ∎

In the special case $c_k = c\,\delta_{1k}$ we have $b_k(t) = (c\log t)^k/k!$ so that in view of (1.37) the condition (2.14) reads as

$$(2.14)'\qquad T_j\circ M_t = t^m\sum_{k=0}^{j} c^{j-k}\,\omega_{j-k}\,T_k\,, \qquad t\in\,]0,+\infty[\,.$$

In this case the following slightly more precise version of Proposition 2.26 holds.

Proposition 2.26'. *For every $c\in\dot{\mathbb{C}}$ the following conditions are equivalent:*

(a) (2.14)' is valid for $j=N$;

(b) (2.14)' is valid for every $j\in\mathbb{N}_N\cup\{0\}$;

(c) T_0 is quasihomogeneous of degree m, and $(\partial_M - m)T_j = c\,T_{j-1}$ for every $j\in\mathbb{N}_N$.

Proof. In view of Proposition 2.26 it suffices to prove the implication "*(a)⇒(b)*".

Using the notation of the proof of Proposition 2.26 we first observe from (2.15) that

$$h_j'(t) = \frac{c}{t} \sum_{k=0}^{j-1} b_{j-1-k}(t) \langle T_k, \varphi \rangle = \frac{c}{t} h_{j-1}(t), \qquad t \in]0,+\infty[.$$

Now suppose that (2.14)' holds for a fixed $j \in \mathbb{N}_N$. This implies (see the proof of Proposition 2.26) that $(\partial_M - m) T_j = c T_{j-1}$ so that

$$g_j'(t) = c\, t^{-m-1} \langle T_{j-1} \circ M_t, \varphi \rangle = \frac{c}{t} g_{j-1}(t).$$

Since (2.14)' means that $g_j = h_j$ and hence $g_j' = h_j'$ it follows that $g_{j-1} = h_{j-1}$ so that (2.14)' is valid with j replaced by j-1, and (b) follows from (a) by induction. ■

To provide a non-trivial example for a meromorphic function of quasihomogeneous distributions we fix a polynomial function $P: V^* \longrightarrow \mathbb{C}$ and recall the method for obtaining *Bernstein's fundamental solution E_P* of the differential polynomial P(D). Suppose that $P \geq 0$. Then by

$$\langle \mathfrak{P}(z), \check{\varphi} \rangle := (2\pi)^{-n} \int_{V^*} P(\xi)^z \hat{\varphi}(\xi)\, d\xi, \qquad \varphi \in \mathcal{S}(V),$$

a holomorphic function $\mathfrak{P}: \{z \in \mathbb{C} ; \operatorname{Re} z > 0\} \longrightarrow \mathcal{S}'(V)$ is defined. As was shown by Bernstein in [2] (see also Björk [3]), it can be extended to a meromorphic function on the whole of $\mathbb{C}$ with values in $\mathcal{S}'(V)$. This extension is denoted by $\mathfrak{P}$, as well. One easily sees that $E_P := a_0(-1; \mathfrak{P})$ is a fundamental solution of P(D), called *Bernstein's fundamental solution.*

Example 2.27. *Suppose that P is quasihomogeneous of degree $\ell \in \mathbb{C}$ and of type M^*. Then*

(i) $\mathfrak{P}(z)$ is quasihomogeneous of degree $-\ell z - \mu$ for every $z \in \mathbb{C} \setminus \mathfrak{Pol}(\mathfrak{P})$;

(ii) setting $N := -\operatorname{ord}(-1; \mathfrak{P})$ we have

$$E_P \circ M_t = t^{\ell-\mu} \Big(E_P + \sum_{j=1}^{N} (-\ell)^j \omega_j(t)\, a_{-j}(-1; \mathfrak{P}) \Big);$$

in particular, if $\ell = 0$ then E_P is quasihomogeneous of degree $-\mu$ while in the case $\ell \neq 0$ E_P is quasihomogeneous (of degree $\ell - \mu$) if and only if $\mathfrak{P}$ is holomorphic at $z = -1$.

<u>Proof</u>. <u>(i)</u>: If $\operatorname{Re} z > 0$ then $\mathcal{F}(\mathfrak{P}(z)) = P^z$ is quasihomogeneous of degree ℓz and of type M^* so that by Proposition 2.19 $\mathfrak{P}(z)$ is quasihomogeneous of degree

$g(z) := -\ell z - \mu$. Hence the assertion follows by Remark 2.24.

(ii): In view of Proposition 2.25 (applied to $g(z) := -\ell z - \mu$) one has to apply Proposition 2.26' to $T_j := a_{j-N}(-1;\mathfrak{P})$, $m := \ell - \mu$, and $c := -\ell$. ■

Further examples of meromorphic functions of quasihomogeneous distributions will come up in Chapters 4 and 6. In fact, using these one can show that for every $T \in \mathfrak{D}'(X)$ which is quasihomogeneous of degree m there is an entire function $h : \mathbb{C} \longrightarrow \mathfrak{D}'(X)$ such that $h(m) = T$ and such that $h(z)$ is quasihomogeneous of degree z for every $z \in \mathbb{C}$.

(d) Almost Quasihomogeneous Distributions

In later chapters a central role is played by distributions T which appear as zero order Laurent coefficients of certain meromorphic functions h of quasihomogeneous distributions where the function g in Proposition 2.25.(a) is the identity map on $\mathbb{C}$. Proposition 2.26' shows that T then satisfies the equation $(\partial_M - m)^N T = 0$ for some $N \in \mathbb{N}$. We call such distributions "*almost quasihomogeneous of degree m*". Their behaviour under the action of M_t is described by (2.14)' for $c = 1$. It is this property - written in a slightly different way - that serves as the basis for the following definition of almost quasihomogeneity. Throughout this section we fix a number $N \in \mathbb{N}$.

Definition 2.28. A distribution $T \in \mathfrak{D}'(X)$ is called *almost quasihomogeneous of degree m (and of type M) and of order* $\leq N$ if and only if there exist distributions $d_1, \ldots, d_N \in \mathfrak{D}'(X)$ satisfying

$$t^{-m}\, T \circ M_t = T + \sum_{k=1}^{N} \omega_k(t)\, d_k . \tag{2.16}$$

In view of (1.42), for every $k \in \mathbb{N}$ the distribution d_k appearing in (2.16) is unique; it is called *the* k^{th} *order deficiency of T*. If $k = 1$ then we also speak of *the deficiency of T*. Setting $d_0 := T$ we call the number $ord_M(T) := \min\{k \in \mathbb{N}_0 ;\ d_k \neq 0\}$ the *(quasihomogeneity) order of T (with respect to M)*.

If X is not supposed to be quasihomogeneous then in the definition of almost quasihomogeneity one has to postulate that

$$(2.16)' \qquad t^{-m-\mu}\langle T, \varphi\circ M_{1/t}\rangle = \langle T,\varphi\rangle + \sum_{k=1}^{N}\omega_k(t)\langle d_k,\varphi\rangle$$

for arbitrary $\varphi \in C_0^\infty(X)$ and $t \in J_M(\operatorname{supp}\varphi; X)$ - see (2.2). By a simple modification of the proof of Proposition 2.5 (compare the proof of Proposition 1.57) one can verify that the assertion of Proposition 1.57 carries over to almost quasihomogeneous distributions. However, we shall not make use of this result in the sequel.

Next we are going to verify that for locally integrable functions the two notions of almost quasihomogeneity introduced in Definitions 1.32 and 2.28 essentially coincide. For this we have to show that similarly as in Remark 1.50 one can compute the deficiencies by purely algebraic means:

Remark 2.29. *Let $T\in\mathcal{D}'(X)$ be almost quasihomogeneous of degree m and of order $\le N$. And let $(t_j)_{1\le j\le N}$ be a sequence of pairwise different numbers in $]0,+\infty[\setminus\{1\}$. Then for every $k\in\mathbb{N}_N$ the kth order deficiency d_k of T is given by*

$$(2.17) \qquad d_k = \sum_{j=1}^{N} L_{kj}\,(t_j^{-m}\, T\circ M_{t_j} - T)$$

where $(L_{kj})_{(k,j)\in\mathbb{N}_N\times\mathbb{N}_N}$ denotes the inverse of the real $N\times N$-matrix (1.66).

Proof. Just as in the proof of Remark 1.50 one has to subtract T from both sides of (2.16), insert t_j for t, and apply (L_{kj}). ∎

Remark 2.30. *Let $f\in\mathcal{L}^1_{loc}(X)$.*

(i) If f is almost quasihomogeneous of degree m in the sense of Definition 1.32 then the distribution T_f induced by f is almost quasihomogeneous of degree m in the sense of Definition 2.28, and for every $k\in\mathbb{N}$ the kth order deficiency q_k of f, being locally integrable by Proposition 1.51.(i), induces the kth order deficiency d_k of T_f.

(ii) Conversely, suppose that T_f is almost quasihomogeneous of degree m in the sense of Definition 2.28; if $V\ne E_M(\sigma_0)$ then there is a function $g: X\longrightarrow\mathbb{C}$ which is almost quasihomogeneous of degree m in the sense of Definition 1.32 and which is equal to f almost everywhere; if f is continuous then f itself is almost quasihomogeneous of degree m in the sense of Definition 1.32.

Proof. *(i)*: in view of (1.39) and Proposition 1.51.(i) a transformation of variables leads to the assertion.

(ii). It follows by Remark 2.29 that for every $k \in \mathbb{N}_N$ there is $f_k \in \mathcal{L}^1_{loc}(X)$ such that the kth order deficiency of T_f is equal to T_{f_k}. We set $f_0 := f$. Since the map $]0,+\infty[\times X \longrightarrow X$, $(t,x) \mapsto M_t x$, is a submersion the map $h:]0,+\infty[\times X \longrightarrow \mathbb{C}$, $(t,x) \mapsto t^{-m} f(M_t x)$, is measurable. The assumption on T_f implies that for every $t \in]0,+\infty[$ we have

$$h(t,\cdot) = \sum_{k=0}^{N} \omega_k(t)\, f_k \quad \text{almost everywhere.}$$

If f is continuous then the f_k can be chosen to be continuous, as well, so that the preceding equations hold everywhere, and the proof is complete. If f is not continous then for any compact subset K of X and for every $t \in]0,+\infty[$ one can only deduce that - since $M_t(K)$ is a compact subset of X - the function $h(t,\cdot)$ is integrable on K, and

$$\int_K |h(t,x)|\, dx \le \sum_{k=0}^{N} |\omega_k(t)| \int_K |f_k(x)|\, dx .$$

Since as a function of t the right-hand side is locally integrable the Fubini-Tonelli theorem shows that h is locally integrable, as well. It follows by Fubini's theorem that for every $\psi \in C_0^\infty(]0,+\infty[\times X)$ we have

$$\int h\psi\, dt\otimes dx = \sum_{k=0}^{N} \int \omega_k \otimes f_k\, \psi\, dt\otimes dx .$$

Consequently, h is equal to $\sum_{k=0}^{N} \omega_k \otimes f_k$ almost everywhere.

Now, we come to the construction of g. First of all, by Proposition 1.80 the assumption on V implies that $X\setminus\Omega(M)$ is a set of measure zero so that we may define $g|_{X\setminus\Omega(M)} :\equiv 0$. Moreover, in view of Notation 1.65 and Lemma 1.64 we find a subset J of $\mathbb{N}$, an open covering $(U_j)_{j\in J}$ of $\Omega(M)$ consisting of non-empty quasihomogeneous open subsets of X, and a family $(\varkappa_j)_{j\in J}$ of C^∞ functions $\varkappa_j : U_j \longrightarrow]0,+\infty[$ which are quasihomogeneous of degree 1. It suffices for every $j \in J$ to find a function $g_j : U_j \longrightarrow \mathbb{C}$ which is almost quasihomogeneous of degree m and coincides with $f|_{U_j}$ almost everywhere. For the sets $X_j := U_j \setminus \bigcup_{i\in J, i<j} U_i$, $j \in J$, are measurable, and defining g on $\Omega(M) = \bigcup_{j\in J} X_j$ by $g|_{X_j} := g_j|_{X_j}$, $j \in J$, completes the construction of g.

So we may assume that X is equipped with a C^∞ function $\varkappa : X \longrightarrow]0,+\infty[$ which is quasihomogeneous of degree 1. We fix $s \in]0,+\infty[$ and define

$$g(M_t M_s \vartheta) := t^m \Big(\sum_{k=0}^{N} \omega_k(t)\, f_k(M_s \vartheta) \Big), \qquad t \in]0,+\infty[,\ \vartheta \in S^{\varkappa}.$$

Applying Remark 1.73'.(i) to $\lambda := \varkappa \circ M_s^{-1}$ instead of $\varkappa$ and observing that $S^{\lambda} = M_s(S^{\varkappa})$ we deduce that g is almost quasihomogeneous of degree m. The remaining problem is to find s in such a way that the functions g and f coincide almost everywhere. By ν we denote the measure $t^{\mu-1}dt$ on $]0,+\infty[$. It follows from Proposition 1.86 that the inverse of the map

$$\Psi :]0,+\infty[\times S^{\varkappa} \times]0,+\infty[\longrightarrow]0,+\infty[\times X,\ (t,\vartheta,s) \mapsto (t, M_s\vartheta),$$

(which is bijective by Proposition 1.72, indeed) maps sets of $dt \otimes dx$-measure zero onto sets of $dt \otimes \tilde{\varkappa} \otimes \nu$-measure zero. Hence, from the equality derived at the beginning of the proof if follows that

$$h \circ \Psi = \sum_{k=0}^{N} (\omega_k \otimes f_k) \circ \Psi \quad dt \otimes \tilde{\varkappa} \otimes \nu\text{-almost everywhere.}$$

This implies that one can find $s \in]0,+\infty[$ such that

$$h \circ \Psi(\cdot, s) = \sum_{k=0}^{N} \omega_k \otimes (f_k \circ M_s|_{S^{\varkappa}}) \quad dt \otimes \tilde{\varkappa}\text{-almost everywhere,}$$

i.e.

$$t^{-m} f(M_t M_s \vartheta) = \sum_{k=0}^{N} \omega_k(t)\, f_k(M_s \vartheta)$$

for $dt \otimes \tilde{\varkappa}$-almost every $(t,\vartheta) \in]0,+\infty[\times S^{\varkappa}$. Since by Proposition 1.86 the mapping $]0,+\infty[\times S^{\varkappa} \longrightarrow X$, $(t,\vartheta) \mapsto (M_t M_s \vartheta)$, maps sets of $dt \otimes \tilde{\varkappa}$-measure zero onto sets of dx-measure zero the number s has the desired property. ∎

The assumption about V in assertion (ii) of Remark 2.30 is perhaps not necessary.

The following proposition gives an infinitesimal characterization for almost quasihomogeneity. It is the analogue of Proposition 2.11 and essentially is a reformulation of Proposition 2.26' for $c = 1$.

Proposition 2.31. *Let $T \in \mathcal{D}'(X)$. Then*

(i) T is almost quasihomogeneous of degree m and of order $\leq N$ if and only if $(\partial_M - m)^{N+1} T = 0$;

(ii) if this is the case then for every $k \in \mathbb{N}_N$ the kth order deficiency d_k of T is equal to $(\partial_M - m)^k T$.

Proof. If (2.16) holds then Proposition 2.26' (applied to $T_N := T$ and $T_{N-k} := d_k$, $k \in \mathbb{N}_N$) shows that $(\partial_M - m)T = d_1$, $(\partial_M - m)d_k = d_{k+1}$ for $k \in \mathbb{N}_{N-1}$, and $(\partial_M - m)d_N = 0$. Hence, by induction we obtain that $(\partial_M - m)^{N+1}T = 0$ and $d_k = (\partial_M - m)^k T$. Conversely, if $(\partial_M - m)^{N+1}T = 0$ then the condition (c) of Proposition 2.26' is valid for $T_j := (\partial_M - m)^{N-j}T$ and $c = 1$. Consequently, (2.16) follows by Proposition 2.26' if we set $d_k := T_{N-k}$. ∎

Proposition 2.31 shows how to compute the polynomial functions $P_{\ell,j}$ associated to any polynomial function P in Remark 1.35:

Corollary 2.32. *Let $P: V \longrightarrow \mathbb{C}$ be a polynomial function. Then for arbitrary $\ell \in \mathfrak{A}(M)$ and $k \in \mathbb{N}$ we have $P_{\ell,k} = (\partial_M - \ell)^k (Q_\ell P)$.* ∎

As an immediate consequence of Proposition 2.31 one obtains the following assertions the first two of which can also easily be derived directly from (2.16) (compare the proof of Lemma 1.48).

Corollary 2.33. *Suppose that T is almost quasihomogeneous of degree m and of order $\leq N$. Then for every $k \in \mathbb{N}_N$ its kth order deficiency d_k has the following properties:*

(i) d_k *is almost quasihomogeneous of degree m and of order $\leq N-k$;*

(ii) $\operatorname{supp} d_k \subset \operatorname{supp} T$;

(iii) $\operatorname{sing\,supp} d_k \subset \operatorname{sing\,supp} T$;

(iv) $WF(d_k) \subset WF(T)$. ∎

In order to prove the analogue of Proposition 2.3 for almost quasihomogeneous distributions we require

Lemma 2.34. *Let $P: X \times V^* \longrightarrow \mathbb{C}$ be a C^∞ function which is a polynomial with respect to the second variable. Then for every $T \in \mathcal{D}'(X)$ we have*

$$\partial_M(P(x,\partial)T) = (\partial_{M \times (-M)^*} P)(x,\partial)T + P(x,\partial)\,\partial_M T.$$

For arbitrary $m, \ell \in \mathbb{C}$ and $k \in \mathbb{N}$ this implies

$$(\partial_M - m - \ell)^k (P(x,\partial)\,T) = \sum_{i=0}^{k} \binom{k}{i} [(\partial_{M\times(-M)^*} - \ell)^i P](x,\partial)\,(\partial_M - m)^{k-i} T.$$

Proof. Working with real coordinates, setting $A := \mathbb{N}_o^n$, writing

$$P(x,\xi) = \sum_{\alpha\in A} P_\alpha(x)\,\xi^\alpha, \qquad x,\xi \in \mathbb{R}^n,$$

and making use of

$$x_k\,\partial^\alpha \partial_j T = \partial^\alpha (x_k \partial_j T) - \alpha_k\,\partial^{\alpha-e_k} \partial_j T$$

we obtain in view of (1.29)' that

$$\partial_M (P(x,\partial)\,T) = \sum_{j,k\in\mathbb{N}_n} M_{jk}\, x_k\, \partial_j \Big(\sum_{\alpha\in A} P_\alpha\, \partial^\alpha T \Big) =$$

$$= \sum_{j,k\in\mathbb{N}_n} \sum_{\alpha\in A} \Big(M_{jk} \big(x_k (\partial_j P_\alpha)\, \partial^\alpha T + P_\alpha\, \partial^\alpha (x_k \partial_j T) - P_\alpha\, \alpha_k\, \partial^{\alpha-e_k} \partial_j T \big) \Big) =$$

$$= \Big[\sum_{j,k\in\mathbb{N}_n} M_{jk}\, x_k\, \partial_{x_j} P - \sum_{j,k\in\mathbb{N}_n} M_{jk}\, \xi_j\, \partial_{\xi_k} P \Big](x,\partial)\,T + P(x,\partial)\,\partial_M T.$$

Since the term in square brackets is equal to $\partial_N P$ where $N := M\times(-M)^*$ the first part of the assertion is proved. Applying this to $((\partial_N - \ell)^i P, (\partial_M - m)^{k-i} T)$ instead of (P,T) we obtain for arbitrary $k\in\mathbb{N}$ and $i\in\{0\}\cup\mathbb{N}_k$

$$(\partial_M - m - \ell)([(\partial_N - \ell)^i P](x,\partial)\,(\partial_M - m)^{k-i} T) =$$

$$= [(\partial_N - \ell)^{i+1} P](x,\partial)\,(\partial_M - m)^{k-i} T + [(\partial_N - \ell)^i P](x,\partial)\,(\partial_M - m)^{k+1-i} T.$$

Since $\binom{k}{i-1} + \binom{k}{i} = \binom{k+1}{i}$ for every $i\in\mathbb{N}_{k+1}$ the second part of the assertion follows by induction on k. ∎

As an immediate consequence of Lemma 2.34 one obtains

Proposition 2.35. *Let $T\in\mathcal{D}'(X)$ be almost quasihomogeneous of degree m and of order $\le N$ with deficiency $d\in\mathcal{D}'(X)$. Moreover, let $P: X\times V^* \longrightarrow \mathbb{C}$ be a C^∞ function which is a polynomial with respect to the second variable. Suppose that P is almost quasihomogeneous of degree $\ell\in\mathbb{C}$, of type $M\times(-M)^*$, and of order $\le N'\in\mathbb{N}_o$ with deficiency Q. Then $P(x,\partial)\,T$ is almost quasihomogeneous of degree $m+\ell$ of order $\le N+N'$ with deficiency $P(x,\partial)\,d + Q(x,\partial)\,T$.* ∎

Two special cases of Proposition 2.35 are formulated as

Corollary 2.36. *Let $T \in \mathcal{D}'(X)$ be almost quasihomogeneous of degree m and of order $\leq N$, let $k \in \mathbb{N}_N$, and let d_k be the kth order deficiency of T.*

(i) If $P: V^* \longrightarrow \mathbb{C}$ is a polynomial function which is quasihomogeneous of degree $\ell \in \mathbb{C}$ and of type M^* then $P(\partial)T$ is almost quasihomogeneous of degree $m-\ell$ with k^{th} order deficiency $P(\partial)d_k$.

(ii) If $q \in C^\infty(X)$ is quasihomogeneous of degree $\ell \in \mathbb{C}$ then qT is almost quasihomogeneous of degree $m+\ell$ with k^{th} order deficiency qd_k. ∎

The next result generalizes Proposition 2.6.

Proposition 2.37. *Let $T \in \mathcal{D}'(X)$ be almost quasihomogeneous. Then*

(i) $\operatorname{supp} T = (\operatorname{supp} T)_M$;

(ii) $\operatorname{sing\,supp} T = (\operatorname{sing\,supp} T)_M$;

(ii)' more generally: $WF(T) = \{(M_t x, M^*_{1/t}\xi) ; (x,\xi) \in WF(T),\ t \in]0,+\infty[\}$.

Proof. We set $d_0 := T$ and for $k \in \mathbb{N}_N$ denote by d_k the kth order deficiency of T.

_(i)_: Let Y be the open subset $X \setminus \operatorname{supp} T$ of X. Then for every $x \in Y_M$ one finds an open neighbourhood U of x in Y_M and a number $t \in]0,+\infty[$ such that $M_{1/t}U \subset Y$. Hence for every $\varphi \in C_0^\infty(U)$ we deduce that $\operatorname{supp}(\varphi \circ M_t) \subset Y$ so that in view of $Y \subset X \setminus \operatorname{supp} d_k$ (see Corollary 2.33.(ii)) it follows from (2.16) that

$$\langle T, \varphi \rangle = t^\mu \langle T \circ M_t, \varphi \circ M_t \rangle = t^{m+\mu} \sum_{k=0}^{N} \omega_k(t) \langle d_k, \varphi \circ M_t \rangle = 0 .$$

This means that $Y_M \subset Y$. Since the converse inclusion is trivial the proof is complete.

_(ii)_. Here we set $Y := X \setminus \operatorname{sing\,supp} T$, and let $f \in C^\infty(Y)$ be such that $T|_Y = T_f$. Since f is almost quasihomogeneous it extends to an almost quasihomogeneous C^∞ function $f_M : Y_M \longrightarrow \mathbb{C}$ by Proposition 1.57. The assertion (i) implies that $T_{(f_M)}$ and T coincide on Y_M. Hence $Y_M \subset Y$, i.e. $Y_M = Y$.

_(ii)'_. By (2.16) and by (2.8)' we have

$$\mathcal{F}(\varphi \circ M_{1/t} T)(M^*_{1/t}\xi) = t^m \sum_{k=0}^{N} \omega_k(t)\, \mathcal{F}((\varphi d_k) \circ M_{1/t})(M^*_{1/t}\xi) =$$

$$= t^{m+\mu} \sum_{k=0}^{N} \omega_k(t)\, \mathcal{F}(\varphi d_k)(\xi), \qquad \xi \in V^*,\ \varphi \in C_0^\infty(X), \text{ and } t \in]0,+\infty[.$$

For arbitrary $x \in X$, $\xi \in V^*$, and $t \in]0,+\infty[$ this implies that $(M_t x, M^*_{1/t}\xi)$ does not belong to WF(T) if (x,ξ) does not belong to $WF(d_k)$ for every $k \in \{0\} \cup \mathbb{N}_N$. In view of Corollary 2.33.(iv) this proves the inclusion "$\supseteq$". As the inverse inclusion is trivial the proof is complete. ■

Carrying over the proof of Corollary 2.7 one deduces from Proposition 2.37.(ii):

Corollary 2.38. *Suppose that (1.14) holds. Let $T \in \mathcal{D}'(V_+)$ be almost quasihomogeneous. If sing supp $T \cap U$ is empty for some neighbourhood U of ker M then T is induced by a C^∞ function.* ■

Combining this with Proposition 1.58 one obtains

Proposition 2.39. *Suppose that $\sigma = \sigma_+$. Let $T \in \mathcal{D}'(V)$ be such that its restriction to $\dot{V}$ is almost quasihomogeneous. Then either $0 \in$ sing supp T or T is induced by a polynomial function.* ■

As for the Fourier transform of almost quasihomogeneous distributions, Lemma 2.21 immediately leads to

Proposition 2.40. *Let $T \in \mathcal{S}'(V)$. Then*

(i) T is almost quasihomogeneous of degree m and of order $\leq N$ if and only if $\hat{T}$ is almost quasihomogeneous of degree $m^* = -m-\mu$, of order $\leq N$ and of type M^*;

(ii) if this is the case then for every $k \in \mathbb{N}_N$ the kth order deficiency of $\hat{T}$ is equal to $(-1)^k \mathcal{F}(d_k)$ where d_k denotes the kth order deficiency of T. ■

Next we generalize Proposition 2.14 and Theorem 2.16 to almost quasihomogeneous distributions.

Proposition 2.41. *If $T \in \mathcal{D}'(X)$ is almost quasihomogeneous then its analytic wave front set $WF_A(T)$ is contained in the set $\Gamma_M(X)$ defined in (2.6).*

Proof. The principal symbol of the differential operator $(\partial_M - m)^{N+1}$ is equal to the function $(iP_M)^{N+1}$ where P_M is defined as in Example 1.21. Hence the characteristic set of the operator $(\partial_M - m)^{N+1}$ is equal to $\Gamma_M(X)$. Consequently, the assertion follows from Proposition 2.31.(i) and Theorem 8.6.1 in Hörmander [11]. ■

Combining Proposition 2.41 and Lemma 2.15 with Corollary 8.2.7 in Hörmander [11] one obtains

Theorem 2.42. *If $T \in \mathcal{D}'(X)$ is almost quasihomogeneous then its restriction $T/_{S^\varkappa}$ to $S^\varkappa$ is well-defined as the pullback by the inclusion map $J: S^\varkappa \longrightarrow X$.* ■

Under special assumptions on X it is possible to present every almost quasihomogeneous distribution as a linear combination of quasihomogeneous distributions where the coefficients are certain distinguished almost quasihomogeneous distributions. For a simple remark illustrating this we suppose that we can fix a distribution $\nu \in \mathcal{D}'(X)$ which is almost quasihomogeneous of degree 0 with deficiency 1. The said remark shows 1° how to construct further almost quasihomogeneous distributions and 2° that in case " $\operatorname{sing\,supp} \nu = \emptyset$ " every almost quasihomogeneous distribution on X of order ≤ 1 has a "quasihomogeneous part" from which it can be reconstructed.

Remark 2.43. *Let $d \in \mathcal{D}'(X)$ be quasihomogeneous of degree m. Suppose that the singular supports of d and ν do not intersect so that νd is a well-defined distribution on X. Moreover, let $T, u \in \mathcal{D}'(X)$ be distributions defining each other by the equation*

(2.18) $\quad T = u + \nu d\,.$

Then under the preceding hypothesis on ν T is almost quasihomogeneous of degree m with deficiency d if and only if u is quasihomogeneous of degree m.

Proof. From (2.18) one deduces by the Leibniz rule that

$$\partial_M T - mT = \partial_M u + (\partial_M \nu)\,d + \nu\,\partial_M d - mu - m\nu d = d + \partial_M u - mu\,.$$

Consequently, $(\partial_M - m)T = d$ if and only if $(\partial_M - m)u = 0$. ■

Under what conditions on X do distributions ν satisfying the assumptions of Remark 2.43 exist? Combining Remark 1.52 with Theorem 3.39 below one sees that such a ν having the additional property "sing supp $\nu = \emptyset$ " exists if and only if X is locally M-bounded (see Definition 3.30 and Proposition 3.31 below). Moreover, if (1.14) holds then such a ν always exists with singular support contained in $X \setminus X_+$: to see this one has to combine Proposition 1.70 with the following corollary to Proposition 1.91.

Remark 2.44. *Suppose that (1.14) holds. If $\varkappa : X_+ \longrightarrow]0,+\infty[$ is continuous and quasihomogeneous of degree 1 then $T_{\log \varkappa}$ extends to an almost quasihomogeneous distribution of degree 0 with deficiency equal to the constant function 1.* ∎

Under stronger assumptions on ν the assertion of Remark 2.43 can be partially generalized:

Proposition 2.45. *Suppose that $\nu : X \longrightarrow \mathbb{C}$ is a C^∞ function which is almost quasihomogeneous of degree 0 with deficiency 1. For every $j \in \mathbb{N}_0$ set $\nu_j := \nu^j / j!$ as in Remark 1.53. Let $T \in \mathcal{D}'(X)$. Then*

(i) T is almost quasihomogeneous of degree m and of order $\leq N$ if and only if there is a sequence of distributions S_ℓ, $0 \leq \ell \leq N$, which are quasihomogeneous of degree m such that

$$(2.19) \qquad T = \sum_{\ell=0}^{N} \nu_\ell \, S_\ell \, ;$$

(ii) if this is the case then the distributions S_ℓ satisfying (2.19) are uniquely determined; in fact, they are given by

$$(2.20) \qquad S_\ell = \sum_{i=0}^{N-\ell} (-1)^i \, \nu_i \, (\partial_M - m)^{i+\ell} \, T, \qquad 0 \leq \ell \leq N;$$

moreover, for every $k \in \mathbb{N}_0$ we have

$$(2.19)_k \qquad T = (\partial_M - m)^k \Big(\sum_{\ell=0}^{N} \nu_{k+\ell} \, S_\ell \Big).$$

Proof. (i), "$\Leftarrow$": this is a consequence of Remark 1.53 and Proposition 2.35.

(ii) and (i), "$\Rightarrow$": Note that Remark 1.53 and Proposition 2.31.(ii) imply

$$(2.21) \qquad \partial_M \nu_j = \nu_{j-1}, \quad j \in \mathbb{N}.$$

To prove uniqueness we deduce from (2.19) making use of Lemma 2.34

$$(\partial_M - m)^k T = \sum_{L=k}^{N} \nu_{L-k} S_L , \qquad 0 \le k \le N .$$

Inserting this for $k = \ell + i$ into the right-hand side of (2.20) and changing the order of summation we obtain

$$\sum_{L=\ell}^{N} \Big[\sum_{i=0}^{L-\ell} (-1)^i \nu_i \nu_{L-\ell-i} \Big] S_L .$$

Taking

$$\text{(2.22)} \qquad \nu_i \nu_{I-i} = \binom{I}{i} \nu_I , \qquad 0 \le i \le I,$$

for $I = L - \ell$ into account we deduce that the term in square brackets is equal to $(1-1)^{L-\ell} \nu_{L-\ell} = \delta_{L\ell}$, i.e. (2.20) is valid.

Conversely, if we define S_ℓ by (2.20) we deduce by Lemma 2.34 and (2.21) that $(\partial_M - m) S_O$ is equal to

$$\sum_{i=1}^{N} (-1)^i \nu_{i-1} (\partial_M - m)^i T + \sum_{i=0}^{N} (-1)^i \nu_i (\partial_M - m)^{i+1} T = (-1)^N \nu_N (\partial_M - m)^{N+1} T.$$

By the assumption on T this implies that S_O is quasihomogeneous of degree m. In view of Corollary 2.33.(i) and Proposition 2.31.(ii) we may apply this argument to $(\partial_M - m)^\ell T$ instead of T. Hence S_ℓ is quasihomogeneous of degree m for every $\ell \in \mathbb{N}$, as well.

Applying Lemma 2.34 and (2.21), inserting (2.20), substituting $I = \ell + i$, changing the order of summation, and taking (2.22) into account we obtain for every $k \in \mathbb{N}_O$ that

$$(\partial_M - m)^k \Big(\sum_{\ell=O}^{N} \nu_{k+\ell} S_\ell \Big) = \sum_{\ell=O}^{N} \nu_\ell S_\ell =$$

$$= \sum_{\ell=O}^{N} \nu_\ell \sum_{i=O}^{N-\ell} (-1)^i \nu_i (\partial_M - m)^{i+\ell} T = \sum_{I=O}^{N} \Big(\sum_{\ell=O}^{I} \binom{I}{\ell} (-1)^{I-\ell} \Big) \nu_I (\partial_M - m)^I T .$$

Since by the binomial formula the term in brackets is equal to $(1-1)^I$ the right-hand side is equal to T, i.e. the equation $(2.19)_k$ is proved. ∎

Note that $\nu := \log$ satisfies the assumptions of Proposition 2.45 in case $M = \mathrm{Id}_{\mathbb{R}}$ and $X =]0,+\infty[$. Hence, combining Proposition 2.45 and Lemma 2.12 one obtains

Corollary 2.46. *A basis for the space of distributions $T \in \mathcal{D}'(]0,+\infty[)$ which are almost homogeneous of degree m and of order $\le N$ is given by the functions $t \mapsto t^m \omega_j(t)$, $0 \le j \le N$.* ∎

Next we formulate a variant of Proposition 1.58.

Proposition 2.47. *Suppose that* $\sigma = \sigma_+$. *Let* $u \in \mathcal{E}'(V)$ *(or more generally* $\mathcal{D}'_C(V)$*) be almost quasihomogeneous of degree* m. *Then there is a polynomial function* $Q: V^* \longrightarrow \mathbb{C}$ *which is almost quasihomogeneous of degree* m^* *and type* M^* *such that* $u = Q(\partial)\,\delta_o$.

Proof. In view of Proposition 2.40.(i) $f := \hat{u}$ satisfies the assumptions of Proposition 1.58 with M replaced by M^* so that the assertion follows. ■

As another consequence of Proposition 1.58 one can state

Proposition 2.48. *Suppose that (1.14) holds. Let* $q: V^* \longrightarrow \mathbb{C}$ *be a hypoelliptic polynomial function. Then the following assertions hold.*

(i) Every distribution $u \in \mathcal{D}'(V)$ *which is almost quasihomogeneous of degree* m *such that* $q(D)u \in C^\infty(V)$ *is a* C^∞ *function being a polynomial with respect to the variables in* $G_M(\sigma_+)$*; moreover, the almost quasihomogeneity order of* u *is bounded by a constant depending on* m*, only; in particular, if* M *is semi-simple then* u *is quasihomogeneous.*

(ii) If M *is semi-simple then for every* $T \in \mathcal{D}'(V)$ *all the almost quasihomogeneous solutions* $S \in \mathcal{D}'(V)$ *of degree* m *of the equation* $q(D)S = T$ *have one and the same deficiency.* ■

Finally, we come back to Example 2.27:

Remark 2.49. *Under the assumptions of Example 2.27 the following holds:*

(i) Bernstein's fundamental solution E_P *of* $P(D)$ *is almost quasihomogeneous of degree* $\ell - \mu$ *and of order* $\leq -\mathrm{ord}(-1;\mathfrak{P})$*; its* 1st *order deficiency is equal to* $-\ell a_{-1}(-1;\mathfrak{P})$*, it is a solution of the equation* $P(D)u = 0$.

(ii) if P *is hypoelliptic and if (1.14) is valid then the deficiency of* E_P *is a* C^∞ *function which is a polynomial with respect to the variables from* $G_M(\sigma_+)$*; in particular, if* $\ell < \mu$ *then* E_P *is quasihomogeneous of degree* $\ell - \mu$.

Proof. *(i)*: this is a reformulation of part of the assertion (ii) of Example 2.27.

(ii): It follows from Example 2.27.(ii) and Corollary 2.33.(i) that $u := a_{-1}(-1;\mathfrak{P})$ is almost quasihomogeneous of degree $\ell-\mu$ of order $< -\mathrm{ord}(-1;\mathfrak{P})$. Since $P(D)u = 0$ Proposition 2.48.(i) implies that $a_{-1}(-1;\mathfrak{P})$ is a C^∞ function, being a polynomial with respect to the variables from $G_M(\sigma_+)$ which must vanish if $\ell-\mu < 0$. ■

In general, E_P need not be quasihomogeneous; in fact, its quasihomogeneity order can be arbitrarily large:

Example 2.50. *Let $k \in \mathbb{N}$ be such that $n = 2k$, and suppose that $M = Id_V$. Set $Q(s,t) := -s^2 - t^2$ and $P := Q \otimes \ldots \otimes Q$ (k times). Then E_Q is almost homogeneous of order 1 but not homogeneous (see e.g. Example 4.13 in Grudzinski [7]) so that $\mathrm{ord}(-1;\mathfrak{P}) = -k$.* ■

(e) Meromorphic Functions of Almost Quasihomogeneous Distributions

In this section, in analogy to Proposition 2.25 we describe meromorphic functions of almost quasihomogeneous distributions by their Laurent series. To this end we let Ω be a connected open subset of $\mathbb{C}$, let $h : \Omega \longrightarrow \mathcal{D}'(X)$ be a meromorphic function, and let $z_o \in \Omega$. Moreover, we fix $N \in \mathbb{N}$, $a \in \dot{\mathbb{C}}$, and $b \in \mathbb{C}$.

Proposition 2.51. *The following conditions are equivalent;*

(a) $h(z)$ is almost quasihomogeneous of degree $az+b$ and of order $\le N$ for every $z \in \Omega \setminus (\{z_o\} \cup \mathfrak{Pol}(h))$;

(b) for every $k \in \mathbb{Z}$ we have

$$a_k(z_o;h) = -\sum_{i=1}^{N+1} \binom{N+1}{i} (-a)^{-i} (\partial_M - az_o - b)^i a_{k+i}(z_o;h). \tag{2.23}$$

Proof. Applying the binomial formula to the sum

$$\partial_M - az - b = (\partial_M - az_o - b) - (az - az_o)$$

of commuting operators we obtain for every $z \in \Omega \setminus \mathfrak{Pol}(h)$

$$(\partial_M - az - b)^{N+1} h(z) = \sum_{i=0}^{N+1} \binom{N+1}{i} (-a)^{N+1-i} (z - z_o)^{N+1-i} (\partial_M - az_o - b)^i h(z).$$

Obviously, for every $i \in \mathbb{N}_0$ the function $g_i : \Omega \longrightarrow \mathcal{D}'(X)$ defined by

$$\Omega \setminus \mathfrak{Pol}(h) \ni z \mapsto (\partial_M - az_o - b)^i h(z)$$

is meromorphic, and from (2.10) one deduces that

$$a_k(z_o; g_i) = (\partial_M - az_o - b)^i a_k(z_o; h), \qquad k \in \mathbb{Z}.$$

It follows that by

$$\Omega \setminus \mathfrak{Pol}(h) \ni z \mapsto (\partial_M - az - b)^{N+1} h(z)$$

a meromorphic function $h_{N+1} : \Omega \longrightarrow \mathcal{D}'(X)$ is defined such that

$$h_{N+1}(z) = \sum_{k \in \mathbb{Z}} (z - z_o)^{N+1+k} \left(\sum_{i=0}^{N+1} \binom{N+1}{i} (-a)^{N+1-i} (\partial_M - az_o - b)^i a_{k+i}(z_o; h) \right).$$

Since a meromorphic function on the connected open set Ω vanishes identically if and only if its Laurent expansion at z_o does so the condition (b) is equivalent to the condition

$$(\partial_M - az - b)^{N+1} h(z) = 0, \qquad z \in \Omega \setminus (\mathfrak{Pol}(h) \cup \{z_o\}),$$

which in view of Proposition 2.31 is equivalent to (a). ∎

In contrast to meromorphic functions of quasihomogeneous distributions the Laurent coefficients of meromorphic functions of almost quasihomogeneous distributions do not have new invariance properties with respect to the action of M_t. In fact they are almost quasihomogeneous:

Remark 2.52. *If one (and hence both) of the conditions of Proposition 2.51 are satisfied then for every $k \in \mathbb{Z}$ satisfying $k \geq \nu := ord(z_o; h)$ the distribution $a_k(z_o; h)$ is almost quasihomogeneous of degree $az_o + b$ and of order $\leq N+k-\nu$.*

Proof. Employing (2.23) with k replaced by k-N-1, multiplying by $(-a)^{N+1}$, rearranging the summands, and applying $(\partial_M - az_o - b)^{k-\nu}$ we obtain

$$(\partial_M - a z_o - b)^{k-\nu+N+1} a_k(z_o;h) =$$

$$= -\left(\sum_{i=0}^{N} \binom{N+1}{i} (-a)^{N+1-i} (\partial_M - a z_o - b)^{k-\nu+i} a_{k-N-1+i}(z_o;h) \right).$$

Since $N+(k-N-1+i)-\nu < k-\nu+i$ and $k-N-1+i \leq k-1$ for $i \leq N$ it follows by induction on k - starting with $k=\nu$ - that

$$(\partial_M - a z_o - b)^{N+k-\nu+1} a_k(z_o;h) = 0, \qquad k \in \mathbb{Z}. \quad \blacksquare$$

If a certain part of the Laurent series of h at z_o vanishes then the condition (2.23) takes on a much simpler form for certain values of k:

Proposition 2.53. *Let $j \in \mathbb{Z}$ such that*

(2.24) $$a_{j-i}(z_o;h) = 0 \quad \text{for every } i \in \mathbb{N}_N.$$

Then for $k = \ell := j-N-1$ the condition (2.23) amounts to

(2.25) $$a_\ell(z_o;h) = -(-a)^{-N-1} (\partial_M - a z_o - b)^{N+1} a_j(z_o;h),$$

and the following conditions are equivalent:

(a) the condition (2.23) holds for every $k \in \mathbb{Z}$ such that $k < \ell$;

(b) for every $i \in \mathbb{N}_0$ we have

(2.26) $$a_{\ell-i}(z_o;h) = \binom{N+i}{N} a^{-i} (\partial_M - a z_o - b)^i a_\ell(z_o;h);$$

(c) for every $i \in \mathbb{N}$ we have

(2.26)' $$a_{\ell-i}(z_o;h) = \frac{N+i}{a i} (\partial_M - a z_o - b) a_{\ell-i+1}(z_o;h).$$

<u>*Proof*</u>. The first part of the assertion is obvious. For the rest of the proof the following equality is required. It is obtained by an application of Lemma 1.76 to $(j,k,\ell) = (L-1,N,N+1)$:

(2.27) $$\sum_{i=1}^{\min(N+1,L)} \binom{N+1}{i} (-1)^{i-1} \binom{N+L-i}{N} = \binom{N+L}{N}, \qquad L \in \mathbb{N}.$$

<u>*(a)⇒(b)*</u>: for $i=0$ the condition (2.26) is trivially valid. For $i \geq 1$ (2.26) is proved by induction: for $I \in \mathbb{N}$ appylying (2.23) to $k = \ell - I$ and taking (2.24) into account one obtains

$$a_{\ell-I}(z_o;h) = -\sum_{i=1}^{\min\{N+1,I\}} \binom{N+1}{i} (-a)^{-i} (\partial_M - az_o - b)^i a_{\ell-I+i}(z_o;h).$$

By the induction hypothesis this is equal to

$$a^{-I}\left(\sum_{i=1}^{\min\{N+1,I\}} \binom{N+1}{i} (-1)^{i-1} \binom{N+I-i}{N}\right) (\partial_M - az_o - b)^I a_\ell(z_o;h).$$

Since by (2.27) the term in brackets is equal to $\binom{N+I}{N}$ the condition (2.26) is valid for $i=I$.

(b)⇒(a): We fix $k\in\mathbb{Z}$ such that $k<\ell$. As an immediate consequence of (2.24) and (2.26) (applied to $i=\ell-k-I$) we obtain

$$\sum_{I=1}^{N+1} \binom{N+1}{I} (-a)^{-I} (\partial_M - az_o - b)^I a_{k+I}(z_o;h) = \sum_{I=1}^{\min\{N+1,\ell-k\}} \cdots =$$

$$= \sum_{I=1}^{\min\{N+1,\ell-k\}} \binom{N+1}{I} (-1)^{-I} a^{-I-(\ell-k-I)} \binom{N+\ell-k-I}{N} (\partial_M - az_o - b)^{I+(\ell-k-I)} a_\ell(z_o;h).$$

Applying (2.27) to $L=\ell-k$ we see that this is equal to

$$-a^{-\ell+k} \binom{N+\ell-k}{N} (\partial_M - az_o - b)^{\ell-k} a_\ell(z_o;h).$$

In view of (2.26) (for $i=\ell-k$) this, in turn, equals $-a_k(z_o;h)$, as desired.

(b)⇒(c): Since $\binom{N+i}{N}/\binom{N+i-1}{N} = \frac{N+i}{i}$ the condition (2.26)' is a combination of (2.26) for i and $i-1$.

(c)⇒(b): this immediately follows by induction. ■

Of course, once one knows that the conditions of Proposition 2.51 are satisfied and hence for each $k\in\mathbb{Z}$ the Laurent coefficient $a_k(z_o;h)$ is almost quasihomogeneous then in order to exhibit the precise behaviour of $a_k(z_o;h)$ under the action of M_t one wants to determine its deficiencies, i.e. - by Proposition 2.31 - the distributions $(\partial_M - az_o - b)^I a_k(z_o;h)$. If $h(z)$ is even quasihomogeneous of degree $az+b$ for every $z\in\Omega\setminus\mathfrak{Pol}(h)$ then Proposition 2.25 says that one just has to compute certain other Laurent coefficients without having to bother about derivatives. However, in general, computing derivatives seems to be unavoidable as a look a (2.23) shows. Nevertheless, if $j\in\mathbb{Z}$ is such that (2.24) holds then for $k=j$ on can do without derivatives at least for sufficiently large I by combining (2.25 and (2.26):

$$(2.28) \qquad (\partial_M - a z_o - b)^I a_j(z_o;h) = (-1)^N a^I \frac{1}{\binom{I-1}{N}} a_{j-I}(z_o;h), \qquad I \geq N+1.$$

By combining (2.23) for $k=j-I$ with (2.24) and substituting i for $i-I$ one obtains

$$(2.29) \qquad (\partial_M - a z_o - b)^I a_j(z_o;h) =$$

$$= \frac{-1}{\binom{N+1}{I}} \sum_{i=1}^{N+1-I} \binom{N+1}{I+i} (-a)^{-i} (\partial_M - a z_o - b)^{I+i} a_{j+i}(z_o;h), \qquad I \in \mathbb{N}_N.$$

However, here the above mentioned difficulty continues to appear as one has to know not only the distributions $a_k(z_o;h)$ for $j<k\leq j+N$, but also their derivatives up to the order $N+1$. Sometimes it is possible to circumvent this difficulty. Namely when one is able to compute the Laurent coefficients of the meromorphic functions $\Omega\setminus\mathfrak{Pol}(h)\ni z\mapsto(\partial_M - az-b)^I h(z)$. In fact, the left-hand side of (2.29) can be described in terms of their Laurent coefficients, as follows.

Remark 2.54. *Suppose that one (and hence each) of the conditions of Proposition 2.51 is valid. Moreover, let $j\in\mathbb{Z}$ be such that (2.24) holds. Then for every $I\in\mathbb{N}_N$ we have*

$$(\partial_M - a z_o - b)^I a_j(z_o;h) = a_j(z_o;h_I)$$

where $h_I:\Omega\longrightarrow\mathcal{D}'(X)$ denotes the meromorphic function defined by

$$(2.30) \qquad h_I(z) := (\partial_M - az - b)^I h(z), \qquad z\in\Omega\setminus\mathfrak{Pol}(h).$$

Proof. As at the beginning of the proof of Proposition 2.51 we apply the binomial formula to obtain for every $z\in\Omega\setminus\mathfrak{Pol}(h)$

$$h_I(z) = \sum_{i=0}^{I} \binom{I}{i} (-a)^{I-i} (z-z_o)^{I-i} (\partial_M - a z_o - b)^i h(z)$$

so that

$$(2.31) \qquad a_k(z_o;h_I) = \sum_{i=0}^{I} \binom{I}{i} (-a)^{I-i} (\partial_M - a z_o - b)^i a_{k+i-I}(z_o;h), \qquad k\in\mathbb{Z}.$$

If $k=j$ and $I\in\mathbb{N}_N$ then (2.24) implies that on the right-hand side of (2.31) the summands indexed by $i<I$ vanish. Hence the assertion is proved. ■

Now, in conclusion we come back to Example 2.27. We would like to weaken the assumption in that we assume of P to be almost quasihomogeneous, only. Here the following analogue of Remark 2.24 will be useful.

Remark 2.24'. *Let $g:\Omega \longrightarrow \mathbb{C}$ be a holomorphic function. Suppose that there is an uncountable subset D of Ω such that $h(z)$ is almost quasihomogeneous of degree $g(z)$ for every $z\in D$. Then there is $N\in\mathbb{N}$ such that $h(z)$ is almost quasihomogeneous of degree $g(z)$ and of order $\leq N$ for every $z\in\Omega\setminus\mathfrak{Pol}(h)$.*

Proof. For every $N\in\mathbb{N}_0$ by $h_N(z) := (\partial_M - g(z))^{N+1} h(z)$, $z\in\Omega\setminus\mathfrak{Pol}(h)$, a meromorphic function $h_N:\Omega\longrightarrow\mathcal{D}'(X)$ is defined. Let $(K_j)_{j\in\mathbb{N}}$ be an increasing sequence of compact subsets of Ω such that $\bigcup_{j\in\mathbb{N}} K_j = \Omega$. Since D is contained in the union of the sets $D_N := \{z\in K_N\,;\, h_N(z) = 0\}$, $N\in\mathbb{N}$, and since D is uncountable we find some $N\in\mathbb{N}$ such that D_N is uncountable. By the principle of analytic continuation it follows that h_N vanishes identically. ■

The following example shows that in case P is only almost quasihomogeneous of degree ℓ but not quasihomogeneous then $\mathfrak{P}(z)$ is not necessarily almost quasihomogeneous where $\mathfrak{P}$ is defined in the text preceding Example 2.27.

Example 2.55. *Let $P:V^*\longrightarrow [0,+\infty[$ be a non-negative polynomial function which is almost quasihomogeneous of degree $\ell\in\mathbb{C}$ and of order ≤ 1. Then for arbitrary $k\in\mathbb{N}$ and $z\in\mathbb{C}$ such that $\mathrm{Re}\, z > k$ we have*

$$(\partial_M - z\ell)^k P^z = \prod_{i=0}^{k-1} (z-i)\ [(\partial_M - \ell)P]^k P^{z-k},$$

and if $(\partial_M - \ell)P$ does not vanish identically then there is a countable subset D of $\mathbb{C}$ such that $\mathfrak{P}(z)$ is not almost quasihomogeneous for every $z\in\mathbb{C}\setminus D$.

Proof. Let $\mathrm{Re}\, z > k+1$. Then by Lemma 2.34 $\partial_M\big([(\partial_M-\ell)P]^k P^{z-k}\big)$ is equal to

$$k\,[(\partial_M-\ell)P]^{k-1}\,[\partial_M(\partial_M-\ell)P]\,P^{z-k} + [(\partial_M-\ell)P]^k\,\partial_M P^{z-k}.$$

Since $\partial_M P^{z-k} = (z-k)\,(\partial_M P - \ell P)\,P^{z-k-1} + (z-k)\,\ell P^{z-k}$ and since by the assumption on P we have $[\partial_M(\partial_M-\ell)P] = \ell\,(\partial_M-\ell)P$ we obtain that

$$(\partial_M - z\ell)\big([(\partial_M-\ell)P]^k P^{z-k}\big) = (z-k)\,[(\partial_M-\ell)P]^{k+1} P^{z-k-1}.$$

Hence the desired equation follows by induction on k. If $(\partial_M-\ell)P\neq 0$ then this equation tells us that for arbitrary $k\in\mathbb{N}$ and $z\in\mathbb{C}$ satisfying $\mathrm{Re}\, z > k+1$ the function P^z is not almost quasihomogeneous of degree $z\ell$ of order $\leq k$. Hence the second part of the assertion follows from Remark 2.24' and Proposition 2.40.(ii). ■

(f) Appendix: (G,σ)-invariant Distributions

Let G be a compact subgroup of GL(V,V) such that $A(X) = X$ for every $A \in G$. Moreover, let $\sigma : G \longrightarrow \dot{\mathbb{C}}$ be a continuous homomorphism of G into the multiplicative group $\dot{\mathbb{C}}$. Note that since G is compact the image of σ is contained in the unit circle S^1. In particular, $|\det A| = 1$ for every $A \in G$. We set $\mathfrak{G} := (G,\sigma)$.

Definition 2.56. A distribution $T \in \mathcal{D}'(X)$ is called *$\mathfrak{G}$-invariant* if and only if

$$T \circ A = \sigma(A)\, T , \qquad A \in G. \tag{2.32}$$

If $\sigma \equiv 1$ we also say that T is *G-invariant.*

Recall that the compactness of G implies that G is unimodular so that the (normalized) *left-invariant Haar measure μ_G on G* is right-invariant, as well.

In the present section we collect the basic material on how to construct $\mathfrak{G}$-invariant distributions by taking the average with respect to μ_G.

Notation 2.57. For any subset L of X we set $L_G := \bigcup_{A \in G} A(L)$.

Lemma 2.58. *(i)* $(\overline{L})_G \cap X = \overline{L_G} \cap X$*; in particular, if L is a closed subset of X so is* L_G.

(ii) If L is compact so is L_G.

Proof. (i): The inclusion "$\subset$" is valid by continuity. To prove "$\supset$" we fix $x \in \overline{L_G} \cap X$ and choose a sequence $(x_k)_{k \in \mathbb{N}}$ in L_G converging to x. Then for every $k \in \mathbb{N}$ we fix $A_k \in G$ and $\ell_k \in L$ such that $x_k = A_k(\ell_k)$. By choosing subsequences if necessary we achieve that (A_k) converges to some $A \in G$ as $k \to \infty$. By continuity it follows that $\ell_k = A_k^{-1}(x_k)$ converges to $\ell := A^{-1}(x)$. Hence $\ell \in \overline{L}$ and $x = A(\ell) \in (\overline{L})_G$.

(ii): L_G is the image of the compact set $G \times L$ under the continuous function $G \times V \longrightarrow V$, $(A,x) \mapsto A(x)$. ■

Proposition 2.59. *Let* $r \in \mathbb{N}_0 \cup \{\infty\}$*, and let* $f \in C^r(X)$*. Then by*

$$f_{\mathfrak{G}}(x) := \int_G \sigma(A^{-1}) f(Ax)\, d\mu_G(A), \qquad x \in X,$$

a $\mathfrak{G}$-invariant C^r function $f_{\mathfrak{G}} : X \longrightarrow \mathbb{C}$ is well-defined having the following properties:

(i) $\operatorname{supp} f_{\mathfrak{G}} \subset (\operatorname{supp} f)_G$;

(ii) $f_{\mathfrak{G}} = f$ *if and only if* f *is* $\mathfrak{G}$*-invariant;*

(iii) the map $C^r(X) \longrightarrow C^r(X)$, $f \mapsto f_{\mathfrak{G}}$, *is linear and continuous.*

Proof. By differentiating under the integral sign and making use of the compactness of G one sees that $f_{\mathfrak{G}}$ belongs to $C^r(X)$ and that there is a constant B_r only depending on r and G such that

(2.33) $$\max\{\,|\partial^\alpha f_{\mathfrak{G}}(x)|\,;\ |\alpha| \le r\} \le B_r \max\{\,|\partial^\beta f(Ax)|\,;\ |\beta| \le r,\ A \in G\}, \qquad x \in X.$$

Since μ_G is invariant and since σ is a homomorphism it follows that

$$f_{\mathfrak{G}}(Bx) = \int_G \sigma(B(AB)^{-1})\, f(ABx)\, d\mu_G(A) = \sigma(B)\, f_{\mathfrak{G}}(x), \qquad x \in X, B \in G.$$

(i): If $x \notin (\operatorname{supp} f)_G$ then for every $A \in G$ we have $Ax \notin \operatorname{supp} f$ and hence $f(Ax) = 0$ so that $f_{\mathfrak{G}}(x) = 0$. Since by Lemma 2.58.(i) the set $(\operatorname{supp} f)_G$ is closed in X the assertion is proved.

(ii): Since $f_{\mathfrak{G}}$ is $\mathfrak{G}$-invariant the implication "$\Rightarrow$" is clear. The converse is obvious in view of $\mu_G(G) = 1$. *(iii)* follows from (2.33). ■

To formulate assertions about the derivatives and the Fourier transform of $f_{\mathfrak{G}}$ we introduce

Notation 2.60. (i) By $G^* := \{A^*;\ A \in G\}$ we denote the subgroup of $GL(V^*, V^*)$ consisting of the transposes of the elements of G ;

(ii) we define a continuous homomorphism $\sigma^* : G^* \longrightarrow \dot{\mathbb{C}}$ by $\sigma^*(A) := \sigma((A^*)^{-1})$;

(iii) we set $\mathfrak{G}^* := (G^*, \sigma^*)$.

Observe that G^* is a compact group, as well, and that its Haar measure μ_{G^*} is described by

$$\int_{G^*} f(A)\, d\mu_{G^*}(A) = \int_G f(A^*)\, d\mu_G(A), \qquad f \in C^0(G^*)$$

Proposition 2.61. *Let $\tau : G \longrightarrow \dot{\mathbb{C}}$ be another continuous homomorphism. Then, writing $\mathfrak{H} := (G,\tau)$, for arbitrary $r \in \mathbb{N}_0 \cup \{\infty\}$ and $f \in C^r(X)$ we have:*

(i) $P(\partial) f_{\mathfrak{G}} = (P(\partial) f)_{(G,\sigma\tau)}$ *for every $\mathfrak{H}^*$-invariant polynomial function $P : V^* \longrightarrow \mathbb{C}$ of degree not larger than r;*

(ii) $q f_{\mathfrak{G}} = (qf)_{(G,\sigma\tau)}$ *for every $\mathfrak{H}$-invariant continuous function $q : X \longrightarrow \mathbb{C}$.*

Proof. *(i)*: The assertion follows from

$$P(\partial)(f \circ A) = \Big((P \circ A^*)(\partial) f\Big) \circ A = \Big(\tau^*(A^*)\, P(\partial) f\Big) \circ A = \tau(A^{-1})\, (P(\partial) f) \circ A, \qquad A \in G.$$

(ii): this is a consequence of $q(x) = \tau(A^{-1})\, q(Ax)$, $A \in G$. ∎

Concerning the Fourier transform one obtains

Proposition 2.62. *Suppose that $X = V$ and $f \in \mathcal{S}(V)$. Then $f_{\mathfrak{G}}$ belongs to $\mathcal{S}(V)$, as well, and*

$$\mathcal{F}(f_{\mathfrak{G}}) = (\hat{f})_{\mathfrak{G}^*}. \tag{2.34}$$

Moreover, the map $\mathcal{S}(V) \longrightarrow \mathcal{S}(V)$, $f \mapsto f_{\mathfrak{G}}$, is linear and continuous.

Proof. Since G is compact there is a constant C such that $1 + |x| \le C\,(1 + |A(x)|)$ for arbitrary $x \in V$ and $A \in G$. Consequently, the first and the third part of the assertion follow from (2.33). For the proof of the second part one observes from (2.8) that $\mathcal{F}(f \circ A) = \hat{f} \circ (A^*)^{-1}$. By Fubini's theorem and by the invariance of μ_G under the transformation $A \mapsto A^{-1}$ one then obtains for every $\xi \in V^*$

$$\mathcal{F}(f_{\mathfrak{G}})(\xi) = \int_G \sigma(A^{-1})\, \mathcal{F}(f \circ A)(\xi)\, d\mu_G(A) =$$

$$= \int_G \sigma^*((A^*)^{-1})\, \hat{f}(A^*\xi)\, d\mu_G(A) = \int_{G^*} \sigma^*(B^{-1})\, \hat{f}(B\xi)\, d\mu_{G^*}(B). \qquad ∎$$

By Fubini's theorem, by a transformation of variables, and by the invariance properties of μ_G we deduce

$$\int_X f_{\mathfrak{G}}(x)\, \varphi(x)\, dx = \int_G \sigma(A^{-1}) \int_X f(Ax)\, \varphi(x)\, dx\, d\mu_G(A) =$$

$$= \int_G \frac{1}{\sigma}(A) \int_X f(x)\, \varphi(A^{-1}x)\, dx\, d\mu_G(A) = \int_X f(x) \int_G \frac{1}{\sigma}(A^{-1})\, \varphi(Ax)\, d\mu_G(A)\, dx =$$

$$= \int_X f(x)\, \varphi_{(G,1/\sigma)}(x)\, dx$$

for arbitrary $f \in C^0(X)$ and $\varphi \in C_0^\infty(X)$. This motivates the following

Definition 2.63. Let $T \in \mathcal{D}'(X)$. We set

$$\langle T_{\mathfrak{G}}, \varphi \rangle := \langle T, \varphi_{\mathfrak{G}'} \rangle, \qquad \varphi \in C_0^\infty(X),$$

where $\mathfrak{G}' := (G, 1/\sigma)$. An alternative way to write this is

$$(2.35) \qquad \langle T_{\mathfrak{G}}, \varphi \rangle = \int_G \sigma(A^{-1}) \langle T \circ A, \varphi \rangle \, d\mu_G(A), \qquad \varphi \in C_0^\infty(X).$$

The result of the computation preceding Definition 2.63 can be rewritten as

$$(2.36) \qquad T_{(f_{\mathfrak{G}})} = (T_f)_{\mathfrak{G}}, \qquad f \in C^0(X).$$

Proposition 2.64. *For every $T \in \mathcal{D}'(X)$ $T_{\mathfrak{G}}$ is a well-defined $\mathfrak{G}$-invariant distribution on X having the following properties:*

(i) *$\operatorname{supp} T_{\mathfrak{G}} \subset (\operatorname{supp} T)_G$;*

(ii) *suppose that $X = V$ and that T is temperate; then $T_{\mathfrak{G}}$ is temperate, as well, the defining equation and (2.35) remain valid for arbitrary $\varphi \in \mathcal{S}(V)$, and*

$$\mathcal{F}(T_{\mathfrak{G}}) = (\hat{T})_{\mathfrak{G}^*};$$

(iii) *$T_{\mathfrak{G}} = T$ if and only if T is $\mathfrak{G}$-invariant.*

Proof. The assertions (i) and (iii) of Proposition 2.59 imply that by $\varphi \mapsto \varphi_{\mathfrak{G}'}$ a continuous linear map $C_0^\infty(K) \longrightarrow C_0^\infty(K_G)$ is well-defined. Hence in view of Lemma 2.58.(ii) $T_{\mathfrak{G}}$ is well-defined and belongs to $\mathcal{D}'(X)$. That $T_{\mathfrak{G}}$ is $\mathfrak{G}$-invariant follows from the $\mathfrak{G}'$-invariance of $\varphi_{\mathfrak{G}'}$ or from (2.35).

(i): If $\varphi \in C_0^\infty(X)$ is such that $\operatorname{supp} \varphi \cap (\operatorname{supp} T)_G = \emptyset$ then for every $A \in G$ $\operatorname{supp} \varphi \circ A = A^{-1}(\operatorname{supp} \varphi)$ does not intersect $\operatorname{supp} T$, i.e. $(\operatorname{supp} \varphi)_G \cap \operatorname{supp} T = \emptyset$ and the assertion follows by Proposition 2.59.(i).

(ii): The first part of the assertion follows from Proposition 2.62. Moreover, using (2.34) we deduce that

$$\langle \mathcal{F}(T_{\mathfrak{G}}), \varphi \rangle = \langle T_{\mathfrak{G}}, \hat{\varphi} \rangle = \langle T, (\hat{\varphi})_{\mathfrak{G}'} \rangle = \langle T, \mathcal{F}(\varphi_{(G^*,1/\sigma^*)}) \rangle =$$

$$= \langle \hat{T}, \varphi_{(G^*,1/\sigma^*)} \rangle = \langle (\hat{T})_{\mathfrak{G}^*}, \varphi \rangle, \qquad \varphi \in \mathcal{S}(V^*)$$

(iii): In view of $\mu_G(G)=1$ this follows from (2.35). ■

The analogue of Proposition 2.61 holds:

Proposition 2.65. *Let τ and $\mathfrak{H}$ be as in Proposition 2.61. Then*

(i) $P(\partial)T_{\mathfrak{G}} = (P(\partial)T)_{(G,\sigma\tau)}$ *for every $\mathfrak{H}^*$-invariant polynomial function $P: V^* \to \mathbb{C}$;*

(ii) $qT_{\mathfrak{G}} = (qT)_{(G,\sigma\tau)}$ *for every $\mathfrak{H}$-invariant C^∞ function $q: X \to \mathbb{C}$.*

Proof. For arbitrary $\varphi \in C_0^\infty(X)$ one obtains by Proposition 2.61 that

$$\langle P(\partial)T_{\mathfrak{G}}, \varphi\rangle = \langle T, (P(-\partial)\varphi)_{(G,1/\sigma)}\rangle =$$

$$= \langle T, P(-\partial)\varphi_{(G,1/(\sigma\tau))}\rangle = \langle (P(\partial)T)_{(G,\sigma\tau)}, \varphi\rangle$$

and

$$\langle qT_{\mathfrak{G}}, \varphi\rangle = \langle T, (q\varphi)_{(G,1/\sigma)}\rangle = \langle T, q\varphi_{(G,1/(\sigma\tau))}\rangle = \langle (qT)_{(G,\sigma\tau)}, \varphi\rangle. \quad ■$$

Finally, we examine when quasihomogeneity is preserved by the operation $T \mapsto T_{\mathfrak{G}}$.

Remark 2.66. *Let $A \in GL(V,V)$. Then*

(i) *A commutes with M if and only if*

(2.37) $$A \circ M_t = M_t \circ A \quad \text{for every } t \in]0,+\infty[;$$

(ii) *if M is of the form (1.1.a) then this is the case if and only if $A_{ij}=0$ for arbitrary $i,j \in \mathbb{N}_n$ satisfying $p_i \neq p_j$.*

Proof. *(i):* "⇒" is clear from the power series presentation of $\exp((\log t)M)$. "⇐" is obtained through differentiating (2.37) at $t=1$ and making use of (1.5).

(ii): For arbitrary $i,j \in \mathbb{N}_n$ we have: $(A\circ M)_{ij} = A_{ij}\,p_j$ and $(M\circ A)_{ij} = p_i A_{ij}$. ■

Note that A^{-1} satisfies (2.37) if A does so. Moreover, (2.37) remains valid if (M,A) is replaced by (A^*,M^*).

Remark 2.67. *(i) Suppose that $A \in GL(V,V)$ commutes with M. Then*

(2.38) $$\partial_M(T\circ A) = (\partial_M T)\circ A, \qquad T \in \mathcal{D}'(X);$$

consequently, if $T \in \mathcal{D}'(X)$ is almost quasihomogeneous of degree m with deficiency $d \in \mathcal{D}'(X)$ then $T \circ A$ is almost quasihomogeneous of degree m with deficiency $d \circ A$; in particular, if T is quasihomogeneous of degree m so is $T \circ A$.

(ii) Suppose that every $A \in G$ commutes with M. Then

$$(2.39) \qquad (\partial_M - m)\, T_{\mathfrak{G}} = ((\partial_M - m)\, T)_{\mathfrak{G}}\,, \qquad T \in \mathcal{D}'(X)\,;$$

consequently, if $T \in \mathcal{D}'(X)$ is almost quasihomogeneous of degree m with deficiency $d \in \mathcal{D}'(X)$ then $T_{\mathfrak{G}}$ is almost quasihomogeneous of degree m with deficiency $d_{\mathfrak{G}}$; in particular, if T is quasihomogeneous of degree m so is $T_{\mathfrak{G}}$.

Proof. *(i)*: For every $\varphi \in C_0^\infty(X)$ the chain rule yields

$$\partial_M(\varphi \circ A)(x) = D\varphi(Ax) \cdot AMx = D\varphi(Ax) \cdot MAx = (\partial_M \varphi)(Ax)\,.$$

In view of Remark 2.8 it follows that $({}^t\partial_M \varphi) \circ A^{-1} = {}^t\partial_M(\varphi \circ A^{-1})$ so that

$$\langle \partial_M(T \circ A), \varphi \rangle = \langle T, ({}^t\partial_M \varphi) \circ A^{-1} \rangle / |\det A| = \langle (\partial_M T) \circ A, \varphi \rangle\,.$$

(ii): this follows from (2.38) and (2.35). ∎

Chapter III

Constructing (Almost) Quasihomogeneous Functions by Taking Quasihomogeneous Averages of Functions with M-bounded Support

Suppose from now on that X is quasihomogeneous. Similarly as in section 2.(f) the simple idea for constructing quasihomogeneous functions out of a given function $f: X \longrightarrow \mathbb{C}$ is to take the average of the functions $t^{-m} f \circ M_t$, $t \in]0,+\infty[$, with respect to the Haar measure of the group $\{M_t;\ t \in]0,+\infty[\}$. Since under the canonical isomorphism of this group onto the multiplicative group $]0,+\infty[$ the Haar measure is induced by $\frac{dt}{t}$, one is lead to introduce the function $f_m: X \longrightarrow \mathbb{C}$ formally given by

$$(3.1) \qquad f_m(x) := \int_0^{+\infty} t^{-m} f(M_t x) \frac{dt}{t}, \qquad x \in X.$$

In order to guarantee that the integral is well-defined, in the present paragraph we impose rather strong conditions, namely that for every $x \in X$ the support of the map defined by $t \mapsto t^{-m} f(M_t x)$ is a compact subset of $]0,+\infty[$; more precisely, we postulate that supp f is an *M-bounded subset of X* in the sense of Definition 3.1 below. In Chapter 5 this assumption will be relaxed.

In section (a) below the theory of the functions f_m is developed. For the sake of applications in Chapter 4 we proceed more generally by introducing the quasihomogeneous averages with respect to measures of the form $w(t)\frac{dt}{t}$ where $w:]0,+\infty[\longrightarrow \mathbb{C}$ is any locally integrable weight function; formally this means that

$$(3.1)' \qquad f_{m,w}(x) := \int_0^{+\infty} t^{-m} f(M_t x)\, w(t) \frac{dt}{t}, \qquad x \in X.$$

Section (b) deals with criteria for the M-boundedness of subsets of X. In particular, it turns out that for some X not every compact subset of X is M-bounded. If X does not have this flaw it is called *locally M-bounded.* Necessary and and sufficient conditions for X to have this property are given in section (c). One of them is the existence of C^{∞} functions $\varkappa : X \longrightarrow]0,+\infty[$ which are quasihomogeneous of degree 1. Another equivalent condition is the existence of a C^{∞} function $\psi : X \longrightarrow \mathbb{C}$ with M-bounded support whose quasihomogeneous average ψ_0 of degree 0 is identically equal to 1.

If X is locally M-bounded then every quasihomogeneous continuous function on X can be written as the quasihomogeneous average of some continuous function f with M-bounded support (section (d)); in fact, finitely many quasihomogeneous functions of pairwise different degrees are described in this manner by means of a single f.

(a) Introducing the Quasihomogeneous Averages $f_{m,w}$

Let $w :]0,+\infty[\longrightarrow \mathbb{C}$ be a weight function belonging to $\mathcal{L}^1_{loc}(]0,+\infty[)$, and set $I := \operatorname{supp} w$. If $f : X \longrightarrow \mathbb{C}$ is a continuous function such that for every $x \in X$ the support of the map $I \longrightarrow \mathbb{C}$, $t \mapsto f(M_t x)$, is contained in a compact subinterval J_x of $]0,+\infty[$ then a function $f_{m,w} : X \longrightarrow \mathbb{C}$ is well-defined by (3.1)'. In order to be able to derive regularity properties of $f_{m,w}$ we require that the intervals J_x can be chosen not to depend too wildly on x; in fact, we are going to postulate that locally one can choose them to be independent from x.

Definition 3.1. A subset L of X is said to be an *(M,I)-bounded subset of X* if and only if for every compact subset K of X the set

(3.2) $$\{t \in I;\ M_t(K) \cap L \neq \emptyset\}$$

is a relatively compact subset of $]0,+\infty[$ (hence of I since I is closed in $]0,+\infty[$). If $I =]0,+\infty[$ one also says *M-bounded* instead of *(M,I)-bounded.*

Note that if, in addition, L is closed in X then the sets (3.2) are compact.

We fix $r \in \mathbb{N}_0 \cup \{\infty\}$ and introduce

Notation 3.2. By $C_I^r(X)$ we denote the set of C^r functions $f: X \longrightarrow \mathbb{C}$ such that $\operatorname{supp} f$ is an (M,I)-bounded subset of X. If $I =]0,+\infty[$ we also write $C_M^r(X)$.

Since finite unions of (M,I)-bounded subsets of X are (M,I)-bounded themselves it follows that $C_I^r(X)$ is a linear subspace of $C^r(X)$.

Now we fix $f \in C_I^r(X)$.

Proposition 3.3. *(i) By (3.1)' a C^r function $f_{m,w}$ is well-defined.*

(ii) The function $\mathfrak{H}_{f,w}: \mathbb{C} \longrightarrow C^r(X)$, $m \mapsto f_{m,w}$, is holomorphic satisfying

$$\mathfrak{H}_{f,w}^{(k)}(m) = k!\,(-1)^k f_{m,w\omega_k}, \qquad k \in \mathbb{N}.$$

<u>Proof.</u> <u>(i)</u>: If I is compact the assertion follows by the rule for differentiating an integral depending on a parameter. In particular, one has

$$(3.3) \qquad (\partial^\alpha f_{m,w})(x) = \int_I t^{-m}\, \partial^\alpha (f \circ M_t)(x)\, \frac{dt}{t}, \qquad x \in X,\ |\alpha| \le r.$$

In general, let $x \in X$ be fixed, and let K be a compact neighbourhood of x in X. Then by the definition of $C_I^r(X)$ we find a compact subinterval J of $]0,+\infty[$ such that $\{t \in I;\ M_t(K) \cap \operatorname{supp} f \neq \emptyset\}$ is contained in J. This means, in particular, that $f_{m,w}$ equals $f_{m,\chi_J w}$ on K where χ_J denotes the characteristic function of J. Consequently, the proof is reduced to the case "I is compact" already dealt with above. Note that (3.3) remains valid.

<u>(ii)</u>: For arbitrary $t \in]0,+\infty[$ and $z \in \mathbb{C}$ we have

$$t^{-z} = t^{-m} \sum_{k=0}^{\infty} \frac{1}{k!} (-1)^k (\log t)^k (z-m)^k,$$

the series converging uniformly if t stays in a compact subset of $]0,+\infty[$ and z in a compact subset of $\mathbb{C}$. Let K be a compact subset of X, and let J be as in the proof of (i). In view of (3.3) it then follows that for every $\alpha \in \mathbb{N}_0^n$ satisfying $|\alpha| \le r$ the series

$$\sum_{k=0}^{\infty} (-1)^k (z-m)^k\, \partial^\alpha f_{m,w\omega_k}$$

converges to $\partial^\alpha f_{z,w}$ uniformly on K and uniformly for z in any compact subset of $\mathbb{C}$. ∎

By exploiting (3.3) one can obtain a convenient formula for the derivatives of $f_{m,w}$. To this end we fix $\ell \in \mathbb{C}$, $N \in \mathbb{N}_0$, and a continuous copolynomial function $P_0 : X \times V^* \longrightarrow \mathbb{C}$ on X of degree $\le r$ (in the sense of Definition 1.16) which is almost quasihomogeneous of degree ℓ, of type $M \times (-M)^*$, and of order $\le N$. Applying Remark 1.50 to $(P, M \times (-M)^*)$ instead of (q_0, M) we conclude that for every $k \in \mathbb{N}_N$ the kth order deficiency P_k of P_0 is a continuous copolynomial function on X, as well. Consequently, for every $k \in \{0\} \cup \mathbb{N}_N$ $P_k(x,\partial)$ is a well-defined differential operator with continuous coefficients, and (1.24) - applied to $A = M_t$ - and (1.65) give

$$(3.4) \qquad P_0(x,\partial)(f \circ M_t) = t^{-\ell} \sum_{i=0}^{N} (-1)^i \omega_i(t)\, (P_i(x,\partial) f) \circ M_t, \qquad t \in]0,+\infty[.$$

Similarly, by applying (1.24) to $(f \circ M_t, M_{1/t})$ one obtains that

$$(3.4)' \qquad (P_0(x,\partial) f) \circ M_t = t^{\ell} \sum_{i=0}^{N} \omega_i(t)\, P_i(x,\partial)(f \circ M_t), \qquad t \in]0,+\infty[.$$

Proposition 3.4. *Under the preceding assumptions on P_0, ℓ, and N we have*

$$(3.5) \qquad P_0(x,\partial) f_{m,w} = \sum_{i=0}^{N} (-1)^i (P_i(x,\partial) f)_{m+\ell, w\omega_i}$$

and

$$(3.5)' \qquad (P_0(x,\partial) f)_{m+\ell,w} = \sum_{i=0}^{N} P_i(x,\partial)\, f_{m, w\omega_i}.$$

In particular, if P_0 is quasihomogeneous of degree ℓ and of type $M \times (-M^*)$ then

$$(3.6) \qquad P_0(x,\partial) f_{m,w} = (P_0(x,\partial) f)_{m+\ell,w}.$$

Proof. Let $x \in X$. Since (3.3) leads to

$$(P_0(x,\partial) f_{m,w})(x) = \int_I t^{-m} P_0(x,\partial)(f \circ M_t)(x)\, w(t) \frac{dt}{t}$$

one obtains (3.5) by inserting (3.4). For the proof of (3.5)' one similarly deduces from (3.4)' that

$$(P_0(x,\partial) f)_{m+\ell,w}(x) = \sum_{i=0}^{N} \int_I t^{-m} P_i(x,\partial)(f \circ M_t)(x)\, w(t)\, \omega_i(t) \frac{dt}{t},$$

and by (3.3), again, the condition (3.5)' follows. ■

In view of Example 1.21 a special case of (3.6) is

$$(3.7) \qquad (\partial_M - z)^j f_{m,w} = ((\partial_M - z)^j f)_{m,w}, \qquad z \in \mathbb{C},\ j \in \mathbb{N}$$

Another special case of Proposition 3.4 worth to be formulated separately is

Corollary 3.5. *Let $\ell \in \mathbb{C}$ and $N \in \mathbb{N}_0$, and let $q \in C^0(X)$ be almost quasihomogeneous of degree ℓ and of order $\leq N$. Then - for every $i \in \mathbb{N}_N$ denoting by q_i the ith order deficiency of q (which is continuous, as well, by Proposition 1.51) - we have*

$$(3.8) \qquad q f_{m,w} = (qf)_{m+\ell,w} + \sum_{i=1}^{N} (-1)^i (q_i f)_{m+\ell,w\omega_i}$$

and

$$(3.8)' \qquad (qf)_{m+\ell,w} = q f_{m,w} + \sum_{i=1}^{N} q_i f_{m,w\omega_i} .$$

In particular, if q is quasihomogeneous of degree ℓ then

$$(3.9) \qquad (qf)_{m+\ell,w} = q f_{m,w} . \quad \blacksquare$$

We now come to the invariance of $f_{m,w}$ under linear changes of variables. Here we require

Lemma 3.6. *Let L be an (M,I)-bounded subset of X.*

(i) Then $A^{-1}(L)$ is an (M,I)-bounded subset of $A^{-1}(X)$ for every $A \in L(V,V)$ commuting with M.

(ii) If G satisfies the assumptions of Remark 2.67.(ii) then L_G (see Notation 2.57) is an (M,I)-bounded subset of X.

Proof. (i): Let K be a compact subset of $A^{-1}(X)$. Then $H := A(K)$ is a compact subset of X such that $\{t \in I ; M_t(K) \cap A^{-1}(L) \neq \emptyset\}$ is equal to $\{t \in I ; M_t(H) \cap L \neq \emptyset\}$, and the assertion follows.

(ii): Let K be a compact subset of X, and let $t \in I$ be such that $M_t(K) \cap L_G = \emptyset$. Then $M_t(K) \cap A(L) \neq \emptyset$ for some $A \in G$ so that $M_t(A^{-1}(K)) \cap L \neq \emptyset$, i.e. $M_t(K_G) \cap L \neq \emptyset$. Since by Lemma 2.58.(ii) K_G is a compact subset of X the assertion follows. ■

Proposition 3.7. *(i) If $A \in L(V,V)$ commutes with M then $f \circ A \in C_I^r(A^{-1}(X))$ and $(f \circ A)_{m,w} = f_{m,w} \circ A$.*

(ii) If $\mathfrak{G}$ satisfies the assumptions of Remark 2.67.(ii) then $f_{\mathfrak{G}} \in C_I^r(X)$ and $(f_{\mathfrak{G}})_{m,w} = (f_{m,w})_{\mathfrak{G}}$.

Proof. *(i)*: Since $\operatorname{supp} f \circ A = A^{-1}(\operatorname{supp} f)$ the first part of the assertion follows from Lemma 3.6.(i). The second part is an immediate consequence of the assumption on A and (3.1)'.

(ii): Since by Proposition 2.59.(i) $\operatorname{supp} f_{\mathfrak{G}}$ is contained in $(\operatorname{supp} f)_G$ the first part follows from Lemma 3.6.(ii). To prove the second part we fix $x \in X$. Since in view of Lemma 3.6.(ii) the set

$$\{t \in I;\ M_t\{x\}_G \cap \operatorname{supp} f \neq \emptyset\} = \{t \in I;\ M_t x \in (\operatorname{supp} f)_G\}$$

is contained in a compact subinterval J of $]0,+\infty[$ we obtain by applying Fubini's theorem that

$$(f_{m,w})_{\mathfrak{G}}(x) = \int_G \sigma(A^{-1}) \int_J t^{-m} f(M_t A x)\, w(t) \frac{dt}{t}\, d\mu_G(A) =$$

$$= \int_J t^{-m} \int_G \sigma(A^{-1})\, (f \circ A)(M_t x)\, d\mu_G(A)\, w(t) \frac{dt}{t} = (f_{\mathfrak{G}})_{m,w}(x)\,. \quad \blacksquare$$

Notation 3.8. For any subset Y of V we set $Y_{M,I} := \bigcup_{t \in I} M_{1/t}(Y)$.

It is obvious that $f_{m,w}$ vanishes outside the set $(f^{-1}(\dot{\mathbb{C}}))_{M,I}$. To determine its closure we require

Lemma 3.9. *Let L be an (M,I)-bounded subset of X. Then* $\overline{L_{M,I}} \cap X = (\overline{L} \cap X)_{M,I}$. *In particular, if L is closed in X so is* $L_{M,I}$.

Proof. The inclusion "$\supset$" is obvious by the continuity of $M_{1/t}$. To prove "$\subset$" we fix $x \in \overline{L_{M,I}} \cap X$, choose a sequence $(x_j)_{j \in \mathbb{N}}$ in $L_{M,I}$ converging to x as $j \to \infty$, and let (t_j) be a sequence in I such that $M_{t_j} x_j \in L$ for every $j \in \mathbb{N}$. Since the set $K := \{x\} \cup \{x_j;\ j \in \mathbb{N}\}$ is a compact subset of X Definition 3.1 shows that there is a compact subset J of I such that $t_j \in J$ for every $j \in \mathbb{N}$. Hence, by choosing subsequences we achieve that (t_j) converges to a number $t \in I$. By continuity it follows that

$$\lim_{j \to \infty} M_{t_j} x_j = M_t x\,.$$

This means that $M_t x \in \overline{L} \cap X$, i.e. $x \in (\overline{L} \cap X)_{M,I}$. $\blacksquare$

Applying Lemma 3.9 to $L = f^{-1}(\dot{\mathbb{C}})$ one obtains in view of $\overline{L} \cap X = \operatorname{supp} f$ that

Proposition 3.10. *The support of $f_{m,w}$ is contained in $(\operatorname{supp} f)_{M,I}$.* ■

The following lemma - which relies on Proposition 3.10 - prepares for the definition of quasihomogeneous averages of distributions on X in Chapter 4 below.

Lemma 3.11. *Let $f \in C_I^o(X)$ and $g \in C_{1/I}^o(X)$ where $1/I := \{1/t \,;\, t \in I\}$. If the sets $(\operatorname{supp} f)_{M,I} \cap \operatorname{supp} g$ and $\operatorname{supp} f \cap (\operatorname{supp} g)_{M,1/I}$ are compact then*

$$\int_X f_{m,w}(x)\, g(x)\, dx = \int_X f(x)\, g_{-m-\mu,v}(x)\, dx$$

where $v(t) := w(1/t)$ (here the integrals are well-defined since the support of each integrand is compact).

Proof. We set $F := \operatorname{supp} f$, $G := \operatorname{supp} g$, $K := F_{M,I} \cap G$, and $L := F \cap G_{M,1/I}$. Since F is an (M,I)-bounded and G an $(M,1/I)$-bounded subset of X the set

$$J := \{t \in I \,;\, M_t(K) \cap F \neq \emptyset\} \cup \{u \in I \,;\, M_{1/u}(L) \cap G \neq \emptyset\}$$

is a compact subset of I. Note that by Proposition 3.10 we have: $\operatorname{supp} f_{m,w} g \subset K$ and $\operatorname{supp} f g_{-m-\mu,v} \subset L$. Applying Fubini's theorem, substituting first $x = M_{1/t} y$ and then $t = 1/s$, taking the inclusions $\operatorname{supp}(f g \circ M_s) \subset F \cap M_{1/s}(K) \subset L$, $s \in 1/I$, into account and applying Fubini's theorem again one verifies that

$$\int_X f_{m,w}(x)\, g(x)\, dx = \int_K \int_J t^{-m} f(M_t x)\, w(t) \frac{dt}{t}\, g(x)\, dx =$$

$$= \int_J t^{-m-\mu} \int_K f(M_t x)\, g(x)\, d(M_t x)\, w(t) \frac{dt}{t} =$$

$$= \int_J t^{-m-\mu} \int_{M_t(K)} f(y)\, g(M_{1/t} y)\, dy\; v(1/t) \frac{dt}{t} =$$

$$= \int_{1/J} s^{m+\mu} \int_{M_{1/s}(K)} f(y)\, g(M_s y)\, dy\; v(s) \frac{ds}{s} =$$

$$= \int_{1/J} s^{m+\mu} \int_L f(y)\, g(M_s y)\, dy\; v(s) \frac{ds}{s} =$$

$$= \int_L f(y) \int_{1/J} s^{m+\mu} g(M_s y)\, v(s) \frac{ds}{s}\, dy = \int_X f(y)\, g_{-m-\mu,v}(y)\, dy .$$ ■

Next we deal with special choices of w.

Lemma 3.12. *Suppose that I is a closed subinterval of $]0,+\infty[$, and let a (resp. b) be its left (resp. right) endpoint. Then for arbitrary $f \in C_I^1(X)$ and $j \in \mathbb{N}_0$ we have*

$$\partial_M f_{m,w_j} = (\partial_M f)_{m,w_j} = m\, f_{m,w_j} - f_{m,w_{j-1}} + B_{b,j} - B_{a,j}$$

where $w_{-1} :\equiv 0$, $w_j := \omega_j \chi_I$, and

$$B_{c,j} := \begin{cases} 0 & \text{if } c \in \{0,+\infty\} \\ c^{-m}\, w_j(c)\, f \circ M_c & \text{if } c \in]0,+\infty[\end{cases} .$$

In particular, if $I =]0,1]$ then $(\partial_M - m)^{j+1} f_{m,w_j} = (-1)^j f$.

Proof. The first equality is a special case of (3.7). To prove the second equality we first observe (see the proof of Proposition 1.22 for $m=0$) that

$$\tag{3.10} \frac{1}{t}(\partial_M f)(M_t x) = \partial_t f(M_t x), \qquad t \in]0,+\infty[,\ x \in X.$$

Moreover, since $\omega_j'(t) = \omega_{j-1}(t)/t$ we have

$$\partial_t(t^{-m}\,\omega_j(t)) = \frac{1}{t}\left(-m\,t^{-m}\,\omega_j(t) + t^{-m}\,\omega_{j-1}(t)\right).$$

Consequently, for arbitrary $a < c < d < b$ partial integration yields:

$$(\partial_M f)_{m,\omega_j \chi_{[c,d]}}(x) = \int_c^d t^{-m}\, \partial_t f(M_t x)\, \omega_j(t)\, dt =$$

$$= B_{d,j}(x) - B_{c,j}(x) + m \int_c^d t^{-m}\, f(M_t x)\, \omega_j(t)\, \frac{dt}{t} - \int_c^d t^{-m}\, f(M_t x)\, \omega_{j-1}(t)\, \frac{dt}{t}.$$

Letting (c,d) tend to (a,b) we derive the first assertion. To prove the second one we observe that in case $I =]0,1]$ the first assertion tells us that

$$(\partial_M - m) f_{m,w_0} = f, \quad \text{and} \quad (\partial_m - m) f_{m,w_j} = -f_{m,w_{j-1}}, \quad j \in \mathbb{N}.$$

Hence the last assertion follows by induction. ■

Now we come to the special case $w = \omega_j$.

Proposition 3.13. *Suppose that $f \in C_M^0(X)$. Then for any $j \in \mathbb{N}_0$ the function f_{m,ω_j} is almost quasihomogeneous of degree m and of order $\leq j$; more precisely, we have*

$$\tag{3.11} t^{-m} f_{m,\omega_j} \circ M_t = \sum_{i=0}^{j} (-1)^i\, \omega_i(t)\, f_{m,\omega_{j-i}}, \qquad t \in]0,+\infty[$$

In particular, by (3.1) a continuous function $f_m : X \longrightarrow \mathbb{C}$ is well-defined (coinciding, of course, with f_{m,ω_0}) which is quasihomogeneous of degree m.

Proof. Substituting $u = ts$ and making use of the binomial formula we obtain

$$\int_0^{+\infty} (ts)^{-m} f(M_{st}x) (\log ts - \log t)^j \frac{ds}{s} =$$

$$= \sum_{i=0}^{j} \binom{j}{i} (-\log t)^i \int_0^{+\infty} u^{-m} f(M_u x) (\log u)^{j-i} \frac{du}{u} .$$

By the definition of ω_j this implies (3.11).

Alternatively, we deduce from Lemma 3.12 that $(\partial_M - m)^i f_{m,w_j} = (-1)^i f_{m,w_{j-i}}$ and $(\partial_M - m) f_{m,w_0} \equiv 0$ so that the assertion follows from Proposition 2.31. ∎

Is every C^r function $q : X \longrightarrow \mathbb{C}$ which is quasihomogeneous of degree m of the form $q = f_m$ for some $f \in C^r_M(X)$? Of course, one cannot take $f = q$ since the support of q is not an M-bounded subset of X unless $q \equiv 0$. So the idea is to multiply q by a cut-off function $\psi \in C^r_M(X)$ whose quasihomogeneous average of degree 0 is identically equal to 1. In this way the question is reduced to the case $m = 0$ and $q \equiv 1$. In a slightly more general formulation this is the content of

Proposition 3.14. *Suppose that for every compact subset K of X there is given a function $\psi_K \in C^o_M(X)$ such that $(\psi_K)_o$ is equal to 1 on K_M. Let $q \in C^o(X)$. Then q is quasihomogeneous of degree m if and only if $q = (\psi_K q)_m$ on K_M for every compact subset K of X.*

Proof. "⇒": by Corollary 3.5 we have $(\psi_K q)_m = q(\psi_K)_o$.

"⇐": by Proposition 3.13 $(\psi_K q)_m$ is quasihomogeneous of degree m. ∎

In section (c) we shall determine under which conditions the assumptions of Proposition 3.14 can be satisfied. In order to prepare this we are going to study (M,I)-bounded subsets of X in the following section.

(b) (M,I)-bounded Subsets of X

If I is compact then, of course, every subset of X is (M,I)-bounded. If J is another closed subset of $]0,+\infty[$ then every subset L of X which is (M,I)- as well as (M,J)-bounded is also $(M,I\cup J)$-bounded. Consequently, if I is an interval there are essentially three non-trivial cases to be distinguished: $I =]0,+\infty[$, $I =]0,1]$, and $I = [1,+\infty[$. Mostly, we shall deal with these cases, but the case that I is not an interval might be of interest, as well, as the following example shows.

Example 3.15. *Let $(s_k)_{k\in\mathbb{N}}$ be a monotone sequence in $]0,+\infty[$ such that*

$$(3.12) \qquad \lim_{k\to\infty} s_{k+1}/s_k = +\infty \quad (\text{resp. } 0).$$

For every $\ell\in\mathbb{N}$ let I_ℓ be a compact subset of $]0,+\infty[\setminus \bigcup_{k\in\mathbb{N}} [1/\ell,\ell]\, s_k$. We set

$$I := \bigcup_{\ell\in\mathbb{N}} I_\ell$$

(note that in view of (3.12) one can easily achieve that I is unbounded - resp. that $0\in\overline{I}$). Finally, let $x\in X$ be such that $\{x\}$ is an M-bounded subset of X. Then

$$L := \{M_{s_k}x \,;\, k\in\mathbb{N}\}$$

is an $(M,]0,1]\cup I)$-(resp. $(M,[1,+\infty[\cup I)$-) bounded but not an $(M,[1,+\infty[)$- (resp. $(M,]0,1])$-) bounded subset of X.

Proof. Since $\{s_k\,;\,k\in\mathbb{N}\}\subset\{t\in]0,+\infty[\,;\, M_t\{x\}\cap L\neq\emptyset\}$ it follows from (3.12) that L is not an $(M,[1,+\infty[)$- (resp. $(M,]0,1])$-) bounded subset of X.

Let K be a compact subset of X. Since $\{x\}$ is an M-bounded subset of X the set $J := \{t\in]0,+\infty[\,;\, x\in M_t(K)\}$ is compact, hence contained in $[1/N,N]$ for some $N\in\mathbb{N}$. Now, let $t\in]0,+\infty[$ be such that $M_t(K)\cap L\neq\emptyset$. Then one can choose $k\in\mathbb{N}$ such that $x\in M_{t/s_k}(K)$, i.e. $t/s_k\in J$ so that

$$t\in s_kJ \subset \bigcup_{k\in\mathbb{N}} [1/\ell,\ell]\, s_k \qquad \text{if } \ell\geq N.$$

This implies, in particular, that $t\geq s_1/N$ (resp. $t\leq N s_1$). Moreover, if, in addition, t belongs to I then it must lie even in the compact set $I_1\cup\ldots\cup I_{N-1}$. ■

We are first going to deal with the question when a given point $x\in X$ is *(M,I)-bounded in X*, i.e. $\{x\}$ is an (M,I)-bounded subset of X.

Proposition 3.16. *Suppose that I is non-compact. Then for every $x\in X$ the following conditions are equivalent:*

(a) x *is (M,I)-bounded in* X;

(b) *the map* $I \longrightarrow X$, $t \mapsto M_{1/t}x$, *induces a homeomorphism onto its image, and*

(3.13) $$\{M_t x\,;\, t\in 1/I\} = \overline{\{x\}_{M,I}} \text{ is a closed subset of } X;$$

(c) $x \in X \setminus E_M(\sigma_0)$, *and (3.13) holds.*

Proof. *(a)⇒(b)*: Since for $K' := i_x(]0,+\infty[)$ we have $\{t\in I\,;\, M_t(K')\cap\{x\}\neq\emptyset\} = I$ and since $\{x\}$ is an (M,I)-bounded subset of X the set K' cannot be compact. Hence Proposition 1.10.(iii) implies that i_x must be injective.

Now let $(t_j)_{j\in\mathbb{N}}$ be a sequence in I such that $(M_{1/t_j}x)_{j\in\mathbb{N}}$ converges to some $y\in X$ as $j\to\infty$. Let K be a compact neighbourhood of y in X. Then for sufficiently large j we have $M_{t_j}(K)\cap\{x\}\neq\emptyset$. Hence a subsequence of (t_j) converges to some $t\in I$ so that by the continuity of i_x we have $y = M_{1/t}x \in \{x\}_{M,I}$. Moreover, for any other subsequence of $(t_j)_{j\in\mathbb{N}}$ there is a subsequence converging to some $s\in I$. Since, again, we have $M_{1/s}x = y$ the injectivity of i_x implies that $t=s$. Of course, this means that $(t_j)_{j\in\mathbb{N}}$ itself converges to t.

(b)⇒(a): Let K be a compact subset of X, and let $(t_\ell)_{\ell\in\mathbb{N}}$ be a sequence in I and $(k_\ell)_{\ell\in\mathbb{N}}$ be a sequence in K such that $M_{t_\ell}k_\ell = x$ for every $\ell\in\mathbb{N}$. By choosing subsequences we achieve that (k_ℓ) converges to some $w\in K$ as $\ell\to\infty$. Consequently,

$$w = \lim_{\ell\to\infty} M_{1/t_\ell}x \in \overline{\{x\}_{M,I}}\,.$$

By (3.13) one finds $t\in I$ such that $w = M_{1/t}x$. The first part of the condition (b) then implies that the sequence (t_ℓ) converges to t.

(b)⇒(c): It suffices to show that

(3.14) $$\overline{\{x\}_M} \subset X \qquad \text{if } x\in X\cap E_M(\sigma_0).$$

Indeed, assuming that $x\in E_M(\sigma_0)$ we are going to derive that I is compact in contradiction to the assumption on I. In fact, let $(t_m)_{m\in\mathbb{N}}$ be any sequence in I. Since the present assumption on x implies that $\overline{\{x\}_M}$ is compact, by choosing a subsequence we achieve that $(M_{1/t_m}x)_{m\in\mathbb{N}}$ converges to some point $y\in\overline{\{x\}_{M,I}}$. By (3.13) and (3.14) the last set is equal to $\{x\}_{M,I}$ so that one finds $t\in I$ such that $y = M_{1/t}x$. Hence, by the first part of (b) the sequence $(t_m)_{m\in\mathbb{N}}$ converges to t, i.e. I is compact, as was to be shown.

For the proof of (3.14) we fix a point y belonging to the left-hand side of (3.14) and choose a sequence $(t_m)_{m\in\mathbb{N}}$ in $]0,+\infty[$ such that the sequence of points $y_m := M_{1/t_m}x$, $m \in \mathbb{N}$, converges to y as $m\to\infty$. Since the endomorphisms M_{t_m} act on $E_M(\sigma_0)$ as isometries (see (1.79) and (1.10)) we observe that

$$\|x - M_{t_m}y\| = \|y_m - y\|, \qquad m \in \mathbb{N},$$

so that $\lim_{n\to\infty} M_{t_m}y = x$. Since X is open this means that $M_{t_m}y \in X$ for sufficiently large m. Since X is quasihomogeneous this implies $y \in X$, as desired.

(c)⇒(b): If x does not belong to $E_M(\sigma_0)$ then by Proposition 1.10.(ii) i_x induces a homeomorphism onto its image and so does its restriction to I. ■

In particular, the points of ker M are never (M,I)-bounded in X if I is non-compact. Sometimes the assumption "$0 \in X$" even implies that there are no non-trivial (M,I)-bounded subsets of X at all:

Remark 3.17. *Suppose that (1.14) holds, and that I is unbounded. Then no point of $X \cap M_0^{-1}(X)$ is (M,I)-bounded in X. In particular, if $X \subset M_0^{-1}(X)$ then no non-empty subset of X is (M,I)-bounded.*

Proof. By Remark 1.8 the assumptions on M and I imply that $M_0 x \in \overline{\{x\}_{M,I}} \setminus \{x\}_{M,I}$ for every $x \in V \setminus \ker M$. If $M_0 x \in X$ it follows by Proposition 3.16 that x is not (M,I)-bounded in X. ■

Note that by Remark 1.8 one has

$$(3.15) \qquad \overline{\{x\}_{M,I}} \setminus \{x\}_{M,I} \subset E_M(\sigma_0).$$

Hence, as another consequence of Proposition 3.16 one obtains

Proposition 3.18. *Suppose that I is non-compact. Then every point of X is (M,I)-bounded if and only if*

$$(3.16) \qquad X \cap E_M(\sigma_0) = \emptyset.$$

In particular, every point of $\dot{V}$ is (M,I)-bounded in $\dot{V}$ if and only if $\sigma_0 = \emptyset$. ■

We now come to the description of general (M,I)-bounded subsets of X.

Remark 3.19. *If L is an (M,I)-bounded subset of X so is* $\overline{L} \cap X$.

Proof. Let K be a compact subset of X, and let U be a compact neighbourhood of K in X. Then the continuity of $M_{1/t}$ implies that $\{t \in I\,;\ M_t(K) \cap \overline{L} \neq \emptyset\}$ is contained in $\{t \in I\,;\ M_t(U) \cap L \neq \emptyset\}$; and the assertion follows. ∎

Corollary 3.20. *Suppose that I is non-compact. If L is an (M,I)-bounded subset of X then*

$$(3.17) \qquad (a)\ \overline{L} \cap X \cap E_M(\sigma_0) = \emptyset, \quad \text{and} \quad (b)\ \overline{L_{M,I}} \cap X \cap E_M(\sigma_0) = \emptyset.$$

Proof. By Remark 3.19 $\overline{L} \cap X$ is an (M,I)-bounded subset of X, as well, so that every point in $\overline{L} \cap X$ is (M,I)-bounded in X. Hence, the condition (3.17.a) is a consequence of Proposition 3.16. Moreover, by Lemma 3.9 the set $\overline{L_{M,I}} \cap X$ coincides with $(\overline{L} \cap X)_{M,I}$. Consequently, since $E_M(\sigma_0)$ is M_t-invariant for every $t \in 1/I$ the condition (3.17.b) follows from (3.17.a). ∎

Preparing for a characterization of (M,I)-bounded subsets we note that (M,I)-boundedness implies some sort of local property.

Remark 3.21. *Suppose that L is an (M,I)-bounded subset of X and that (3.17.a) holds. Then*

$$(3.18.a) \qquad L_I := \{\,(M_{1/t}x, x)\,;\ x \in \overline{L} \cap X,\ t \in I\}$$

is a closed subset of $X \times X$, *and*

$$(3.18.b) \qquad \text{the map } L_I \longrightarrow \,]0,+\infty[,\ (M_{1/t}x, x) \mapsto t, \text{ is continuous.}$$

Proof. First of all we observe that by (3.17.a) and Proposition 1.10.(ii) the map in (3.18.b) is well-defined, indeed. Moreover, by Remark 3.19 we may assume that L is closed in X. Now, let $(x_m)_{m\in\mathbb{N}}$ (resp. $(t_m)_{m\in\mathbb{N}}$) be a sequence in L (resp. I), and let $(x,y) \in X \times X$ be such that

$$\lim_{m\to\infty} (M_{1/t_m}x_m, x_m) = (y,x).$$

This implies that the set $K := \{y\} \cup \{M_{1/t_m}x_m\,;\ m \in \mathbb{N}\}$ is a compact subset of X and that the numbers t_m belong to the set (3.2). Consequently, we find $t \in I$ such that a suitable subsequence of $(t_m)_{m\in\mathbb{N}}$ converges to t. Since any subsequence of $(x_m)_{m\in\mathbb{N}}$ tends to x, as well, and since the map $(r,y)\mapsto M_r y$ is continuous it follows that $y = M_{1/t}x$. Since L is closed in X this shows that $(y,x)\in L_I$, i.e. (3.18.a) is proved.

Moreover, by the same argument as above we see that any subsequence of $(t_m)_{m\in\mathbb{N}}$ has a subsequence converging to some number $s\in I$ which by the continuity of the map $(r,y)\mapsto M_r y$, again, satisfies the equation $M_{1/s}x = M_{1/t}x$. Since by (3.17.a) and Proposition 1.10.(ii) this implies $s = t$ we conclude that the whole sequence (t_m) converges to t as $m\to\infty$. Hence (3.18.b) is proved, as well. ∎

The following characterization of (M,I)-bounded subsets of X is essential for Chapters 4 and 6.

Proposition 3.22 . *Suppose that I is non-compact. Let L be a subset of X satisfying (3.17.a) and (3.18.b). Then the following conditions are equivalent:*

(a) L is an (M,I)-bounded subset of X;

(b) every subset H of the set L_I defined in (3.18.a) such that $\pi_1(H)$ is a relatively compact subset of X is relatively compact in L_I (here $\pi_1 : X\times X \longrightarrow X$ denotes the projection onto the first factor);

(c) $\overline{L}\cap K_{M,1/I}$ is compact for every compact subset K of X.

Note that for the converse implication it does not suffice to postulate that $\overline{L}\cap\overline{K_{M,1/I}}\cap X$ is compact (see Example 3.24 below).

Proof. *(a)⇒(b):* Let $(\ell_m)_{m\in\mathbb{N}}$ be a sequence in $\overline{L}\cap X$ and (t_m) any sequence in I such that $k_m := M_{1/t_m}\ell_m$ converges to some $x\in X$ as $m\to\infty$. Then the subset $K := \{x\}\cup\{k_m;\ m\in\mathbb{N}\}$ of X is compact. Since the numbers t_m lie in the set (3.2) with L replaced by $\overline{L}\cap X$ and since by Remark 3.19 $\overline{L}\cap X$ is (M,I)-bounded in X, as well, we achieve by choosing subsequences that (t_m) converges to some $t\in I$. By continuity we deduce that $\ell_m = M_{t_m}k_m$ converges to $M_t x =: \ell\in\overline{L}\cap X$ as $m\to\infty$. Hence the sequence of pairs $(M_{1/t_m}\ell_m, \ell_m)$ converges to $(M_{1/t}\ell, \ell)\in L_I$.

(b)⇒(c): Let $(t_m)_{m\in\mathbb{N}}$ be a sequence in I and (k_m) be a sequence in K such that $\ell_m := M_{t_m}k_m$ belongs to $\overline{L}$ for every $m\in\mathbb{N}$. Choosing subsequences we achieve that

(k_m) converges to some $x \in K$ as $m \to \infty$. Applying (b) to the set $H = \{(k_m, \ell_m);\ m \in \mathbb{N}\}$ we find $\ell \in \overline{L} \cap X$ and $t \in I$ such that (k_m, ℓ_m) converges to $(M_{1/t}\ell, \ell)$ so that ℓ_m converges to $\ell = M_t x \in \overline{L} \cap K_{M,1/I}$.

<u>*(c)⇒(a)*</u>: Let K be a compact subset of X. And let $(t_m)_{m \in \mathbb{N}}$ be a sequence in the set (3.2). It suffices to show that (t_m) has a subsequence converging to some $t \in]0, +\infty[$. For every $m \in \mathbb{N}$ we fix $k_m \in K$ such that $\ell_m := M_{t_m} k_m$ belongs to L. By choosing subsequences we achieve that (k_m) converges to some $x \in K$ as $m \to \infty$. Hence, $H := \{x\} \cup \{k_m;\ m \in \mathbb{N}\}$ is a compact subset of X. Since ℓ_m lies in $L \cap H_{M,1/I}$ for every $m \in \mathbb{N}$ the condition (c) shows that by choosing subsequences, again, we achieve that ℓ_m tends to some $\ell \in \overline{L} \cap H_{M,1/I}$ as $m \to \infty$. Then we fix $h \in H$ and $t \in I$ such that $\ell = M_t h$. If $h = x$ then it follows by (3.18.b) that $(t_m)_{m \in \mathbb{N}}$ converges to t as $m \to \infty$, as desired. If $h \neq x$ then for every $N \in \mathbb{N}$ repeating the preceding argument with H replaced by $\{x\} \cup \{k_m;\ m \geq N\}$ we deduce that a subsequence of $(k_m)_{m \in \mathbb{N}}$ belongs to $\{\ell\}_{M,I}$ so that $x \in \overline{\{\ell\}_{M,I}}$. In view of (3.15) the assumption (3.17.a) shows that $x \in \{\ell\}_{M,I}$, i.e. $\ell = M_t x$ for some $t \in I$. As above we deduce that $(t_m)_{m \in \mathbb{N}}$ converges to t. ■

By applying Proposition 3.22 one can easily determine the (M,I)-bounded subsets of $\dot{V}$ in case $\sigma = \sigma_+$:

Remark 3.23 . *Suppose that $V = G_M(\sigma_+) + E_M(\sigma_0)$ and that $X \cap E_M(\sigma_0) = \emptyset$. Let L be a subset of X.*

(i) If L is bounded then L is an $(M, [1, +\infty[)$-bounded subset of X;

(ii) if $\mathrm{dist}(L, E_M(\sigma_0)) > 0$ then L is an $(M,]0,1])$-bounded subset of X;

(iii) if L is a relatively compact subset of $V \setminus E_M(\sigma_0)$ then L is an M-bounded subset of X;

(iv) if $\sigma = \sigma_+$ and if $X = \dot{V}$ then the converse implications are valid.

<u>*Proof.*</u> By Remark 1.8 the assumptions on M and X imply that $\lim_{t \to \infty} |M_t x| = +\infty$ and $\lim_{t \to 0} \mathrm{dist}(M_t x, E_M(\sigma_0)) = 0$ uniformly if x stays in any compact subset of $V \setminus E_M(\sigma_0)$. From this the assertions <u>*(i) - (iii)*</u> are easily deduced. Moreover, if the assumptions of <u>*(iv)*</u> are satisfied then Proposition 1.70 shows that we can choose the norm $|\cdot|$ on V in such a way that for every $x \in \dot{V}$ the function $t \mapsto |M_t x|$

is strictly increasing. Hence, it follows that the compact subset $K := \{x \in V;\ |x| = 1\}$ of $\dot{V}$ satisfies the equations

$$K_{M,]0,1]} = V \setminus B, \quad K_{M,[1,+\infty[} = B \setminus \{0\}, \quad \text{and} \quad K_M = \dot{V}$$

where $B := \{x \in V;\ |x| \leq 1\}$. Hence the assertion (iv) is obtained by an application of Proposition 3.22. ∎

If the assumption on V and M is dropped then the implications (i) - (iii) are false in general; in fact, not every compact subset of $\dot{V}$ is M-bounded as is illustrated by the following

Example 3.24. *Suppose that $n=2$ and $p \in]0,+\infty[\times]-\infty,0[$. Let $\varepsilon > 0$, and set*

$$L := [0, \varepsilon^{p_1}] \times \{1\} \quad \text{and} \quad K := \{1\} \times [0, \varepsilon^{-p_2}] .$$

Then $\{t \in]0,+\infty[;\ M_t(K) \cap L \neq \emptyset\} =]0,\varepsilon]$, and $L \cap K_{M,1/]0,\varepsilon]} = L \setminus \{(0,1)\}$. In particular, L is not a $(p,]0,1])$-bounded and K not a $(p,[1,+\infty[)$-bounded subset of $\mathbb{R}^2 \setminus \{0\}$; and consequently no neighbourhood of $(0,1)$ (resp. $(1,0)$) is a $(p,]0,1])$- (resp. $(p,[1,+\infty[)$-) bounded subset of $\mathbb{R}^2 \setminus \{0\}$. ∎

Since by Proposition 3.18 the assumptions of Example 3.24 imply that the sets $\{(0,1)\}$ and $\{(1,0)\}$ are p-bounded subsets of $\mathbb{R}^2 \setminus \{0\}$ the Example 3.24 shows, in particular, that (p,I)-bounded subsets of X do not necessarily have (p,I)-bounded neighbourhoods.

The following lemma is relevant for Lemma 4.5 below.

Lemma 3.25. *Suppose that I is a closed subinterval of $]0,+\infty[$. Let A be an (M,I)-bounded and B be an $(M,1/I)$-bounded subset of X.*

(i) If I is non-compact then both $A_{M,I} \cap B$ and $A \cap B_{M,1/I}$ are M-bounded subsets of X;

(ii) if I is compact then $A_{M,I} \cap B$ is an M-bounded subset of X if and only if $A \cap B_{M,1/I}$ is one.

For the proof of this and other assertions the following lemma is useful.

Lemma 3.26. *Suppose that I is a proper closed subinterval of $]0,+\infty[$. If L is an (M,I)-bounded subset of X so is $L_{M,I}$.*

Proof. Let K be a compact subset of X. Since the assumptions on L and I imply that L is an (M,J)-bounded subset of X where $J := \{st\,;\ s,t \in I\}$ it follows that the set $\{u \in J\,;\ M_u(K) \cap L \neq \emptyset\}$ is contained in a compact subinterval $[c,d]$ of $]0,+\infty[$. Now, let $t \in I$ be such that $M_t(K) \cap L_{M,I} \neq \emptyset$, and choose $k \in K$, $\ell \in L$ and $s \in I$ such that $M_t k = M_{1/s}\ell$, i.e. $M_{st}k = \ell$. Hence $c \leq st \leq d$. If $b \in]0,+\infty[$ is such that I equals $]0,b]$ (resp. $[b,+\infty[$) then this implies that $t \in [c/b,b]$ (resp. $[b,d/b]$). ∎

Proof of Lemma 3.25. *(i)*: If $I =]0,+\infty[$ then A and B are M-bounded subsets of X, and there is nothing to be proved. So we suppose that $I \neq]0,+\infty[$. Since I is non-compact the closure J of $]0,+\infty[\setminus I$ differs from $1/I$ by a relatively compact subset of $]0,+\infty[$ so that B is an (M,J)-bounded subset of X, as well. Since by Lemma 3.26 $A_{M,I}$ is an (M,I)-bounded and $B_{M,1/I}$ an $(M,1/I)$-bounded and hence (M,J)-bounded subset of X the assertion follows.

(ii), "⇒": Let K be a compact subset of X, and choose $t \in]0,+\infty[$ and $k \in K$ such that $a := M_t k \in A \cap B_{M,1/I}$. Then we find $s \in 1/I$ such that $b := M_s a \in B$. This means that $M_t(M_s k) = M_s a \in A_{M,I} \cap B$. Since by the assumption on I the set $\{M_s k\,;\ k \in K, s \in 1/I\}$ is a compact subset of X the assertion follows.

"⇐": one has to interchange the roles of (A,I) and $(B,1/I)$ and apply the implication "⇒". ∎

The following lemma is required for the proof of Theorem 4.8 below.

Lemma 3.27. *Let I and J be closed subintervals of $]0,+\infty[$ such that $I \cap J$ is compact. Let L be an (M,I)-bounded subset of X. Then for every compact subset K of X the set $\overline{L}_{M,I} \cap K_{M,J}$ is a compact subset of X.*

Proof. If $I =]0,+\infty[$ then the assumption on J implies that J is compact so that $K_{M,J}$ is a compact subset of X, and the assertion follows by Lemma 3.9. Hence we suppose that $I \neq]0,+\infty[$. Then the assumption on J and I means that I and $1/J$ differ by a relatively compact subset of $]0,+\infty[$. Since by Remark 3.19, Lemma 3.9, and Lemma 3.26 $\overline{L}_{M,I} \cap X$ is an (M,I)-bounded subset of X it is $(M,1/J)$-

bounded, as well. Since by Lemma 3.9 $\overline{L}_{M,I} \cap X$ is a closed subset of X an application of Corollary 3.20 and Proposition 3.22 leads to the desired assertion. ■

In some sense (M,I)-boundedness is a local property:

Remark 3.28. *Let $\mathfrak{U}$ be a locally finite covering of X consisting of quasihomogeneous open subsets of X. Let L be a subset of X.*

(i) If there exists a family $(L_U)_{U \in \mathfrak{U}}$ of (M,I)-bounded subsets of X such that $L_U \subset U$ for every $U \in \mathfrak{U}$ and $L = \bigcup_{U \in \mathfrak{U}} L_U$ then L is an (M,I)-bounded subset of X.

(ii) L is an (M,I)-bounded subset of X if and only if $L \cap U$ is an (M,I)-bounded subset of U for every $U \in \mathfrak{U}$.

Note that in (i) it is not sufficient to assume that each L_U is an M-bounded subset of U: For example, if $\sigma_0 = \emptyset$ then in view of Proposition 3.34.A below every compact subset of $X_\pm$ is an M-bounded subset of $X_\pm$, but in case $X_\pm \neq X_+ \cup X_-$ not every compact subset of $X_+ \cup X_-$ is an M-bounded subset of $X_+ \cup X_-$.

Proof. Let K be a compact subset of X. Since every $U \in \mathfrak{U}$ is quasihomogeneous we have $(K \cap U)_M = K_M \cap U$. Since $\mathfrak{U}$ is locally finite we can find a finite subset $\mathfrak{V}$ of $\mathfrak{U}$ such that $K_M \cap U = \emptyset$ for every $U \in \mathfrak{U} \setminus \mathfrak{V}$. Hence, under the assumptions of *(i)*, for every $t \in I$ we have

$$M_t(K) \cap L = \bigcup_{U \in \mathfrak{V}} (M_t(K) \cap L_U)$$

from which the conclusion of (i) follows.

To prove *(ii)* we choose a family $(K_U)_{U \in \mathfrak{V}}$ of compact subsets K_U of $K \cap U$ such that $\bigcup_{U \in \mathfrak{V}} K_U = K$ and observe that then

$$M_t(K) \cap L = \bigcup_{U \in \mathfrak{V}} M_t(K_U) \cap (L \cap U) \ . \quad ■$$

We close this section with an elementary but important criterion for (M,I)-boundedness. It requires the existence of positive quasihomogeneous functions on X.

Lemma 3.29. *Let $x: X \longrightarrow]0,+\infty[$ be a continuous function which is quasihomogeneous of degree 1. Then a subset L of X is (M,I)-bounded if and only if for every compact subset K of X the following condition holds:*

(3.19) *$I \cap J\varkappa(L \cap K_M)$ is a relatively compact subset of I for every compact subset J of $]0,+\infty[$.*

Note that if I is an interval then (3.19) is equivalent to

(3.19)' *$I \cap \varkappa(L \cap K_M)$ is a relatively compact subset of I.*

Proof. "$\Leftarrow$": If K is a compact subset of X then $J := 1/\varkappa(K)$ is a compact subset of $]0,+\infty[$. Since for arbitrary $t \in]0,+\infty[$ and $k \in K$ the condition "$\ell := M_t k \in L$" implies "$t = \varkappa(\ell)/\varkappa(k) \in J\varkappa(L \cap K_M)$" the set (3.2) is contained in $I \cap J\varkappa(L \cap K_M)$ and hence is relatively compact in I by (3.19).

"$\Rightarrow$": We fix a compact subset K of X, and let J be a compact subset of $]0,+\infty[$. Note that by the implication "$\Leftarrow$", already proved above, the set $\varkappa^{-1}(1/J)$ is an M-bounded subset of X. Moreover, note that the existence of $\varkappa$ implies (3.16) and the continuity of the map $\{(x, M_t x) ; x \in X,\ t \in]0,+\infty[\}$, $(x, M_t x) \mapsto t$. Hence, it follows by Proposition 3.22 that $H := \varkappa^{-1}(1/J) \cap K_M$ is a compact subset of X. Now we let $j \in J$ and $\ell \in L \cap K_M$, set $t := j\varkappa(\ell)$, and choose $s \in]0,+\infty[$ and $k \in K$ such that $\ell = M_s k$. Then $\varkappa(M_{s/t} k) = \varkappa(\ell)/t = 1/j \in 1/J$, i.e. $M_{s/t} k \in H$ so that $\ell = M_t(M_{s/t} k) \in M_t(H)$. Hence we conclude that $I \cap J\varkappa(L \cap K_M)$ is contained in the set $\{t \in I ; M_t(H) \cap L \neq \emptyset\}$ which is relatively compact in I by the assumption on L. ∎

Note that in view of Proposition 1.70 the implications (i) - (iii) of Remark 3.23 are special cases of Lemma 3.29.

(c) When is every Compact Subset of X M-bounded?

Definition 3.30. X is called *locally M-bounded* if and only if every compact subset of X is an M-bounded subset of X.

Proposition 3.31. *Let I be a non-compact closed subset of $]0,+\infty[$. Then the following conditions are equivalent:*

(a) *X is locally M-bounded;*

(b) *every compact subset of X is (M,I)-bounded;*

(c) *the condition (3.16) holds,* $X_I := \{(x, M_t x)\,;\ x \in X\,,\ t \in I\}$ *is a closed subset of X×X, and the map* $X_I \longrightarrow]0,+\infty[$, $(x, M_t x) \mapsto t$, *is continuous;*

(d) *for every compact subset K of X there exists an M-bounded subset L of X such that* $L_M \supset K$ *;*

(e) *there is an M-bounded subset L of X satisfying* $L_M = X$ *.*

For the proof of the implication "(b)⇒(a)" in case I is not an interval we require two lemmata, the first one elaborating on Example 3.24.

Lemma 3.32. *Suppose that* $\sigma_0 = \emptyset$*. Let* $y_- \in G_M(\sigma_-)$ *and* $y_+ \in G_M(\sigma_+)$*, and let* $U_\pm$ *be a neighbourhood of* $y_\pm$ *. Then there exists* $a \in]0,+\infty[$ *such that* $M_t(U_-) \cap U_+ \neq \emptyset$ *for every* $t \in [a,+\infty[$ *. In particular, there is a compact subset K of* U_M *such that* $M_t(K) \cap U \neq \emptyset$ *for every* $t \in]0,+\infty[$ *where* $U := U_+ \cup U_-$ *.*

Proof. We fix $\varepsilon > 0$ so small that $K_\pm := \{x \in V\,;\ P_\pm(x) = P_\pm(y_\pm)$, and $|P_\mp(x)| \leq \varepsilon\}$ is contained in $U_\pm$. Setting $s_+ := 0$ and $s_- := +\infty$ we observe that by Remark 1.8.(i) we have $\lim_{t \to s_\pm} M_t y_\pm = 0$. Hence we can choose $a \in]1,+\infty[$ such that $|M_t y_\pm| \leq \varepsilon$ for every $t \in I_\pm$ where $I_- := [a,+\infty[$ and $I_+ := 1/I_- =]0,1/a]$. It follows that for every $t \in I_-$ by

$$P_+(x) := M_{1/t}\, y_+$$

a unique element x of K_- is well-defined such that $M_t x \in K_+$. This shows that $M_t(K_-) \cap K_+ \neq \emptyset$ for every $t \in I_-$. So the first part of the assertion is proved.
From this it follows that $M_t(L) \cap L \neq \emptyset$ for every $t \in]0,+\infty[\setminus I'$ where $L := K_+ \cup K_-$ and $I' := [1/a\,, a]$. Since $M_t(M_{a/t}(L)) \cap L = M_a(L) \cap L \neq \emptyset$ we see that $M_t(K) \cap L \neq \emptyset$ for every $t \in I'$ where K is the compact subset of X defined by

$$K := \bigcup_{t \in I'} M_{a/t}(L)\,.$$

Since K contains L we conclude that the set (3.2) is equal to $]0,+\infty[$. In view of $K \subset L_M \subset U_M$ the proof is complete. ∎

Lemma 3.33. *Let* $y, z \in G_M(\sigma_0)$ *be such that z is M-connected to y in the sense* *of Definition 1.81. Then for every neighbourhood U of z and for every neighbourhood W of y there is a finite subset R of* $]1,+\infty[$ *and a compact subinterval J of* $]0,+\infty[$ *such that*

$$M_t\Big(\bigcup_{r\in R} M_r(W)\Big)\cap U\neq\emptyset \quad \textit{for every } t\in\,]0,+\infty[\,\setminus J.$$

Proof. We set $\tau:=\sigma_O$. By Lemma 1.84 we find a point $w\in T(\tau)\subset\mathbb{C}^\tau$ (see Definition 1.79.(iv)) and a continuous function $\delta:]0,+\infty[\setminus\{1\}\longrightarrow G_M(\sigma_O)$ such that the condition (1.101) is valid for every $s\in\{0,+\infty\}$ with x replaced by z. Observing that $H_v^+(x):=(v_\lambda x_\lambda)_{\lambda\in\tau}$ is continuous as a function of $(v,x)\in\mathbb{C}^\tau\times G_M(\tau)$ we find open neighbourhoods U_τ of w in $\mathbb{C}^\tau$ and Z of $H_w^-(z):=(z_\lambda/w_\lambda)_{\lambda\in\tau}$ in $G_M(\tau)$ such that $H_v^+(x)\in U$ for every $(v,x)\in U_\tau\times Z$. Then by Lemma 1.11 we find a finite subset R of $]1,+\infty[$ such that

(3.20) for every $v\in T(\tau)$ there is $r\in R$ satisfying $(r^\lambda v_\lambda)_{\lambda\in\tau}\in U_\tau$.

Now, by (1.101) we can choose a compact subset J of $]0,+\infty[$ such that $\delta(rt)\in W$ and $N_{rt}\delta(rt)\in Z$ for arbitrary $t\in I:=]0,+\infty[\setminus J$ and $r\in R$ where N_t is defined by (1.102). Let $t\in I$. Since by the definition of $T(\tau)$ the point $\gamma(t):=(t^\lambda)_{\lambda\in\tau}$ belongs to $T(\tau)$, by (3.20) we find a number $r\in R$ such that $\gamma(rt)\in U_\tau$. It follows that $H^+_{\gamma(rt)}(N_{rt}\delta(rt))\in U$. Since by (1.102) we have $H^+_{\gamma(rt)}\circ N_{rt}=M_{rt}=M_t\circ M_r$ the assertion is proved. ■

Proof of Proposition 3.31. *(b)⇒(a)*: We observe that for arbitrary compact subsets K and L of X and for every $t\in I$ we have: $M_t(K)\cap L=M_t(K\cap M_{1/t}(L))\neq\emptyset$ if and only if $K\cap M_{1/t}(L)\neq\emptyset$. Consequently, the condition (b) remains valid if I is replaced by $1/I$ and hence by $J:=I\cup 1/I$. If I is a non-compact interval then J and $]0,+\infty[$ differ by a compact subinterval of $]0,+\infty[$ so that (a) follows in this case. Moreover, for the proof of the general case we may suppose that I is unbounded.

The proof of the general case is done by contraposition. So we assume that there are compact subsets K and L of X such that $J:=\{t\in\,]0,+\infty[;\ M_t(K)\cap L\neq\emptyset\}$ is not compact. Hence, in view of the observation at the beginning of the proof, by interchanging the roles of K and L if necessary we achieve that J is unbounded. Consequently, we can choose a sequence $(t_m)_{m\in\mathbb{N}}$ converging to $+\infty$ as $m\to\infty$ and a sequence $(k_m)_{m\in\mathbb{N}}$ in K such that $\ell_m:=M_{t_m}k_m$ belongs to L for every

$m\in\mathbb{N}$. By choosing subsequences if necessary we achieve that (k_m) (resp. (ℓ_m)) converges to some $k\in K$ (resp. $\ell\in L$) as $m\to\infty$. In view of Remark 1.8.(i) we conclude that

$$\text{(3.21.a)} \qquad P_-(\ell) = \lim_{m\to\infty} P_-(\ell_m) = \lim_{m\to\infty} M_{t_m}P_-(k_m) = 0$$

and

$$\text{(3.21.b)} \qquad P_+(k) = \lim_{m\to\infty} P_+(k_m) = \lim_{m\to\infty} M_{1/t_m}P_+(\ell_m) = 0 .$$

Setting $y_+ := P'(\ell)$ and $y_- := P'(k)$ (where $P' := P_+ + P_-$) we can choose compact neighbourhoods $U'_\pm$ of $y_\pm$ in $V' := G_M(\sigma_+ \cup \sigma_-)$ and U of $z := P_0(\ell)$ and W of $y := P_0(k)$ in $G_M(\sigma_0)$ such that $U'_+ + U$ and $U'_- + W$ are contained in X. Note that in view of (3.21) the points y_+ and y_- satisfy the assumptions of Lemma 3.32 if (V,M) is replaced by (V',M') where $M' := M_{V'}$. Let a be a positive number such that the conclusion of Lemma 3.32 holds. Moreover, noting that z is M-connected to y we fix J and R as in the conclusion of Lemma 3.33. Finally, we fix $b\in\,]0,+\infty[$ such that $]b,+\infty[\,\cap J = \emptyset$ and $rb\ge a$ for every $r\in R$. Consequently, for every $t\in\,]b,+\infty[$ we find $r\in R$, $w\in W$, and $u_-\in U'_-$ such that $M_{rt}u_-\in U'_+$ and $M_tM_rw\in U$, i.e. $M_t(M_r(u_-+w))\in L := U'_+ + U$. Since $K' := \bigcup_{r\in R} M_r(U'_- + W)$ is a compact subset of X and since I is assumed to be unbounded this shows that L is not (M,I)-bounded in X. Since L is compact the condition (b) is violated.

(b)⇒(c): The first part of (c) is a consequence of Proposition 3.18. To deduce the other parts we fix sequences $(x_m)_{m\in\mathbb{N}}$ in X and $(t_m)_{m\in\mathbb{N}}$ in I such that $(x_m, M_{t_m}x_m)$ converges to some $(x,y)\in X\times X$ as $m\to\infty$. Then, in particular, $L := \{y\}\cup\{M_{t_m}x_m;\, m\in\mathbb{N}\}$ is a compact and hence – by (b) – an (M,I)-bounded subset of X. Consequently, Remark 3.21 implies 1.° that $y = M_tx$ for some $t\in I$, i.e. $(x,y)\in X_I$, and 2.° that (t_m) converges to t as $m\to\infty$. This proves the second and the third part of the condition (c).

(c)⇒(b): Let K and L be compact subsets of X, and let $(t_m)_{m\in\mathbb{N}}$ be a sequence in the set (3.2). For every $m\in\mathbb{N}$ we fix $k_m\in K$ and $\ell_m\in L$ such that $M_{t_m}k_m = \ell_m$. By choosing subsequences if necessary we achieve that (k_m) converges to some $k\in K$ and (ℓ_m) to some $\ell\in L$ as $m\to\infty$. By the second part of the condition (c) we deduce that $\ell = M_tk$ for some $t\in I$. And by the last part of (c) we conclude that $t_m\to t$ as $m\to\infty$, as desired.

(a)⇒(e): We choose a sequence of compact subsets X_m, $m\in\mathbb{N}_0$, of X such tha

$$(3.22) \qquad \text{(a)}\ \emptyset = X_0 = X_1 \subset \ldots \subset X_m \subset \overset{\circ}{X}_{m+1},\ m \in \mathbb{N}; \text{ and } \quad \text{(b)}\ \bigcup_{m\in\mathbb{N}} X_m = X.$$

Since the quasihomogeneous hull of an open set is a union of open sets the sets

$$(3.23) \qquad L_m := X_{m+1} \setminus (\overset{\circ}{X}_m)_M, \qquad m \in \mathbb{N},$$

are compact subsets of X. We set $L := \bigcup_{m\in\mathbb{N}} L_m$. Then

$$(3.24) \qquad L_M = \bigcup_{m\in\mathbb{N}} (L_m)_M \supset \bigcup_{m\in\mathbb{N}} (X_{m+1})_M \setminus (X_m)_M = \bigcup_{m\in\mathbb{N}} (X_{m+1})_M = X_M = X.$$

To show that L is an M-bounded subset of X we fix a compact subset K of X, choose $m \in \mathbb{N}$ such that $K \subset X_m$ and observe that for every $t \in]0,+\infty[$ we have

$$M_t(K) \cap L \subset M_t(X_m) \cap (X_m)_M \cap L = M_t(X_m) \cap \bigcup_{k=1}^{m} L_k.$$

Since $\bigcup_{k=1}^{m} L_k$ is compact it follows from (a) that the set (3.2) is contained in a compact subinterval of $]0,+\infty[$.

(d)⇒(a): Let Z and K be compact subsets of X. Choose an M-bounded subset L of X such that $K \subset L_M$. For every $Y \in \{Z,K\}$ set $J_Y := \{s \in]0,+\infty[\,;\ M_s(Y) \cap L \neq \emptyset\}$. Let $t \in]0,+\infty[$ satisfy $M_t(K) \cap Z \neq \emptyset$, and choose $k \in K$ such that $z := M_t k \in Z$. Since $K_M \subset L_M$ we find $s \in]0,+\infty[$ such that $M_s z \in L$. Then 1.° $s \in J_Z$ and 2.° $M_{st}k \in L$, i.e. $st \in J_K$. Consequently, $t \in J := \{r/s\,;\ r \in J_K,\ s \in J_Z\}$. Since J_K and J_Z are relatively compact subsets of $]0,+\infty[$ so is J. Hence Z is an M-bounded subset of X.

Since the implications "*(e)⇒(d)*" and "*(a)⇒(b)*" are trivial the proof is complete. ∎

The following proposition gives a fairly explicit description of locally M-bounded open sets. We divide it into parts A and B.

Proposition 3.34.A. *Suppose that $\sigma_0 = \emptyset$. Then the following conditions are equivalent:*

(a) X is locally M-bounded;

(b) there exists a continuous (resp. real analytic) function $x : X \longrightarrow]0,+\infty[$ which is quasihomogeneous of degree 1;

(c) $X = X_+$ or $X = X_-$.

Proof. *(a)⇒(c):* If (c) does not hold then we find $y_- \in X \setminus X_+ \subset G_M(\sigma_-)$ and

$y_+ \in X \setminus X_- \subset G_M(\sigma_+)$. Since X is open we can choose a compact neighbourhood U of $\{y_+, y_-\}$ which is contained in X. By Lemma 3.32 we find a compact subset K of $U_M \subset X$ such that $M_t(K) \cap U \neq \emptyset$ for every $t \in]0,+\infty[$. This shows that U is not an M-bounded subset of X.

(c) ⇒ (b): see Example 1.68.(ii) and Proposition 1.70.

(b) ⇒ (a): Let K be a compact subset of X. If $\varkappa$ is continuous then $\varkappa(K)$ is a compact subset of $]0,+\infty[$. Hence it follows by Lemma 2.29 applied to K instead of L that K is an M-bounded subset of X. ■

In order to handle the case "$\sigma_0 \neq \emptyset$" we introduce

Notation 3.35. For every $y \in G_M(\sigma_0)$ we set $X_y := \{x \in X;\ P_0(x) = y\}$.

Proposition 3.34.B. *Suppose that $\sigma_0 \neq \emptyset$. Then X is locally M-bounded if and only if for arbitrary $y, z \in G_M(\sigma_0)$ such that z is M-connected to y (see Definition 1.81) we have*

(3.25) $\quad$ (a) $X_y \cup X_z \subset V_+$; or (b) $X_y \cup X_z \subset V_-$; or (c) $X_y \subset V_+ \cap V_-$; or (d) $X_z \subset V_+ \cap V_-$.

Proof. First of all we convince ourselves that (3.25) can be replaced by

(3.25)' $\quad X_y \subset V_+$, or $X_z \subset V_-$.

In fact, (3.25) means that (3.25)' is valid for both (y,z) and (z,y) so that our claim follows in view of Corollary 1.83.(i).

"⇒": Let y,z be points in $G_M(\sigma_0)$ such that z is M-connected to y and such that (3.25)' is violated. Then we find points $x_- \in X_y \setminus V_+$ and $x_+ \in X_z \setminus V_-$. We set $y_\pm := P'(x_\pm)$ and $M' := M_{V'}$ where $P' := P_+ + P_-$ and $V' := G_M(\sigma_+ \cup \sigma_-)$. By the choice of $x_\pm$ we have $y_\pm \in G_{M'}(\sigma'_\pm)$. We choose a compact neighbourhood $U'_\pm$ of $y_\pm$ in V' and for every $w \in \{y,z\}$ a compact neigbourhood U^o_w of w in $G_M(\sigma_0)$ such that $K := U'_- + U^o_y$ and $L := U'_+ + U^o_z$ are contained in X. Let (t_m) and (y_m) be as in Definition 1.81.(i) such that (1.97) is valid with x replaced by z. Then by Lemma 3.3 there is $N \in \mathbb{N}$ such that for every $m \geq N$ we can find $k'_m \in U'_-$ satisfying $M_{t_m} k'_m \in U'_+$. By making N larger if necessary we achieve that $y_m \in U^o_y$ and $M_{t_m} y_m \in U^o_z$ if $m \geq N$

This means that $M_{t_m}(k'_m+y_m) \in L$ for $m \geq N$, i.e. the sequence $(t_m)_{m\geq N}$ is contained in the set (3.2) so that L is not an $(M,[1,+\infty[)$-bounded subset of X, and X is not locally M-bounded by Proposition 3.31.

"$\Leftarrow$": Suppose otherwise. Then we find compact subsets K and L of X and a sequence $(t_m)_{m\in\mathbb{N}}$ in $]0,+\infty[$ converging to $+\infty$ as $m\to\infty$ such that $M_{t_m}(K)\cap L \neq \emptyset$. For every $m\in\mathbb{N}$ we fix $k_m\in K$ such $\ell_m := M_{t_m}k_m \in L$. Since K and L are compact, by choosing subsequences if necessary we achieve that (k_m) converges to some $k\in K$ and (ℓ_m) to some $\ell\in L$ as $m\to\infty$. In particular, the point $z := P_0(\ell)$ is M-connected to $y := P_0(k)$. Moreover, in view of Remark 1.8.(i) we conclude that the equations (3.21) are valid. Consequently, k belongs to $X_y\backslash V_+$ and ℓ to $X_z\backslash V_-$, i.e. (3.25)' is violated. ■

If $V_- = \emptyset$ or $V_+ = \emptyset$ the assertion of Proposition 3.34.B becomes much simpler:

Corollary 3.36. *Suppose that $\sigma_- = \emptyset$ or $\sigma_+ = \emptyset$. Then X is locally M-bounded if and only if X does not intersect the set*

$$(3.26) \qquad \mathscr{C}_M(X) := \{z \in P_0(X)\,;\ z \text{ is } M\text{-connected to some } y\in P_0(X)\}. \quad ■$$

Special assumptions on $G_M(\sigma_0)$ lead to other simplifications of the condition of Proposition 3.34.B. For example, if $\mathscr{C}_M(X) \subset E_M(\sigma_0)$ or, more generally, if the asumption of Corollary 1.83.(iii) holds then the following consequence of Proposition 3.34.B applies.

Corollary 3.37. *Suppose that $\sigma_0 \neq \emptyset$ and that M-connectedness defines an equivalence relation on the set $\mathscr{C}_M(X)$ defined in (3.26). Then X is locally M-bounded if and only if for every M-connectedness equivalence class $R\subset \mathscr{C}_M(X)$ the open quasihomogeneous subset $X'_R := \{P'(x)\,;\ x\in X,\ P_0(x)\in R\}$ of the vector space $V' := G_M(\sigma_+\cup\sigma_-)$ is locally M'-bounded where $P' := P_+ + P_-$ and $M' := M_{V'}$.*

Proof. Let R be such an equivalence class. First of all we observe directly from Definition 1.81.(i) that R is quasihomogeneous so that X'_R is quasihomogeneous (of type M and hence of type M'), indeed. By Proposition 3.34.A the set X'_R is locally M'-bounded if and only if

(3.27) $X'_R \subset V_+$ or $X'_R \subset V_-$.

Now, obviously the condition (3.27) implies that (3.25)' is valid for arbitrary $y, z \in R$. Conversely, suppose that the latter is the case. Then if $X'_R \not\subset V_+$ then we choose $y \in R$ such that $X_y \not\subset V_+$ so that (3.25)' implies that $X_z \subset V_-$ for every $z \in R$, i.e.

$$X'_R = \bigcup_{z \in R} P'(X_z) \subset V_- .$$

It follows that X'_R is locally M'-bounded if and only if (3.25) is satisfied for arbitrary $y, z \in R$. Since $X_y = \emptyset$ for every $y \in \mathfrak{C}_M \setminus \mathfrak{C}_M(X)$ the assertion follows by Proposition 3.34.B. ■

Example 3.38. *Let $p \in \mathbb{R}^3$ such that $p_1 > 0$, $p_2 < 0$, and $p_3 = 0$. We define*

$$X := \bigcup_{\alpha \in \mathbb{R}} R_\alpha(Q) \times \{\alpha\} \subset \mathbb{R}^3$$

where $Q :=]0, +\infty[^2$ and where $R_\alpha \in GL(2;\mathbb{R})$ denotes the rotation by the angle α. Then X is open, connected, quasihomogeneous of type p, and locally p-bounded, but $X \neq X_\pm$. ■

Now we come to the main result on locally M-bounded open sets. In particular, it includes the analogue of the condition (b) in Proposition 3.34.A for the case "$\sigma_0 \neq \emptyset$".

Theorem 3.39. *The following conditions are equivalent:*

(a) X is locally M-bounded;

(b) there is a continuous (resp. C^∞) function $\varkappa : X \longrightarrow]0, +\infty[$ which is quasihomogeneous of degree 1;

(c) there exists a (non-negative) function $\psi \in C^\infty_M(X)$ such that $\psi_0 \equiv 1$;

(c)' for every family $\mathcal{U}$ of open subsets of X such that $(U_M)_{U \in \mathcal{U}}$ is a locally finite covering of X there exists a family of non-negative functions $\psi_U \in C^\infty_M(X)$, $U \in \mathcal{U}$, such that

(3.28) $$\operatorname{supp} \psi_U \subset U, \qquad U \in \mathcal{U},$$

and

(3.29) $$\sum_{U \in \mathcal{U}} (\psi_U)_0 \equiv 1 ;$$

(c)" for every compact subset K of X there exists $\psi_K \in C^o_M(X)$ such that $(\psi_K)_o \neq 0$ on K.

The first step in the proof is the construction of a suitable covering of X. Since it is used in section (d) below, as well, we formulate it as a separate lemma.

Lemma 3.40. *Suppose that X is locally M-bounded. Let $\mathfrak{U}$ be a family of open subsets of X such that $(U_M)_{M\in\mathfrak{U}}$ covers X. Then there exist sequences $(K_m)_{m\in\mathbb{N}}$ of compact subsets of X, $(V_m)_{m\in\mathbb{N}}$ of relatively compact open subsets of X, and $(U_m)_{m\in\mathbb{N}}$ in $\mathfrak{U}$ having the following properties:*

(3.30) $$K_m \subset V_m \subset \overline{V_m} \subset U_m\,, \qquad m\in\mathbb{N}\,;$$

(3.31) $$\bigcup_{m\in\mathbb{N}} (K_m)_M = X\,;$$

and

(3.32) *for every $m\in\mathbb{N}$ only finitely many of the $(V_j)_M$, $j\in\mathbb{N}$, intersect $(V_m)_M$.*

Proof. First of all we replace $\mathfrak{U}$ by an open covering $\mathfrak{W}$ of $Y := \bigcup\mathfrak{U}$ such that for every $W\in\mathfrak{W}$ there is $U\in\mathfrak{U}$ satisfying $\overline{W}\subset U$. Then we fix a sequence of compact subsets X_m, $m\in\mathbb{N}$, satisfying (3.22). For every $m\in\mathbb{N}$ we define L_m by (3.23). For every $y\in Y$ we choose $W_y\in\mathfrak{W}$ and $\varepsilon_y>0$ satisfying $\overline{K}(y,2\varepsilon_y)\subset W_y$. Since $\bigcup_{W\in\mathfrak{W}} W_M = X$ and since L_m is compact we find a finite subset A_m of $]0,+\infty[\times Y$ such that

(3.33) $$L_m \subset \bigcup_{(t,y)\in A_m} M_{1/t}(K(y,\varepsilon_y))\,.$$

For $a=(t,y)\in A_m$ we set $K_{m,a} := \overline{K}(y,\varepsilon_y)\cap M_t(L_m)$ and

$$V_{m,a} := K(y,2\varepsilon_y)\cap\Big(M_t(\overset{o}{X}_{m+2})\setminus(X_{m-1})_M\Big)\,.$$

By the assumption and by Lemma 3.9 it follows that $(X_{m-1})_M$ is a closed subset of X. Hence $V_{m,a}$ is an open neighbourhood of the compact subset $K_{m,a}$. Since $(K_{m,a})_M = (M_{1/t}(\overline{K}(y,\varepsilon_y))\cap L_m)_M$ it follows from (3.33) and (3.24) that

$$\bigcup_{m\in\mathbb{N}}\ \bigcup_{a\in A_m} (K_{m,a})_M \supset \bigcup_{m\in\mathbb{N}} (L_m)_M = X\,.$$

Since $(V_{m,a})_M \subset Z_m := (\overset{o}{X}_{m+2})_M\setminus(X_{m-1})_M$, since $\{k\in\mathbb{N};\ Z_k\cap Z_m\neq\emptyset\}$ is contained in the set $\{k\in\mathbb{N};\ m-2\le k\le m+2\}$, and since the sets A_m, $m\in\mathbb{N}$, are finite it follows that every set $(V_{m,a})_M$ intersects with finitely many of the sets $(V_{k,b})_M$, $k \in \mathbb{N}$ and $b \in A_k$, only. Since the index set $\bigcup_{m\in\mathbb{N}} \{m\} \times A_m$ is countable we can replace it by $\mathbb{N}$ and arrive at the desired assertion. ■

Proof of Theorem 3.39. *(a)⇒(c)'*: We choose K_m, V_m, and U_m as in Lemma 3.40. Then we fix $\varphi_m \in C_0^\infty(V_m)$ such that $\varphi_m \geq 0$ and $\varphi_m \equiv 1$ on K_m and for every $U \in \mathfrak{U}$ set

$$\chi_U := \sum_{U_k = U} \varphi_k .$$

Since by (3.32) for every $m \in \mathbb{N}$ the set $I_{U,m} := \{ k \in \mathbb{N} ;\ (V_k)_M \cap (V_m)_M \neq \emptyset ,\ U_k = U\}$ is finite it follows that for every $m \in \mathbb{N}$ the sum is finite on $(V_m)_M$ so that χ_U is a well-defined C^∞ function on X satisfying

$$\operatorname{supp} \chi_U \cap (V_m)_M \subset \bigcup_{k \in I_{U,m}} \operatorname{supp} \varphi_k .$$

Since the right-hand side is compact the condition (a) implies that it is an M-bounded subset of X. Hence Remark 3.28.(ii) (applied to the covering consisting of the sets $(V_m)_M$) implies that $\operatorname{supp} \chi_U$ is an M-bounded subset of X. Moreover, from (3.30) it follows that $\operatorname{supp} \chi_U \subset U$. Finally, for every $k \in \mathbb{N}$ satisfying $U_k = U$ we have $\chi_U \geq \varphi_k \geq 0$ so that $(\chi_U)_o \geq (\varphi_k)_o$ which is positive on $(K_k)_M$. Consequently,

$$q := \sum_{U \in \mathfrak{U}} (\chi_U)_o \quad \text{is positive on} \quad \bigcup_{k \in \mathbb{N}} (K_k)_M = X \quad (\text{see } (3.31)).$$

Since by Proposition 3.10 $\operatorname{supp}(\chi_U)_o$ is contained in U_M and since $(U_M)_{U\in\mathfrak{U}}$ is locally finite q is a well-defined C^∞ function on X which is quasihomogeneous of degree 0. Hence in view of Corollary 3.5 the functions $\psi_U := \chi_U / q$ have the desired properties.

(c)'⇒(c): we apply (c)' to $\mathfrak{U} = \{X\}$. *(c)⇒(c)"*: is trivial.

(c)"⇒(a): If $(\psi_K)_o \neq 0$ on K then $(\operatorname{supp} \psi_K)_M \supset K$. Hence the assertion follows by the implication "(d)⇒(a)" in Proposition 3.31.

(c)⇒(b): Let ψ be as in (c). Then by Proposition 3.13 $\varkappa := \psi_1$ is a C^∞ function which is quasihomogeneous of degree 1. If ψ is non-negative then $\varkappa$ is positive on the set $(\psi^{-1}(]0,+\infty[))_M$ which in view of the condition "$\psi_o \equiv 1$" coincides with X.

To show that we may assume that ψ is non-negative we first observe that ψ can be replaced by $\mathrm{Re}\,\psi$ so that we may suppose that ψ is real-valued. Then we set $\chi := \psi^2$. Since the condition "$\psi_O \equiv 1$" implies that χ_O is positive everywhere the function $\Psi := \chi/\chi_O$ is non-negative and belongs to $C^\infty_M(X)$. Moreover, since by Proposition 3.13 χ_O is quasihomogeneous of degree 0 Corollary 3.5 implies that $\Psi_O \equiv 1$, indeed.

(b)⇒(a) : since $\varkappa$ is continuous the set $\varkappa(K)$ is compact in $]0,+\infty[$ for every compact subset K of X. Hence the assertion follows by Lemma 3.29. ■

In view of Notation 1.65 and Lemma 1.64 the following assertion is a consequence of Theorem 3.39.

Corollary 3.41. *Every point in $\Omega(M)$ possesses a quasihomogeneous open neighbourhood which is locally M-bounded.* ■

We are now ready to prove the following supplement to Proposition 3.31.

Proposition 3.42. *Suppose that I is a non-compact closed subset of $]0,+\infty[$. Then X is locally M-bounded if and only if every point of X is (M,I)-bounded in X and every (M,I)-bounded subset of X has an open neighbourhood which is an (M,I)-bounded subset of X, as well.*

Proof. "⇐": The assumptions mean that every point of X has an (M,I)-bounded open neighbourhood. Since every compact subset of X is covered by finitely many of such neighbourhoods it is (M,I)-bounded as finite unions of (M,I)-bounded subsets of X are (M,I)-bounded.

"⇒": We choose a continuous function $\varkappa$ as in Theorem 3.39.(b) and by Lemma 3.40 fix a sequence of relatively compact open subsets V_m of X satisfying (3.32) and

$$(3.34) \qquad \bigcup_{m\in\mathbb{N}} (V_m)_M = X .$$

Now, let L be an (M,I)-bounded subset of X. For every $m\in\mathbb{N}$ the set $I_m :=]1/2,2[\cdot\varkappa(L\cap(\overline{V_m})_M)$ is an open neighbourhood of $\varkappa(L\cap(\overline{V_m})_M)$ in $]0,+\infty[$. By Lemma 3.29 and in view of $I\cap JI_m \subset I\cap\big(J\cdot[1/2,2]\cdot\varkappa(L\cap(\overline{V_m})_M)\big)$ the set $I\cap JI_m$ is a relatively compact subset of I for every compact subset J of $]0,+\infty[$. We set

$$U := \bigcup_{m\in\mathbb{N}} \varkappa^{-1}(I_m) \cap (V_m)_M .$$

Let K be any compact subset of X. Then one finds a finite subset A of $\mathbb{N}$ such that

$$K \subset \bigcup_{\ell\in A} (V_\ell)_M .$$

Moreover, let J be a compact subset of $]0,+\infty[$. Then

$$I \cap J\varkappa(U \cap K_M) \subset \bigcup_{\ell\in A} I \cap J\varkappa(U \cap (V_\ell)_M) \subset$$

$$\subset \bigcup_{\ell\in A} \bigcup_{m\in B_\ell} I \cap J\varkappa(\varkappa^{-1}(I_m) \cap (V_m)_M \cap (V_\ell)_M) \subset \bigcup_{\ell\in A} \bigcup_{m\in B_\ell} I \cap J I_m$$

where $B_\ell := \{m\in\mathbb{N};\ (V_m)_M \cap (V_\ell)_M \neq \emptyset\}$. Since by (3.32) the sets B_ℓ, $\ell\in\mathbb{N}$, are finite it follows that $I \cap J\varkappa(U \cap K_M)$ is relatively compact in I. Hence, by Lemma 3.29 U is an (M,I)-bounded subset of X. Since U contains the set $\bigcup_{m\in\mathbb{N}} L \cap (V_m)_M$ which is equal to L by (3.34) we see that U is the desired open neighbourhood of L. ∎

In the proof of the implication "(c)⇒(b)" of Theorem 3.39 the function $\varkappa$ is explicitly constructed out of ψ. Here is a different way to do this:

Remark 3.43. *Let $\psi \in C^r_M(X)$ be such that $\psi_0 \equiv 1$. Then by*

(3.35) $$\varkappa := \exp\circ(-\psi_{0,\log})$$

a nowhere vanishing C^r function $\varkappa : X \longrightarrow \mathbb{C}$ is defined which is quasihomogeneous of degree 1; $\varkappa$ is positive everywhere if ψ is real-valued.

Proof. That $\varkappa$ is a C^r function follows from Proposition 3.3.(i). As a special case of (3.11) we note that

(3.36) $$\psi_{0,\log} \circ M_t = \psi_{0,\log} - \log t\, \psi_0 , \qquad t \in]0,+\infty[$$

Since $\psi_0 \equiv 1$ it follows by Remark 1.52.(i) that $\varkappa$ is quasihomogeneous of degree 1 indeed. ∎

The preceding relation between $\varkappa$ and ψ will be relevant in Chapter 8 below. Here we use it in combination with Theorem 3.39 to derive the following supplement to Lemma 3.29.

Lemma 3.44. *Suppose that X is locally M-bounded. Let I be a non-compact proper closed subinterval of $]0,+\infty[$, and let L be a subset of X. Then L is an (M,I)-bounded subset of X if and only if there is a C^∞ function $x: X \longrightarrow]0,+\infty[$ which is quasihomogeneous of degree 1 such that $x(L) \cap I$ is a relatively compact subset of $]0,+\infty[$.*

A part of the proof is formulated as

Lemma 3.45. *Suppose that X is locally M-bounded. Let I be a non-compact closed subinterval of $]0,+\infty[$, and let L be an (M,I)-bounded subset of X. Then there exists an M-bounded open subset Y of X such that $Y_M = X$ and*

(3.37) $$\{t \in]0,+\infty[;\ M_t(L) \cap Y \neq \emptyset\} \subset I .$$

<u>Proof.</u> By Lemma 3.40 we choose a sequence $(V_m)_{m\in\mathbb{N}}$ as in the proof of the implication "$\Rightarrow$" in Proposition 3.42.

<u>Case 1</u>: $I = [a,+\infty[$ for some $a > 0$. Since L is $(M,[1/a,+\infty[)$-bounded, as well, and since for every $m \in \mathbb{N}$ $\overline{V_m}$ is a compact subset of X we find $\varepsilon_m \in]0,1]$ such that $\{s \in]0,a];\ M_s(L) \cap \overline{V_m} \neq \emptyset\} \subset [\varepsilon_m a, a]$. It follows that

(3.38) $$\{t \in]0,+\infty[;\ M_t(L) \cap Y_m \neq \emptyset\} \subset [a,+\infty[$$

where $Y_m := M_{1/\varepsilon_m}(V_m)$, $m \in \mathbb{N}$. Since X is locally M-bounded the sets Y_m are (M,I)-bounded open subsets of X. Since in view of (3.32) and (3.34) the covering consisting of the sets $(V_m)_M$ satisfies the requirements of Remark 3.28.(i) it follows that $Y := \bigcup_{m\in\mathbb{N}} Y_m$ is an M-bounded subset of X. By (3.34) and (3.38) Y has the desired properties.

<u>Case 2</u>: $I =]0,b]$ for some $b \in]0,+\infty[$. Since L is $(M,]0,1/b])$-bounded, as well, one finds $\varepsilon_m \in]0,1]$ such that $\{s \in [b,+\infty[;\ M_s(L) \cap \overline{V_m} \neq \emptyset\} \subset [b, b/\varepsilon_m]$. Consequently,

(3.38)' $$\{t \in]0,+\infty[;\ M_t(L) \cap Y_m \neq \emptyset\} \subset]0,b]$$

where $Y_m := M_{\varepsilon_m}(V_m)$, $m \in \mathbb{N}$. Similarly as in Case 1 it follows that the set $Y := \bigcup_{m\in\mathbb{N}} Y_m$ has the desired properties. ∎

<u>Proof of Lemma 3.44</u>. The implication "$\Leftarrow$" is a special case of Lemma 3.29. To

prove "$\Rightarrow$" we choose Y as in Lemma 3.45, and by Theorem 3.39 we fix $\psi \in C^\infty(X)$ with support contained in Y such that $\psi \geq 0$ and $\psi_0 \equiv 1$. Then we define $\varkappa$ by (3.35). In view of what was said at the beginning of section (b) we may assume that $I =]0,1]$ or $I = [1,+\infty[$. If $I =]0,1]$ (resp. $[1,+\infty[$) then by (3.37) it follows that $-\psi_{0,\log} \geq 0$ (resp. ≤ 0) on L so that $\varkappa \geq 1$ (resp. ≤ 1) on L. In view of Remark 3.43 the proof is complete. ■

(d) Describing Quasihomogeneous Functions on X as Quasihomogeneous Averages

Throughout this section we suppose that X is locally M-bounded so that each of the conditions of Theorem 3.39 is valid. Then families of functions ψ_K, as required in Proposition 3.14, exist. In fact, one can find a single non-negative function $\psi \in C_M^\infty(X)$ such that $\psi_0 \equiv 1$ on the whole of X. Moreover, choosing any family $\mathfrak{U}$ of open subsets of X such that the sets U_M, $U \in \mathfrak{U}$, constitute a locally finite covering of X one can fix a family $(\psi_U)_{U \in \mathfrak{U}}$ in $C^\infty(X)$ having the properties of condition (c)' in Theorem 3.39. With these data one obtains the following variant of Proposition 3.14.

Proposition 3.46. *Let $q \in C^0(X)$. Then q is quasihomogeneous of degree m if and only if*

$$q = \sum_{U \in \mathfrak{U}} (\psi_U \, q)_m .$$

Here the sum is locally finite since $\operatorname{supp} (\psi_U q)_m \subset U_M$ for every $U \in \mathfrak{U}$. ■

An immediate consequence is

Theorem 3.47. *Let $r \in \mathbb{N}_0 \cup \{\infty\}$, and let $q \in C^r(X)$ be quasihomogeneous of degree m. Then there exists a family $(f_U)_{U \in \mathfrak{U}}$ in $C_M^r(X)$ such that $\operatorname{supp} f_U \subset U$ for every $U \in \mathfrak{U}$ and*

$$q = \sum_{U \in \mathfrak{U}} (f_U)_m . \quad ■$$

The following result shows that finitely many (almost) quasihomogeneous functions of pairwise different degrees can be described as quasihomogeneous averages of a single function.

Theorem 3.48. *Let Λ be a finite subset of $\mathbb{C}$, and let $(j_m)_{m\in\Lambda}$ be a family in $\mathbb{N}_0$. For every $m\in\Lambda$ let functions $q_{m,0},\dots,q_{m,j_m}\in C^r(X)$ be given. Moreover, let Z be a closed quasihomogeneous subset of X such that*

$$\text{(3.39)}\qquad \operatorname{supp} q_{m,j}\subset Z,\qquad m\in\Lambda \text{ and } j\in\mathbb{N}_{j_m}\cup\{0\}.$$

Finally, let Y be an open subset of X such that $Y_M=X$. Then the following conditions are equivalent:

(a) there exists $f\in C^r_M(X)$ with support contained in $Y\cap Z$ such that

$$\text{(3.40)}\qquad f_{m,\omega_j}=q_{m,j},\qquad m\in\Lambda,\ j\in\mathbb{N}_{j_m}\cup\{0\};$$

(b) for every $m\in\Lambda$ $q_{m,0}$ is quasihomogeneous of degree m and

$$\text{(3.41)}\qquad t^{-m}q_{m,j}\circ M_t=\sum_{i=0}^{j}(-1)^i\,\omega_i(t)\,q_{m,j-i}\,,\qquad t\in]0,+\infty[,\ j\in\mathbb{N}_{j_m}.$$

Proof. *(a)* $\Rightarrow$ *(b)*: this is valid by Proposition 3.13.

(b) $\Rightarrow$ *(a)*: We fix a C^∞ function $\varkappa$ as in condition (b) of Theorem 3.39 and proceed in two steps.

Step 1: we suppose first that $Y=\varkappa^{-1}(I)$ for some relatively compact open subset I of $]0,+\infty[$. We define $w_{m,j}:]0,+\infty[\longrightarrow\mathbb{C}$ by $t\mapsto t^{-m}\omega_j(t)$ and fix $\chi\in C_0^\infty(I)$ such that $\chi\neq 0$. It follows from (1.42) that the functions $\chi_{m,j}:=\chi w_{m,j}$, $m\in\mathbb{C}$ and $j\in\mathbb{N}_0$, are linearly independent over $\mathbb{C}$. Let H be the subspace of $C_0^\infty(I)$ generated by the functions $\chi_{m,j}$, $0\le j\le j_m$ and $m\in\Lambda$. Equipped with the scalar product defined by

$$\langle\varphi,\psi\rangle:=\int_0^{+\infty}\varphi(t)\,\overline{\psi(t)}\,\frac{dt}{t}$$

H becomes a Hilbert space. By $(\beta^{m,j})$ we denote the basis of the dual space H^* which is dual to the basis $(\chi_{m,j})$ of H. Then by the Riesz lemma we can choose $\psi_{m,j}\in H$ such that $\langle\cdot,\psi_{m,j}\rangle=\beta^{m,j}$. This means that the functions $\varphi_{m,j}:=\chi\,\overline{\psi_{m,j}}$, belonging to $C_0^\infty(I)$, satisfy

(3.42) $$\int_0^{+\infty} t^{-m}\varphi_{z,\ell}(t)\,\omega_j(t)\frac{dt}{t} = \delta_{mz}\delta_{j\ell}, \qquad z,m\in\Lambda,\ \ell\in\mathbb{N}_{j_z}\cup\{0\},\ j\in\mathbb{N}_{j_m}\cup\{0\}.$$

Now we can define

(3.43) $$f := \sum_{z\in\Lambda}\sum_{\ell=0}^{j_z} q_{z,\ell}\circ\rho_\varkappa\ \varphi_{z,\ell}\circ\varkappa$$

where the C^∞ function $\rho_\varkappa$ is defined in Notation 1.71. Then f is a C^r function on X with support contained in $Z_1\cap Z_2$ where $Z_1 := \rho_\varkappa^{-1}(Z) = Z_M = Z$ and $Z_2 := \varkappa^{-1}(\operatorname{supp}\chi) \subset \varkappa^{-1}(I) = Y$. Since by Lemma 3.29 $\varkappa^{-1}(I)$ is an M-bounded subset of X f belongs to $C^r_M(X)$. To prove (3.40) we fix $x\in S^\varkappa$. Then $\rho_\varkappa(M_t x) = x$ and $\varkappa(M_t x) = t$, $t\in\,]0,+\infty[$, so that

$$f_{m,\omega_j}(x) = \sum_{z\in\Lambda}\sum_{\ell=0}^{j_z}\int_0^{+\infty} t^{-m} q_{z,\ell}(x)\,\varphi_{z,\ell}(t)\,\omega_j(t)\frac{dt}{t}, \qquad m\in\Lambda,\ 0\le j\le j_m.$$

By (3.42) this coincides with $q_{m,j}(x)$, i.e. (3.40) holds on $S^\varkappa$. By (3.11) and (b) the equality (3.40) is then valid at the points $M_t x$, $x\in S^\varkappa$ and $t\in\,]0,+\infty[$.

Step 2: the general case. Let $x\in Y$. Since by Proposition 1.72 the map $\Phi_\varkappa$ (see Notation 1.71) is a homeomorphism we find a relatively compact open subinterval J_x of $]0,+\infty[$ and an open subset V_x of $S^\varkappa$ such that $Y_x := \{M_t\vartheta\ ;\ \vartheta\in V_x,\ t\in J_x\}$ is a relatively compact open neighbourhood of x in Y. One observes that $Y_x =$ $= (Y_x)_M\cap\varkappa^{-1}(J_x)$. Applying Lemma 3.40 to the family $\mathfrak{U} := (Y_x)_{x\in Y}$ one obtains a sequence of relatively compact open subsets V_ℓ, $\ell\in\mathbb{N}$, of Y such that the sets $(V_\ell)_M$, $\ell\in\mathbb{N}$, form a locally finite open covering of X and such that for every $\ell\in\mathbb{N}$ one can fix $x_\ell\in Y$ satisfying $\overline{V_\ell}\subset Y_{x_\ell}$. We define $I_\ell := J_{x_\ell}$ and $U_\ell := (V_\ell)_M\cap\varkappa^{-1}(I_\ell)$. Then U_ℓ is open in Y_{x_ℓ} satisfying $(U_\ell)_M = (V_\ell)_M$. Consequently, $\mathfrak{U} := \{U_\ell\ ;\ \ell\in\mathbb{N}\}$ satisfies the assumptions of condition (c)' of Theorem 3.39. Hence we find a sequence of functions $\psi_\ell\in C^\infty_M(X)$ with support contained in U_ℓ such that

(3.44) $$\sum_{\ell=1}^{\infty}(\psi_\ell)_0 \equiv 1\ .$$

By the first step of the proof, for every $\ell\in\mathbb{N}$ we choose a function $f_\ell\in C^r_M(X)$ with support contained in $\varkappa^{-1}(I_\ell)$ such that (3.40) is valid for f_ℓ instead of f. We define

$$f := \sum_{\ell=1}^{\infty}(\psi_\ell)_0\, f_\ell\ .$$

Since by the definition of U_ℓ we have $(U_\ell)_M\cap\varkappa^{-1}(I_\ell) = U_\ell$ it follows by Proposition 3.10 and by the local finiteness of the family (U_ℓ) that f is a well-defined

C^r function on X with support contained in

$$\bigcup_{\ell \in \mathbb{N}} U_\ell \cap Z \subset Y \cap Z .$$

Since by Lemma 3.29 every U_ℓ is an M-bounded subset of X and since the sets $(U_\ell)_M$, $\ell \in \mathbb{N}$, form a locally finite open covering of X Remark 3.28.(ii) implies that supp f is an M-bounded subset of X, as well. Moreover, since by Proposition 3.13 the functions $(\psi_\ell)_0$ are quasihomogeneous of degree 0 it follows by Corollary 3.5 and by (3.44) that f satisfies (3.40). ■

A special case of Theorem 3.48 is the fact that every function $\varkappa$ as in Theorem 3.39.(b) is of the form (3.35):

Corollary 3.49. *Let $\varkappa : X \longrightarrow]0,+\infty[$ be a C^r function which is quasihomogeneous of degree 1. If Y is as in Theorem 3.48 then there exists a function $\psi \in C^r_M(X)$ with support contained in Y such that $\psi_0 \equiv 1$ and (3.35) holds.*

Proof. Since $-\log \varkappa(M_t x) = -\log \varkappa(x) - \log t$ one obtains the assertion by applying Theorem 3.48 to $q_{0,0} :\equiv 1$ and $q_{0,1} := -\log \varkappa$. ■

Next we are going to have a look at certain spaces of almost quasihomogeneous functions required in section 4.(e). To this end we fix $r \in \mathbb{N}_0 \cup \{\infty\}$ and $k \in \mathbb{N}_0$ and introduce

Notation 3.50. By $\mathcal{U}\mathfrak{Q}^r_{m,k}(X)$ we denote the space of all C^r functions $q : X \longrightarrow \mathbb{C}$ which are almost quasihomogeneous of degree m and of order $\leq k$ such that

(3.45) $\quad \operatorname{supp} q \subset K_M$ for some compact subset K of X.

Proposition 3.51. *(i) $\mathcal{U}\mathfrak{Q}^r_{m,k}(X) = \{ f_{m,\omega_k} ; f \in C^r_0(X) \}$.*

(ii) Let $\varkappa : X \longrightarrow]0,+\infty[$ be a C^∞ function which is quasihomogeneous of degree 1. Then, $S^\varkappa$ being a C^∞ submanifold of X, the space $C^\infty_0(S^\varkappa)$ is well-defined, and by

$$(3.46) \qquad q \mapsto \left(((\partial_M - m)^i q)|_{S^\varkappa} \right)_{0 \leq i \leq k}$$

a linear isomorphism $\mathcal{U}\mathfrak{Q}^\infty_{m,k}(X) \longrightarrow C^\infty_0(S^\varkappa)^{k+1}$ is well-defined, its inverse being given by

(3.46)' $$(g_i)_{0\le i\le k} \mapsto \sum_{i=0}^{k} \eta_i\, x^m\, g_i \circ \rho_x$$

where the η_i are defined in Remark 1.74 and ρ_x is defined in Notation 1.71.

Proof. *(i).* "$\supset$": this is a consequence of Propositions 3.13 and 3.10.

"$\subset$": By Propositions 3.31 and 3.42 we find an M-bounded open subset Y of X such that $Y_M = X$. We fix $q \in \mathfrak{A}\mathfrak{Q}^r_{m,k}(X)$. Then by Proposition 1.51.(ii) its deficiencies q_j, $j \in \mathbb{N}_k$, are C^r functions, as well. Moreover, since the equations (1.39) and $(1.39)_k$ (see Lemma 1.48) are valid for $q_0 := q$ the condition (b) in Theorem 3.48 is satisfied for $\Lambda = \{m\}$, $j_m := k$ and $q_{m,i} := (-1)^{k-i} q_{k-i}$. Consequently, by Theorem 3.48 we find a C^r function $f: X \longrightarrow \mathbb{C}$ such that $q = f_{m,\omega_k}$ and such that $\operatorname{supp} f \subset Y \cap \operatorname{supp} q_0$. Choosing a compact subset K of X satisfying (3.45) we conclude that supp f is contained in the set $Y \cap K_M$ which is relatively compact in X by Proposition 3.22.

(ii): Since by Lemma 3.29 S^x is an M-bounded subset of X it follows by Proposition 3.22 that $S^x \cap K_M$ is compact for every compact subset K of X. Consequently, the map given by (3.46) has values in $C_0^\infty(S^x)^{1+k}$. On the other hand, since $\operatorname{supp} g_i \circ \rho_x = (\operatorname{supp} g_i)_M$ it follows by Remark 1.73'.(i) that the map (3.46)' is well-defined and has values in $\mathfrak{A}\mathfrak{Q}^\infty_{m,k}(X)$. That these maps are inverse to each other is also a consequence of Remark 1.73'. ∎

The fact that in Theorem 3.48 the case $|\Lambda| \ge 2$ is admitted, as well, might be useful, for example, for recovering the quasihomogeneous parts of finite sums of quasihomogeneous functions:

Remark 3.52. *Let Λ be a finite subset of $\mathbb{C}$, and for every $m \in \Lambda$ let $q_m \in C^0(X)$ be quasihomogeneous of degree m. We set*

$$q := \sum_{m\in\Lambda} q_m .$$

If $\psi \in C^0_M(X)$ is such that $\psi_z \equiv \delta_{0z}$ for every $z \in \Lambda - \Lambda$ then $(\psi q)_m = q_m$ for every $m \in \Lambda$.

Proof. Corollary 3.5 shows that $(\psi q_m)_\ell = \psi_{\ell - m}\, q_m$ for arbitrary $m, \ell \in \Lambda$. ∎

For example, if $\sigma = \sigma_+$ and if M is semi-simple then Remark 3.52 can be applied to any polynomial function q and to $\Lambda = \{ m \in \mathfrak{A}(M) ;\ Q_m q \ne 0\}$ and $q_m := Q_m q$.

Chapter IV

Constructing (Almost) Quasihomogeneous Distributions by Taking Quasihomogeneous Averages. The Case: X is Locally M-bounded

In section (a) the method of taking quasihomogeneous averages is extended to distributions. Here it is required that the support of every test function $\varphi \in C_0^\infty(X)$ is an M-bounded subset of X. Consequently, we assume throughout the present chapter that X is locally M-bounded in the sense of Definition 3.30. A few basic facts about the singular support and the wave front set of the quasihomogeneous averages appear in section (d).

In section (b) it is established that every quasihomogeneous distribution T on X appears as the quasihomogeneous average u_m of some distribution u with M-bounded support. In fact, as with functions, any finite collection of quasihomogeneous distributions of different degrees can be written in this way by means of a single u. The corresponding result for almost quasihomogeneous distributions is valid, as well (see Theorem 4.14). A consequence of these facts is the result that for every quasihomogeneous real analytic function $q: X \longrightarrow \mathbb{C}$ the equation $qS = T$ has an (almost) quasihomogeneous solution $S \in \mathcal{D}'(X)$ for any (almost) quasihomogeneous right-hand side $T \in \mathcal{D}'(X)$ (see Theorem 4.15 below).

In section (c) the *inhomogeneous Euler equation* $(\partial_M - m)S = T$ is shown to be solvable for every $T \in \mathcal{D}'(X)$; in particular, every (almost) quasihomogeneous distribution on X appears as the deficiency of some other almost quasihomogeneous distribution.

Finally, in section (e) it will be proved that every quasihomogeneous distribution on X is of the form $\varkappa^m \rho_\varkappa^* \nu$ for some $\nu \in \mathcal{D}'(S^\varkappa)$ where $\varkappa: X \longrightarrow]0,+\infty[$ is any

C^∞ function which is quasihomogeneous of degree 1 and where $\rho_\varkappa$ is the canonical submersion from X onto $S^\varkappa$ associated with $\varkappa$ (see Notation 1.71). The proof relies on the quasihomogeneous continuations $\nu_m \in \mathcal{D}'(X)$ which are introduced in section (e), as well.

(a) Introducing the Quasihomogeneous Averages $u_{m,w}$

Before coming to the main definition we recall an elementary fact from the theory of distributions (see for example Theorem 2.2.5 in Hörmander [11]).

Remark 4.1. *Let $u \in \mathcal{D}'(X)$. Then for any open subset Y of V there exists a unique linear extension $\tilde{u}$ of $u|_{X \cap Y}$ to the space*

$$D_u(Y) := \{\psi \in C^\infty(Y);\ \operatorname{supp}\psi \cap \operatorname{supp} u \text{ is compact}\}$$

vanishing on the subspace $\{\psi \in C^\infty(Y);\ \operatorname{supp}\psi \cap \operatorname{supp} u = \emptyset\}$. The following properties hold for every $\psi \in D_u(Y)$:

(i) $\langle (\partial^\alpha u)^\sim, \psi \rangle = \langle \tilde{u}, (-\partial)^\alpha \psi \rangle, \qquad \alpha \in \mathbb{N}_o^n;$

(ii) $\langle \widetilde{fu}, \psi \rangle = \langle \tilde{u}, f\psi \rangle, \qquad f \in C^\infty(X);$

(iii) if $Y \cap \operatorname{supp} u$ is a closed subset of Y then $C_o^\infty(Y)$ is contained in $D_u(Y)$, and the restriction of $\tilde{u}$ to $C_o^\infty(Y)$ is a well-defined distribution on Y. ■

Below we shall omit the superscript and denote $\tilde{u}$ by u, as well. Moreover, we shall not specify the choice of Y; mostly, we have to deal with the case $Y = X$.

As in Chapter 3 we fix a closed subset I of $]0,+\infty[$ and a locally integrable function $w:]0,+\infty[\longrightarrow \mathbb{C}$ with support contained in I. We set

(4.1) (a) $1/I := \{1/t;\ t \in I\}$; (b) $v(t) := w^*(t) := w(1/t)$; (c) $m^* := -m-\mu$.

We observe that v is locally integrable on $]0,+\infty[$, as well, and that the support of v is contained in 1/I. To motivate the main definition we let $f \in C_I^o(X)$ and $\varphi \in C_o^o(X)$ and observe that $(\operatorname{supp} f)_{M,I} \cap \operatorname{supp}\varphi$ (by Lemma 3.9) and

$\operatorname{supp} f \cap (\operatorname{supp}\varphi)_{M,1/I}$ (by Proposition 3.22) are compact so that Lemma 3.11 implies

$$(4.2)\qquad \int_X f_{m,w}(x)\,\varphi(x)\,dx = \int_X f(x)\,\varphi_{m^*,v}(x)\,dx\ .$$

Notation 4.2. By $\mathcal{D}'_I(X)$ we denote the space of distributions $u \in \mathcal{D}'(X)$ such that $\operatorname{supp} u$ is an (M,I)-bounded subset of X. If $I =]0,+\infty[$ we also write $\mathcal{D}'_M(X)$.

We now fix $u \in \mathcal{D}'_I(X)$. Then by Propositions 3.10 and 3.22 and by Remark 4.1 it makes sense to define

$$(4.3)\qquad \langle u_{m,w}\,, \varphi\rangle := \langle u, \varphi_{m^*,v}\rangle\,, \qquad \varphi \in C_0^\infty(X).$$

Note that (4.2) now reads as

$$(4.2)'\qquad T_{f_{m,w}} = (T_f)_{m,w}\,, \qquad f \in C_I^0(X).$$

Proposition 4.3. *(i)* $u_{m,w}$ *is a well-defined distribution on* X *with support contained in* $(\operatorname{supp} u)_{M,I}$.

(ii) If $\varphi \in C^\infty_{1/I}(X)$ *is such that* $\operatorname{supp}\varphi \cap (\operatorname{supp} u)_{M,I}$ *is compact then the set* $\operatorname{supp}\varphi_{m^*,v} \cap \operatorname{supp} u$ *is compact, as well, and the defining equation in (4.3) remains valid; moreover,*

$$(4.4)\qquad \langle u_{m,w}\,, \varphi\rangle = \int_I t^{-m}\langle u\circ M_t\,, \varphi\rangle\, w(t)\,\frac{dt}{t}$$

(here the integrand on the right-hand side is a well-defined continuous function on I *with compact support).*

(iii) The function $\mathfrak{H}_{u,w} : \mathbb{C} \longrightarrow \mathcal{D}'(X)$, $m \mapsto u_{m,w}$, *is holomorphic satisfying*

$$\mathfrak{H}^{(k)}_{u,w}(m) = k!\,(-1)^k\, u_{m,w\omega_k}\,, \qquad k \in \mathbb{N}.$$

Proof. *(i)*: We set $L := \operatorname{supp} u$. Let K be a compact subset of X. Then by Proposition 3.22 $L \cap K_{M,1/I}$ is a compact subset of X. We let W be a compact neighbourhood of this set in X and fix $\chi \in C_0^\infty(W)$ such that $\chi \equiv 1$ near $L \cap K_{M,1/I}$. Since by Proposition 3.10 the support of $\varphi_{m^*,v}$ is contained in $K_{M,1/I}$ for every $\varphi \in C_0^\infty(K)$ one sees that the definition of $u_{m,w}$ amounts to

$$(4.5)\qquad \langle u_{m,w}\,, \varphi\rangle = \langle u\,, \chi\varphi_{m^*,v}\rangle\,, \qquad \varphi \in C_0^\infty(K).$$

Since K is an M-bounded subset of X there is a compact subinterval J of $]0,+\infty[$ such that $\{t \in]0,+\infty[\,;\, M_t(W) \cap K \neq \emptyset\} \subset J$. Hence for every $\varphi \in C_0^\infty(K)$ the functions $\varphi_{m^*,v}$ and $\varphi_{m^*,v}\chi_J$ coincide on W so that in view of (3.5) it follows that the linear map $C_0^\infty(K) \longrightarrow C_0^\infty(W)$, $\varphi \mapsto \chi\varphi_{m^*,v}$, is continuous. Consequently, by (4.5) $u_{m,w}$ is continuous on $C_0^\infty(K)$.

To prove the assertion about $\operatorname{supp} u_{m,w}$ we first observe that $K \cap (\operatorname{supp} u)_{M,I} = \emptyset$ if and only if $\operatorname{supp} u \cap K_{M,1/I} = \emptyset$. Since $\operatorname{supp} \varphi_{m^*,v} \subset K_{M,1/I}$ for every $\varphi \in C_0^\infty(K)$ this implies in view of Remark 4.1 that $u_{m,w}$ vanishes on $X \setminus (\operatorname{supp} u)_{M,I}$. Since by Lemma 3.9 the last set is open in X the proof of (i) is complete.

(iii). Choosing K, W, and χ as above and fixing $\varphi \in C_0^\infty(K)$ we deduce from Proposition 3.3.(ii) that the map $\mathbb{C} \longrightarrow C_0^\infty(W)$, $m \mapsto \chi\varphi_{m^*,v}$, is holomorphic. Since u maps $C_0^\infty(W)$ continuously into $\mathbb{C}$ this implies in view of (3.5) that the map $\mathbb{C} \longrightarrow \mathbb{C}$, $m \mapsto \langle u_{m,w}, \varphi \rangle$, is holomorphic. The formula about the derivatives of $\mathfrak{H}_u$ follows from

$$\langle \mathfrak{H}_{u,w}^{(k)}(m), \varphi \rangle = (-1)^k \langle u, \mathfrak{H}_{\varphi,v}^{(k)}(m^*) \rangle, \qquad k \in \mathbb{N},$$

from the corresponding formula in Proposition 3.3.(ii), and from (1.65).

(ii): We set $S := \operatorname{supp} u$ and $\Phi := \operatorname{supp} \varphi$. Then by the assumption on Φ and by Proposition 3.22 the set $\Phi_{M,1/I} \cap S = S \cap (\Phi \cap S_{M,I})_{M,1/I}$ is a compact subset of X. Since by Proposition 3.10 the support of $\varphi_{m^*,v}$ is contained in $\Phi_{M,1/I}$ the first part of the assertion is proved.

To prove the other parts of the assertion we are first going to find an open neighbourhood U of S such that U is an (M,I)-bounded subset of X and such that $K := \overline{\Phi \cap U_{M,I}}$ is a compact subset of X. Indeed, since it was proved above that $L := \Phi_{M,1/I} \cap S$ is a compact subset of X we can choose $\varepsilon > 0$ such that $L + \overline{K}(0,\varepsilon) \subset X$. Moreover, by Proposition 3.42 we find an open neighbourhood W of S which is an (M,I)-bounded subset of X. Then by Lemma 3.9 $U := (W \setminus \Phi_{M,1/I}) \cup (L + K(0,\varepsilon))$ is an open (M,I)-bounded subset of X containing S such that $L' := \overline{U \cap \Phi_{M,1/I}}$ (which is equal to $(L + \overline{K}(0,\varepsilon)) \cap \Phi_{M,1/I}$) is a compact subset of X. Since by Proposition 3.22 the set $\Phi \cap L'_{M,I}$ is a compact subset of X containing the set $\Phi \cap (U \cap \Phi_{M,1/I})_{M,I}$ which coincides with $\Phi \cap U_{M,I}$ we see that U has the desired properties, indeed.

We now choose $\chi \in C_0^\infty(X)$ equal to 1 near K. If $x \in U$ then for every $t \in J_x :=$

$\{t \in 1/I\ ;\ M_t x \in \Phi\}$ we have $M_t x \in \Phi \cap U_{M,I}$ so that $\chi(M_t x) = 1$ and

$$\varphi_{m^*,v}(x) = \int_{J_x} t^{-m^*} (\chi\varphi)(M_t x)\, v(t)\, \frac{dt}{t} = (\chi\varphi)_{m^*,v}(x),$$

i.e. $\varphi_{m^*,v}$ and $(\chi\varphi)_{m^*,v}$ coincide on U. And since $\operatorname{supp} u_{m,w}$ is contained in $U_{M,I}$ it follows in view of Remark 4.1 that

$$\langle u, \varphi_{m^*,v} \rangle = \langle u, (\chi\varphi)_{m^*,v} \rangle = \langle u_{m,w}, \chi\varphi \rangle = \langle u_{m,w}, \varphi \rangle .$$

To prove the last part of the assertion we define J to be the closure in I of the set $\{t \in I\ ;\ M_{1/t}x \in \Phi$ for some $x \in U\}$. As a subset of $\{t \in I\ ;\ M_t K \cap \overline{U} \neq \emptyset\}$ J is compact. Hence we can fix a function $\lambda \in C_0^\infty(]0,+\infty[)$ equal to 1 near J. We set

$$\psi(x,t) := t^{m^*} \lambda(t)\, (\chi\varphi)(M_{1/t}x), \qquad (x,t) \in X \times]0,+\infty[.$$

Since the support of ψ is contained in $(\operatorname{supp}\chi)_{M,1/J'} \times J'$ where $J' := \operatorname{supp}\lambda$ ψ belongs to $C_0^\infty(X \times]0,+\infty[)$.

Now we observe that by the definition of J we have $\varphi\, u \circ M_t \equiv 0$ for every $t \in I \setminus J$. Moreover, from the definition of K we deduce that $\operatorname{supp}(\varphi\, u \circ M_t) \subset K$ for every $t \in J$. Consequently,

$$\int_I t^{-m} \langle u \circ M_t, \varphi \rangle\, w(t)\, \frac{dt}{t} = \int_J t^{-m} \langle u \circ M_t, \chi\varphi \rangle\, w(t)\, \lambda(t)\, \frac{dt}{t} =$$

$$= \int_I t^{m^*} \langle u, (\chi\varphi) \circ M_{1/t} \rangle\, w(t)\, \lambda(t)\, \frac{dt}{t} = \langle u \otimes T, \psi \rangle$$

where $T \in \mathcal{D}'(]0,+\infty[)$ is defined by

$$\langle T, \eta \rangle := \int_I \eta(t)\, w(t)\, \frac{dt}{t}, \qquad \eta \in C_0^\infty(]0,+\infty[).$$

On the other hand, since for every $x \in U$ the support of the function $I \longrightarrow \mathbb{C}$, $t \mapsto (\chi\varphi)(M_{1/t}x)$, is contained in J and since $\chi_I = \lambda$ on J it follows that

$$\langle T, \psi(x,\cdot) \rangle = \int_I t^{m^*} (\chi\varphi)(M_{1/t}x)\, w(t)\, \frac{dt}{t} = (\chi\varphi)_{m^*,v}(x), \qquad x \in U,$$

so that $\langle u \otimes T, \psi \rangle = \langle u, (\chi\varphi)_{m^*,v} \rangle$. In view of the preceding equalities the proof of (4.4) is complete. ∎

Next we are going to prove the analogues of Propositions 3.4 and 3.7 and of Corollary 3.5 for distributions.

Proposition 4.4. *(i) Let $N \in \mathbb{N}_0$ and $\ell \in \mathbb{C}$, and let $P_0 : X \times V^* \longrightarrow \mathbb{C}$ be a C^∞ copolynomial function which is almost quasihomogeneous of degree ℓ, of type $M \times (-M)^*$, and of order $\leq N$. Then we have*

$$(4.6) \qquad P_0(x,\partial)\, u_{m,w} = \sum_{i=0}^{N} (-1)^i \, (P_i(x,\partial)\, u)_{m+\ell, w\omega_i}$$

and

$$(4.6)' \qquad (P_0(x,\partial)\, u)_{m+\ell,w} = \sum_{i=0}^{N} P_i(x,\partial)\, u_{m,w\omega_i}$$

where

$$(4.7) \qquad P_i := (\partial_{M\times(-M)^*} - \ell)^i \, P, \qquad i \in \mathbb{N}.$$

(ii) Let $\ell \in \mathbb{C}$ and $N \in \mathbb{N}_0$, and let $q \in C^\infty(X)$ be almost quasihomogeneous of degree ℓ and of order $\leq N$; then

$$(4.8) \qquad q\, u_{m,w} = (qu)_{m+\ell,w} + \sum_{i=1}^{N} (-1)^i \, (q_i u)_{m+\ell, w\omega_i}$$

and

$$(4.8)' \qquad (qu)_{m+\ell,w} = q\, u_{m,w} + \sum_{i=1}^{N} q_i \, u_{m,w\omega_i}$$

where

$$(4.9) \qquad q_i := (\partial_M - \ell)^i \, q\,, \qquad i \in \mathbb{N}.$$

(iii) $(u \circ A)_{m,w} = u_{m,w} \circ A$ for every $A \in GL(V,V)$ commuting with M.

(iv) $(u_{\mathfrak{G}})_{m,w} = (u_{m,w})_{\mathfrak{G}}$ if $\mathfrak{G}$ satisfies the assumptions of Remark 2.67.(ii).

Proof. Note first that $P_i(x,\partial)u$, qu, and – by Lemma 3.6.(ii) – $u_{\mathfrak{G}}$ belong to $\mathcal{D}'_I(X)$ and that by Lemma 3.6.(i) $u \circ A$ belongs to $\mathcal{D}'_I(A^{-1}(X))$, indeed.

(i): Let $\varphi \in C_0^\infty(X)$. Then by Proposition 3.4 and by (1.65) we have

$$\langle P_0(x,\partial)\, u_{m,w}, \varphi \rangle = \langle u, (P_0(x,-\partial)\, \varphi)_{m^*,v} \rangle =$$
$$= \sum_{i=0}^{N} \langle u, P_i(x,-\partial)\, \varphi_{m^*-\ell, v\omega_i} \rangle = \sum_{i=0}^{N} (-1)^i \langle (P_i(x,\partial)\, u)_{m+\ell, w\omega_i}, \varphi \rangle$$

and

$$\langle (P_0(x,\partial)\, u)_{m+\ell,w}, \varphi \rangle = \langle u, P_0(x,-\partial)\, \varphi_{m^*-\ell,v} \rangle =$$
$$= \sum_{i=0}^{N} (-1)^i \langle u, (P_i(x,-\partial)\, \varphi)_{m^*, v\omega_i} \rangle = \sum_{i=0}^{N} \langle P_i(x,\partial)\, u_{m,w\omega_i}, \varphi \rangle .$$

(ii): This is a special case of (i). It also follows from (3.8) and (3.8)'.

(iv): From Proposition 3.7.(ii) we deduce

$$\langle (u_{\mathfrak{S}})_{m,w}, \varphi \rangle = \langle u, (\varphi_{m^*,v})_{\mathfrak{S}'} \rangle = \langle u, (\varphi_{\mathfrak{S}'})_{m^*,v} \rangle = \langle (u_{m,w})_{\mathfrak{S}}, \varphi \rangle .$$

Finally, *(iii)* follows from Proposition 3.7.(i) in a similar way.

By making use of (4.4) one can prove the assertions in a perhaps more direct way without recourse to the defining equations (4.3). ■

In view of Example 1.21 we obtain as a special case of (4.6):

$$(4.10) \qquad (\partial_M - m)^j u_{m,w} = ((\partial_M - m)^j u)_{m,w}, \qquad j \in \mathbb{N}.$$

The following lemma is required for the proof of Proposition 4.9 below.

Lemma 4.5. *Suppose that I is an interval. Let $f \in C^{\infty}_{1/I}(X)$ and $u \in \mathcal{D}'_I(X)$, and assume that $\operatorname{supp} f \cap (\operatorname{supp} u)_{M,I}$ is an M-bounded subset of X. Then for every $\ell \in \mathbb{C}$ the distributions $f u_{m,w}$ and $f_{\ell-m,v} u$ belong to $\mathcal{D}'_M(X)$ satisfying*

$$(f u_{m,w})_{\ell,\omega_k} = \sum_{i=0}^{k} (f_{\ell-m,v\omega_i} u)_{\ell,\omega_{k-i}}, \qquad k \in \mathbb{N}_0 .$$

Note that by Lemma 3.25.(i) the assumption on the supports of f and u is automatically satisfied if I is non-compact.

Proof. The first part of the assertion follows by Propositions 4.3.(i) and 3.10 and by Lemma 3.25. Let $\varphi \in C^{\infty}_{0}(X)$. Then by Propositions 4.3.(i) and 3.10 we have

$$\operatorname{supp} u_{m,w} \cap \operatorname{supp}(f\varphi_{\ell^*,\omega_k}) \subset (\operatorname{supp} u)_{M,I} \cap \operatorname{supp} f \cap (\operatorname{supp} \varphi)_M$$

and

$$\operatorname{supp} u \cap \operatorname{supp}(f_{\ell-m,v\omega_i} \varphi_{\ell^*,\omega_{k-i}}) \subset \operatorname{supp} u \cap (\operatorname{supp} f)_{M,1/I} \cap (\operatorname{supp} \varphi)_M .$$

By Proposition 3.22 and Lemma 3.25 these sets are compact. Hence, taking (1.65) and Remark 4.1 into account and applying Corollary 3.5 to the function $q = \varphi_{\ell^*,\omega_k}$ (note that by Proposition 3.13 its ith order deficiency q_i is equal to $(-1)^i \varphi_{\ell^*,\omega_{k-i}}$) we deduce that

$$\langle (f u_{m,w})_{\ell,\omega_k}, \varphi \rangle = (-1)^k \langle u, (f\varphi_{\ell^*,\omega_k})_{m^*,v} \rangle =$$

$$= \sum_{i=0}^{k} (-1)^{k+i} \langle u, f_{\ell-m,v\omega_i} \varphi_{\ell^*,\omega_{k-i}} \rangle = \sum_{i=0}^{k} \langle (f_{\ell-m,v\omega_i} u)_{\ell,\omega_{k-i}}, \varphi \rangle . \quad ■$$

(b) Describing Quasihomogeneous Distributions in Terms of Quasihomogeneous Averages

For the whole section we fix a number $k \in \mathbb{N}_0$.

Notation 4.6. For any $u \in \mathcal{D}'_I(X)$ we denote u_{m,χ_I} by $u_{m,I}$; if $I =]0,+\infty[$ then we also write u_m instead of $u_{m,I}$.

Proposition 4.7. *Let $u \in \mathcal{D}'_M(X)$. Then u_{m,ω_k} is almost quasihomogeneous of degree m and of order $\leq k$. In particular, u_m is quasihomogeneous of degree m. Moreover, for every $i \in \mathbb{N}_k$ we have*

$$(4.11) \qquad (\partial_M - m)^i u_{m,\omega_k} = (-1)^i u_{m,\omega_{k-i}}.$$

Proof. Let $\varphi \in C_0^\infty(X)$. Applying (3.11) to (φ, m^*) instead of (f,m), making use of Proposition 3.7.(i) and taking (1.65) into account we conclude that for every $t \in]0,+\infty[$ we have

$$\begin{aligned}
&\langle u_{m,\omega_k} \circ M_t, \varphi \rangle = t^{-\mu} \langle u_{m,\omega_k}, \varphi \circ M_{1/t} \rangle = \\
&= (-1)^k t^{-\mu} \langle u, (\varphi \circ M_{1/t})_{m^*,\omega_k} \rangle = (-1)^k t^{-\mu} \langle u, \varphi_{m^*,\omega_k} \circ M_{1/t} \rangle = \\
&= t^{-\mu} (1/t)^{m^*} \sum_{i=0}^{k} \omega_i(1/t)\, (-1)^{k-i} \langle u, \varphi_{m^*,\omega_{k-i}} \rangle = \\
&= t^m \sum_{i=0}^{k} (-1)^i\, \omega_i(t) \langle u_{m,\omega_{k-i}}, \varphi \rangle .
\end{aligned}$$

Hence, the assertion follows by Proposition 2.31. An alternative proof can be based on (4.4). For still another alternative proof one verifies $(\partial_M - m)\, u_{m,\omega_i} = -u_{m,\omega_{i-1}}$, $i \in \mathbb{N}$, and $(\partial_M - m)\, u_m = 0$ by making use of (2.5) and Lemma 3.12. ■

We can now formulate the main theorem of this section.

Theorem 4.8. *Let $T \in \mathcal{D}'(X)$. Then the following conditions are equivalent:*

(a) *T is almost quasihomogeneous of degree m and of order $\leq k$;*

(b) *$\langle T, \varphi \rangle = 0$ for every $\varphi \in C_0^\infty(X)$ such that $\varphi_{m^*,\omega_k} \equiv 0$;*

(c) *$T = u_{m,\omega_k}$ for some $u \in \mathcal{D}'_M(X)$.*

Proof. *(a)⇒(b)*: Let $\varphi \in C_0^\infty(X)$, and set $v_k := \omega_k \chi_{]0,1]}$ and $\phi := \varphi_{m^*,v_k}$. Suppose that $\varphi_{m^*,\omega_k} \equiv 0$. This means that $\phi = -\varphi_{m^*,\omega_k}\chi_{[1,+\infty[}$. In view of Proposition 3.10 this implies that

$$\operatorname{supp}\phi \subset (\operatorname{supp}\varphi)_{M,]0,1]} \cap (\operatorname{supp}\varphi)_{M,[1,+\infty[}\,.$$

Since by Lemma 3.27 the right-hand side is a compact subset of X this means that $\phi \in C_0^\infty(X)$. Note that by Lemma 3.12 and (2.5) we have

$$\varphi = -\,{}^t(\partial_M - m)^{k+1}\phi\,.$$

Hence by (a) and Proposition 2.31.(i) we deduce that

$$\langle T,\varphi\rangle = -\langle T, {}^t(\partial_M - m)^{k+1}\phi\rangle = -\langle(\partial_M - m)^{k+1}T, \phi\rangle = 0\,.$$

(b)⇒(a): Let $\varphi \in C_0^\infty(X)$. Applying (2.5) and (3.7) we obtain

$$({}^t(\partial_M - m)^i\varphi)_{m^*,\omega_k} = (-1)^i\,(\partial_M - m^*)^i\,\varphi_{m^*,\omega_k}\,, \qquad i \in \mathbb{N}_0\,.$$

Hence, setting

$$\chi := t^{m^*}\varphi\circ M_{1/t} - \sum_{i=0}^{k}\omega_i(t)\,{}^t(\partial_M - m)^i\varphi$$

we obtain by Proposition 3.7.(i) and in view of (1.65) that

$$\chi_{m^*,\omega_k} = (1/t)^{-m^*}\varphi_{m^*,\omega_k}\circ M_{1/t} - \sum_{i=0}^{k}\omega_i(1/t)\,(\partial_M - m^*)^i\,\varphi_{m^*,\omega_k}\,.$$

Since by Proposition 3.13 φ_{m^*,ω_k} is almost quasihomogeneous of degree m^* and of order $\leq k$ the right-hand side of the preceding equation vanishes, and the condition (b) - applied to χ - implies that

$$0 = \langle T,\chi\rangle = t^{-m}\langle T\circ M_t, \varphi\rangle - \sum_{i=0}^{k}\omega_i(t)\,\langle(\partial_M - m)^i T, \varphi\rangle,$$

i.e. (a) holds.

(c)⇒(a): this follows from Proposition 4.7.

For the proof of the implication "*(a)⇒(c)*" we require the following partial generalization of Corollary 3.5 to distributions. The proof makes use of the implication "(a)⇒(b)" of Theorem 4.8.

Proposition 4.9. *Let $\ell \in \mathbb{C}$, $N \in \mathbb{N}_0$, and let $T \in \mathcal{D}'(X)$ be almost quasihomogeneous of degree $\ell \in \mathbb{C}$ and of order $\leq N$. Then setting $T_i := (\partial_M - m)^i T$ we have for* *every* $f \in C_I^\infty(X)$:

$$(4.12) \qquad f_{m,w}T = (fT)_{m+\ell,w} + \sum_{i=1}^{N} (-1)^i (fT_i)_{m+\ell,w\omega_i}$$

and

$$(4.12)' \qquad (fT)_{m+\ell,w} = f_{m,w}T + \sum_{i=1}^{N} f_{m,w\omega_i} T_i \,.$$

<u>*Proof.*</u> Let $\varphi \in C_0^\infty(X)$. For the proof of <u>*(4.12)*</u> we first observe that

$$\langle f_{m,w}T, \varphi \rangle = \langle T, \psi \rangle \quad \text{where} \quad \psi := f_{m,w}\varphi$$

and – in view of (1.65) –

$$\langle (fT)_{m+\ell,w} + \sum_{i=1}^{N} (-1)^i (fT_i)_{m+\ell,w\omega_i}, \varphi \rangle = \langle T, \chi \rangle$$

where

$$\chi := \sum_{i=0}^{N} {}^t(\partial_M - \ell)^i (f\varphi_{-m-\ell-\mu,v\omega_i}) \,.$$

Hence, by the equivalence "(a)$\Longleftrightarrow$(b)" of Theorem 4.8 it suffices for the proof of (4.12) to show that $(\psi - \chi)_{\ell^*,\omega_N} = 0$. To this end we deduce by making use of (2.5) and (3.7) and of Proposition 3.13 that

$$({}^t(\partial_M - \ell)^i (f\varphi_{-m-\ell-\mu,v\omega_i}))_{\ell^*,\omega_N} =$$

$$= (-1)^i (\partial_M - \ell^*)^i (f\varphi_{-m-\ell-\mu,v\omega_i})_{\ell^*,\omega_N} = (f\varphi_{-m-\ell-\mu,v\omega_i})_{\ell^*,\omega_{N-i}} \,.$$

Applying Lemma 4.5 to $(\ell^*, \varphi, T_f, N)$ instead of (ℓ, f, u, k) and taking (4.2)' into account we conclude that

$$\psi_{\ell^*,\omega_N} = \sum_{i=0}^{N} (f\varphi_{-\ell-\mu-m,v\omega_i})_{\ell^*,\omega_{N-i}} = \chi_{\ell^*,\omega_N} \,,$$

as desired.

<u>*(4.12)'*</u>: Note that

$$\langle (fT)_{m+\ell,w}, \varphi \rangle = \langle T, \psi \rangle \quad \text{where} \quad \psi := f\varphi_{-m-\ell-\mu,v}$$

and

$$\langle f_{m,w}T + \sum_{i=1}^{N} f_{m,w\omega_i} T_i, \varphi \rangle = \langle T, \chi \rangle$$

where

$$\chi := \sum_{i=0}^{N} {}^t(\partial_M - \ell)^i (f_{m,w\omega_i}\varphi) \,.$$

Hence, by Theorem 4.8, "(a)$\Longleftrightarrow$(b)", again, it suffices to show that $(\psi - \chi)_{\ell^*,\omega_N}$

vanishes identically. To this end we make use of (2.5) and (3.7) and of Proposition 3.13 to obtain

$$({}^t(\partial_M-\ell)^i(f_{m,w\omega_i}\varphi))_{\ell^*,\omega_N}=(-1)^i(\partial_M-\ell^*)^i(f_{m,w\omega_i}\varphi)_{\ell^*,\omega_N}=(f_{m,w\omega_i}\varphi)_{\ell^*,\omega_{N-i}}.$$

Applying Lemma 4.5 to $(\ell^*-m,\ell^*,T_\varphi,v,N)$ instead of (m,ℓ,u,w,k) and taking (4.2)' into account we conclude that

$$\psi_{\ell^*,\omega_N}=\sum_{i=0}^{N}(f_{m,w\omega_i}\varphi)_{\ell^*,\omega_{N-i}}=\chi_{\ell^*,\omega_N}. \quad \blacksquare$$

Corollary 4.10. *Let $T\in\mathcal{D}'(X)$ be almost quasihomogeneous of degree m and of order $\leq k$. Then*

$$T=\sum_{i=0}^{k}(-1)^i(\psi(\partial_M-m)^iT)_{m,\omega_i} \tag{4.13}$$

for any $\psi\in C_M^\infty(X)$ satisfying $\psi_0\equiv 1$. ■

End of the proof of Theorem 4.8 : "(a)⇒(c)". Setting $u_i:=\psi(\partial_M-m)^iT$ we deduce from (4.11) and (4.10) that

$$(u_i)_{m,\omega_i}=(-1)^{k-i}(\partial_M-m)^{k-i}(u_i)_{m,\omega_k}=(-1)^{k-i}((\partial_M-m)^{k-i}u_i)_{m,\omega_k}.$$

Hence in view of (4.13) the equation $T=u_{m,\omega_k}$ is valid for

$$u=\sum_{i=0}^{k}(-1)^k(\partial_M-m)^{k-i}(\psi(\partial_M-m)^iT). \quad \blacksquare \tag{4.14}$$

As a consequence of Corollary 4.10 we note in passing

Corollary 4.11. *Let $T\in\mathcal{D}'(X)$ be almost quasihomogeneous of degree m and of order $\leq k$. Then there exists a sequence $(q_j)_{j\in\mathbb{N}}$ in $C^\infty(X)$ converging weakly to T as $j\to\infty$ such that for every $j\in\mathbb{N}$ q_j is almost quasihomogeneous of degree m and of order $\leq k$, the support of q_j being contained in K_M for some compact subset K of X.*

Proof. Let $(f_j)_{j\in\mathbb{N}}$ be a sequence in $C_0^\infty(X)$ converging weakly to T as $j\to\infty$, let $\psi\in C_M^\infty(X)$ be such that $\psi_0\equiv 1$, and set

$$g_{i,j}:=(\psi(\partial_M-m)^if_j)_{m,\omega_i}.$$

Then in view of (4.2)' for every $i\in\mathbb{N}_k\cup\{0\}$ the sequence of distributions

$$T_{g_{i,j}} = (\psi\,(\partial_M - m)^i\, T_{f_j})_{m,\omega_i}$$

converges weakly to $(\psi\,(\partial_M - m)^i T)_{m,\omega_i}$ as $j\to\infty$. The assertion follows from Corollary 4.10 and from Proposition 3.10. ∎

Our next aim is to find more general distributions $u\in\mathcal{D}'_M(X)$ than those given by (4.14) which satisfy the equation $T = u_{m,\omega_k}$. First we are going to rewrite (4.14). Applying Lemma 2.34 and substituting $I=k-s$ we obtain

$$(\partial_M - m)^{k-i}\,(\psi(\partial_M - m)^i T) = \sum_{s=0}^{k-i} \binom{k-i}{s}\, \partial_M^s\psi\,(\partial_M - m)^{k-s}\, T =$$

$$= \sum_{I=i}^{k} \binom{k-i}{k-I}\, \partial_M^{k-I}\psi\,(\partial_M - m)^I\, T$$

so that (4.14) becomes

$$u = \sum_{I=0}^{k} c_{k,I}\,(-\partial_M)^{k-I}\psi\,(\partial_M - m)^I\, T \tag{4.14'}$$

where the coefficients $c_{k,i}$ are defined by $c_{k,0} := 1$ and by

$$c_{k,i} := (-1)^i \sum_{I=0}^{i} \binom{k-I}{k-i}, \qquad i\in\mathbb{N}_k. \tag{4.15}$$

In Proposition 4.13 below the coefficients $c_{k,i}$ appear via a property described in assertion (i) of the following lemma. The assertion (ii) of this lemma shows that the functions

$$\psi_i = c_{k,i}\,(-\partial_M)^{k-i}\psi, \qquad i\in\{0\}\cup\mathbb{N}_k, \tag{4.16}$$

appearing in (4.14)' satisfy the assumption of Proposition 4.13 below.

Lemma 4.12. *(i) If we set $c_{k,0} := 1$ then the unique solution of the system of equations*

$$\sum_{i=0}^{j} \binom{k+j-i}{k} c_{k,i} = 0, \qquad j\in\mathbb{N}_k, \tag{4.17}$$

is given by (4.15); moreover, the numbers $c_{k,i}$ satisfy

$$\sum_{i=0}^{k} \binom{k+j-i}{k} c_{k,i} = (-1)^k \binom{j-1}{k}, \qquad j\in\mathbb{N}\setminus\mathbb{N}_k \tag{4.17'}$$

(ii) If $\psi\in C_M^\infty(X)$ satisfies $\psi_0\equiv 1$ then by (4.16) a sequence $(\psi_i)_{0\le i\le k}$ in $C_M^\infty(X)$ is defined such that

$$(\psi_i)_{0,\omega_{k+j-i}} = c_{k,i}\,(\psi_0)_{0,\omega_{k+j}}, \qquad i\in\mathbb{N}_k,\ j\in\mathfrak{J} \tag{4.18}$$

is valid for $\mathfrak{J}=\mathbb{N}_0$ *and such that*

(4.19) $$(\psi_0)_{0,\omega_k}\equiv 1\,.$$

Proof. (i): We fix $j\in\mathbb{N}$ and set $m:=\min\{j,k\}$. Inserting (4.15), changing the order of summation and substituting $i:=i-I$ we obtain

$$\sum_{i=0}^{m}\binom{k+j-i}{k}c_{k,i}=\sum_{i=0}^{m}(-1)^i\binom{k+j-i}{k}\sum_{I=0}^{i}\binom{k-I}{k-i}=$$

$$=\sum_{I=0}^{m}\sum_{i=I}^{m}(-1)^i\binom{k+j-i}{k}\binom{k-I}{k-i}=\sum_{I=0}^{m}(-1)^I\Big[\sum_{i=0}^{m-I}(-1)^i\binom{k+j-i-I}{k}\binom{k-I}{i}\Big].$$

Computing the sum in square brackets by applying Lemma 1.76 to $\ell=k-I$ we conclude that this is equal to

$$\sum_{I=0}^{m}(-1)^I\binom{j}{I}.$$

Since in case "$j\le k$" by the binomial formula the last sum is equal to $(1-1)^j=0$ we see that the k-tuple $(c_{k,1},\dots,c_{k,k})$ is a solution of (4.17). Since in the equation (4.17) the coefficient of $c_{k,j}$ is equal to 1 the system of equations (4.17) can be rewritten in such a way that it gives a recursive definition of the $c_{k,i}$. This shows that the solution is unique.

Finally, for the derivation of (4.17)' making use of

$$\binom{j}{I}=\binom{j-1}{I-1}+\binom{j-1}{I},\qquad I\in\mathbb{N}_{j-1},$$

we conclude that

(4.20) $$\sum_{I=0}^{k}(-1)^{k+I}\binom{j}{I}=(-1)^k+\sum_{I=1}^{k}(-1)^{k+I}\binom{j-1}{I-1}+\sum_{I=1}^{k}(-1)^{k+I}\binom{j-1}{I}=$$

$$=(-1)^k-\sum_{I=0}^{k-1}(-1)^{k-I}\binom{j-1}{I}+\sum_{I=1}^{k}(-1)^{k-I}\binom{j-1}{I}=\binom{j-1}{k},\qquad j\in\mathbb{N}\setminus\mathbb{N}_k,$$

and (4.17)' follows by the computations above.

(ii): Since $(-\partial_M)^i\psi_i=c_{k,i}\psi_0$ it follows by (3.7) and by Proposition 3.13 that

$$(c_{k,i}\psi_0)_{0,\omega_{k+j}}=(-\partial_M)^i(\psi_i)_{0,\omega_{k+j}}=(\psi_i)_{0,\omega_{k+j-i}},\qquad j\in\mathbb{N}_0,$$

i.e. (4.18) is valid. Moreover, the case $i=k$ and $j=0$ shows that $(\psi_0)_{0,\omega_k}=(\psi)_0$ so that (4.19) is a consequence of the assumption on ψ. ∎

The following proposition provides more general distributions u satisfying $T=u_{m,\omega_k}$, thereby giving another proof of the implication "(a)⇒(c)" of Theorem 4.8.

Proposition 4.13. *Let $N \in \mathbb{N}_0$ such that $N \geq k$, and let $T \in \mathcal{D}'(X)$ be almost quasihomogeneous of degree m and of order $\leq N$. Then for any sequence $(\psi_i)_{0 \leq i \leq k}$ in $C_M^\infty(X)$ satisfying (4.18) for $\mathfrak{J} = \{0\} \cup \mathbb{N}_N$ and (4.19) we have*

$$(4.21) \qquad T = \sum_{i=0}^{k} (\psi_i (\partial_M - m)^i T)_{m,\omega_k} - (-1)^k \sum_{j=k+1}^{N} \binom{j-1}{k} (\psi_0)_{0,\omega_{k+j}} (\partial_M - m)^j T .$$

Note that in case N = k and if the ψ_i are chosen as in Lemma 4.12.(ii) then one obtains (4.21) by inserting (4.14)' into the equation $T = u_{m,\omega_k}$.

Proof. For every $i \in \mathbb{N}_k \cup \{0\}$ we apply (4.12)' to $((\partial_M - m)^i T, m, 0, \psi_i, \omega_k)$ instead of (T, ℓ, m, f, w) and obtain

$$(\psi_i (\partial_M - m)^i T)_{m,\omega_k} = \sum_{I=0}^{N-i} (\psi_i)_{0,\omega_k \omega_I} (\partial_M - m)^{i+I} T .$$

Combining this with the equation (1.38) and taking (4.18) into account we see that the first sum on the right-hand side of (4.21) is equal to

$$\sum_{i=0}^{k} \sum_{I=0}^{N-i} \binom{k+I}{k} c_{k,i} (\psi_0)_{0,\omega_{k+i+I}} (\partial_M - m)^{i+I} T$$

which via the substitution j = i+I turns out to be equal to

$$\sum_{j=0}^{N} d_{k,j} (\psi_0)_{0,\omega_{k+j}} (\partial_M - m)^j T$$

where

$$d_{k,j} := \sum_{i=0}^{\min(j,k)} \binom{k+j-i}{k} c_{k,i} .$$

Since by Lemma 4.12.(i) we have $d_{k,0} = 1$, $d_{k,j} = 0$ for $j \in \mathbb{N}_k$, and $d_{k,j} = (-1)^k \binom{j-1}{k}$ for j > k the equality (4.21) follows in view of (4.19). ∎

Proposition 4.13 enables us to prove the following generalization of Theorem 3.48 to distributions.

Theorem 4.14. *Let Λ be a finite subset of $\mathbb{C}$, let $(j_m)_{m \in \Lambda}$ be a family in $\mathbb{N}_0$, and for every $m \in \Lambda$ let $T_m \in \mathcal{D}'(X)$ be almost quasihomogeneous of degree m and of order $\leq j_m$. Then for every open subset Y of X satisfying $Y_M = X$ there exist $u \in \mathcal{D}'_M(X)$ with support contained in Y such that $u_{m,\omega_{j_m}} = T_m$ for every $m \in \Lambda$.*

Proof. Let $\varkappa : X \longrightarrow]0, +\infty[$ be a C^∞ function which is quasihomogeneous of degree 1. We fix $z \in \Lambda$, set $k := j_z$, and fix $i \in \mathbb{N}_k \cup \{0\}$, and set $J := \{(m, \ell) \in \Lambda \times \mathbb{N}_0$

$\ell \le j_m + k - i\}$. For every $(m,\ell) \in J$ we define

$$(4.22.a) \qquad q_{m,\ell} := \begin{cases} c_{k,i}\, \eta_{\ell-k+i}\, \varkappa^z & \text{in case } m = z \text{ and } \ell \ge k-i \\ 0 & \text{otherwise} \end{cases}$$

where $\eta_j := (-1)^j \omega_j \circ \varkappa$. Then by Lemma 2.34 and by Proposition 2.31 and Remark 1.74 we have $(\partial_M - m)^j q_{m,\ell} = (-1)^j q_{m,\ell-j}$ for arbitrary $(m,\ell) \in J$. In view of Proposition 2.31 this means that the condition (b) of Theorem 3.48 is satisfied. Hence, there exists $\psi_{z,i} \in C_M^\infty(X)$ with support contained in Y such that

$$(4.22.b) \qquad (\psi_{z,i})_{m,\omega_\ell} = q_{m,\ell}, \qquad (m,\ell) \in J.$$

Once in this way for arbitrary $z \in \Lambda$ and $i \in \mathbb{N}_{j_z} \cup \{0\}$ the functions $\psi_{z,i}$ are constructed the distribution

$$u := \sum_{m\in\Lambda} \varkappa^{-m} \sum_{i=0}^{j_m} \psi_{m,i} (\partial_M - m)^i T_m$$

is well-defined, its support lying in Y. In particular, u belongs to $\mathcal{D}'_M(X)$. In order to verify the other desired properties of u we fix $z \in \Lambda$, again, and use the abbreviations introduced above. Now, for arbitrary $m \in \Lambda$ and $i \in \mathbb{N}_{j_m} \cup \{0\}$ we apply (4.12)' to $((\partial_M - m)^i T_m, j_m - i, m, z-m, \varkappa^{-m}\psi_{m,i}, \omega_k)$ instead of (T, N, ℓ, m, f, w) and obtain

$$(\varkappa^{-m} \psi_{m,i} (\partial_M - m)^i T_m)_{z,\omega_k} = \sum_{I=0}^{j_m - i} (\varkappa^{-m} \psi_{m,i})_{z-m,\omega_k \omega_I} (\partial_M - m)^{i+I} T_m .$$

Applying (3.9) to the function $q = \varkappa^{-m}$ (which is quasihomogeneous of degree $-m$) and taking (1.38) into account we deduce that

$$(\varkappa^{-m} \psi_{m,i})_{z-m,\omega_k\omega_I} = \binom{k+I}{k} \varkappa^{-m} (\psi_{m,i})_{z,\omega_{k+I}} .$$

Note that by (4.22) the right-hand side vanishes in case $m \ne z$. More precisely, the condition (4.22) leads to

$$(\varkappa^{-m} \psi_{m,i} (\partial_M - m)^i T_m)_{z,\omega_k} = \delta_{mz} \sum_{I=0}^{k-i} \binom{k+I}{k} c_{k,i}\, \eta_{i+I} (\partial_M - z)^{i+I} T_z .$$

Summing over $m \in \Lambda$ and $i \in \mathbb{N}_{j_m} \cup \{0\}$ we obtain T_z by a similar argument as in the proof of Proposition 4.13 or by applying Proposition 4.13 to T_z instead of T and to the functions $\psi_i := \varkappa^{-z} \psi_{z,i}$. This shows that $u_{z,\omega_k} = T_z$. ■

The following division theorem is obtained by an application of Theorem 4.8.

Theorem 4.15. *Let $\ell \in \mathbb{C}$, and let $q \in C^\infty(X)$ be quasihomogeneous of degree ℓ.*

Suppose that for every $T \in \mathcal{D}'(X)$ the equation

(4.23) $$qS = T$$

has a solution $S \in \mathcal{D}'(X)$. Then for every $T \in \mathcal{D}'(X)$ which is almost quasihomogeneous of degree m and of order $\leq k$ the equation (4.23) has a solution $S \in \mathcal{D}'(X)$ which is almost quasihomogeneous of degree $m-\ell$ and of order $\leq k$.

Note that by the Lojasiewicz division theorem (see [14]) the assumption is satisfied if q is real analytic.

<u>*Proof*</u>. Let $T \in \mathcal{D}'(X)$ be as in the theorem. Then by Theorem 4.8 we find $u \in \mathcal{D}'_M(X)$ such that $u_{m,\omega_k} = T$. By the assumption we choose a solution $v \in \mathcal{D}'(X)$ of the equation $qv = u$. In view of Proposition 3.42 we can multiply v by a suitable cut-off function with M-bounded support and still obtain a solution of $qv = u$. Hence we may assume that v belongs to $\mathcal{D}'_M(X)$. It then follows by Propositions 4.4.(ii) and 4.7 that $S := v_{m-\ell,\omega_k}$ is the desired solution of (4.23). ■

(e) Solving the Equation $(\partial_M - m)S = T$

Let $\varkappa : X \longrightarrow]0,+\infty[$ be a continuous function which is quasihomogeneous of degree 1, and set $U_0 := \varkappa^{-1}(]0,2[)$ and $U_\infty := \varkappa^{-1}(]1,+\infty[)$. Then we choose $\chi \in C^\infty(X)$ such that $\operatorname{supp}\chi \subset U_0$ and $\operatorname{supp}(1-\chi) \subset U_\infty$. Hence, by Lemma 3.2 $\operatorname{supp}\chi$ is an $(M,[1,+\infty[)$-bounded and $\operatorname{supp}(1-\chi)$ an $(M,]0,1])$-bounded subset of X. Finally, we fix $T \in \mathcal{D}'(X)$. Then $(1-\chi)T$ belongs to $\mathcal{D}'_{]0,1]}(X)$ and χT to $\mathcal{D}'_{[1,+\infty[}(X)$ so that it makes sense to define (see Notation 4.6)

(4.24) $$\tilde{T}_m := ((1-\chi)T)_{m,]0,1]} - (\chi T)_{m,[1,+\infty[}\,.$$

Theorem 4.16. *$\tilde{T}_m$ is a well-defined distribution on X having the following properties:*

(i) $(\partial_M - m)\tilde{T}_m = T$;

(ii) *supp* $\widetilde{T}_m \subset (\operatorname{supp} T)_M$;

(iii) *let* $r \in \mathbb{N}_0 \cup \{\infty\}$; *if* T *is induced by a* C^r *function so is* $\widetilde{T}_m$;

(iv) *if* $q \in C^\infty(X)$ *is quasihomogeneous of degree* $\ell \in \mathbb{C}$ *then* $q\widetilde{T}_m = (\widetilde{qT})_{m+\ell}$.

Proof. (i): Let $\varphi \in C_0^\infty(X)$. We observe that by Lemma 3.12 we have

$$(\partial_M \varphi)_{m^*,]0,1]} = m^* \varphi_{m^*,]0,1]} + \varphi$$

and

$$(\partial_M \varphi)_{m^*,[1,+\infty[} = m^* \varphi_{m^*,[1,+\infty[} - \varphi \, .$$

Hence, using (2.5) we obtain

$$\langle \partial_M \widetilde{T}_m, \varphi \rangle = \langle \widetilde{T}_m, {}^t\partial_M \varphi \rangle =$$

$$= -\mu \langle \widetilde{T}_m, \varphi \rangle + \langle \chi T, (\partial_M \varphi)_{m^*,]0,1]} \rangle - \langle (1-\chi)T, (\partial_M \varphi)_{m^*,[1,+\infty[} \rangle =$$

$$= -\mu \langle \widetilde{T}_m, \varphi \rangle - m^* \langle \widetilde{T}_m, \varphi \rangle + \langle \chi T, \varphi \rangle - \langle (1-\chi)T, (-\varphi) \rangle =$$

$$= m \langle \widetilde{T}_m, \varphi \rangle + \langle T, \varphi \rangle \, .$$

(ii): is a consequence of Proposition 4.3.(i).

(iii): follows from Proposition 3.3.(i) and (4.2)'.

(iv): follows from Proposition 4.4.(ii). ■

Corollary 4.17. *For every* $d \in \mathcal{D}'(X)$ *which is almost quasihomogeneous of degree* m *there exists* $u \in \mathcal{D}'(X)$ *which is almost quasihomogeneous of degree* m *with deficiency* d. ■

As another consequence of Theorem 4.16 one obtains the following generalization of the Division Theorem 4.15.

Theorem 4.18. *Let* q *and* ℓ *satisfy the assumptions of Theorem 4.15, and let* $T, c \in \mathcal{D}'(X)$. *If* $qc = (\partial_M - m)T$ *then there exists a solution* $S \in \mathcal{D}'(X)$ *of the equation (4.23) such that* $(\partial_M - (m-\ell))S = c$.

In particular, if $T \in \mathcal{D}'(X)$ *is almost quasihomogeneous of degree* m *with deficiency* d *and if* $c \in \mathcal{D}'(X)$ *is a solution of the equation* $qc = d$ *then there exists a solution* $S \in \mathcal{D}'(X)$ *of (4.23) which is almost quasihomogeneous of degree* $m-\ell$ *with deficiency* c.

Proof. By Lemma 2.34 and by Theorem 4.16.(i) (applied to $(c,m-\ell)$ instead of (T,m)) we have $(\partial_M - m)(q\tilde{c}_{m-\ell}) = q(\partial_M-(m-\ell))\tilde{c}_{m-\ell} = qc = (\partial_M - m)T$, i.e. $T - q\tilde{c}_{m-\ell}$ is quasihomogeneous of degree m. By Theorem 4.15 we find a distribution $R \in \mathcal{D}'(X)$ which is quasihomogeneous of degree $m-\ell$ such that $qR = T - q\tilde{c}_{m-\ell}$. Hence in view of the preceding application of Theorem 4.16.(i) the distribution $S := R + \tilde{c}_{m-\ell}$ is the desired solution. ■

(d) Singular Support and Wave Front Sets of the Distributions $u_{m,w}$

Proposition 4.19. *Sing supp $u_{m,w} \subset (\text{sing supp } u)_{M,I}$ for every $u \in \mathcal{D}'_I(X)$.*

Proof. Let K be a compact subset of $X \setminus (\text{sing supp } u)_{M,I}$. Then $K_{M,1/I}$ does not intersect sing supp u. Since by Proposition 3.22 $L := K_{M,1/I} \cap \text{supp } u$ is a compact subset of X we can fix a function $\chi \in C_0^\infty(X)$ such that $\chi \equiv 1$ on a neighbourhood of L and such that supp χ is contained in the open set $Y := X \setminus \text{sing supp } u$. We choose $f \in C^\infty(Y)$ such that $T_f = u|_Y$. Then χf extends to a function $g \in C_0^\infty(X)$ satisfying $\chi u = T_g$. Let $\varphi \in C_0^\infty(K)$. Then by Proposition 3.10 L contains the set $\text{supp } u \cap \text{supp } \varphi_{m*,v}$ so that

$$\langle u_{m,w}, \varphi \rangle = \langle \chi u, \varphi_{m*,v} \rangle = \langle T_g, \varphi_{m*,v} \rangle = \langle (T_g)_{m,w}, \varphi \rangle .$$

Since by (4.2)' we have $(T_g)_{m,w} = T_{g_{m,w}}$ and since by Proposition 3.3.(i) $g_{m,w}$ is a C^∞ function it follows that on K the distribution $u_{m,w}$ is induced by a C^∞ function. Since by Proposition 3.10 $(\text{sing supp } u)_{M,I}$ is a closed subset of X the assertion follows. ■

For the wave front sets of $u_{m,w}$ a similar result is valid:

Theorem 4.20. *For every $u \in \mathcal{D}'_I(X)$ the set $WF(u)_{M,I} := \bigcup_{t\in I} M_t^* WF(u)$ – where $M_t^* WF(u) := \{(M_{1/t}y, M_t^*\eta);\ (y,\eta) \in WF(u)\}$ – is a closed subset of $X \times \dot{V}^*$ containing $WF(u_{m,w})$. The assertion remains valid if WF is replaced by WF_L.*

As for the proof. The assertions are special cases of more general results that will be proved in Chapter 9 (see Remark 9.10 and Theorems 9.11 and 9.28 applied to (M, Id_{V^*}) instead of (N, M)). ∎

Suppose that $w \equiv 1$. Then the assertion of Theorem 4.20 neglects the smoothing effect which comes about by the integration along the quasihomogeneous rays $\{M_t x;\ t \in]0,+\infty[\}$, $x \in X$. In fact, since u_m is quasihomogeneous Proposition 2.14 says that $WF_A(u_m)$ is contained in $\Gamma_M(X)$. For WF instead of WF_A we are going to prove this inclusion more directly, thereby, in addition, obtaining continuity properties which are required for the proof of Theorem 4.25 below.

Recall that for any closed conic subset Γ of $X \times \dot{V}^*$ the space $\mathcal{D}'_\Gamma(X)$ is defined as the set of all distributions $T \in \mathcal{D}'(X)$ such that $WF(T) \subset \Gamma$. The topology of $\mathcal{D}'_\Gamma(X)$ is defined by the semi-norms of the weak topology of $\mathcal{D}'(X)$ and the semi-norms of the form

$$(4.25) \qquad T \mapsto \sup\{\tau^N\ |\mathcal{F}(\varphi T)(\tau\xi)|\ ;\ \tau \in]0,+\infty[,\ \xi \in H\}$$

where $N \in \mathbb{N}$ and $\varphi \in C_0^\infty(X)$ and where H is any compact subset of $\dot{V}^*$ such that $(\operatorname{supp}\varphi) \times H \cap \Gamma = \emptyset$ (see Definition 8.2.2 in Hörmander [11]).

Moreover, for any closed subset Z of X we denote by $\mathcal{D}'(X;Z)$ the space of all distributions on X with support contained in Z. Equipped with the topology of uniform convergence on the bounded subsets of $C_0^\infty(X)$ it is a closed subspace of $\mathcal{D}'(X)$.

Theorem 4.21. *Suppose that w is a C^∞ function. Let L be a closed M-bounded subset of X. Then the map*

$$\mathcal{D}'(X;L) \longrightarrow \mathcal{D}'_{\Gamma_M(X)}(X), \quad u \mapsto u_{m,w},$$

is well-defined and continuous.

Note that the assumption that w be a C^∞ function cannot be omitted. For example, the conclusion is false if w equals $w_1 := \chi_{]0,1]}$ or $w_2 := \chi_{[1,+\infty[}$. Indeed, since the conclusion is valid for $w \equiv 1$ and since $u_m = u_{m,w_1} + u_{m,w_2}$ it suffices to see that it cannot be valid for both w_1 and w_2. But this, in turn, is clear since there exist distributions $u \in \mathcal{D}'(X;L)$ such that $WF(u)$ is not contained in $\Gamma_M(X)$ and since by the assertion (i) of Theorem 4.16 one has

$$\mathrm{WF}(u) \subset \mathrm{WF}(\tilde{u}_m) \subset \mathrm{WF}\big(((1-\chi)u)_{m,]0,1]}\big) \cup \mathrm{WF}\big((\chi u)_{m,[1,+\infty[}\big).$$

Proof of Theorem 4.21. Let K (resp. H) be a compact subset of X (resp. $\dot{V}^*$) such that $K\times H \cap \Gamma_M(X) = \emptyset$. Since $K\times H$ is compact, as well, one finds a compact neighbourhood K' of K in X such that

$$(4.26) \qquad \inf\{\,|\langle \xi, Mx\rangle|\,;\ (x,\xi) \in K'\times H\} > 0\,.$$

By the assumption on L and by Proposition 3.22 the set $L \cap K_{M,1/I}$ is a compact subset of X. Let W be a compact neighbourhood of $L\cap K_{M,1/I}$ in X, and let $\chi \in C_0^\infty(W)$ be equal to 1 near $L \cap K_{M,1/I}$. Then by Proposition 3.10 and Remark 4.1 we have for every $u \in \mathcal{D}'(X;L)$

$$(4.27) \qquad \mathcal{F}(\varphi u_{m,w})(\zeta) = \langle u, \chi\,(\varphi e^{-i\langle\zeta,\,\cdot\,\rangle})_{m*,v}\rangle, \qquad \zeta \in V,\ \varphi \in C_0^\infty(K).$$

In order to derive suitable estimates for $(\varphi e^{-i\langle\zeta,\,\cdot\,\rangle})_{m*,v}$, $\zeta \in \bigcup_{\tau\ge 1}\tau H$, we fix $\varepsilon > 0$ such that $W_\varepsilon := W + \overline{K}(0,\varepsilon)$ is contained in X. Since the map $(t,x) \mapsto M_t x$ is continuous one finds for every $y \in W$ constants $\delta_y \in]0,\varepsilon[$ and $\gamma_y > 0$ such that

$$\{M_t x;\ t \in I_y + [-2\gamma_y, 2\gamma_y],\ x \in \overline{K}(y,\delta_y)\} \subset K'$$

where $I_y := \{t \in]0,+\infty[;\ M_t y \in K\}$. Since the sets I_x, $x \in W_\varepsilon$, are all contained in the compact subset $\{t \in]0,+\infty[;\ M_t(W_\varepsilon)\cap K \ne \emptyset\}$ of $]0,+\infty[$ we may assume (after having made δ_y smaller if necessary) that $I_x \subset I_y +]-\gamma_y, \gamma_y[$ for arbitrary $y \in W$ and $x \in K(y,\delta_y)$. Since W is compact it is covered by finitely many of the balls $K(y,\delta_y)$. Consequently, one obtains a finite open covering $\mathfrak{U}$ of W of open subsets of W_ε and a family $(I_U)_{U\in\mathfrak{U}}$ consisting of finite unions I_U of compact subintervals of $]0,+\infty[$ such that for every $U \in \mathfrak{U}$

$$(4.28) \qquad (a)\ \{M_t x;\ t \in I_U,\ x \in U\} \subset K'; \qquad (b)\ \bigcup_{x\in U} I_x \subset \mathring{I}_U\,.$$

Now, in view of (4.28.b) it follows for arbitrary $\varphi \in C_0^\infty(K)$, $\xi \in H$, $\tau \in [1,+\infty[$, $U \in \mathfrak{U}$, and $x \in U$ that

$$(4.29) \qquad (\varphi e^{-i\langle\tau\xi,\,\cdot\,\rangle})_{m*,v}(x) = \int_{I_U} t^{-m^*}\varphi(M_t x)\,\exp(-i\langle\tau\xi, M_t x\rangle)\,v(t)\,\frac{dt}{t}\,.$$

By combining (4.28.a) and (4.26) one observes that on a neighbourhood of the set $D := \bigcup_{U\in\mathfrak{U}} I_U \times U \times H$ by

$$g(t,x,\xi) := \frac{-it}{\langle \xi, M_t M x \rangle}$$

a C^∞ function g is well-defined, all of whose derivatives are bounded on D. Note that

$$\exp(-i\langle \tau\xi, M_t x\rangle) = -\frac{1}{\tau} g(t,x,\xi)\ \partial_t \exp(-i\langle \tau\xi, M_t x\rangle), \qquad (t,x,\xi) \in D,\ \tau > 0.$$

We insert this equation into the right-hand side of (4.29) and do partial integration; here the boundary terms vanish since by (4.28.b) for arbitrary $x \in U$ and $t \in \partial I_U$ the point $M_t x$ does not belong to K. Repeating this procedure we see that for arbitrary $U \in \mathfrak{U}$, $x \in U$, and $N \in \mathbb{N}$ the right-hand side of (4.29) is equal to

$$\tau^{-N} \int_{I_U} Q^N\big(t^{-m^*-1} v(t)\, \varphi(M_t x)\big) \exp(-i\langle \tau\xi, M_t x\rangle)\, dt$$

where $Q = Q(t,x,\xi,\partial_t)$ denotes the differential operator $f \mapsto \partial_t(gf)$. Using the Leibniz rule, making use of (3.5), and applying the preceding considerations to the derivatives of φ, as well, we conclude that for arbitrary $\varphi \in C_0^\infty(K)$ and $N \in \mathbb{N}$ the set

$$\mathfrak{B}_N := \{\tau^N \chi\,(\varphi e^{-i\langle \tau\xi, \cdot\rangle})_{m^*,v}\ ;\ \tau \in [1,+\infty[,\ \xi \in H\}$$

is a bounded subset of $C_0^\infty(W)$. In view of (4.27) this shows that if π_N is the semi-norm on $\mathcal{D}'_{\Gamma_M(X)}(X)$ defined by (4.25) then $u \mapsto \pi_N(u_{m,w})$ defines a continuous semi-norm on $\mathcal{D}'(X;L)$. The assertion follows. ■

Corollary 4.22. *The sequence (q_j) in Corollary 4.11 can be chosen in such a way that (T_{q_j}) converges to T in the topology of $\mathcal{D}'_{\Gamma_M(X)}(X)$ as $j \to \infty$.*

Proof. Let f_j and ψ be as in the proof of Corollary 4.11. Then for every $i \le k$ the sequence of distributions induced by the functions $\psi(\partial_M - m)^i f_j$ converges to $\psi(\partial_M - m)^i T$ in the space $\mathcal{D}'(X;L)$ where $L := \operatorname{supp}\psi$. Consequently, in view of the proof of Corollary 4.11 the assertion follows from Theorem 4.21. ■

(e) The Quasihomogeneous Continuations ν_m and the Distributions $\varkappa^m \rho_\varkappa^*(\nu)$, $\nu \in \mathcal{D}'(S^\varkappa)$

We fix a C^∞ function $\varkappa : X \longrightarrow]0,+\infty[$ which is quasihomogeneous of degree 1 (see Theorem 3.39). To motivate the main definition we first prove the following lemma in a form which is suitable for applications in Chapter 8, as well.

Lemma 4.23. *Let $k \in \mathbb{N}$, let $q_0 \in \mathcal{L}^1_{loc}(X)$ be almost quasihomogeneous of degree m and of order $\leq k$, and let $f \in C^0(X)$ be such that*

$$(4.30) \qquad \overline{(\operatorname{supp} f)_M} \cap \operatorname{supp} q_i \cap S^\varkappa \text{ is compact for every } i \in \mathbb{N}_k \cup \{0\}$$

where q_i denotes the ith order deficiency of q_0. Suppose, moreover, that $|f|_{\operatorname{Re} m^*, |w\omega_k|}$ is well-defined by (3.1)' (with (m, w) replaced by $(\operatorname{Re} m^*, |w\omega_k|)$) and continuous. Then for every $i \in \mathbb{N}_k \cup \{0\}$ $f_{m^*, w\omega_i}$ is well-defined by (3.1)' (with m and w replaced by m^* and $w\omega_i$) and is measurable on $S^\varkappa$, and the function $q_i f (w\omega_i) \circ \varkappa$ belongs to $\mathcal{L}^1(X)$ such that

$$(4.31) \qquad \int_X q_0(x)\, f(x)\, (w \circ \varkappa)(x)\, dx = \sum_{i=0}^{k} \int_{S^\varkappa} q_i(\vartheta)\, f_{m^*, w\omega_i}(\vartheta)\, d\tilde{\varkappa}(\vartheta)$$

and

$$(4.31)' \qquad \int_{S^\varkappa} q_0(\vartheta)\, f_{m^*, w}(\vartheta)\, d\tilde{\varkappa}(\vartheta) = \sum_{i=0}^{k} (-1)^i \int_X q_i(x)\, f(x)\, ((w\omega_i) \circ \varkappa)(x)\, dx\, .$$

Proof. We denote the union of the compact sets in (4.30) by K. Since by Lemma 1.90 the functions $q_i|_{S^\varkappa}$ belong to $\mathcal{L}^1_{loc}(\tilde{\varkappa})$ and since the functions $|f|_{\operatorname{Re} m^*, |w\omega_i|}$ are well-defined and continuous with support contained in $\overline{(\operatorname{supp} f)_M}$ it follows by (4.30) that the functions $h_i := (|q_i|\, |f|_{\operatorname{Re} m^*, |w\omega_i|})|_{S^\varkappa}$ belong to $\mathcal{L}^1(\tilde{\varkappa})$. Hence, in particular, the right-hand side of (4.31) is well-defined. In order to prove the same about the left-hand side of (4.31) we first observe that in view of Proposition 1.86 and by the Fubini and the Fubini-Tonelli theorem it suffices to show that the function

$$g :]0,+\infty[\times S^\varkappa \longrightarrow \mathbb{C}, \quad (t, \vartheta) \mapsto (q_0 f\, w \circ \varkappa)(M_t \vartheta)\, t^{\mu - 1},$$

belongs to $\mathcal{L}^1(dt \otimes \tilde{\varkappa})$. In view of the equality

$$g(t,\vartheta) = \sum_{i=0}^{k} q_i(\vartheta)\, t^{-m^*} \omega_i(t)\, f(M_t\vartheta)\, w(t)/t\ , \qquad \vartheta \in S^{\varkappa},$$

the assumptions on f imply that the function $S^{\varkappa} \longrightarrow \mathbb{R}$, $\vartheta \mapsto \int_0^{+\infty} |g(t,\vartheta)|\, dt$, is well-defined and majorized by the sum of the functions h_i defined above. Consequently, the Fubini-Tonelli theorem leads to the desired conclusion and, moreover, to the equation (4.31).

For the proof of (4.31)' we first observe that by what was proved above the integrands of both sides of (4.31)' are absolutely integrable. Hence in view of Proposition 1.86 and the equation

$$q_0(\vartheta)\, t^{-m^*} f(M_t\vartheta)\, w(t) = q_0(M_{1/t}(M_t\vartheta))\, t^{m+\mu}\, f(M_t\vartheta)\, (w\circ\varkappa)(M_t\vartheta) =$$
$$= \sum_{i=0}^{k} q_i(M_t\vartheta)\, t^{\mu}\, (-1)^i\, (\omega_i\circ\varkappa)(M_t\vartheta)\, f(M_t\vartheta)\, (w\circ\varkappa)(M_t\vartheta)$$

(which is valid for arbitrary $t \in]0,+\infty[$ and $\vartheta \in S^{\varkappa}$) the equation (4.31)' follows by the Fubini theorem. ■

Recall now that $S^{\varkappa}$ is a C^{∞} submanifold of X (see section 1.(f)) and that the space $\mathcal{D}'(S^{\varkappa})$ of distributions on $S^{\varkappa}$ can be defined as the dual of the space of C_0^{∞} densities on $S^{\varkappa}$ (see for example § 6.3 in Hörmander [11]). In the following we shall identify $\mathcal{D}'(S^{\varkappa})$ with the dual of $C_0^{\infty}(S^{\varkappa})$ via the strictly positive density $\omega_M|_{S^{\varkappa}}$ (see Notation 1.85):

(4.32) $$\langle \nu, \psi \rangle := \langle \nu, \psi\, \omega_M|_{S^{\varkappa}} \rangle, \qquad \nu \in \mathcal{D}'(S^{\varkappa}),\ \psi \in C_0^{\infty}(S^{\varkappa}).$$

Now we fix $k \in \mathbb{N}$, a (1+k)-tuple $\nu = (\nu_0, \ldots, \nu_k) \in \mathcal{D}'(S^{\varkappa})^{1+k}$, and $m \in \mathbb{C}$. In view of the preceding identification it makes sense to define

(4.33) $$\langle \nu_m, \varphi \rangle := \sum_{i=0}^{k} \langle \nu_i, \varphi_{m^*,\omega_i}|_{S^{\varkappa}} \rangle, \qquad \varphi \in C_0^{\infty}(X).$$

Note that when Lemma 4.23 is applied to $f = \varphi \in C_0^{\infty}(X)$ and $w \equiv 1$ the formula (4.31) reads as

(4.34) $$T_{q_0} = \left(T_{q_0}|_{S^{\varkappa}}, \ldots, T_{q_k}|_{S^{\varkappa}} \right)_m .$$

Proposition 4.24. *(i)* ν_m *is a well-defined distribution on X with support contained in* $\bigcup_{i=0}^{k} (\operatorname{supp} \nu_i)_M$ *;*

(ii) ν_m *is almost quasihomogeneous of degree m and of order* $\leq k$ *;*

(iii) for every $j \in \{0\} \cup \mathbb{N}_k$ we have

$$(\partial_M - m)^j \nu_m = x^m \sum_{i=0}^{k-j} \eta_i \, \rho_x^*(\nu_{i+j}) \tag{4.35}$$

where ρ_x is defined in Notation 1.71 and the η_i are defined in Lemma 1.74;

(iv) the function $\mathbb{C} \longrightarrow \mathcal{D}'(X)$, $m \mapsto \nu_m$, is holomorphic.

Proof. (i): Let K be a compact subset of X. Since by Lemma 3.29 S^x is an M-bounded subset of X it follows by Proposition 3.22 that $L := K_M \cap S^x$ is a compact subset of S^x. By Proposition 3.10 we see that if $\varphi \in C_0^\infty(X)$ then the functions $\varphi_{m^*,\omega_i}|_{S^x}$ belong to $C_0^\infty(L)$. Since K is an M-bounded subset of X we find a compact subinterval J of $]0,+\infty[$ such that

$$\varphi_{m^*,\omega_i}(\vartheta) = \int_J t^{m+\mu} \varphi(M_t \vartheta)\, \omega_i(t)\, \frac{dt}{t}, \qquad \vartheta \in L,\ \varphi \in C_0^\infty(K).$$

Consequently, in view of (3.5) the linear maps $C_0^\infty(K) \longrightarrow C_0^\infty(L)$, $\varphi \mapsto \varphi_{m^*,\omega_i}|_{S^x}$, are well-defined and continuous. Hence ν_m is well-defined and continuous on the space $C_0^\infty(K)$.

The assertion about supp ν_m is proved in a similar way as the corresponding part of Proposition 4.3.(i).

(iv): Let K and L be as above. We fix $\varphi \in C_0^\infty(K)$ and $i \in \{0\} \cup \mathbb{N}_k$. Since the restriction map $C^\infty(X) \longrightarrow C^\infty(S^x)$, $f \mapsto f|_{S^x}$, is linear and continuous Proposition 3.3.(ii) implies that the map $\mathbb{C} \longrightarrow C_0^\infty(L)$, $m \mapsto \varphi_{m^*,\omega_i}|_{S^x}$, is holomorphic. Since $\nu_i : C_0^\infty(S^x) \longrightarrow \mathbb{C}$ is linear and continuous the assertion follows.

(ii): this is a consequence of (iii) or can be proved directly with the help of the almost quasihomogeneity properties of the functions φ_{m^*,ω_i}.

(iii): If $\nu = (T_{g_0}, \dots, T_{g_k})$ for some $g_0, \dots, g_k \in C^0(S^x)$ then by Remark 1.73' the functions q_k defined by (1.82) satisfy (1.39) (with N replaced by k) so that in view of (1.83) and because of $q_i|_{S^x} = g_i$ the equality (4.35) for $j=0$ is identical to (4.34) in this case. In general ν_i can be approximated (in the weak topology) by a sequence of functions $g_{i,j} \in C_0^\infty(S^x)$, $j \in \mathbb{N}$. Since the maps ρ_x^* and

$$\mathcal{D}'(S^x)^{1+k} \longrightarrow \mathcal{D}'(X), \quad \nu \mapsto \nu_m,$$

are (weakly) continuous the equality (4.35) follows for $j=0$. In order to derive (4.35) for $j \in \mathbb{N}_k$ we conclude by Remark 1.74 and Theorem 2.17.(i) and by changing the order of summation and substituting $I = i-j$ that

$$\nu_m \circ M_t = t^m \varkappa^m \sum_{i=0}^{k} \Big(\sum_{j=0}^{i} \omega_j(t)\, \eta_{i-j} \Big) \rho_\varkappa^*(\nu_i) =$$

$$= t^m \sum_{j=0}^{k} \omega_j(t) \Big(\varkappa^m \sum_{I=0}^{k-j} \eta_I\, \rho_\varkappa^*(\nu_{j+I}) \Big), \qquad t \in]0,+\infty[.$$

By Proposition 2.31 this implies (4.35). ■

We can now complete the description of almost quasihomogeneous distributions on X.

Theorem 4.25. *Let $T \in \mathcal{D}'(X)$ be almost quasihomogeneous of degree m and of order $\leq k$. Then the (1+k)-tuple* d(T) *of distributions $(\partial_M - m)^i T/_{S^\varkappa}$, $0 \leq i \leq k$, (which are well-defined by Proposition 2.31.(ii), Corollary 2.33.(i), and Theorem 2.42) has the following properties:*

(i) $d(T)_m = T$;

(ii) if $\nu \in \mathcal{D}'(S^\varkappa)^{1+k}$ satisfies the equation $\nu_m = T$ then $\nu = d(T)$.

Proof. *(i)*: If for every $i \in \{0\} \cup \mathbb{N}_k$ there exists $q_i \in C^0(X)$ such that $(\partial_M - m)^i T = T_{q_i}$ then the assertion is identical with (4.34). In general, by Corollary 4.22 we choose a sequence of functions $q_j \in C^\infty(X)$ which are almost quasihomogeneous of degree m and of order $\leq$ k such that (T_{q_j}) converges to T in the topology of the space $\mathcal{D}'_{\Gamma_M(X)}(X)$ as $j \to \infty$. Then (by Theorem 2.42 and Lemma 2.15 and by Theorem 8.2.4 in Hörmander [11]) the sequence of distributions $d(T_{q_j})$ converges to d(T) as $j \to \infty$, and hence $T_{q_j} = d(T_{q_j})_m$ converges to $d(T)_m$ as $j \to \infty$.

(ii): Since $\eta_i|_{S^\varkappa} \equiv \delta_{i0}$ this follows from (4.35) and Theorem 2.17.(iv). ■

Corollary 4.26. *Let $T \in \mathcal{D}'(X)$ be almost quasihomogeneous of degree m and of order $\leq k$, and let $S_0, \ldots, S_k$ be the quasihomogeneous (of degree m) distributions defined by (2.20) with (N, ν_i) replaced by (k, η_i). Then we have*

$$(4.36) \qquad S_\ell = \varkappa^m \rho_\varkappa^*\big(((\partial_M - m)^\ell T)/_{S^\varkappa}\big), \qquad \ell \in \{0\} \cup \mathbb{N}_k .$$

Proof. Inserting the equation (i) in Theorem 4.25 into the left-hand side of the equation (4.35) for $\nu = d(T)$ and $j = 0$ we see that the equation (2.19) is valid for $\nu_i = \eta_i$ if the distributions S_ℓ are given by (4.36). Since by Theorem 2.17.(i) and

Corollary 2.4.(ii) the distributions given by (4.36) are quasihomogeneous of degree m the assertion follows by the uniqueness part of Proposition 2.45. ■

As another consequence of Theorem 4.25 we note in passing:

Corollary 4.27. *Let* $T \in \mathcal{D}'(X)$ *be almost quasihomogeneous of degree* m *and of order* $\leq k$. *Let* $(f_j)_{j\in\mathbb{N}}$ *be a sequence in* $C^k(X) \cap \mathcal{D}'_{\Gamma_M(X)}(X)$ *converging to* T *in the topology of* $\mathcal{D}'_{\Gamma_M(X)}(X)$ *as* $j \to \infty$ *then the sequence of distributions induced by the functions*

$$\sum_{i=0}^{k} \varkappa^m((\partial_M - m)^i f_j) \circ \rho_\varkappa$$

converges (weakly) to T *as* $j \to \infty$.

Proof. We set $g_{i,j} := ((\partial_M - m)^i f_j)|_{S^\varkappa}$. The assumption on (f_j) implies that $(T_{g_{i,j}})$ converges to $(\partial_M - m)^i T|_{S^\varkappa}$ as $j \to \infty$ so that

$$\sum_{i=0}^{k} \varkappa^m((\partial_M - m)^i f_j) \circ \rho_\varkappa = (T_{g_{0,j}}, \ldots, T_{g_{k,j}})_m$$

converges to $(T|_{S^\varkappa}, \ldots, (\partial_M - m)^k T|_{S^\varkappa})_m = d(T)_m = T$ as $j \to \infty$. ■

Combining Theorem 4.25 and Proposition 4.24.(ii) one sees that a distribution $T \in \mathcal{D}'(X)$ is almost quasihomogeneous of degree m and of order $\leq k$ if and only if $T = \nu_m$ for some $\nu \in \mathcal{D}'(S^\varkappa)^{1+k}$. Introducing

Notation 4.28. By $\mathcal{UQ}'_{m,k}(X)$ we denote the complex vector space of all distributions $T \in \mathcal{D}'(X)$ which are almost quasihomogeneous of degree m and of order $\leq k$

we reformulate Theorem 4.25 and Proposition 4.24.(ii) as follows.

Theorem 4.25'. *The map* $\mathcal{UQ}'_{m,k}(X) \longrightarrow \mathcal{D}'(S^\varkappa)^{1+k}$, $T \mapsto d(T)$, *is a linear isomorphism, its inverse being given by* $\nu \mapsto \nu_m$. ■

We end this section by rewriting the equality (i) in Theorem 4.25.

Theorem 4.29. *Let $T \in \mathfrak{UQ}'_{m,k}(X)$ and $q \in \mathfrak{UQ}^{\infty}_{m^*,k}(X)$ (see Notation 3.50). And let $(\psi_i)_{0 \le i \le k}$ be any sequence of functions in $C^{\infty}_M(X)$ satisfying (4.18) for $\mathfrak{J} = \{0\} \cup \mathbb{N}_k$ and (4.19). Then*

$$(4.37) \qquad \sum_{i=0}^{k} (-1)^i \langle (\partial_M - m)^i T_{/S^\varkappa}, ((\partial_M - m^*)^{k-i} q)_{/S^\varkappa} \rangle = \sum_{i=0}^{k} \langle (\partial_M - m)^i T, \psi_i q \rangle$$

where the left-hand side is well-defined since by Proposition 3.51.(ii) the functions $((\partial_M - m)^i q)_{/S^\varkappa}$, $i \in \{0\} \cup \mathbb{N}_k$, belong to $C^{\infty}_O(S^\varkappa)$.

In particular, it follows that the left-hand (resp. right-hand) side of (4.37) is independent of the choice of $\varkappa$ (resp. $\psi_O, \ldots, \psi_k$).

Proof. Let K be a compact subset of X such that (3.45) is valid. By Theorem 3.48 we find a C^∞ function $\varphi: X \longrightarrow \mathbb{C}$ with support contained in $L := K_M \cap \varkappa^{-1}([1,2])$ such that $q = \varphi_{m^*,\omega_k}$. By Lemma 3.29 and Proposition 3.22 the set L is compact. Hence φ belongs to $C^{\infty}_O(X)$. By Proposition 4.13 we have

$$\langle T, \varphi \rangle = (-1)^k \sum_{i=0}^{k} \langle (\partial_M - m)^i T, \psi_i q \rangle .$$

By Theorem 4.25.(i) the left-hand side of this equation is equal to

$$\langle d(T)_m, \varphi \rangle = \sum_{i=0}^{k} \langle ((\partial_M - m)^i T)|_{S^\varkappa}, \varphi_{m^*,\omega_i}|_{S^\varkappa} \rangle .$$

Since by Propositions 3.13 and 2.31 we have $\varphi_{m^*,\omega_i} = (-1)^{k-i} (\partial_M - m^*)^{k-i} q$ the assertion follows. ■

This result can be interpreted in terms of duality brackets as follows.

Remark 4.30. *Defining $\langle T, q \rangle$ by the right-hand side of (4.37) one obtains a (non-degenerate) duality bracket $\mathfrak{UQ}'_{m,k}(X) \times \mathfrak{UQ}^{\infty}_{m^*,k}(X) \longrightarrow \mathbb{C}$. In view of (4.37) this duality bracket is isomorphic - via the isomorphisms in Theorem 4.25' and Proposition 3.51.(ii) - to the duality bracket*

$$(4.38) \qquad \mathcal{D}'(S^\varkappa)^{1+k} \times C^{\infty}_O(S^\varkappa)^{1+k} \longrightarrow \mathbb{C}, \quad (\nu, \chi) \mapsto \sum_{i=0}^{k} (-1)^i \langle \nu_i, \chi_{k-i} \rangle ,$$

where $\langle \cdot, \cdot\cdot \rangle$ is defined by (4.32). ■

Of course, via the isomorphisms in Theorem 4.25' and Proposition 3.51.(ii) the spaces $\mathfrak{UQ}'_{m,k}(X)$ and $\mathfrak{UQ}^{\infty}_{m,k}(X)$ can be equipped with locally convex topologies in such a way that each of these spaces is identified as the topological dual of the other via the duality bracket defined in Remark 4.30. At present we refrain from giving an intrinsic description of these locally convex topologies. However, under the assumptions of Chapter 5 we shall have to do this for the analogous spaces defined there (see § 5.g and § 6.f).

Chapter V

Constructing (Almost) Quasihomogeneous Functions by Taking Quasihomogeneous Averages of Functions Not Necessarily Having M-bounded Support

The purpose of the present chapter is to extend part of the theory of Chapter 3 to the case where X is not locally M-bounded. Here one wants to admit functions f such that the support of the functions $]0,+\infty[\longrightarrow \mathbb{C}$, $t \mapsto f(M_t x)$, is no longer a compact subset of $]0,+\infty[$. This leads to two kinds of difficulties: it has to be ensured that the integrand on the right-hand side of (3.1)' is integrable 1.° near $t = +\infty$ and 2.° near $t = 0$. The first difficulty is easily overcome by suitable growth conditions on f and w at infinity. However, the second difficulty is more serious: since $t \mapsto t^{-m-1}$ is only integrable near $t=0$ if $\mathrm{Re}\, m < 0$, in general it is possible to define $f_{m,w}$ by (3.1)' only if m belongs to the half plane $\mathbb{C}(-\infty,d) := \{z \in \mathbb{C}\, ;\, \mathrm{Re}\, z < d\}$ for some $d \in \mathbb{R}$. This is elaborated in section (a) - eventually under the assumption that (1.14) holds.

Under very special assumptions on w, in section (b) for every $f \in C^\infty(X)$ a *functional equation* is derived which makes it possible to extend the holomorphic function $\mathbb{C}(-\infty,d) \longrightarrow C^\infty(X)$, $m \mapsto f_{m,w}$, to a meromorphic function on the whole of $\mathbb{C}$. This leads to a natural definition of $f_{m,w}$ for every $m \in \mathbb{C}$. It is for the derivation of the functional equation that the assumption (1.14) is essential. In section (c) the negative Laurent coefficients at the poles of the meromorphic extension are computed while section (d) contains explicit formulas for $f_{m,w}$.

In section (e), as a combination of the results of the preceding sections a reasonable definition of f_m for arbitrary $f \in \mathcal{S}(V)$ and arbitrary $m \in \mathbb{C}$ is obtained. These functions f_m, however, do not always have the desired properties: in general they are defined on X_+, only, and if $m \in \mathfrak{U}(M)$ then it happens that f_m is not quasihomogeneous but only almost quasihomogeneous of degree m, its deficiency extending to an (almost) quasihomogeneous C^∞ function on the whole of X.

In section (f) locally convex spaces of functions of the latter kind are introduced, while section (g) is devoted to locally convex spaces of functions of the form f_m. Their topological standard properties are of importance in Chapter 7.

(a) Defining $f_{m,w}$ by (3.1)' when Supp f is Not Necessarily M-bounded

As in Chapters 3 and 4 let $w:]0,+\infty[\longrightarrow \mathbb{C}$ be a locally integrable weight function. For various choices of subspaces E and F of $C^0(X)$ and open subsets D of $\mathbb{C}$ we are going to verify the following combination of assertions:

(5.1) *(i) for arbitrary $m \in D$ and $f \in E$ a function $f_{m,w} \in F$ is well-defined by (3.1)';*

(ii) the map $D \times E \longrightarrow F$, $(m,f) \mapsto f_{m,w}$, is continuous;

(iii) for every $k \in \mathbb{N}$ the conditions (i) and (ii) remain valid if w is replaced by $w\omega_k$;

(iv) for every $f \in E$ the map $D \longrightarrow F$, $m \mapsto f_{m,w}$, is holomorphic, for every $k \in \mathbb{N}$ its derivative of order k being given by $m \mapsto k!\,(-1)^k f_{m,w\omega_k}$.

In order to introduce the relevant subspaces of $C^0(X)$ we fix a continuous weight function $W: X \longrightarrow [0,+\infty[$.

Notation 5.1. (i) By $C^0(W)$ we denote the Banach space of continuous functions $f: X \longrightarrow \mathbb{C}$ such that $\|f\|_W := \sup\{|f(x)|/W(x);\ x \in X \setminus W^{-1}(0)\}$ is finite and such that $|f| \le \|f\|_W W$. Note that this definition makes sense for any weight function $W: X \longrightarrow [0,+\infty]$, but then $C^0(W)$ need not be a Banach space.

(ii) Let $r,s \in \mathbb{N}_0 \cup \{\infty\}$; referring to Definition 1.18.(i) (where the space $C^{r,s}(X)$ was defined with respect to some pre-fixed M-invariant decomposition of V of the form (1.26)) we denote by $C^{r,s}(W)$ the space of all functions $f \in C^{r,s}(X)$

such that for every polynomial function $P: V^* \longrightarrow \mathbb{C}$ of degree $\leq (r,s)$ the function $P(\partial)f$ belongs to $C^o(W)$. Of course, if $V_1 = \{0\}$ we write $C^r(W)$ instead of $C^{r,s}(W)$. Note that equipped with its natural norms $C^{r,s}(W)$ is a Banach space in case $r,s < +\infty$ and a Fréchet space otherwise.

Now we fix another continuous weight function $U: X \longrightarrow [0,+\infty[$ and constants $c,d \in \mathbb{R}$ such that $c < d$. We suppose that there exists $\gamma \in \mathcal{L}^1(]0,+\infty[; dt/t)$ such that

(5.2) $$W(M_t x)\,|w(t)| \leq U(x)\,\gamma(t)\,\min\{t^c, t^d\}, \qquad x \in X,\ t \in]0,+\infty[.$$

Lemma 5.2. *Under the preceding assumptions the assertion (5.1) is valid for $E = C^o(W)$, $F = C^o(U)$, and $D = \mathbb{C}(c,d) := \{z \in \mathbb{C} :\ c < \mathrm{Re}\, z < d\}$.*

Proof. *(i)*: Let $m \in \mathbb{C}(c,d)$. Then $|t^{-m}| = t^{-\mathrm{Re}\, m} \leq \max\{t^{-c}, t^{-d}\} = 1/\min\{t^c, t^d\}$. Hence (5.2) implies that

(5.3) $$|t^{-m} w(t)|\, W(M_t x) \leq U(x)\,\gamma(t), \qquad x \in X,\ t \in]0,+\infty[.$$

Consequently, for every $f \in C^o(W)$ one deduces that $f_{m,w}$ is a well-defined function on X satisfying

(5.4) $$\| f_{m,w} \|_U \leq C \| f \|_W \qquad \text{where} \qquad C := \int_0^{+\infty} \gamma(t) \frac{dt}{t} < +\infty.$$

In order to show that $f_{m,w}$ is continuous we fix $x \in X$ and a compact neighbourhood K of x in X and deduce from (5.3) that

(5.5) $$|t^{-m} f(M_t y)\, w(t)| \leq C'\gamma(t), \qquad y \in K,\ t \in]0,+\infty[,$$

where $C' := \| f \|_W \sup\{U(y) ;\ y \in K\}$ is finite. Hence Lebesgue's Dominated Convergence Theorem shows that $f_{m,w}(y)$ tends to $f_{m,w}(x)$ as $y \to x$.

(iii): First of all we fix $\varepsilon \in]0, \frac{d-c}{2}[$ and observe that

$$|\omega_k(t)| \min\{t^c, t^d\} \leq C_{\varepsilon,k} \min\{t^{c+\varepsilon}, t^{d-\varepsilon}\}, \qquad t \in]0,+\infty[,\ k \in \mathbb{N},$$

where $C_{\varepsilon,k} := \inf\{t^{-\varepsilon}\omega_k(t) ;\ t \in [1+\infty[\}$. Consequently,

(5.6) (5.2) remains valid if (w,γ,c,d) is replaced by $(w\omega_k, C_{\varepsilon,k}\gamma, c+\varepsilon, d-\varepsilon)$

so that everything we did in the proof of (i) remains valid for $w\omega_k$ instead of w provided that m belongs to $\mathbb{C}(c+\varepsilon, d-\varepsilon)$.

It follows that the assertion (5.1.i) remains valid for $D=\mathbb{C}(c+\varepsilon,d-\varepsilon)$ with w replaced by $w\omega_k$. Since ε can be made arbitrarily small this is also true for $D=\mathbb{C}(c,d)$.

For the assertion (5.2.ii) the same argument applies once it is proved in its original form.

(ii): We fix $m\in\mathbb{C}(c,d)$, choose $\varepsilon>0$ so small that $\operatorname{Re}m+[-2\varepsilon,2\varepsilon]\subset\,]c,d[$, and let $h\in\mathbb{C}(-\varepsilon,\varepsilon)$. Note that then $m+h\in\mathbb{C}(c+\varepsilon,d-\varepsilon)$. By the main theorem of calculus we obtain for every $f\in C^o(W)$ that

$$(5.7)\qquad (f_{m+h,w}-f_{m,w})(x)=\int_0^{+\infty}(t^{-h}-1)\,t^{-m}\,f(M_t x)\,w(t)\,\frac{dt}{t}=$$

$$=-h\int_0^{+\infty}\log t\int_0^1 t^{-sh}\,ds\;t^{-m}\,f(M_t x)\,w(t)\,\frac{dt}{t}\,.$$

Consequently, $\|f_{m+h,w}-f_{m,w}\|_U\le|h|\,N(f)$ where the number

$$N(f):=2\max\{\,\|\,|f|_{\operatorname{Re}(m+sh),v}\|_U\,;\;s\in\{0,1\}\,\}$$

is well-defined and finite for $v:=|w\log|$ by the first part of condition (iii) already completely proved above. Applying the inequality (5.4) to $(|f|,\operatorname{Re}(m+sh),v)$ instead of (f,m,w) we deduce that $N(f)\le C\,C_{\varepsilon,1}\|f\|_W$. Combining the preceding estimates with

$$\|f_{m+h,w}-g_{m,w}\|_U\le\|f_{m+h,w}-f_{m,w}\|_U+\|(f-g)_{m,w}\|_U\,,\qquad g\in C^o(W),$$

and with (5.4) (applied to $f-g$ instead of f) we derive the continuity of the map under consideration at the point $(m,g)\in\mathbb{C}(c,d)\times C^o(W)$.

(iv): Let m, ε, and h be as in the proof of (ii). Since by (5.6) the estimate (5.5) is valid with (w,C') replaced by $(w\log,C'C_{\varepsilon,1})$ we can apply Fubini's theorem to the right-hand side of (5.7) to obtain

$$\frac{1}{h}(f_{m+h,w}-f_{m,w})(x)=-\int_0^1 f_{m+sh,w\log}(x)\,ds\,,\qquad x\in X.$$

Since by (iii) the function $\mathbb{C}(-\varepsilon,\varepsilon)\longrightarrow C^o(U)$, $h\mapsto f_{m+h,w\log}$, is continuous one deduces that the map $\mathfrak{H}:\mathbb{C}(c,d)\longrightarrow C^o(U)$, $z\mapsto f_{z,w}$, is holomorphic at the point $z=m$ with derivative $-f_{m,w\log}$. In view of (5.6), for every $k\in\mathbb{N}$ we can apply this to $w\omega_k$ instead of w and in this way obtain the formula for the kth order derivative of $\mathfrak{H}$ by induction. ■

Next we are going to have a look at the standard properties of $f_{m,w}$. As for the multiplication by quasihomogeneous functions, the following assertion is easily verified directly from the definitions.

Remark 5.3. *Let $\ell \in \mathbb{C}$ and $N \in \mathbb{N}$, and let $q: X \longrightarrow \mathbb{C}$ be almost quasihomogeneous of degree ℓ and of order $\leq N$. For every $j \in \mathbb{N}_N$ let q_j be its jth order deficiency. Then for arbitrary $f \in C^o(X)$ and $m \in \mathbb{C}$ such that $(qf)_{m+\ell,w}$, $(q_j f)_{m+\ell,w}$, and $f_{m,w}$ are well-defined by (3.1)' the equations (3.8) and (3.8)' hold. In particular, if q is quasihomogeneous of degree ℓ then (3.9) is valid.* ■

We now formulate an assertion about derivatives. For simplicity we restrict ourselves to first order derivatives. Of course, one can deduce a corresponding assertion for higher order derivatives by induction if one carefully formulates suitable conditions on the set of admissable m .

Lemma 5.4. *Let $f \in C^{1,o}(W)$ (see Notation 5.1.ii), and let $P: V^* \longrightarrow \mathbb{R}$ be a linear function such that $\ker P \supset V_2^*$. Then for every $m \in \bigcap_{\ell \in \Lambda(P)} \mathbb{C}(c + \mathrm{Re}\,\ell, d + \mathrm{Re}\,\ell)$ the function $P(\partial) f_{m,w}$ is well-defined and equal to*

$$\sum_{\ell \in \Lambda(P)} \sum_{j \in \mathbb{N}_o} (P_{\ell,j}^*(\partial) f)_{m-\ell, w\omega_j}$$

where the polynomials $P_{\ell,j}^*$ are defined in the text preceding Remark 1.42. Moreover, we have

$$(5.8) \qquad (P(\partial) f_{m,w})(x) = \int_0^{+\infty} t^{-m} P(\partial)(f \circ M_t)(x)\, w(t) \frac{dt}{t}, \qquad x \in X.$$

Proof. First of all we observe that the assumption on P means that P is a polynomial function of degree $\leq (1,0)$ such $P(\partial)$ is the directional derivative operator with respect to a unique vector $y \in V_1$. In particular, the polynomial functions $P_{\ell,j}^*$ are of degree $\leq (1,0)$, too, so that the derivatives $P_{\ell,j}^*(\partial) f$, $\ell \in \Lambda(P)$ and $j \in \mathbb{N}_o$, are well-defined and belong to $C^o(W)$.

We fix $x \in X$. Then by the main theorem of calculus, by (1.49), and by Fubini's theorem - which can be applied in view of (5.5) with (f, w) replaced by $(\partial_i f, w\omega_j)$ or $(P_{\ell,j}^*(\partial) f, w\omega_j)$ - we deduce for every sufficiently small $h \in \mathbb{R}$

$$(5.9) \qquad \frac{1}{h}\Big(f_{m,w}(x + hy) - f_{m,w}(x)\Big) =$$

$$= \int_0^{+\infty} t^{-m} \frac{1}{h}\Big((f\circ M_t)(x+hy) - (f\circ M_t)(x)\Big)\, w(t)\, \frac{dt}{t} =$$

$$= \int_0^{+\infty} t^{-m} \int_0^1 P(\partial)(f\circ M_t)(x+shy)\, ds\; w(t)\, \frac{dt}{t} =$$

$$= \sum_{\ell\in\Lambda(P)} \int_0^{+\infty} t^{-m} \int_0^1 \Big[t^{\ell} \sum_{j\in\mathbb{N}_0} \omega_j(t)\, P^*_{\ell,j}(\partial) f\Big](M_t(x+shy))\, ds\; w(t)\, \frac{dt}{t} =$$

$$= \sum_{\ell\in\Lambda(P)} \sum_{j\in\mathbb{N}_0} \int_0^1 (P^*_{\ell,j}(\partial) f)_{m-\ell, w\omega_j}(x+shy)\, ds\,.$$

Since by Proposition 5.2 the functions $(P^*_{\ell,j}(\partial) f)_{m-\ell,w\omega_j}$ are continuous the first part of the assertion follows.

From this one obtains the condition (5.8) by another application of (1.49). ■

The proof of the following assertion is trivial.

Remark 5.5. *Let $f\in C^o(W)$. If $A\in L(V,V)$ commutes with M then $f\circ A$ belongs to $C^o(W\circ A)$, and $(f\circ A)_{m,w} = f_{m,w}\circ A$ for every $m\in\mathbb{C}(c,d)$.* ■

Now let G and $\mathfrak{G}$ satisfy the assumptions of Remark 2.67.(ii). First we observe that if $f\in C^o(W)$ then $f_{\mathfrak{G}}$ belongs to $C^o(W_G)$ where $W_G : X \longrightarrow [0,+\infty[$ is the continuous weight function defined by

$$W_G(x) := \sup\{\, W(Ax)\,;\; A\in G\}\,, \qquad x\in X.$$

Furthermore, we note that (5.2) remains valid if the pair (W,U) is replaced by (W_G, U_G). By applying Fubini's theorem (compare the proof of Proposition 3.7.(ii)) we derive

Remark 5.6. *If $f\in C^o(W)$ then $(f_{m,w})_{\mathfrak{G}}$ belongs to $C^o(U_G)$ and is equal to $(f_{\mathfrak{G}})_{m,w}$.* ■

As for the support of $f_{m,w}$, one immediately obtains

Remark 5.7. *Let $f\in C^o(X)$ be such that by (3.1)' a function $f_{m,w} : X \longrightarrow \mathbb{C}$ is well-defined. Then $\operatorname{supp} f_{m,w} \subset \overline{(\operatorname{supp} f)_{M,I}}$ where $I := \operatorname{supp} w$.* ■

Note that the set $(\operatorname{supp} f)_{M,I}$ need not be closed even if $\operatorname{supp} f$ is compact:

Example 5.8. *(i) Suppose that (1.14) holds. If* $x \in V_+$ *then* $0 \in \overline{L} \setminus L$ *where* $L := \{x\}_{M,[1,+\infty[}$.

(ii) Suppose that $p \in]0,+\infty[^2$. *Let* $\beta > p_2/p_1$, *and set*

$$K := \{(x,y) \in [-1,1] \times [0,1];\ |x|^\beta \le y\}.$$

Then $L := K_{p,]0,1]} = \mathbb{R} \times]0,+\infty[\cup \{(0,0)\}$, *and* $\overline{L} \setminus L = \dot{\mathbb{R}} \times \{0\}$.

Proof. *(i)*: The assumption (1.14) implies that $M_{1/t}x$ tends to 0 as $t \to +\infty$.

(ii): Let $(u,v) \in \mathbb{R} \times]0,+\infty[$. Then we can choose $t \in]0,1]$ so small that $x := t^{p_1}u \in [-1,1]$, $y := t^{p_2}v \in]0,1]$, and $t^{p_2 - \beta p_1} \ge |u|^\beta / v$, i.e. $y \ge |x|^\beta$. ∎

In general, for $\overline{K_{M,I}} \setminus K_{M,I}$ the following inclusion is valid:

Lemma 5.9. *Suppose that (1.14) holds. Let I be a closed subset of* $]0,+\infty[$, *and let K be a compact subset of V. Then* $\overline{K_{M,I}} \setminus K_{M,I} \subset K_0 \cup K_\infty$ *where*

$$K_0 := \begin{cases} \emptyset & \text{if } 1/I \text{ is bounded} \\ M_0^{-1}(K \cap M_0(K)) & \text{if } 1/I \text{ is unbounded} \end{cases}$$

and

$$K_\infty := \begin{cases} \emptyset & \text{if } I \text{ is bounded} \\ M_0(K) & \text{if } I \text{ is unbounded} \end{cases} .$$

Proof. Let $x \in \overline{K_{M,I}}$. Then we find sequences $(t_j)_{j \in \mathbb{N}}$ in $1/I$ and $(k_j)_{j \in \mathbb{N}}$ in K such that $M_{t_j}k_j$ converges to x as $j \to \infty$. By choosing subsequences if necessary we achieve that (k_j) converges to some $k \in K$ and (t_j) to some $s \in \{0,+\infty\} \cup 1/I$. If $s < +\infty$ then by continuity we obtain that $x = M_s k$. This means: if $s = 0$ then $x \in M_0(K)$ and I is unbounded; if $0 < s < +\infty$ then $s \in 1/I$ and $x \in K_{M,I}$. If $s = +\infty$ then $1/I$ is unbounded, and the condition

$$\lim_{j \to \infty} t_j\, \dot{x}_+(k_j) = \dot{x}_+(x)$$

(which is valid since $\dot{x}_+$ is continuous - see Propositions 1.70 and 1.61) implies that

$$\dot{x}_+(k) = \lim_{j \to \infty} \dot{x}_+(k_j) = 0 ,$$

i.e. $k \in K \cap M_0(V) = K \cap M_0(K)$. Since M_0 is continuous it follows that

$$M_o x = \lim_{j\to\infty} M_o(M_{t_j} k_j) = \lim_{j\to\infty} M_o k_j = M_o k = k\,, \quad \text{i.e.} \quad x \in M_o^{-1}(K \cap M_o(K)).$$ ∎

From now on we suppose that (1.14) holds. Moreover, we assume that the norm $|\cdot|$ is induced by a scalar product on V which satisfies the condition (1.79). In order to apply the preceding results we are going to choose special W, U, and w satisfying (5.2). Our choices depend on whether the support of w stays away from 0 or from $+\infty$. In order to deal with the first case we fix a continuous function $u : M_o(X) \longrightarrow]0,+\infty[$, and for $N \in \mathbb{R}$ we define a continuous function $W_N : X \longrightarrow]0,+\infty[$ by

$$W_N(x) := u(M_o x)\,(1+|P_+(x)|)^N, \qquad x \in X, \tag{5.10}$$

where P_+ is defined in Notation 1.6.B.(iii). Moreover, for every $\eta > 0$ we define $W_{N,\eta} : X_+ \longrightarrow]0,+\infty]$ by

$$W_{N,\eta}(x) := \begin{cases} W_N(x) & \text{if } x \in \Omega_{[\eta,+\infty[} \\ +\infty & \text{otherwise} \end{cases}, \tag{5.11}$$

employing the following

Notation 5.10. For any subset J of $[0,+\infty[$ we set $\Omega_J := \{x \in V;\ |P_+(x)| \in J\}$ where P_+ is defined in Notation 1.6.B.(iii).

Proposition 5.11. *Suppose that* $a := \inf \operatorname{supp} w > 0$.

(i) Let $L \in \mathbb{R}$ *be such that*

$$\int_a^{+\infty} |w(t)|\, t^L \frac{dt}{t} < +\infty\,. \tag{5.12}$$

Then for every $N \in [0,+\infty[$ *the assertion (5.1) is valid for the Banach space* $E = C^o(W_{-N}/X_+)$, *for the Fréchet space* $F = \bigcap_{\eta \in]0,1]} C^o(W_{-N,\eta})$ *and for the set* $D = \mathbb{C}(-N\lambda_{min}-L,+\infty)$ *where* $\lambda_{min} := \min\{\operatorname{Re}\lambda\,;\ \lambda \in \sigma_+\}$; *moreover, for arbitrary* $f \in C^o(W_{-N}/X_+)$ *and* $m \in \mathbb{C}(-N\lambda_{min}-L+1,+\infty)$ *we have*

$$\lim_{\varepsilon\to 0} f_{m,w_\varepsilon} = f_{m,w} \text{ in the topology of } F \text{ where } w_\varepsilon(t) := e^{-\varepsilon t} w(t). \tag{5.13}$$

(ii) If (5.12) is satisfied for every $L \in]0,+\infty[$ *then for every* $N \in \mathbb{R}$ *the assertion (5.1) holds for* $E = F = C^o(W_N)$ *and* $D = \mathbb{C}$.

Proof. *(i):* We fix $\Lambda \in]0, \lambda_{min}[$ and set $c := -N\Lambda - L$ and $\delta := \lambda_{min} - \Lambda$. Moreover, we choose $B \in]0, +\infty[$ such that

$$B\,t^{\Lambda} \le I(t) := \min\{t^{\operatorname{Re}\lambda-\delta}, t^{\operatorname{Re}\lambda+\delta} ;\ \lambda \in \sigma_+\}, \qquad t \in [a, +\infty[\,.$$

Since by Lemma 1.69 we may assume that $|P_+(M_t x)| \ge I(t)\,|P_+(x)|$ it follows that

$$(1 + |P_+(M_t x)|)^{-N}\,|w(t)| \le B^{-N}\,|P_+(x)|^{-N}\,|w(t)|\,t^L\,t^c, \qquad x \in X_+,\ t \in [a, +\infty[\,.$$

Since

$$t^c \le \max\{1, a^{c-d}\}\,\min\{t^c, t^d\}, \qquad t \in [a, +\infty[\,,\ d \in]c, +\infty[\,, \tag{5.14}$$

we deduce that (5.2) (with X replaced by X_+) is satisfied for $W = W_{-N}\big|_{X_+}$, $U(x) = |P_+(x)|^{-N} u(M_0 x)$, and $\gamma(t) = B^{-N} \max\{1, a^{c-d}\}\,|w(t)|\,t^L$. Note that $U \le (\eta/2)^{-N}\,W_{N,\eta}$ on $\Omega_{[\eta, +\infty[}$. Since by (5.12) γ belongs to $\mathscr{L}^1(]0, +\infty[; \frac{dt}{t})$ the first part of the assertion follows by Lemma 5.2.

In view of $|e^{-\varepsilon t} - 1| \le \varepsilon t$, $t \in]0, +\infty[$, we have

$$|f_{m,w_\varepsilon} - f_{m,w}| \le \varepsilon\,|f|_{\operatorname{Re} m - 1, |w|}\,. \tag{5.15}$$

Since the first part of the assertion remains valid if w is replaced by $|w|$ the assertion (5.13) follows.

(ii): We set $\hat{N} := \max\{N\lambda_{max}, 0\}$ where $\lambda_{max} := \max\{\operatorname{Re}\lambda ;\ \lambda \in \sigma_+\}$. Then by Lemma 1.69 we may assume that there exists a constant C_N such that

$$(1 + |P_+(M_t x)|)^N \le C_N\,t^{\hat{N}+1}\,(1 + |P_+(x)|)^N, \qquad x \in V,\ t \in [a, +\infty[\,.$$

We fix $c < 0 < d$. Then (5.2) is satisfied for $W = U = W_N$ and - in view of (5.14) - $\gamma(t) = \max\{1, a^{c-d}\}\,C_N\,|w(t)|\,t^{\hat{N}+1-c}$. ∎

Next we deal with the case that the support of w is bounded.

Proposition 5.12. *Suppose that* $b := \sup \operatorname{supp} w$ *is finite and that*

$$\int_0^b |w(t)|\,t^{-d}\,\frac{dt}{t} < +\infty\,. \tag{5.16}$$

Then for every $N \in [0, +\infty[$ *the assertion (5.1) is valid for* $E = F = C^o(W_N)$ *and* $D = \mathbb{C}(-\infty, d)$. *Moreover, for every* $f \in C^o(W_N)$ *the functions* f_{m,w_ε} *converge to* $f_{m,w}$ *in the norm of* $C^o(W_N)$ *as* $\varepsilon \to 0$ *where* $w_\varepsilon(t) := e^{-\varepsilon t}\,w(t)$.

Proof. Since $N \geq 0$ and $b < +\infty$ it follows from the last assertion of Proposition 1.70 and from Lemma 1.69 that there is a constant C such that $W_N(M_t x) \leq C\, W_N(x)$ for arbitrary $x \in X$ and $t \in]0,b]$. Consequently, for any $c \in]-\infty, d[$ the estimate (5.2) holds for $W = U = W_N$ and $\gamma(t) = C \max\{1, b^{d-c}\}\, |w(t)|\, t^{-d}$, and the first assertion follows by Lemma 5.2.

The second assertion is an immediate consequence of (5.15). ■

Working with general weight functions W requires an additional assumption on X: from now on for the rest of Chapter 5 with the exception of section (g) we assume that X satisfies

(5.17) $$X \supset M_0(X), \text{ i.e. } X = G_M(\sigma_+) + X'' \text{ where}$$
$$X'' := \{x'' \in \ker M \;;\; x' + x'' \in X \text{ for some } x' \in G_M(\sigma_+)\}.$$

Proposition 5.13. *Suppose, in addition, that the assumptions of Proposition 5.12 are valid. Then the assertion (5.1) holds for $E = F = C^o(X)$ and $D = \mathbb{C}(-\infty, d)$. More precisely, if $W : X \longrightarrow [0, +\infty[$ is any continuous weight function then by*

(5.18) $$W_{(b)}(x) := \sup\{\, W(M_t x) \;;\; t \in [0, b]\}$$

a continuous weight function $W_{(b)} : X \longrightarrow [0, +\infty[$ is defined such that the assertion (5.1) holds for $E = C^o(W)$, $F = C^o(W_{(b)})$, and $D = \mathbb{C}(-\infty, d)$.

Proof. That $W_{(b)}$ is well-defined and continuous follows from the continuity of W, from (5.17) and from the continuity of $[0,b] \times V \longrightarrow V$, $(t,x) \mapsto M_t x$, (see Corollary 1.9). Obviously, for any $c \in]-\infty, d[$ the condition (5.2) is satisfied for $V = W_{(b)}$ and $\gamma(t) := \max\{1, b^{d-c}\}\, t^{-d}\, w(t)$. Since by (5.16) γ belongs to $\mathcal{L}^1(]0, +\infty[\,; \frac{dt}{t})$ the second part of the assertion follows from Lemma 5.2. One obtains the first part by first fixing $f \in C^o(X)$ and then applying the second part to $W := |f| + 1$. ■

If the support of f stays away from ker M then $f_{m,w}$ is well-defined by (3.1)':

Remark 5.14. *Suppose that $b := \sup \operatorname{supp} w < +\infty$. Let $f \in C^o(X)$ be such that*

(5.19) $$\operatorname{supp} f \subset X_+ .$$

Then $f_{m,w}: X \longrightarrow \mathbb{C}$ is well-defined by (3.1)' for every $m \in \mathbb{C}$, and by $m \mapsto f_{m,w}$ a holomorphic function on $\mathbb{C}$ with values in $C^o(X)$ is defined.

Proof. By Proposition 3.3 it suffices to show that supp f is an $(M,]0,b])$-bounded subset of X.

Let $y \in X$. Since supp f is a closed subset of X the assumptions (5.17) and (5.19) imply that $\mathrm{dist}(M_0 y, \mathrm{supp}\, f)$ is positive. Hence, by the continuity of the map $(t,x) \mapsto M_t x$ at $(0,y)$ (see Corollary 1.9) we find $\varepsilon_y > 0$ such that $M_t x \notin \mathrm{supp}\, f$ for arbitrary $t \in]0, \varepsilon_y]$ and $x \in K(y, \varepsilon_y)$. If K is a compact subset of X then we find a finite subset N of K such that K is covered by the balls $K(y, \varepsilon_y)$, $y \in N$. Consequently, $\{ t \in]0,b] \,;\, M_t(K) \cap \mathrm{supp}\, f \neq \emptyset \}$ is contained in $[\varepsilon, b]$ where $\varepsilon := \min\{ \varepsilon_y \,;\, y \in N\}$ is positive. ∎

In the following section our aim is to get rid of the assumption (5.19).

(b) Meromorphic Extensions of the Maps $f \mapsto f_{m,w}$

We continue to suppose that (1.14) and (5.17) are valid. Moreover, we assume that $b := \sup \mathrm{supp}\, w$ is finite and that

(5.20) $$\delta(w) := \sup\{ d \in \mathbb{R} \,;\, (5.16) \text{ holds}\} \geq 0 \,.$$

We fix a function $f \in C^o(X)$. Recall that by Proposition 5.13 the prescription $m \mapsto f_{m,w}$ defines a holomorphic function from the half plane $\mathbb{C}(-\infty, 0)$ into $C^o(X)$. Under additional assumptions on w and f we are going to extend it to a meromorphic function on the whole of $\mathbb{C}$. In view of

(5.21) $$t^{-m-1} = -\frac{1}{m}\, \partial_t t^{-m}$$

the idea is to do partial integration with the defining integral of $f_{m,w}$ in (3.1)'. Consequently, we are lead to postulate conditions ensuring that the integrand in (3.1)' is differentiable with respect to t. So first of all we suppose that $w|_{]0,b]}$ is a C^1 function. Moreover, we require f to be continuously differentiable with respect to the variables in $G_M(\sigma_o)$; more precisely, we suppose that f belongs to

$C^{1,0}(X)$ where from now on we take the decomposition (1.14) as the basis for the definition of $C^{r,s}(X)$, i.e. in Definition 1.18 we have

$$V_1 = G_M(\sigma_+) \text{ and } V_2 = \ker M .$$

Then it follows by the chain rule and by (1.5) that for every $x \in X$ the function $]0,b] \longrightarrow \mathbb{C}$, $t \mapsto f(M_t x)\, w(t)$, is differentiable satisfying

$$\partial_t\Big(f(M_t x)\, w(t)\Big) = \frac{1}{t}\,(D_1 f(M_t x)\cdot P_+ M M_t x)\, w(t) + f(M_t x)\, w'(t) \tag{5.22.a}$$

for every $t \in]0,b]$ where by D_1 we denote differentiation with respect to the variables in $G_M(\sigma_+)$. In order to compute the derivative with respect to coordinates we suppose that mixed real-complex coordinates are chosen according to Convention 1.24. This means, in particular, that the basis A of $V_{\mathbb{R}}$ contains a basis of $\ker M$. By J_O we are going to denote the set of indices describing the coordinates of the points in $\ker M$. More precisely, employing the notation from Convention 1.24 we introduce

Notation 5.15. (i) $J_O := \{ j \in \mathbb{N}_c ;\ \rho_j = 0 \}$;

(ii)$^+$ $\mathfrak{A}^+ := \{ \alpha = (\beta,\gamma,\delta) \in \mathfrak{A} ;\ \beta_j = 0 \text{ for every } j \in J_O \}$;

(ii)^O $\mathfrak{A}^O := \{ \alpha = (\beta,\gamma,\delta) \in \mathfrak{A} ;\ \beta_j = 0 \text{ for every } j \in \mathbb{N}_c \setminus J_O,\ \gamma = 0 = \delta \}$;

(iii) For any $\alpha \in \mathfrak{A}$ we define $\alpha_+ \in \mathfrak{A}^+$ and $\alpha_O \in \mathfrak{A}^O$ by the equation $\alpha = \alpha_+ + \alpha_O$.

(iv) For arbitrary $r,s \in \mathbb{N}_O \cup \{\infty\}$ we set $\mathfrak{A}^{r,s} := \{ \alpha \in \mathfrak{A} ;\ |\alpha_+| \le r,\ |\alpha_O| \le s \}$.

Note that in view of (1.32) and Remark 1.30 we have

$$D_1 f(M_t x)\cdot P_+ M M_t x = \sum_{\alpha \in \mathfrak{A}_1} t^{\alpha_M} \sum_{j \in \mathbb{N}_O} \omega_j(t)\, R_{\alpha,j}(x)\, (\partial^\alpha f)(M_t x) \tag{5.22.b}$$

where $\mathfrak{A}_1 := \{ \alpha \in \mathfrak{A}^+ ;\ |\alpha| = 1 \}$ and where the $R_{\alpha,j} : V \longrightarrow \mathbb{C}$ are suitable polynomial functions of degree $\le (1,0)$ which are almost quasihomogeneous of degree α_M; of course, only finitely many of them do not vanish identically. Note that in case M is semi-simple we have

$$R_{\alpha,j}(x) = \delta_{Oj}\, \lambda^\alpha x^\alpha , \qquad j \in \mathbb{N}_O,\ \alpha \in \mathfrak{A}_1 . \tag{5.23}$$

Combining (5.22.a) and (5.22.b) we see, in particular, that the derivative of $t \mapsto f(M_t x)\, w(t)$ belongs to $\mathcal{L}^1(]0,b], t^{-d}dt)$ for every $d \in]-\infty,0[$ provided that

(5.20)' $\delta(w') \geq -1$.

Consequently, if we assume, in addition, that

(5.24) $$\lim_{t\to 0} w(t)\, t^{-d} = 0 \quad \text{for every } d \in]-\infty,0[$$

then for every $m \in \mathbb{C}(-\infty,0)$ the equation (5.21) and partial integration lead to

(5.25) $$f_{m,w}(x) = \frac{1}{m}\int_0^b t^{-m}\partial_t\big(f(M_t x)\, w(t)\big)\, dt - \frac{1}{m} b^{-m} f(M_b x)\, w(b)$$

which in view of (5.22.a) and (5.22.b) can be written as

(5.26) $$f_{m,w} = \frac{1}{m}\Big(\sum_{\alpha\in\mathfrak{A}_1} \sum_{j\in\mathbb{N}_0} R_{\alpha,j}\, (\partial^\alpha f)_{m-\alpha_M, w\omega_j} + f_{m-1,w'} - b^{-m}\, w(b)\, f\circ M_b \Big).$$

Now, assuming that not only (5.20)' but even

(5.20)" $\delta(w') \geq 0$

is valid, then in view of $\delta(w\omega_j) = \delta(w) \geq 0$ we observe that the right-hand side of (5.26) is well-defined if $\operatorname{Re} m < c_w$ where

(5.27) $$c_w := \begin{cases} \lambda_{min} := \min\{\operatorname{Re}\lambda\, ;\, \lambda \in \sigma_+\} & \text{if } w' \equiv 0 \\ \min\{1, \lambda_{min}\} & \text{if } w' \not\equiv 0 \end{cases};$$

in fact, it defines a meromorphic function on $\mathbb{C}(-\infty, c_w)$ with values in $C^0(X)$ and possibly with a (simple) pole at $m = 0$. Consequently, (5.26) shows that $m \mapsto f_{m,w}$ extends to a meromorphic function on $\mathbb{C}(-\infty, c_w)$ with values in $C^0(X)$.

In order to extend this meromorphic extension further one wants to iterate the above procedure. For this a good choice for w seems to be the function w_0 defined by

(5.28) $$w_0(t) := e^{-\varepsilon t}\chi_{]0,b]}(t), \qquad t \in]0,+\infty[,$$

with $\varepsilon \in \mathbb{C}$ being a fixed constant. Since on $]0,b]$ we have $w_0' = -\varepsilon w_0$ the assumptions (5.20) and (5.24) are satisfied, indeed. However, when going to repeat the extension procedure we have to work with the weight functions $w_k := w_0\,\omega_k$, $k \in \mathbb{N}$, as well, which in view of

(5.29) $$w_k'(t) = -\varepsilon w_k(t) + \frac{1}{t} w_{k-1}(t) \quad (\text{where } w_{-1} :\equiv 0), \qquad t \in]0,b],$$

no longer satisfy (5.20)". To overcome this difficulty we observe that the w_k

still satisfy (5.20) and (5.24) so that (5.26) may be applied to $w = w_k$. In view of (5.29) and (1.38) the equation (5.26) then becomes

$$(5.26)_k \qquad f_{m,w_k} = \frac{1}{m}\Big(\sum_{\alpha\in\mathfrak{A}_1} \sum_{j\in\mathbb{N}_0} \binom{k+j}{k} R_{\alpha,j} (\partial^\alpha f)_{m-\alpha_M, w_{k+j}} + f_{m,w_{k-1}} - \\ - \varepsilon f_{m-1,w_k} - b^{-m} w_k(b)\, f\circ M_b \Big).$$

The only term on the right-hand side which is not defined for every $m\in\mathbb{C}(-\infty,c_w)$ is $f_{m,w_{k-1}}$. Consequently, we insert first the equation $(5.26)_{k-1}$ and then $(5.26)_{k-2},\dots,(5.26)_1$ and finally (5.26) into $(5.26)_k$ until all the terms containing f_{m,w_i}, $i\in\{0\}\cup\mathbb{N}_k$, are eliminated. This leads to

$$(5.30) \qquad f_{m,w_k} = \sum_{i=0}^{k} m^{i-k-1}\Big(\sum_{\alpha\in\mathfrak{A}_1} \sum_{j\in\mathbb{N}_0} \binom{i+j}{i} R_{\alpha,j} (\partial^\alpha f)_{m-\alpha_M, w_{i+j}} - \\ - \varepsilon f_{m-1,w_i} - b^{-m} w_i(b)\, f\circ M_b \Big).$$

Since here the right-hand side is well-defined for every $m\in\mathbb{C}(-\infty,c_w)$ we see that the function $\mathbb{C}(-\infty,0)\longrightarrow C^0(X)$, $m\mapsto f_{m,w_k}$, extends to a meromorphic function on $\mathbb{C}(-\infty,c_w)$ with values in $C^0(X)$ and possibly with a pole of order $\le k+1$ at $m=0$.

In order to employ the above extension procedure for a second time we have to assume that the derivatives $\partial^\alpha f$, $\alpha\in\mathfrak{A}_1$, belong to $C^{1,0}(X)$ themselves. In fact, if we fix $r\in\mathbb{N}\cup\{\infty\}$ and assume that f belongs to $C^{r,0}(X)$ we obtain by induction a meromorphic extension to $\mathbb{C}(-\infty,rc_w)$ with poles lying in the set

$$(5.31) \qquad \mathfrak{A}_\varepsilon(M) := \begin{cases} \mathfrak{A}(M) & \text{if } \varepsilon=0 \\ \mathfrak{A}(M)+\mathbb{N}_0 & \text{if } \varepsilon\ne 0 \end{cases}.$$

It is desirable to keep track of the derivatives with respect to the variables in $\ker M$, as well. To this end for the rest of this and the following two sections we fix $r\in\mathbb{N}\cup\{\infty\}$, $s\in\mathbb{N}_0\cup\{\infty\}$, and $f\in C^{r,s}(X)$. Moreover, we choose $k\in\mathbb{N}_0$ and set $w=w_k$. Instead of c_w (see (5.27)) from now on we write c_ε.

Proposition 5.16. *The function $\mathbb{C}(-\infty,0)\ni m\mapsto f_{m,w}$ has a (unique) meromorphic extension $\mathfrak{M}_{f,w}:\mathbb{C}(-\infty,rc_\varepsilon)\longrightarrow C^{r,s}(X)$ having the following properties:*

(i) the poles of $\mathfrak{M}_{f,w}$ are contained in the set $\mathfrak{A}_\varepsilon(M)$ defined by (5.31); they are simple if $k=0$ and if M is semi-simple;

(ii) $\operatorname{supp} a_i(m;\mathfrak{M}_{f,w}) \subset \overline{(\operatorname{supp} f)_{M,]0,b]}}$ for arbitrary $m\in\mathbb{C}(-\infty,rc_\varepsilon)$ and $i\in\mathbb{Z}$;

(iii) for every polynomial function $P: V^* \longrightarrow \mathbb{C}$ of degree $\leq (r,s)$ we have

$$(5.32) \qquad P(\partial)\, a_i(m; \mathfrak{M}_{f,w}) = \sum_{\ell \in \Lambda(P)} \sum_{j \in \mathbb{N}_o} \binom{k+j}{j} a_i(m-\ell; \mathfrak{M}_{g_{\ell,j}, w_{k+j}})$$

for arbitrary $m \in \mathbb{C}(-\infty, r c_\varepsilon)$ and $i \in \mathbb{Z}$ where $g_{\ell,j} := P^*_{\ell,j}(\partial) f$, the polynomial functions $P^*_{\ell,j}$ being defined in the text preceding Remark 1.42.

Note that for arbitrary $m \in \mathbb{C}(-\infty, r c_\varepsilon)$ and $\ell \in \Lambda(P)$ the point $m-\ell$ belongs to the domain of definition of $\mathfrak{M}_{g_{\ell,j}, w_{k+j}}$, indeed. In fact, since $P^*_{\ell,j}$ is almost quasihomogeneous of degree ℓ this is seen from the following simple remark which is given a separate formulation for the sake of easy reference.

Remark 5.17. *Let $\ell \in \mathfrak{A}(M)$ be such that the set $A := \{\alpha \in \mathfrak{A}^{r,s};\ \alpha_M = \ell\}$ is non-empty, and define*

$$\rho := \max\{|\alpha_+|;\ \alpha \in A\} \qquad \text{and} \qquad \sigma := \max\{|\alpha_0|;\ \alpha \in A\}.$$

(i) For every polynomial function $P: V^* \longrightarrow \mathbb{C}$ of degree $\leq (r,s)$ which is almost quasihomogeneous of degree ℓ and of type M^* the function $P(\partial) f$ belongs to $C^{r-\rho, s-\sigma}(X)$.

(ii) For every $m \in \mathbb{C}(-\infty, r c_\varepsilon)$ we have $m - \ell \in \mathbb{C}(-\infty, (r-\rho) c_\varepsilon)$.

Proof. *(i)*: By the very definition of ρ and σ the functions $\partial^\alpha f$, $\alpha \in A$, belong to $C^{r-\rho, s-\sigma}(X)$. Since the assumption on P means that $P(\partial) f$ is a linear combination of the $\partial^\alpha f$, $\alpha \in A$, the assertion follows.

(ii): We choose $\alpha \in A$ such that $|\alpha_+| = \rho$. Then

$$\operatorname{Re} \ell = \operatorname{Re} \alpha_M \geq |\alpha_+| \lambda_{min} = \rho \lambda_{min} \geq \rho c_\varepsilon$$

so that $\qquad \operatorname{Re}(m-\ell) < r c_\varepsilon - \operatorname{Re} \ell \leq r c_\varepsilon - \rho c_\varepsilon \quad$ for every $m \in \mathbb{C}(-\infty, r c_\varepsilon)$. ∎

Proof of Proposition 5.16. By the reasoning preceding the formulation of the proposition we obtain a meromorphic extension $\mathcal{F}_{f,w}: \mathbb{C}(-\infty, r c_\varepsilon) \longrightarrow C^0(X)$ of $m \mapsto f_{m,w}$ satisfying (i). In order to verify the differentiability properties of the functions in the image of $\mathcal{F}_{f,w}$ we first deal with the case

$$(5.33) \qquad P = \langle \cdot, y \rangle \quad \text{for some } y \in G_M(\sigma_+) \cup \ker M.$$

Here the functions $g_{\ell,j}$ belong to $C^{r-q, s+1-q}(X)$ where $q := 1$ if $y \in G_M(\sigma_+)$ and

$q := 0$ if $y \in \ker M$. It follows that $\mathcal{F}_{g_{\ell,j},w_{k+j}}$, $\ell \in \Lambda(P)$ and $j \in \mathbb{N}_0$, are well-defined meromorphic functions from $\mathbb{C}(-\infty,(r-q)c_\varepsilon)$ into $C^0(X)$. If, in addition, we fix $x \in X$ and $h \in \dot{\mathbb{R}}$ sufficiently small then in view of (1.38) the formula (5.9) reads as

$$\text{(5.9)'} \qquad \frac{1}{h}\Big(\mathcal{F}_{f,w_k}(m)(x+hy) - \mathcal{F}_{f,w_k}(m)(x)\Big) =$$

$$= \sum_{\ell \in \Lambda(P)} \int_0^1 \sum_{j \in \mathbb{N}_0} \binom{j+k}{j} \mathcal{F}_{g_{\ell,j},w_{k+j}}(m-\ell)(x+shy)\,ds$$

for every $m \in \mathbb{C}(-\infty,0)$. Note that $m-\ell$ belongs to $\mathbb{C}(-\infty,(r-q)c_\varepsilon)$ if $m \in \mathbb{C}(-\infty,rc_\varepsilon)$ and $\ell \in \Lambda(P)$. Hence, by the principle of analytic continuation the formula (5.9)' remains valid for every $m \in D_r := \mathbb{C}(-\infty,rc_\varepsilon) \setminus \mathfrak{A}_\varepsilon(M)$. Since $\mathcal{F}_{g_{\ell,j},w_{k+j}}(m-\ell)$ is continuous it follows for every $m \in D_r$ that $P(\partial)\,\mathcal{F}_{f,w}(m)$ exists and is equal to

$$\sum_{\ell \in \Lambda(P)} \sum_{j \in \mathbb{N}_0} \binom{j+k}{j} \mathcal{F}_{g_{\ell,j},w_{k+j}}(m-\ell).$$

By induction one derives that $\mathcal{F}_{f,w}$ induces a meromorphic function $\mathfrak{M}_{f,w}$ defined on $\mathbb{C}(-\infty,rc_\varepsilon)$ with values in $C^{r,s}(X)$. Moreover, the formula

$$\text{(5.32)'} \qquad P(\partial)\,\mathfrak{M}_{f,w}(m) = \sum_{\ell \in \Lambda(P)} \sum_{j \in \mathbb{N}_0} \binom{k+j}{j} \mathfrak{M}_{g_{\ell,j},w_{k+j}}(m-\ell), \qquad m \in D_r,$$

is already proved for the polynomial functions P of the form (5.33). Consequently, making use of Corollary 1.37.(ii) we deduce by induction that (5.32)' is valid if P is a product of polynomial functions of the form (5.33). Since an arbitrary P is a linear combination of such products we conclude that (5.32)' holds for arbitrary P.

In order to complete the proof of <u>*(iii)*</u> we fix $m \in \mathbb{C}(-\infty,rc_\varepsilon)$, choose $a > 0$ so small that $\overline{K}(m,a) \subset \mathbb{C}(-\infty,rc_\varepsilon)$ and $\overline{K}(z,a) \cap \mathfrak{A}_\varepsilon(M) \subset \{z\}$, $z \in \{m\} \cup (m-\Lambda(P))$, and define $\gamma: [0,2\pi] \longrightarrow \mathbb{C}$ by $\gamma(t) := a e^{it}$. Then for arbitrary $x \in X$ and $I \in \mathbb{Z}$ we obtain

$$\text{(5.32)''} \qquad [P(\partial)\,a_I(m;\mathfrak{M}_{f,w})](x) = \frac{1}{2\pi i}\int_\gamma P(\partial)\,[\mathfrak{M}_{f,w}(m+z)](x)\, z^{-I-1}\,dz$$

and

$$\frac{1}{2\pi i}\int_\gamma \mathfrak{M}_{g_{\ell,j},w_{k+j}}(m-\ell+z)(x)\, z^{-I-1}\,dz = a_I(m-\ell;\mathfrak{M}_{g_{\ell,j},w_{k+j}})(x)$$

Herce, the equality (5.32) follows from (5.32)'.

For the proof of <u>*(ii)*</u> we observe (see Remark 5.7) that in case $\operatorname{Re} m < 0$ we have

$$\mathfrak{M}_{f,w}(m)(x) = 0, \qquad x \in X \setminus \overline{(\operatorname{supp} f)_{M,]0,b]}}$$

By the principle of analytic continuation this remains valid for every $m \in D_r$. From this the assertion follows since for $m \in \mathfrak{U}_\varepsilon(M) \cap \mathbb{C}(-\infty, rc_\varepsilon)$ and $I \in \mathbb{Z}$ we have

$$a_I(m;\mathfrak{M}_{f,w})(x) = \frac{1}{2\pi i}\int_\gamma \mathfrak{M}_{f,w}(m+z)(x)\, z^{-I-1}\, dz$$

where γ is chosen as above. ■

An explicit formula for $\mathfrak{M}_{f,w}$ is given in Proposition 5.28 below.

Notation 5.18. $f_{m,w} := a_0(m;\mathfrak{M}_{f,w})$, $\qquad m \in \mathbb{C}(-\infty, rc_\varepsilon)$.

If f satisfies (5.19) then this notation is compatible with (3.1)' as is seen from Remark 5.14 by the principle of analytic continuation.

At first glance it may seem to be surprising that $f_{m,w}$ is differentiable of the same orders as f although for its very definition derivatives are used up. However, a look at formula (5.32) and at Remark 5.17 shows that for $(g_{\ell,j})_{m-\ell,w_{k+j}}$ to be well-defined for $m \in \mathbb{C}(-\infty, rc_\varepsilon)$ requires lower order differentiability of $g_{\ell,j}$ since $m-\ell$ lies to the left of m.

As for almost quasihomogeneous differential operators with variable coefficients the analogue of Proposition 3.4 is valid. As for the assumptions compare also the text preceding that proposition.

Proposition 5.19. *Let $\ell \in \mathbb{C}$ and $N \in \mathbb{N}$, and let $P_0 : X \times V^* \longrightarrow \mathbb{C}$ be a continuous copolynomial function of degree $\leq (r,s)$ which is almost quasihomogeneous of degree ℓ, of order $\leq N$, and of type $M \times (-M)^*$. For every $j \in \mathbb{N}_N$ let $P_j : X \times V^* \longrightarrow \mathbb{C}$ be its jth order deficiency which is a continuous copolynomial function of degree $\leq (r,s)$, as well. Moreover, let $\rho \in \mathbb{N}_0$ be such that $P_j(x,\partial) f \in C^{\rho,s}(X)$ for every $j \in \{0\} \cup \mathbb{N}_N$. Finally, let $m \in \mathbb{C}(-\infty, \min\{rc_\varepsilon, \rho c_\varepsilon - \mathrm{Re}\,\ell\})$. Then for every $i \in \mathbb{Z}$ we have*

$$(5.34) \qquad P_0(x,\partial)\, a_i(m;\mathfrak{M}_{f,w}) = \sum_{j=0}^{N} (-1)^j \binom{k+j}{j}\, a_i(m+\ell;\mathfrak{M}_{P_j(x,\partial)f,\,w_{k+j}})$$

and

$$(5.34)' \qquad a_i(m+\ell;\mathfrak{M}_{P_0(x,\partial)f,\,w}) = \sum_{j=0}^{N} \binom{k+j}{j}\, P_j(x,\partial)\, a_i(m;\mathfrak{M}_{f,\,w_{k+j}}).$$

In view of Notation 5.18, for $i=0$ the equations (5.34) and (5.34)' take the familiar form (3.4) and (3.4)' with $w=w_k$.

Proof. If Re m is sufficiently small then by Lemma 5.4 in the defining integral for $f_{m,w}$ we may interchange differentiation and integration. Hence performing the computations done in the proof of Proposition 3.4 and taking (1.38) into account we obtain for sufficiently small Re m

$$P_o(x,\partial)\,f_{m,w} = \sum_{j=0}^{N} \binom{k+j}{j} (-1)^j\,(P_j(x,\partial)f)_{m+\ell,w_{k+j}}$$

and

$$(P_o(x,\partial)\,f)_{m+\ell,w} = \sum_{j=0}^{N} \binom{k+j}{j} P_j(x,\partial)\,f_{m,w_{k+j}}\,.$$

By the principle of analytic continuation this remains valid for every $m\in \mathbb{C}(-\infty,R)\setminus(\mathfrak{U}_\varepsilon(M)\cap(\mathfrak{U}_\varepsilon(M)-\ell))$ where $R:=\min\{rc_\varepsilon,\rho c_\varepsilon-\mathrm{Re}\,\ell\}$.

Now, for any $m\in\mathbb{C}(-\infty,R)$ let $a>0$ be such that $\overline{K}(m,a)\subset\mathbb{C}(-\infty,R)$ and $\overline{K}(z,a)\cap\mathfrak{U}_\varepsilon(M)\subset\{z\}$ for every $z\in\{m,m+\ell\}$, and define $\gamma:[0,2\pi]\longrightarrow\mathbb{C}$ by $t\mapsto ae^{it}$. Then for arbitrary $I\in\mathbb{Z}$ and $x\in X$ it follows that

$$[P_j(x,\partial)\,a_I(m;\mathfrak{M}_{f,w})](x) = \frac{1}{2\pi i}\int_\gamma [P_j(x,\partial)\,f_{m+z,w}](x)\,z^{-I-1}\,dz$$

and

$$a_I(m+\ell;\mathfrak{M}_{P_j(x,\partial)f,w_{k+j}})(x) = \frac{1}{2\pi i}\int_\gamma [P_j(x,\partial)\,f]_{m+\ell+z,w_{k+j}}(x)\,z^{-I-1}\,dz\,.$$

Combining this with the equations above one obtains (5.34) and (5.34)'. ■

A special case is the following analogue of Corollary 3.5.

Corollary 5.20. *Let $\ell\in\mathbb{C}$ and $N\in\mathbb{N}$, and let $q_o:X\longrightarrow\mathbb{C}$ be almost quasihomogeneous of degree ℓ and of order $\le N$. Moreover, let $\rho\in\mathbb{N}_o$ be such that for every $j\in\{0\}\cup\mathbb{N}_N$ the function q_jf belongs to $C^{\rho,o}(X)$ where in case $j\ge 1$ $q_j:X\longrightarrow\mathbb{C}$ be the jth order deficiency of q_o. Then for arbitrary $m\in\mathbb{C}(-\infty,\min\{rc_\varepsilon,\rho c_\varepsilon-\mathrm{Re}\,\ell\}$ and $i\in\mathbb{Z}$ we have*

(5.35) $$q_o\,a_i(m;\mathfrak{M}_{f,w}) = \sum_{j=0}^{N}(-1)^j\binom{k+j}{j}\,a_i(m+\ell;\mathfrak{M}_{q_jf,w_{k+j}})$$

and

(5.35)' $$a_i(m+\ell;\mathfrak{M}_{q_o f,w}) = \sum_{j=0}^{N}\binom{k+j}{j}\,q_j\,a_i(m;\mathfrak{M}_{f,w_{k+j}})\,. \quad ■$$

Next we prove the analogue of Proposition 3.7, resp. Remarks 5.5 and 5.6.

Proposition 5.21. *(i) If $A \in L(V,V)$ commutes with M then $f \circ A \in C^{r,s}(A^{-1}(X))$, and for arbitrary $m \in \mathbb{C}(-\infty, rc_\varepsilon)$ and $i \in \mathbb{Z}$ we have $a_i(m;\mathfrak{M}_{f\circ A,w}) = a_i(m;\mathfrak{M}_{f,w}) \circ A$.*

(ii) If $\mathfrak{G}$ satisfies the assumptions of Remark 2.67.(ii) then $f_{\mathfrak{G}} \in C^{r,s}(X)$, and for arbitrary $m \in \mathbb{C}(-\infty, rc_\varepsilon)$ and $i \in \mathbb{Z}$ we have $a_i(m;\mathfrak{M}_{f_{\mathfrak{G}},w}) = (a_i(m;\mathfrak{M}_{f,w}))_{\mathfrak{G}}$.

Proof. First of all we note that every $A \in L(V,V)$ commuting with M leaves every subspace of the form $\ker Q(M)$ invariant where Q is any complex polynomial in one variable. Consequently, the generalized eigenspaces $G_M(\lambda)$, $\lambda \in \sigma$, and hence $G_M(\sigma_+)$ and $\ker M$ are A-invariant. This shows that $f \circ A \in C^{r,s}(A^{-1}(X))$ and - under the assumptions of (ii) - $f_{\mathfrak{G}} \in C^{r,s}(X)$. Moreover, every A commuting with M commutes with M_0 so that $A^{-1}(X)$ satisfies the condition (4.17), as well.

Now, by Remarks 5.5 and 5.6 the assertions (i) and (ii) are valid if $m \in \mathbb{C}(-\infty, 0)$. By the principle of analytic continuation they remain valid for every $m \in \mathbb{C}(-\infty, rc_\varepsilon)$ not belonging to $\mathfrak{A}_\varepsilon(M)$. We fix an arbitrary $m \in \mathbb{C}(-\infty, rc_\varepsilon)$, choose $a > 0$ such that $\overline{K}(m,a) \subset \mathbb{C}(-\infty, rc_\varepsilon)$ and $\overline{K}(m,a) \cap \mathfrak{A}_\varepsilon(M) \subset \{m\}$, and define $\gamma : [0, 2\pi] \longrightarrow \mathbb{C}$ by $\gamma(t) := a e^{it}$. Then for arbitrary $I \in \mathbb{Z}$ and $x \in X$ we conclude that

$$a_I(m;\mathfrak{M}_{f\circ A,w})(x) = \frac{1}{2\pi i}\int_\gamma (f\circ A)_{m+z,w}(x)\, z^{-I-1} dz =$$

$$= \frac{1}{2\pi i}\int_\gamma f_{m+z,w}(Ax)\, z^{-I-1} dz = a_I(m;\mathfrak{M}_{f,w})(Ax)$$

and - by making use of Fubini's theorem -

$$a_I(m;\mathfrak{M}_{f_{\mathfrak{G}},w})(x) = \frac{1}{2\pi i}\int_\gamma (f_{\mathfrak{G}})_{m+z,w}(x)\, z^{-I-1} dz =$$

$$= \frac{1}{2\pi i}\int_\gamma (f_{m+z,w})_{\mathfrak{G}}(x)\, z^{-I-1} dz =$$

$$= \int_G \sigma(A^{-1}) \frac{1}{2\pi i}\int_\gamma f_{m+z,w}(Ax)\, z^{-I-1} dz\, d\mu_G(A) =$$

$$= (a_I(m;\mathfrak{M}_{f,w}))_{\mathfrak{G}}(x). \quad \blacksquare$$

(c) Computing the Residues of $\mathfrak{M}_{f,w}$

The following example is basic.

Example 5.22. *Suppose that* $\varepsilon = 0$. *Let* $\ell \in \mathbb{C}$ *and* $N \in \mathbb{N}$, *and suppose that* f *is almost quasihomogeneous of degree* ℓ *and of order* $\le N$. *Let* $(q_i)_{1\le i\le N}$ *be the sequence of its deficiencies.*

(i) If $b = 1$ *then* $f_{m,w}$ *is equal to*

$$(5.36) \qquad -(m-\ell)^{-k-1} f - \sum_{i\in\mathbb{N}} (m-\ell)^{-k-i-1} \binom{k+i}{i} q_i , \qquad m \in \mathbb{C}\setminus\{\ell\}.$$

(ii) In general, $\mathfrak{M}_{f,w}$ *is holomorphic on* $\mathbb{C}\setminus\{\ell\}$, *and the negative part of its Laurent series at* $m = \ell$ *is given by (5.36).*

Proof. *(i):* Suppose that $\operatorname{Re} m < \min\{0, \operatorname{Re}\ell\}$. Inserting (1.39) (for $q_0 = f$) and taking (1.38) into account we obtain

$$f_{m,w}(x) = \int_0^1 t^{-m} f(M_t x)\, \omega_k(t)\, \frac{dt}{t} = f(x)\, I_k(m-\ell) + \sum_{i\in\mathbb{N}} q_i(x) \binom{k+i}{i} I_{k+i}(m-\ell)$$

where

$$I_j(m) := \int_0^1 t^{-m}\, \omega_j(t)\, \frac{dt}{t} .$$

Performing partial integration on the basis of (5.21) and using the equation $\omega_j'(t) = \omega_{j-1}(t)/t$ we obtain $I_j(m) = I_{j-1}(m)/m$ so that in view of $I_0(m) = -1/m$ we conclude that

$$(5.37) \qquad I_j(m) = -m^{-j-1} .$$

Inserting this into the equation above leads to (5.36).

(ii): Since $\mathbb{C} \longrightarrow C^0(X)$, $m \mapsto f_{m,v_k}$, is holomorphic where $v := \chi_{]1,b]}$ the assertion follows from (i). ∎

It is easy to compute the derivatives of $f_{m,w}$ at the points of $M_0(X)$ if $\varepsilon = 0$:

Remark 5.23. *Suppose that* $\varepsilon = 0$. *Let* $\ell \in \mathbb{C}$, $N \in \mathbb{N}$, *and* P_i, $0 \le i \le N$, *satisfy the* *assumptions of Proposition 5.19. Then for every $x \in M_o(X)$ the following assertions hold:*

(i) If $b=1$ then

$$(5.38)\qquad (P_o(x,\partial) f_{m,w})(x) = \sum_{i=0}^{N} (-1)^{i+1} \binom{k+i}{i} [P_i(x,\partial)f](x)\,(m+\ell)^{-k-i-1}$$

(ii) The meromorphic function on $\mathbb{C}(-\infty, r\lambda_{min})$ defined by

$$\mathbb{C}(-\infty, r\lambda_{min}) \setminus \mathfrak{A}(M) \ni m \mapsto (P_o(x,\partial) f_{m,w})(x)$$

extends to a meromorphic function on $\mathbb{C}$ which is holomorphic outside the point $m=-\ell$, and the negative part of its Laurent expansion at this point is given by (5.38).

Proof. *(i)*: Suppose that $\operatorname{Re} m < \min\{0, -\operatorname{Re} \ell\}$ and $b=1$. Since by the second part of Lemma 5.4 one can interchange differentiation and integration one obtains by making use of (3.4) and (1.38) and in view of $M_t x = x$ that

$$(P_o(x,\partial) f_{m,w_k})(x) = \int_0^1 t^{-m} P_o(x,\partial)(f \circ M_t)(x)\, \omega_k(t)\, \frac{dt}{t} =$$

$$= \sum_{i=0}^{N} (-1)^i \binom{k+i}{i} [P_i(x,\partial)f](x)\, I_{k+i}(m+\ell)$$

where $I_{k+i}(m+\ell)$ is defined as in the proof of Example 5.22. Hence the first assertion follows by (5.37) and by the principle of analytic continuation.

(ii): Since $\mathbb{C} \longrightarrow C^o(X)$, $m \mapsto f_{m,v_k}$, is holomorphic where $v := \chi_{]1,b]}$ the assertion follows from (i). ■

Notation 5.24. (i) For any $m \in \mathbb{C}$ we set $\mathfrak{A}^+_m := \{\alpha \in \mathfrak{A}^+ ;\ \alpha_M = m\}$ (for the definition of $\mathfrak{A}^+$ see Notation 5.15). Of course, $\mathfrak{A}^+_m \neq \emptyset$ if and only if $m \in \mathfrak{A}(M)$.

(ii) For every $m \in \mathbb{C}(-\infty, r\lambda_{min})$ the so-called *(almost) quasihomogeneous part $Q_m f$ of degree m (and of type M) in f* is defined by

$$(Q_m f)(x) := \sum_{\alpha \in \mathfrak{A}^+_m} \frac{1}{\alpha!} f^{(\alpha)}(M_o x)\, x^\alpha, \qquad x \in X.$$

Proposition 5.25. *If $m \in \mathfrak{A}_\varepsilon(M) \cap \mathbb{C}(-\infty, rc_\varepsilon)$ then*

(i) $a_{-i}(m; \mathfrak{M}_{f,w}) = 0$ for every $i \in \mathbb{N}_k$;

$$(ii)\quad a_{-k-1-i}(m;\mathfrak{M}_{f,w}) = -\binom{k+i}{k}\sum_{j=0}^{r}\frac{1}{j!}(-\varepsilon)^j(\partial_M - m + j)^i Q_{m-j}f, \qquad i\in\mathbb{N}_0.$$

Proof. By freezing the variables in ker M one reduces the proof to the case $X = G_M(\sigma_+)$. Instead of giving a direct computational proof by induction based on the main theorem of calculus, I prefer to give a slightly less elementary proof which seems to be more suitable to explain the result.

The case "$\varepsilon = 0$": We fix $i \in \mathbb{N}$, set $q := a_{-i}(m;\mathfrak{M}_{f,w})$, and let $d \in]0,+\infty[$. Then substituting $t = \tau/d$ and making use of (1.64) and (1.65) one obtains for every $z \in \mathbb{C}(-\infty,0)$ that

$$f_{z,w}(M_d x) = \int_0^b t^{-z} f(M_{td}x)\,\omega_k(t)\,\frac{dt}{t} =$$

$$= d^z \int_0^{bd} \tau^{-z} f(M_\tau x)\,\omega_k(\tau/d)\,\frac{d\tau}{\tau} = d^z \sum_{I=0}^{k} (-1)^I \omega_I(d)\, f_{z,v_{k-I}}(x)$$

where $v := \chi_{]0,bd]}$. By the principle of analytic continuation this equation remains valid for every $z \in \mathbb{C}(-\infty, rc_\varepsilon)\setminus\mathfrak{A}(M)$. Consequently, expanding d^{z-m}, comparing coefficients, changing the order of summation, taking (1.38) into account, substituting $j = I + \ell$, and re-changing the order of summation one deduces

$$(5.39)\qquad d^{-m}\, q\circ M_d = \sum_{I=0}^{k}(-1)^I\omega_I(d)\sum_{\ell=0}^{\infty}\omega_\ell(d)\,a_{-i-\ell}(m;\mathfrak{M}_{f,v_{k-I}}) =$$

$$= \sum_{j=0}^{\infty}\omega_j(d)\sum_{\ell=\max\{0,j-k\}}^{j}(-1)^{j+\ell}\binom{j}{\ell}\,a_{-i-\ell}(m;\mathfrak{M}_{f,v_{k-j+\ell}}).$$

Since $\operatorname{supp}(v-w)$ is a compact subset of $]0,+\infty[$ the function $\mathfrak{M}_{f,v_k-w_k}$ is holo morphic on $\mathbb{C}(-\infty, rc_\varepsilon)$ so that

$$(5.40)\qquad a_{-\ell}(m;\mathfrak{M}_{f,v_j}) = a_{-\ell}(m;\mathfrak{M}_{f,w_j}), \qquad \ell\in\mathbb{N},\ j\in\mathbb{N}_0$$

Inserting this into the equation above one sees that the summand for $j = 0$ is equa to q. Hence, (5.39) means that q is almost quasihomogeneous of degree m. Sinc q is a C^r function it must be a polynomial by Proposition 1.58, i.e.

$$q(x) = \sum_{\alpha_M = m}\frac{1}{\alpha!}q^{(\alpha)}(0)\,x^\alpha.$$

Note that by (5.32)" and by the integral formula for the Laurent coefficients w have

$$q^{(\alpha)}(0) = a_{-i}(m;h_\alpha)$$

where the meromorphic function $h_\alpha : \mathbb{C}(-\infty, r\lambda_{min}) \longrightarrow \mathbb{C}$ is defined by $h_\alpha(z) := (f_{z,w})^{(\alpha)}(0)$ for $z \in \mathbb{C}(-\infty, r\lambda_{min}) \setminus \mathfrak{A}(M)$. Consequently, Remark 5.23 shows that

$$q^{(\alpha)}(0) = \begin{cases} 0 & \text{if } i < k+1 \\ -f^{(\alpha)}(0) & \text{if } i = k+1 \\ (-1)^{i-k}\binom{i-1}{k}[P^*_{\alpha,i-k-1}(\partial)f](0) & \text{if } i > k+1 \end{cases}$$

where $P^*_{\alpha,j}$ denotes the jth order deficiency of the polynomial P_α satisfying (1.53). This implies the assertion (i) as well as the equality in (ii) for $i=0$.

Now, suppose that $Q_m f \equiv 0$, i.e. $f^{(\beta)}(0) = 0$ for every $\beta \in \mathfrak{A}^+_m$. Since $P^*_{\alpha,j}$ is almost quasihomogeneous of degree α_M so that by Remark 1.42 $P^*_{\alpha,j}(\partial)$ is a linear combination of the operators ∂^β, $\beta \in \mathfrak{A}^+_m$, this implies that $[P^*_{\alpha,j}(\partial)f](0) = 0$ for arbitrary $\alpha \in \mathfrak{A}^+_m$ and $j \in \mathbb{N}_0$. Consequently, the preceding formulas show that the assertion (ii) turns out to be valid for every $i \in \mathbb{N}_0$ in case $Q_m f \equiv 0$.

In the general case one obtains the assertion (ii) by applying the special case to $g := f - Q_m f$ (note that $Q_m g \equiv 0$) and taking Example 5.22 with f replaced by $Q_m f$ into account. Note that in case $k=0$ in view of Proposition 2.31 the assertion (ii) is also a consequence of the conditions (5.39) and (5.40).

<u>*The case "$\varepsilon \neq 0$"*</u>: By $R : \mathbb{R} \longrightarrow \mathbb{C}$ we denote the C^∞ function defined by

$$e^{-\varepsilon t} = \sum_{j=0}^{r} \frac{1}{j!}(-\varepsilon)^j t^j + R(t)\, t^{r+1}, \qquad t \in \mathbb{R}.$$

Setting $v := \chi_{]0,b]}$ and $u := vR$ one deduces for every $z \in \mathbb{C}(-\infty, 0)$ that

$$f_{z,w} = \sum_{j=0}^{r} \frac{1}{j!}(-\varepsilon)^j f_{z-j,v_k} + f_{z-r-1,u_k}.$$

Since by Proposition 5.13 the function $z \mapsto f_{z-r-1,u_k}$ is holomorphic on $\mathbb{C}(-\infty, r+1)$ the principle of analytic continuation shows that

$$\mathfrak{M}_{f,w} - \sum_{j=0}^{r} \frac{1}{j!}(-\varepsilon)^j \mathfrak{M}_{f,v_k}((\cdot)-j)$$

is holomorphic on $\mathbb{C}(-\infty, rc_\varepsilon)$. Consequently,

$$a_{-i}(m; \mathfrak{M}_{f,w}) = \sum_{j=0}^{r} \frac{1}{j!}(-\varepsilon)^j a_{-i}(m-j; \mathfrak{M}_{f,v_k}), \qquad i \in \mathbb{N}.$$

Hence, the assertion (i) is a simple consequence of the coresponding assertion for $\varepsilon = 0$. Since by the assertion (ii) for the case $\varepsilon = 0$ $a_{-k-1-i}(m-j; \mathfrak{M}_{f,v_k})$ is equal to $-\binom{k+i}{i}(\partial_M - (m-j))^i Q_{m-j} f$ the assertion (ii) follows. ■

Applying Proposition 5.25.(ii) to $\varepsilon = 0$ and $i = 0$ one obtains

Corollary 5.26. *For every $m \in \mathbb{C}(-\infty, r\lambda_{min})$ the $C^{\infty,s}$ function $Q_m f$ does not depend on the choice of coordinates.* ∎

It is now easy to compute $(\partial_M - m) f_{m,w}$:

Proposition 5.27. *For every $m \in \mathbb{C}(-\infty, rc_\varepsilon)$ we have*

$$(\partial_M - m) f_{m,w} = \varepsilon f_{m-1,w} - f_{m,w_{k-1}} - \delta_{k0} \sum_{j \in \mathbb{N}_0} \frac{1}{j!} (-\varepsilon)^j Q_{m-j} f + b^{-m} w(b) f \circ M_b .$$

Proof. We note first that by Example 1.21 and by Propositions 5.16 and 5.19 we have

$$\partial_M f_{m,w} = (\partial_M f)_{m,w} . \tag{5.41}$$

Moreover, we observe that (5.22.a) can be written as

$$\partial_t\Big(f(M_t x)\, w(t)\Big) = \frac{1}{t} (\partial_M f)(M_t x)\, w(t) + f(M_t x)\, w'(t) . \tag{5.22.a)'}$$

Hence, if $\operatorname{Re} m < 0$ then in view of (5.29) the equation (5.25) becomes

$$m f_{m,w} = (\partial_M f)_{m,w} - \varepsilon f_{m-1,w} + f_{m,w_{k-1}} - b^{-m} w(b) f \circ M_b . \tag{5.25)'}$$

By the principle of analytic continuation the equation (5.25)' remains valid for every $m \in \mathbb{C}(-\infty, rc_\varepsilon) \setminus \mathfrak{A}_\varepsilon(M)$, and in view of (5.41) the desired equation is proved for these m.

Now let $m \in \mathfrak{A}_\varepsilon(M)$. Then by the first part of the proof and by (5.41) one obtains for every $z \in \mathbb{C}(-\infty, rc_\varepsilon) \setminus \mathfrak{A}_\varepsilon(M)$ that

$$(\partial_M f)_{z,w} = m f_{z,w} + (z-m) f_{z,w} + \varepsilon f_{z-1,w} - f_{z,w_{k-1}} + b^{-z} e^{-\varepsilon b} f \circ M_b .$$

Hence, using (5.41) again we conclude that

$$\begin{aligned} \partial_M f_{m,w} &= a_0(m; \mathfrak{M}_{\partial_M f,w}) = \\ &= m\, a_0(m; \mathfrak{M}_{f,w}) + a_{-1}(m; \mathfrak{M}_{f,w}) + \varepsilon\, a_0(m-1; \mathfrak{M}_{f,w}) - \\ &\qquad - a_0(m; \mathfrak{M}_{f,w_{k-1}}) + b^{-m} e^{-\varepsilon b} f \circ M_b \end{aligned}$$

In view of Proposition 5.25 this is the desired equation. ∎

(d) A Formula for $f_{m,w}$ if Re m ≥ 0

First we are going to derive an explicit formula for $f_{m,w}$ in case m does not belong to $\mathfrak{A}_\varepsilon(M)$. A few abbreviations are required.

We set $\lambda_O := 1$, and for every $\alpha \in \mathfrak{A}^{1,O} \setminus \{0\}$ (see Notation 5.15) we set $\lambda_\alpha := \alpha_M$ and $p_\alpha := \mathrm{Re}\,\lambda_\alpha$. We fix a number $\ell \in]0, rc_\varepsilon]$ and define

$$I_\ell := \{(N,\beta)\,;\ N \in \mathbb{N},\ \beta \in (\mathfrak{A}^{1,O})^N,\ p_{\beta_1}+\ldots+p_{\beta_{N-1}} < \ell \leq p_{\beta_1}+\ldots+p_{\beta_N}\}$$

and

$$J_\ell := \{(N,\beta)\,;\ N \in \mathbb{N},\ \beta \in (\mathfrak{A}^{1,O})^N,\ p_{\beta_1}+\ldots+p_{\beta_N} < \ell\}\,.$$

For $\beta \in (\mathfrak{A}^{1,O})^N$ we set

$$\tilde{\beta} := \sum_{j=1}^{N} \beta_j \in \mathfrak{A}\,.$$

Note that $|\tilde{\beta}| \leq r$ if $(N,\beta) \in I_\ell \cup J_\ell$ (for $|\tilde{\beta} - \beta_N|\, c_\varepsilon \leq p_{\beta_1}+\ldots+p_{\beta_{N-1}} < \ell \leq rc_\varepsilon$ and $\beta_N \neq 0$ imply $|\tilde{\beta}|-1 < r$, and $p_{\beta_1}+\ldots+p_{\beta_N} < \ell$ similarly implies $|\tilde{\beta}| \leq r$).

Finally, we define

$$\mathfrak{B}_\ell := \bigcup_{N\in\mathbb{N}} \{(\beta,\sigma)\,;\ (\beta,N) \in I_\ell,\ \sigma \in \mathbb{N}_O \times \mathbb{N}^N\}$$

and

$$\mathfrak{C}_\ell := \bigcup_{N\in\mathbb{N}} \{(\beta,\sigma)\,;\ (\beta,N) \in J_\ell,\ \sigma \in \mathbb{N}_O \times \mathbb{N}^{N+1}\}\,.$$

Proposition 5.28. *There is a (unique) family of polynomial functions $R_{\beta,\sigma} : V \longrightarrow \mathbb{C}$, $(\beta,\sigma) \in \mathfrak{B}_\ell \cup \mathfrak{C}_\ell$, of degree $\leq (r,0)$, only finitely many of them not vanishing identically, having the following properties:*

(i) for arbitrary $\varepsilon \in \mathbb{C}$ and $m \in \mathbb{C}(-\infty, rc_\varepsilon) \setminus \mathfrak{A}_\varepsilon(M)$ the function $f_{m,w}$ is equal to

$$\sum_{(N,\beta)\in I_\ell} (-\varepsilon)^{N-|\tilde{\beta}|} \sum_{\sigma\in\mathbb{N}_O\times\mathbb{N}^N} R_{\beta,\sigma} \prod_{\nu=1}^{N} \left(m-\lambda_{\beta_1}-\ldots-\lambda_{\beta_{\nu-1}}\right)^{-\sigma_\nu} (\partial^{\tilde{\beta}} f)_{m-\lambda_{\beta_1}-\ldots-\lambda_{\beta_N},\, w_{\sigma_O}} -$$

$$- b^{-m}\Big(\sum_{j=O}^{k} w_j(b)\, m^{j-k-1}\Big) f\circ M_b - \sum_{(N,\beta)\in J_\ell} (-\varepsilon)^{N-|\tilde{\beta}|}\, b^{(\lambda_{\beta_1}+\ldots+\lambda_{\beta_N}-m)}$$

$$\sum_{\sigma\in\mathbb{N}_O\times\mathbb{N}^{N+1}} w_{\sigma_O}(b)\, R_{\beta,\sigma} \prod_{\nu=O}^{N} \left(m-\lambda_{\beta_1}-\ldots-\lambda_{\beta_\nu}\right)^{-\sigma_{\nu+1}} (\partial^{\tilde{\beta}} f)\circ M_b\,.$$

(ii) For every $(\beta,\sigma)\in\mathfrak{B}_\ell\cup\mathfrak{C}_\ell$ the function $R_{\beta,\sigma}$ is almost quasihomogeneous of degree $(\tilde{\beta})_M$.

(iii) Suppose that M is semi-simple, and let $(\beta,\sigma)\in\mathfrak{B}_\ell$ (resp. $\mathfrak{C}_\ell$); then $R_{\beta,\sigma}$ vanishes identically if and only if σ does not belong to the set

$$S_k(N) := \{\sigma\in\mathbb{N}_0\times\mathbb{N}^N;\ \textstyle\sum_{\nu=0}^{N}\sigma_\nu = k+N\}$$

(resp. $S_k(N+1)$) where N is defined by the condition $(N,\beta)\in I_\ell\cup J_\ell$; moreover, if $R_{\beta,\sigma}\neq 0$ then $R_{\beta,\sigma}(x) = \lambda^{\tilde{\beta}}x^{\tilde{\beta}}$.

Proof. By the principle of analytic continuation and in view of Remark 5.17 it suffices to prove the equality in (i) for $\operatorname{Re} m < 0$. We fix $q\in\{0\}\cup\mathbb{N}_r$ and a finite sequence $(\ell_i)_{0\le i\le q}$ in $]0,+\infty[$ such that $\ell_0 < c_\varepsilon$, $\ell_q = \ell$, and

$$\ell_{i-1} < \ell_i \le \ell_{i-1} + c_\varepsilon\,, \qquad i\in\mathbb{N}_q\,. \tag{5.42}$$

The proof is done by induction on q, starting with $q=0$, the assertion then being valid by (5.30). In fact, if $q=0$ then $\ell < c_\varepsilon$ so that $I_\ell = \{1\}\times\mathfrak{A}^{1,0}$ and $J_\ell = \emptyset$, and via the change of summation variables $\sigma_1 = k+1-i$ and $i+j=\sigma_0$ the equation (5.30) becomes the formula in the proposition if we set

$$R_{\alpha,(\sigma_1,\sigma_2)} := \begin{cases} \binom{\sigma_0}{k+1-\sigma_1} R_{\alpha,\sigma_0+\sigma_1-k-1} & \text{if } k+1-\sigma_0\le\sigma_1\le k+1 \text{ and } \alpha\neq 0 \\ 1 & \text{if } \sigma_1\in\mathbb{N}_{k+1},\ \sigma_0 = k+1-\sigma_1, \text{ and } \alpha = 0\,. \\ 0 & \text{otherwise} \end{cases} \tag{5.43}$$

For the induction step we fix $i\in\mathbb{N}_{q-1}\cup\{0\}$, and for every $(N,\beta)\in I_{\ell_i}\setminus I_{\ell_{i+1}}$ and $\sigma_0\in\mathbb{N}_0$ we apply (5.30) to $(\partial^{\tilde{\beta}}f, m-\lambda_{\beta_1}-\ldots-\lambda_{\beta_N},\sigma_0)$ instead of (f,m,k). Defining polynomial functions $P_{i,\alpha,j}$ of degree $\le (1,0)$ by

$$P_{i,\alpha,j} := \begin{cases} \delta_{j0} & \text{if } \alpha = 0 \\ \binom{i+j}{i} R_{\alpha,j} & \text{if } \alpha\neq 0 \end{cases} \tag{5.44}$$

we obtain

$$(\partial^{\tilde{\beta}} f)_{m-\lambda_{\beta_1}-\ldots-\lambda_{\beta_N},w_{\sigma_0}} = \tag{5.45}$$

$$= \sum_{i=0}^{\sigma_0} (m-\lambda_{\beta_1}-\ldots-\lambda_{\beta_N})^{i-\sigma_0-1}\Big(\sum_{\alpha\in\mathfrak{A}^{1,0}} (-\varepsilon)^{\delta_{0\alpha}} \sum_{j\in\mathbb{N}_0} P_{i,\alpha,j}$$

$$(\partial^{\tilde{\beta}+\alpha} f)_{m-\lambda_{\beta_1}-\ldots-\lambda_{\beta_N}-\lambda_\alpha,w_{i+j}} - b^{(\lambda_{\beta_1}+\ldots+\lambda_{\beta_N}-m)} w_i(b)\,(\partial^{\tilde{\beta}} f)\circ M_b\Big)$$

Note that the functions $P_{i,\alpha,j}$ are almost quasihomogeneous of degree λ_α and that only finitely many of them do not vanish identically.

Inserting (5.45) into the formula in assertion (i) for $\ell = \ell_i$ one obtains the formula for $\ell = \ell_{i+1}$ by taking into account the conditions $\delta_{0\alpha} = 1 - |\alpha|$,

(5.46) $$I_{\ell_i} \setminus I_{\ell_{i+1}} = J_{\ell_{i+1}} \setminus J_{\ell_i},$$

and

(5.47) $$I_{\ell_{i+1}} \setminus I_{\ell_i} = \{(N+1, (\beta_1, \dots, \beta_N, \alpha)) ; (N,\beta) \in I_{\ell_i} \setminus I_{\ell_{i+1}}, \ \alpha \in \mathfrak{A}^{1,0}\}$$

to be proved below. In fact, denoting the summands in (i) following the summation over $(N,\beta) \in I_\ell$ (resp. J_ℓ) by $T_\ell(N,\beta)$ (resp. $U_\ell(N,\beta)$) we obtain by inserting (5.45) that

$$\sum_{(N,\beta) \in I_{\ell_i} \setminus I_{\ell_{i+1}}} T_{\ell_i}(N,\beta) = B - C$$

where

$$B := \sum_{(N,\beta) \in I_{\ell_i} \setminus I_{\ell_{i+1}}} (-\varepsilon)^{N-|\tilde{\beta}|} \sum_{\sigma \in \mathbb{N}_0 \times \mathbb{N}^N} \sum_{i=0}^{\sigma_0} \sum_{\alpha \in \mathfrak{A}^{1,0}} (-\varepsilon)^{\delta_{0\alpha}} \sum_{j=1}^{\infty} \prod_{\nu=1}^{N} (m - \lambda_{\beta_1} - \dots - \lambda_{\beta_{\nu-1}})^{-\sigma_\nu}$$
$$(m - \lambda_{\beta_1} - \dots - \lambda_{\beta_N})^{-(\sigma_0+1-i)} R_{\beta,\sigma} P_{i,\alpha,j} (\partial^{\tilde{\beta}+\alpha} f)_{m-\lambda_{\beta_1}-\dots-\lambda_{\beta_N}-\lambda_\alpha, w_{i+j}}$$

and

$$C := \sum_{(N,\beta) \in I_{\ell_i} \setminus I_{\ell_{i+1}}} (-\varepsilon)^{N-|\tilde{\beta}|} \sum_{\sigma \in \mathbb{N}_0 \times \mathbb{N}^N} \sum_{i=0}^{\sigma_0} \prod_{\nu=1}^{N} (m - \lambda_{\beta_1} - \dots - \lambda_{\beta_{\nu-1}})^{-\sigma_\nu}$$
$$(m - \lambda_{\beta_1} - \dots - \lambda_{\beta_N})^{-(\sigma_0+1-i)} \, b^{(\lambda_{\beta_1} + \dots + \lambda_{\beta_N} - m)} w_i(b) R_{\beta,\sigma} (\partial^{\tilde{\beta}} f) \circ M_b .$$

In order to compute B we introduce new summation variables $\sigma_{N+1} := \sigma_0 + 1 - i$ and $\sigma_0' := i + j = \sigma_0 + 1 - \sigma_{N+1} + j$ and verify the following summation rule:

$$\sum_{\sigma_0 \in \mathbb{N}_0} \sum_{i=0}^{\sigma_0} \sum_{j=0}^{\infty} \dots = \sum_{\sigma_0 \in \mathbb{N}_0} \sum_{\sigma_{N+1}=1}^{\sigma_0+1} \sum_{j=0}^{\infty} \dots =$$
$$= \sum_{\sigma_{N+1}=1}^{\infty} \sum_{\sigma_0=\sigma_{N+1}-1}^{\infty} \sum_{\sigma_0'=\sigma_0+1-\sigma_{N+1}}^{\infty} \dots = \sum_{\sigma_{N+1}=1}^{\infty} \sum_{\sigma_0'=0}^{\infty} \sum_{\sigma_0=\sigma_{N+1}-1}^{\sigma_0'+\sigma_{N+1}-1} \dots .$$

Consequently, defining

(5.48) $$R_{(\beta_1,\dots,\beta_N,\alpha),(\sigma_0',\sigma_1,\dots,\sigma_{N+1})} :=$$
$$:= \sum_{\sigma_0=\sigma_{N+1}-1}^{\sigma_0'+\sigma_{N+1}-1} R_{\beta,(\sigma_0,\sigma_1,\dots,\sigma_N)} P_{\sigma_0+1-\sigma_{N+1},\alpha,\sigma_0'+\sigma_{N+1}-\sigma_0-1}$$

and taking $(\beta_1,\ldots,\beta_N,\alpha)^\sim = \tilde{\beta}+\alpha$ and $N-|\tilde{\beta}|+\delta_{0\alpha} = N+1-|\tilde{\beta}+\alpha|$ into account one obtains in view of (5.47) that

$$B = \sum_{(N,\beta)\in I_{\ell_{i+1}}\setminus I_{\ell_i}} T_{\ell_{i+1}}(N,\beta)\,.$$

For the computation of C we introduce the summation variable $\sigma_{N+1} := \sigma_0+1-i$, rename i as σ_0', and verify

$$\sum_{\sigma_0\in\mathbb{N}_0}\sum_{i=0}^{\sigma_0}\ldots = \sum_{i=0}^{\infty}\sum_{\sigma_0=i}^{\infty}\ldots = \sum_{\sigma_0'=0}^{\infty}\sum_{\sigma_{N+1}=1}^{\infty}\ldots\,.$$

Hence, defining

$$(5.49)\qquad R_{\beta,(\sigma_0',\sigma_1,\ldots,\sigma_{N+1})} := R_{\beta,(\sigma_{N+1}+\sigma_0'-1,\sigma_1,\ldots,\sigma_N)}$$

we obtain in view of (5.46) that

$$C = \sum_{(N,\beta)\in J_{\ell_{i+1}}\setminus J_{\ell_i}} U_{\ell_{i+1}}(N,\beta)\,.$$

Since $T_{\ell_i}(N,\beta) = T_{\ell_{i+1}}(N,\beta)$ for every $(N,\beta)\in I_{\ell_i}\cap I_{\ell_{i+1}}$ and $U_{\ell_i}(N,\beta) = U_{\ell_{i+1}}(N,\beta)$ for every $(N,\beta)\in J_{\ell_i}\subset J_{\ell_{i+1}}$ we arrive at the formula for $\ell=\ell_{i+1}$. The induction step is complete.

The assertion *(ii)* is immediately obtained from (5.43), (5.44), (5.48) and (5.49) by induction.

Under the assumption of *(iii)* we deduce from (5.43) and (5.23) that

$$(5.43)'\qquad R_{\alpha,\sigma}(x) = \delta_{\sigma_0+\sigma_1,k+1}\lambda^\alpha x^\alpha\,,\qquad \alpha\in\mathfrak{A}^{1,0},\ \sigma\in\mathbb{N}_0\times\mathbb{N}\,.$$

Similarly, one obtains that

$$(5.44)'\qquad P_{i,\alpha,j}(x) = \delta_{0,j}\lambda^\alpha x^\alpha\,,\qquad 0\le i\le\sigma_0,\ \alpha\in\mathfrak{A}^{1,0},\ j\in\mathbb{N}_0\,.$$

Following now the lines of the induction step above, one derives the assertion (iii) in its full generality from (5.48) and (5.49) by induction.

Proof of (5.46). "$\subset$": If $(N,\beta)\in I_{\ell_i}$ then $(N,\beta)\notin J_{\ell_i}$ und $p_{\beta_1}+\ldots+p_{\beta_{N-1}} < \ell_i < \ell_{i+1}$ so that if, in addition, $(N,\beta)\notin I_{\ell_{i+1}}$ it follows that $p_{\beta_1}+\ldots+p_{\beta_N} < \ell_{i+1}$, i.e. $(N,\beta)\in J_{\ell_{i+1}}$.

"$\supset$": If $(N,\beta)\in J_{\ell_{i+1}}$ then $(N,\beta)\notin I_{\ell_{i+1}}$ and $p_{\beta_1}+\ldots+p_{\beta_N} < \ell_{i+1}$ so that by (5.42) we deduce that $p_{\beta_1} + \dots + p_{\beta_{N-1}} < \ell_{i+1} - c_\varepsilon \leq \ell_i$. If, in addition, $(N,\beta) \notin J_{\ell_i}$ then this implies that $(N,\beta) \in I_{\ell_i}$.

Proof of (5.47). In view of (5.46) we first observe that on the right-hand side of (5.47) the set $I_{\ell_i} \setminus I_{\ell_{i+1}}$ can be replaced by $J_{\ell_{i+1}} \setminus J_{\ell_i}$.

"$\subset$": If $(N,\beta) \in I_{\ell_{i+1}} \setminus I_{\ell_i}$ then

(5.50) (a) $p_{\beta_1} + \dots + p_{\beta_N} \geq \ell_{i+1}$, and (b) $p_{\beta_1} + \dots + p_{\beta_{N-1}} < \ell_{i+1}$,

and

(5.51) (a) $p_{\beta_1} + \dots + p_{\beta_N} < \ell_i$, or (b) $p_{\beta_1} + \dots + p_{\beta_{N-1}} \geq \ell_i > 0$.

Since $\ell_{i+1} > \ell_i$ it follows in view of (5.50.a) that (5.51.a) cannot hold so that (5.51.b) is valid which implies that $N-1 \geq 1$. Moreover, from (5.50.b) we see that $(N-1,(\beta_1,\dots,\beta_{N-1}))$ belongs to $J_{\ell_{i+1}} \setminus J_{\ell_i}$.

"$\supset$": That (N,β) belongs to $J_{\ell_{i+1}} \setminus J_{\ell_i}$ means that

(5.52) (a) $p_{\beta_1} + \dots + p_{\beta_N} < \ell_{i+1}$, and (b) $p_{\beta_1} + \dots + p_{\beta_N} \geq \ell_i$.

From (5.52.b) and (5.42) it follows for every $\alpha \in \mathfrak{A}^{1,0}$ that

$$p_{\beta_1} + \dots + p_{\beta_N} + p_\alpha \geq \ell_i + c_\varepsilon \geq \ell_{i+1}$$

which in combination with (5.52.a) implies that $(N+1,(\beta_1,\dots,\beta_N,\alpha))$ belongs to $I_{\ell_{i+1}}$. In view of (5.52.b) it does not belong to I_{ℓ_i}. ■

Combining Proposition 5.28 and Remark 5.17 with Propositions 5.13 and 5.16 one obtains

Corollary 5.29. *The condition (ii) in assertion (5.1) is satisfied for* $E = F = C^{r,s}(X)$ *and* $D = \mathbb{C}(-\infty, rc_\varepsilon) \setminus \mathfrak{A}_\varepsilon(M)$.

More precisely, let $W : X \longrightarrow [0,+\infty[$ *be any continuous weight function; then the condition (ii) in assertion (5.1) holds for* $E = C^{r,s}(W)$, $D = \mathbb{C}(-\infty, rc_\varepsilon) \setminus \mathfrak{A}_\varepsilon(M)$, *and* $F = C^{r,s}(U)$ *where* $U := (1 + |P_+(\cdot)|)^\nu \, W_{(b)}$ *with*

(5.53) $\nu := \max\{\deg R_{\beta,\sigma} \; ; \; (N,\beta) \in \mathfrak{B}_\ell \cup \mathfrak{C}_\ell ,\; 0 < \ell \leq rc_\varepsilon\}$

and where $W_{(b)}$ *is defined by (5.18).* ■

Our next aim is to find explicit formulas for $f_{z,w}$ if z belongs to $\mathbb{C}(-\infty, rc_\varepsilon) \cap \mathfrak{U}_\varepsilon(M)$. We are going to deal with the case $z=0$ first.

Proposition 5.30. *(i) For arbitrary $k \in \mathbb{N}_0$ and $x \in X$ we have*

$$f_{0,w}(x) = -\int_0^b \partial_t\Big(f(M_t x)\Big)\, w_{k+1}(t)\, dt + \varepsilon f_{-1,w_{k+1}} + w_{k+1}(b)\, f(M_b x);$$

(ii) the integral in (i) is equal to

$$\sum_{\alpha \in \mathfrak{U}_1} \sum_{j=0}^{\infty} \binom{k+j+1}{j} [R_{\alpha,j}\, (\partial^\alpha f)_{-\alpha_M, w_{k+j+1}}](x)$$

where $R_{\alpha,j}$ are the polynomial functions from (5.22.b).

Proof. *(ii):* We have to apply (5.22) to $w \equiv 1$ and take (1.38) into account.

(i): Combining (5.25) and (5.29) we obtain for arbitrary $m \in \mathbb{C}(-\infty,0)$ and $x \in X$ that

$$f_{m,w_k}(x) = \frac{1}{m} \int_0^b t^{-m} \partial_t\Big(f(M_t x)\, e^{-\varepsilon t}\Big)\, \omega_k(t)\, dt + \frac{1}{m} f_{m,w_{k-1}} - \frac{1}{m} b^{-m} f(M_b x)\, w_k(b).$$

By induction on k this leads to

$$f_{m,w_k}(x) = m^{-k-1}\Big(\int_0^b \partial_t\Big(f(M_t x)\, e^{-\varepsilon t}\Big) S_k(t,m)\, dt - f(M_b x)\, e^{-\varepsilon b} S_k(b,m)\Big)$$

where

$$S_k(t,m) := t^{-m} \sum_{i=0}^{k} \omega_i(t)\, m^i .$$

Since by the main theorem of calculus and by the continuity of $t \mapsto f(M_t x)$ we have

$$f(M_0 x) = -\int_0^b \partial_t\Big(f(M_t x)\, e^{-\varepsilon t}\Big)\, dt + f(M_b x)\, e^{-\varepsilon b}, \qquad x \in X,$$

it follows that

$$f_{m,w_k}(x) + m^{-k-1} f(M_0 x) = \tag{5.54}$$

$$= m^{-k-1}\Big(-\int_0^b \partial_t\Big(f(M_t x)\, e^{-\varepsilon t}\Big)\, (1 - S_k(t,m))\, dt + f(M_b x)\, e^{-\varepsilon b} (1 - S_k(b,m))\Big)$$

By Taylor's formula we have

$$1 - S_k(t,m) = t^{-m}\Big(t^m - \sum_{i=0}^{k} \omega_i(t)\, m^i\Big) =$$

$$= t^{-m} m^{k+1} \omega_{k+1}(t) \int_0^1 (k+1)\, (1-\tau)^k\, t^{\tau m}\, d\tau =$$

$$= m^{k+1}\, \omega_{k+1}(t) \int_0^1 (k+1)\, \tau^k\, t^{-\tau m}\, d\tau .$$

Since $|t^{-\tau m}| \leq \max\{1, b^{-\operatorname{Re} m}\}$ where the right-hand side tends to 1 as $m \to 0$ the integral converges to 1 uniformly for $t \in]0,b]$ as $m \to 0$. Since in view of (5.22) the function $]0,b] \ni t \mapsto \partial_t(f(M_t x)\, e^{-\varepsilon t})\, \omega_{k+1}(t)$ is integrable it follows by Lebesgue's Dominated Convergence Theorem from (5.54) that

$$\lim_{\substack{m \to \infty \\ \operatorname{Re} m < 0}} \left(f_{m,w_k}(x) + m^{-k-1} f(M_0 x) \right) =$$

$$= - \int_0^b \partial_t\left(f(M_t x)\, e^{-\varepsilon t}\right) \omega_{k+1}(t)\, dt + f(M_b x)\, e^{-\varepsilon b}\, \omega_{k+1}(b), \qquad x \in X .$$

This implies (i). Note that it also gives an independent proof of the assertion of Proposition 5.25 in case $m = 0$. ■

Combining Proposition 5.30 with Proposition 5.16.(iii) and taking $R_{\alpha,j} \circ M_0 \equiv 0$ into account one obtains

Corollary 5.31. *For every $\ell \in \mathfrak{A}(M)$ and every polynomial function $P: V^* \longrightarrow \mathbb{C}$ which is almost quasihomogeneous of degree ℓ and of type M^* we have*

$$(P(\partial) f_{\ell,w}) \circ M_0 =$$

$$= \sum_{j \in \mathbb{N}_0} \binom{k+j}{j} \left[\varepsilon\, (P_j(\partial) f)_{-1, w_{k+j+1}} \circ M_0 + w_{k+j+1}(b)\, (P_j(\partial) f) \circ M_0 \right]$$

where $P_j := (\partial_{M^*} - \ell)^j P$. ■

The following proposition gives an explicit formula for $f_{z,w}$ in case z belongs to $\mathfrak{A}_\varepsilon(M)$. It generalizes the one for $z = 0$ in Proposition 5.30.

Proposition 5.32. *Let $z \in \mathbb{C}(-\infty, rc_\varepsilon) \cap \mathfrak{A}_\varepsilon(M)$. Then there is a (unique) family of polynomial functions $P_{j,\alpha,\rho}: V \longrightarrow \mathbb{C}$, $j, \rho \in \mathbb{N}_0$ and $\alpha \in \mathfrak{A}^+$, of degree $\leq (r,0)$, only finitely many of them not vanishing identically, having the following properties:*

(i) for every $\varepsilon \in \mathbb{C}$ we have

$$f_{z,w} = \sum_{j=0}^{+\infty} (-\varepsilon)^j\, (-H_j + T_j - B_j)$$

where

$$H_j(x) := \sum_{\alpha\in\mathfrak{A}^+_{z-j}} \sum_{\rho\in\mathbb{N}_0} P_{j,\alpha,\rho}(x)\Big(\int_0^b \partial_t\big((\partial^\alpha f)(M_t x)\,e^{-\varepsilon t}\big)\,\omega_{\rho+1}(t)\,dt - w_{\rho+1}(b)\,f(M_b x)\Big),$$

$$T_j := \sum_{\alpha\in\mathfrak{A}_j(z)} \sum_{\rho\in\mathbb{N}_0} P_{j,\alpha,\rho}\,(\partial^\alpha f)_{z-j-\alpha_M,\,w_\rho}$$

with

$$\mathfrak{A}_j(z) := \{\alpha\in\mathfrak{A}^+;\ \alpha_M\neq z-j,\ \operatorname{Re}\alpha_M\geq \ell-j\} \quad \text{with} \quad \ell := \operatorname{Re} z,$$

and where

$$B_j := \sum_{\substack{\alpha\in\mathfrak{A}^+\\ \operatorname{Re}\alpha_M<\ell-j}} \sum_{\rho\in\mathbb{N}_0} P_{j,\alpha,\rho}\, b^{(\alpha_M+j-z)}\, w_\rho(b)\,(\partial^\alpha f)\circ M_b;$$

(ii) for arbitrary $j,\rho\in\mathbb{N}_0$ and $\alpha\in\mathfrak{A}^+$ the function $P_{j,\alpha,\rho}$ is almost quasihomogeneous of degree α_M;

(iii) let $(j,\alpha)\in\mathbb{N}_0\times\mathfrak{A}^+$ be such that $\alpha_M+j=z$ and let $\rho\in\mathbb{N}_0$; if $\rho\leq k-1$ or $j\geq r+1$ then $P_{j,\alpha,\rho}\equiv 0$; if $j\leq r$ and if, in addition, M is semi-simple then

$$P_{j,\alpha,\rho}(x) = \delta_{\rho k}\,\frac{1}{j!}\,\frac{1}{\alpha!}\,x^\alpha.$$

Note that by Proposition 5.30.(ii) we have

$$(5.55)\qquad \int_0^b \partial_t\big((\partial^\alpha f)(M_t x)\,e^{-\varepsilon t}\big)\,\omega_{\rho+1}(t)\,dt =$$

$$= \Big[\sum_{\beta\in\mathfrak{A}_1}\sum_{i=0}^{\infty}\binom{\rho+i+1}{i} R_{\beta,i}\,(\partial^{\alpha+\beta}f)_{-\beta_M,\,w_{\rho+i+1}}\Big](x) - \varepsilon\,(\partial^\alpha f)_{-1,\,w_{\rho+1}}(x).$$

Proof. The case $z=0$ is an obvious reformulation of Proposition 5.30. So we suppose that $z\neq 0$. Then $\ell>0$. We set $L := \{(N,\beta);\ N\in\mathbb{N},\ \beta\in(\mathfrak{A}^{1,0})^N,\ \lambda_{\beta_1}+\ldots+\lambda_{\beta_N}=z\}\subset I_\ell$ (see p. 205). Note first that for arbitrary $N\in\mathbb{N}$ and $\beta\in(\mathfrak{A}^{1,0})^N$ we have

$$(5.56)\qquad (\tilde\beta)_M + N - |\tilde\beta| = \lambda_{\beta_1}+\ldots+\lambda_{\beta_N}.$$

For arbitrary $j\in\mathbb{N}_0$ and $\alpha\in\mathfrak{A}^+$ we set

$$K_{j,\alpha} := \{(N,\beta);\ N=j+|\alpha|,\ \beta\in(\mathfrak{A}^{1,0})^N,\ \tilde\beta=\alpha\}.$$

Hence it follows from (5.56) that

$$L = \bigcup_{j\in\mathbb{N}_0}\ \bigcup_{\alpha\in\mathfrak{A}^+_{z-j}} K_{j,\alpha}, \qquad I_\ell\setminus L = \bigcup_{(j,\alpha)\in D} K_{j,\alpha}$$

where $D := \{(j,\alpha)\in\mathbb{N}_0\times\mathfrak{A}^+;\ \operatorname{Re}\alpha_M\geq\ell-j,\ \alpha_M\neq z-j,\ \exists\,(N,\beta)\in I_\ell:\ \alpha=\tilde\beta\}$, and

$$J_\ell = \bigcup\big\{K_{j,\alpha};\ j\in\mathbb{N}_0,\ \alpha\in\mathfrak{A}^+,\ \operatorname{Re}\alpha_M<\ell-j\big\}$$

where all the unions are disjoint.

Our aim is to let m tend to z from the left in every term in the formula of Proposition 5.28.(i). We do this for each summand separately.

If $(N,\beta) \in L$ and $\sigma = (\sigma_0, \ldots, \sigma_N) \in \mathbb{N}_0 \times \mathbb{N}^N$ then

$$\lim_{m \to z} \prod_{\nu=1}^{N} (m - \lambda_{\beta_1} - \ldots - \lambda_{\beta_{\nu-1}})^{-\sigma_\nu} =$$

$$= \prod_{\nu=1}^{N} (z - \lambda_{\beta_1} - \ldots - \lambda_{\beta_{\nu-1}})^{-\sigma_\nu} = \prod_{\nu=1}^{N} (\lambda_{\beta_\nu} + \ldots + \lambda_{\beta_N})^{-\sigma_\nu},$$

and - by Propositions 5.25 and 5.30 -

$$\lim_{m \to z} \left((\partial^{\tilde\beta} f)_{m-z, w_{\sigma_0}}(x) + (m-z)^{-\sigma_0 - 1} (\partial^{\tilde\beta} f)(M_0 x) \right) = (\partial^{\tilde\beta} f)_{0, w_{\sigma_0}}(x) =$$

$$= - \int_0^b \partial_t \left((\partial^{\tilde\beta} f)(M_t x)\, e^{-\varepsilon t} \right) \omega_{\sigma_0 + 1}(t)\, dt + (\partial^{\tilde\beta} f)(M_b x)\, w_{\sigma_0+1}(b) .$$

If $(N,\beta) \in I_\ell \setminus L$ then $\lambda_{\beta_1} + \ldots + \lambda_{\beta_N} \neq z$ so that - using Corollary 5.29 - we conclude for every $\sigma \in \mathbb{N}_0 \times \mathbb{N}^N$ that

$$\lim_{m \to z} \prod_{\nu=1}^{N} (m - \lambda_{\beta_1} - \ldots - \lambda_{\beta_{\nu-1}})^{-\sigma_\nu} (\partial^{\tilde\beta} f)_{m - \lambda_{\beta_1} - \ldots - \lambda_{\beta_N}, w_{\sigma_0}}(x) =$$

$$= \prod_{\nu=1}^{N} (z - \lambda_{\beta_1} - \ldots - \lambda_{\beta_{\nu-1}})^{-\sigma_\nu} (\partial^{\tilde\beta} f)_{z - \lambda_{\beta_1} - \ldots - \lambda_{\beta_N}, w_{\sigma_0}}(x) .$$

Finally, if $(N,\beta) \in J_\ell$ we have $p_{\beta_1} + \ldots + p_{\beta_N} < \ell$ so that for $\sigma \in \mathbb{N}^{N+1}$ we have

$$\lim_{m \to z} \prod_{\nu=0}^{N} (m - \lambda_{\beta_1} - \ldots - \lambda_{\beta_\nu})^{-\sigma_\nu + 1}\, b^{(\lambda_{\beta_1} + \ldots + \lambda_{\beta_N} - m)} =$$

$$= \prod_{\nu=0}^{N} (z - \lambda_{\beta_1} - \ldots - \lambda_{\beta_\nu})^{-\sigma_\nu + 1}\, b^{(\lambda_{\beta_1} + \ldots + \lambda_{\beta_N} - z)} .$$

Now, if $\mathrm{Re}\, \alpha_M \geq \ell - j$ then we define

$$P_{j,\alpha,\rho} := \sum_{(N,\beta) \in K_{j,\alpha}} \sum_{\sigma \in \mathbb{N}^N} \prod_{\nu=1}^{N} (z - \lambda_{\beta_1} - \ldots - \lambda_{\beta_{\nu-1}})^{-\sigma_\nu} R_{\beta,(\rho,\sigma_1,\ldots,\sigma_N)} ;$$

on the other hand, if $(j,\alpha) \neq (0,0)$ and $\mathrm{Re}\, \alpha_M < \ell - j$ then we set

$$P_{j,\alpha,\rho} := \sum_{(N,\beta) \in K_{j,\alpha}} \sum_{\sigma \in \mathbb{N}^{N+1}} \prod_{\nu=0}^{N} (z - \lambda_{\beta_1} - \ldots - \lambda_{\beta_\nu})^{-\sigma_\nu + 1} R_{\beta,(\rho,\sigma_1,\ldots,\sigma_{N+1})} .$$

Finally, if $(j,\alpha) = (0,0)$ then we set

$$P_{0,0,\rho} :\equiv \begin{cases} 0 & \text{if } \rho \geq k+1 \\ z^{\rho - k - 1} & \text{if } 0 \leq \rho \leq k \end{cases} .$$

Combining the computations above with Proposition 5.28.(i) we first of all deduce that the negative Laurent coefficients of $\mathfrak{M}_{f,w}$ at $m=z$ are given by

$$(5.57)\qquad a_{-\rho-1}(z;\mathfrak{M}_{f,w}) = -\sum_{j\in\mathbb{N}_0}(-\varepsilon)^j \sum_{\alpha\in\mathfrak{A}^+_{z-j}} P_{j,\alpha,\rho}\,(\partial^\alpha f)\circ M_0\,.$$

Moreover, from the formula in Proposition 5.28.(i) we obtain the main assertion of the proposition by subtracting the negative part of the Laurent series of $\mathfrak{M}_{f,w}$ at $m=z$ and taking the limit as $m\to z$.

Finally, comparing (5.57) with the formulas in Proposition 5.25 we conclude that for every pair (j,α) satisfying $\alpha_M+j=z$ the polynomial function $P_{i,\alpha,\rho}$ vanishes identically in case $\rho\le k-1$ or $j\ge r+1$ and is given by

$$(5.58)\qquad P_{j,\alpha,\rho} = \binom{\rho}{k}\frac{1}{j!}\frac{1}{\alpha!}(\partial_M - z+j)^{\rho-k}(\cdot)^\alpha \qquad \text{if } \rho\ge k \text{ and } j\le r.$$

If M is semi-simple then $(\cdot)^\alpha$ is quasihomogeneous of degree $\alpha_M=z-j$ so that $P_{j,\alpha,\rho}$ vanishes for $\rho>k$, as well, and for $\rho=k$ is equal to $\frac{1}{j!}\frac{1}{\alpha!}(\cdot)^\alpha$. ■

Combining Proposition 5.32 with Propositions 5.13 and 5.16 and Remark 5.17 one obtains

Corollary 5.33. *For every $m\in\mathbb{C}(-\infty,rc_\varepsilon)\cap\mathfrak{A}_\varepsilon(M)$ the map $C^{r,s}(X)\longrightarrow C^{r,s}(X)$, $f\mapsto f_{m,w}$, is well-defined and continuous.*

More precisely, let $W:X\longrightarrow[0,+\infty[$ be any continuous weight function; then the map $C^{r,s}(W)\longrightarrow C^{r,s}(U)$, $f\mapsto f_{m,w}$, is well-defined and continuous where $U:=(1+|P_+(\cdot)|)^\nu\,W_{(b)}$ with

$$(5.59)\qquad \nu := 1+\max\{\deg P_{j,\alpha,\rho}\,;\ j,\rho\in\mathbb{N}_0,\ \alpha\in\mathfrak{A}^+,\ z\in\mathfrak{A}_\varepsilon(M)\cap\mathbb{C}(-\infty,rc_\varepsilon)\}$$

and with $W_{(b)}$ being defined by (5.18). ■

Notation 5.34. For every $N\in[0,+\infty[$ we denote by $\mathcal{S}^N_M(V)$ the Fréchet space of all C^∞ functions $g:V\longrightarrow\mathbb{C}$ such that

$$\sup\{\,|g^{(\alpha)}(x)|\,(1+|P_+(x)|)^{-N}\,(1+|M_0(x)|)^L\,;\ x\in V\}$$

is finite for arbitrary $\alpha\in\mathfrak{A}$ and $L\in\mathbb{N}$. We set

$$\mathcal{S}^\infty_M(V) := \operatorname*{ind}_{N\to+\infty}\mathcal{S}^N_M(V)\,.$$

Recall that the space $\mathcal{O}_{\mathfrak{M}}(V)$ of all multiplication operators on the space $\mathcal{S}(V)$ consists of all C^{∞} functions $\chi : V \longrightarrow \mathbb{C}$ such that for every $\alpha \in \mathfrak{A}$ there is a constant $N \in \mathbb{N}$ such that $|\partial^{\alpha} f|(1+|\cdot|)^{-N}$ is bounded.

Remark 5.35. *Let $\chi \in \mathcal{O}_{\mathfrak{M}}(V)$ be with support contained in Ω_J (see Notation 5.10) for some bounded subinterval J of $[0,+\infty[$. Then the map $\mathcal{S}_M^{\infty}(V) \longrightarrow \mathcal{S}(V)$, $g \mapsto \chi g$, is well-defined and continuous.*

Proof. By the assumption on J there is a constant C_J such that for every $x \in \Omega_J$ the estimate $(1+|x|) \le C_J(1+|M_O(x)|)$ is valid. Hence the assertion follows by the Leibniz rule. ■

Proposition 5.36. *For every $m \in \mathbb{C}$ the function $\mathcal{O}_{\mathfrak{M}}(V) \longrightarrow \mathcal{O}_{\mathfrak{M}}(V)$, $f \mapsto f_{m,w}$, is well-defined, linear and continuous; for every $N \in [0,+\infty[$ it maps the space $\mathcal{S}_M^N(V)$ continuously into $\mathcal{S}_M^{\nu+N}(V)$ where ν is defined by (5.53), resp. (5.59), if $r \in \mathbb{N}$ is chosen such that $\mathrm{Re}\, m < r c_{\varepsilon}$.*

Moreover, for every $f \in \mathcal{O}_{\mathfrak{M}}(V)$ we have

$$\lim_{\varepsilon \to o} f_{m,w} = f_{m,\chi_{]0,b]}\omega_k} \quad \text{in the topology of } \mathcal{O}_{\mathfrak{M}}(V).$$

Proof. We apply Proposition 5.12 to $u = (1+|\cdot|)^N \circ M_O$ (see (5.10)) and take Propositions 5.16, 5.28 and 5.32 into account. ■

(e) Introducing f_m for Arbitrary $m \in \mathbb{C}$ and $f \in \mathcal{S}(V)$

If no restricting condition on the support of w is assumed we decompose w according to $w = \chi_{]0,1]} w + \chi_{]1,+\infty[} w$. If $f_{m,w}$ is well-defined by (3.1)' then this decomposition immediately leads to

$$(5.60) \qquad f_{m,w} = f_{m,u} + f_{m,v} \quad \text{where} \quad u := \chi_{]0,1]} w \quad \text{and} \quad v := \chi_{[1,+\infty[} w .$$

In general, $f_{m,w}$ can be defined by (5.60) if $f_{m,u}$ and $f_{m,v}$ are well-defined. Conditions ensuring this are obtained by combining the results of sections (a) and (b). So we assume in this section that (5.17) holds and that for a fixed integer $k \in \mathbb{N}_O$ we have

(5.61) $$w(t) = \omega_k(t)\, e^{-\varepsilon t}, \qquad t \in \,]0,+\infty[.$$

We are going to deal with the cases "$\varepsilon = 0$" and "$\varepsilon > 0$" separately.

<u>*The case "ε = 0"*</u>. In this case the condition (5.12) with $a = 1$ is satisfied for $L < 0$ only so that the domain of definition of $f_{m,v}$ cannot be the whole of X, but is X_+ only. In order that $f_{m,u}$ be defined for every $m \in \mathbb{C}$ f has to be a C^∞ function with respect to the variables in $G_M(\sigma_+)$. So we fix $s \in \mathbb{N}_0 \cup \{\infty\}$ and $f \in C^{\infty,s}(X)$. Finally, we require that the growth of f is restricted as follows:

(5.62) $$\sup\{\,|f^{(\alpha)}(x)|\,(1+|P_+(x)|)^N\,;\ x \in M_0^{-1}(K)\} < +\infty$$

for arbitrary $\alpha \in \mathfrak{A}^{\infty,s}$ and $N \in \mathbb{N}_0$ and for every compact subset K of X.

We can now state and prove the main result on the quasihomogeneous averages f_{m,ω_k}.

Theorem 5.37. *Under the preceding assumptions for every $m \in \mathbb{C}$ by (5.60) (restricted to X_+) a function $f_{m,w} \in C^{\infty,s}(X_+)$ is well-defined having the following properties:*

(i) If $m \in \mathbb{C}(-\infty,0)$ or if (5.19) holds then $f_{m,w}$ is given by (3.1)'.

(ii) $\operatorname{supp} f_{m,w} \subset \overline{(\operatorname{supp} f)_M}$.

(iii) $f_{m,w}$ is almost quasihomogeneous of degree m; more precisely, we have

(5.63) $$(\partial_M - m)^i f_{m,w} = \begin{cases} (-1)^i f_{m,\omega_{k-i}} & \text{if } i \in \mathbb{N}_k \\ (-1)^{k+1} (\partial_M - m)^{i-k-1} Q_m f|_{X_+} & \text{if } i \in \mathbb{N}\setminus\mathbb{N}_k \end{cases};$$

here the term in the second line on the right-hand side of (5.63) vanishes in case $m \notin \mathfrak{A}(M)$, i.e. in this case (3.11) is valid.

(iv) By $\mathfrak{M}_{f,w}(m) := f_{m,w}$, $m \in \mathbb{C}\setminus\mathfrak{A}(M)$, a meromorphic function $\mathfrak{M}_{f,w}$ on $\mathbb{C}$ with values in $C^{\infty,s}(X)$ is defined, its poles lying in $\mathfrak{A}(M)$; moreover, if $m \in \mathfrak{A}(M)$ then $a_0(m;\mathfrak{M}_{f,w}) = f_{m,w}$ and

(5.64) $$a_{-i}(m;\mathfrak{M}_{f,w}) = \begin{cases} 0 & \text{if } i \in \mathbb{N}_k \\ -\binom{i-1}{k} (\partial_M - m)^{i-k-1} Q_m f|_{X_+} & \text{if } i \in \mathbb{N}\setminus\mathbb{N}_k \end{cases}.$$

(v) Let $\ell \in \mathbb{C}$ and $N \in \mathbb{N}$, and let $P_0 : X\times V^* \longrightarrow \mathbb{C}$ be a continuous copolynomial function of degree $\le (\infty,s)$ which is almost quasihomogeneous of degree ℓ, of

order $\leq N$, and of type $M\times(-M)^*$. For every $j \in \mathbb{N}_N$ let $P_j : X\times V^* \longrightarrow \mathbb{C}$ be its jth order deficiency which is a continuous copolynomial function of degree $\leq (r,s)$, as well. Moreover, suppose that for every $j \in \{0\}\cup\mathbb{N}_N$ the function $P_j(x,\partial)f$ belongs to $C^{\infty,s}(X)$ and satisfies the assumptions of Theorem 5.37, as well. Then on X_+ the formulas (5.34) and (5.34)' are valid.

(vi) Let $\ell \in \mathbb{C}$ and $N \in \mathbb{N}$, and let $q_0 : X \longrightarrow \mathbb{C}$ be almost quasihomogeneous of degree ℓ and of order $\leq N$ such that for every $j \in \{0\}\cup\mathbb{N}_N$ the function $q_j f$ belongs to $C^{\infty,s}(X)$ and satisfies the assumptions of Theorem 5.37, as well, where in case $j \geq 1$ $q_j : X \longrightarrow \mathbb{C}$ be the jth order deficiency of q_0. Then on X_+ the formulas (5.35) and (5.35)' hold.

(vii) $f_{m,w}\circ A = (f\circ A)_{m,w}$ for every $A \in L(V,V)$ commuting with M.

(viii) $(f_{m,w})_{\mathfrak{G}} = (f_{\mathfrak{G}})_{m,w}$ if $\mathfrak{G}$ satisfies the assumptions of Remark 2.67.(ii); in particular, if f is $\mathfrak{G}$-invariant so is $f_{m,w}$.

(ix) The formulas of Propositions 5.28 and 5.32 remain valid if $\int_0^b \ldots$ is replaced by $\int_0^{+\infty} \ldots$ and if all the other terms containing b and ε explicitly are deleted.

Proof. The assumptions on the growth of f mean that for arbitrary $\alpha \in \mathfrak{A}^{\infty,s}$ and $N \in \mathbb{N}_0$ there is a continuous function $u_{\alpha,N} : M_0(X) \longrightarrow]0,+\infty[$ such that

$$\sup\{\,|f^{(\alpha)}(x)|(1+|P_+(x)|)^N / u_{\alpha,N}(M_0 x)\,;\ x \in X\,\}$$

is finite. Then one obtains the first part of the assertion by combining Propositions 5.11.(i) and 5.12 and Lemma 5.4 with Proposition 5.16 and Corollaries 5.29 and 5.33.

(i): this is clear from Propositions 5.11.(i) and 5.12 and from Remark 5.14.

(ii): follows by Remark 5.7 and Proposition 5.16.(ii).

(iv): since by Proposition 5.11.(i) the function $\mathbb{C} \ni m \mapsto f_{m,v}$ (where $v := \chi_{[1,+\infty[}w$) is holomorphic the assertion follows from Propositions 5.16 and 5.25.

(iii): Under the assumptions of property (i) the condition (3.11) is derived as in the proof of Proposition 3.13. By the principle of analytic continuation it remains valid for every $m \in \mathbb{C}\setminus\mathfrak{A}(M)$. Hence $h := \mathfrak{M}_{f,w}$ satisfies the condition (a) of Proposition 2.51 for $(a,b) = (1,0)$. By Proposition 2.31 the condition (3.11) means that

$$(\partial_M - m)^i f_{m,w} = (-1)^i f_{m,\omega_{k-i}}, \qquad i \in \mathbb{N}_k,$$

holds for every $m \in \mathbb{C} \setminus \mathfrak{A}(M)$. Since by (iv) the assumptions of Remark 2.54 are satisfied for $N = k$ and $j = 0$ it follows that the preceding equations hold for arbitrary $m \in \mathfrak{A}(M)$, as well. Moreover, since by property (iv), again, the assumptions of Proposition 2.53 are satisfied for $N = k$ and $j = 0$, i.e. $\ell = -k-1$, we conclude that

$$(\partial_M - m)^{k+1} f_{m,w} = (-1)^k a_{-k-1}(m; \mathfrak{M}_{f,w})$$

and

$$(\partial_M - m)^{i+k+1} f_{m,w} = (-1)^k (\partial_M - m)^i a_{-k-1}(m; \mathfrak{M}_{f,w}) =$$
$$= (-1)^k a_{-k-1-i}(m; \mathfrak{M}_{f,w}) / \binom{k+i}{k} \qquad i \in \mathbb{N}.$$

Hence the condition (5.63) follows by property (iv).

(v): by Proposition 5.19 the assertion is valid for $\mathfrak{M}_{f,u}$. To prove it for $\mathfrak{M}_{f,v}$ we observe that by Lemma 5.4 in the defining integral for $f_{m,v}$ we may interchange differentiation and integration. Hence performing the computations done in the proof of Proposition 3.4 and taking (1.38) into account we arrive at the desired formulas.

(vi): this is a special case of (v) or can similarly be proved directly.

(vii) and (viii): see Remarks 5.5 and 5.6 and Proposition 5.21.

(ix): One easily verifies that under the assumptions of Theorem 5.37 the results of sections (b), (c), and (d) remain valid if in (5.28) the term $\chi_{]0,b]}$ is deleted, if $\int_0^b$ is replaced by $\int_0^{+\infty}$ and if all the other terms containing b explicitly are omitted provided that $\varepsilon \geq 0$. Alternatively, one could argue that $f_{m,w}\chi_{]0,b]}$ converges pointwise to $f_{m,w}$ as $b \to \infty$. ■

Notation 5.38. If $k = 0$ we write f_m instead of $f_{m,w}$.

Note that under the assumptions of Theorem 5.37.(i) f_m is given by (3.1). Moreover, the assertion of Theorem 5.37.(iii) means that f_m is quasihomogeneous of degree m in case $m \notin \mathfrak{A}(M)$ but may not be quasihomogeneous in case $m \in \mathfrak{A}(M)$.

Estimates for $f_{m,w}$ can be obtained from the results in sections (a) - (d). For the purposes in the following chapters, it suffices to have estimates if f belongs to $\mathcal{S}(V)$. Observe that in this case the growth conditions (5.62) are satisfied.

Notation 5.39. For every $N \in [0,+\infty[$ we denote by $\mathfrak{X}_M^N$ the Fréchet space of all C^∞ functions $g : V_+ \longrightarrow \mathbb{C}$ such that

$$\sup\{\, |g^{(\alpha)}(x)|\,(1+|P_+(x)|)^{-N}\,(1+|M_0 x|)^L \ ;\ x \in \Omega_{[\eta,+\infty[}\}$$

is finite for arbitrary $\alpha \in \mathfrak{A}$, $L \in \mathbb{N}$, and $\eta > 0$. We define

$$\mathfrak{X}_M^\infty := \operatorname*{ind}_{N \to +\infty} \mathfrak{X}_M^N .$$

Similarly as Remark 5.35 one proves

Remark 5.40. *Let $\chi \in \mathcal{O}_{\mathfrak{M}}(V)$ be such that its support is contained in Ω_J (see Notation 5.10) for some compact subinterval J of $]0,+\infty[$. Then the map $\mathfrak{X}_M^\infty \longrightarrow \mathcal{S}(V)$, $g \mapsto \chi g$, is well-defined and continuous.* ■

Applying Propositions 5.11.(i) and 5.12 to $u = (1+|\cdot|)^{-L} \circ M_0$ and taking Propositions 5.28 , 5.32 and 5.36 into account one obtains

Proposition 5.41. *Let $f \in \mathcal{S}(V)$. Then for every $m \in \mathbb{C}$ the function $f_{m,w}$ belongs to $\mathfrak{X}_M^\infty$. More precisely, let $r \in \mathbb{N}$ and choose ν according to (5.53), resp. (5.59); then*

(i) the map $\mathfrak{M}_{f,w}$, defined in Theorem 5.37.(iv), maps the set $\mathbb{C}(-\infty, rc_\varepsilon)$ meromorphically into $\mathfrak{X}_M^\nu$;

(ii) by $f \mapsto f_{m,w}$ a continuous linear map from $\mathcal{S}(V)$ into $\mathfrak{X}_M^\nu$ is defined provided that $\mathrm{Re}\, m < rc_\varepsilon$. ■

<u>*The case "$\varepsilon > 0$"*</u>. Here (5.12) is valid for every $L \in \mathbb{R}$ so that the domain of definition of $f_{m,w}$ will be the whole of X. As for the assumptions on f, we content ourselves with stating the results for $f \in \mathcal{O}_{\mathfrak{M}}(V)$.

Theorem 5.42. *Suppose that $\varepsilon > 0$. Let $f \in \mathcal{O}_{\mathfrak{M}}(V)$. Then for every $m \in \mathbb{C}$ by (5.60) the function $f_{m,w}$ (where w is given by (5.61)) is well-defined and belongs to $\mathcal{O}_{\mathfrak{M}}(V)$. If f belongs to $\mathcal{S}_M^\infty(V)$ (see Notation 5.34) so does $f_{m,w}$.*

By $\mathfrak{M}_{f,w}(m) := f_{m,w}$, $m \in \mathbb{C} \setminus \mathfrak{A}_\varepsilon(M)$, a meromorphic function $\mathfrak{M}_{f,w}$ on $\mathbb{C}$ with values in $\mathcal{O}_{\mathfrak{M}}(V)$ is defined, its poles lying in $\mathfrak{A}_\varepsilon(M)$; if $m \in \mathfrak{A}_\varepsilon(M)$ then

$a_0(m;\mathfrak{M}_{f,w}) = f_{m,w}$, and for every $i \in \mathbb{N}$ $a_{-i}(m;\mathfrak{M}_{f,w})$ is given by the formulas of Proposition 5.25.

The assertions (i), (ii), (vii), and (viii) of Theorem 5.37 remain valid. The assertions (v) and (vi) carry over if in their assumptions one postulates that the functions $P_j(x,\partial)f$ and $q_j f$ belong to $\mathcal{O}_{\mathfrak{M}}(V)$. Moreover, the formulas of Propositions 5.27, 5.28, and 5.32 remain valid if $\int_0^b \ldots$ is replaced by $\int_0^{+\infty} \ldots$ and if all the other terms containing b explicitly are deleted.

Finally, by $f \mapsto f_{m,w}$ continuous linear maps $\mathcal{O}_{\mathfrak{M}}(V) \longrightarrow \mathcal{O}_{\mathfrak{M}}(V)$ and $\mathcal{S}_M^\infty(V) \longrightarrow \mathcal{S}_M^\infty(V)$ are defined.

Proof: is analogous to the proofs of Theorem 5.37 and Proposition 5.41. ■

We close this section with an approximation result required in Chapter 8.

Proposition 5.43. *For arbitrary $f \in \mathcal{S}(V)$ and $m \in \mathbb{C}$ $f_{m,w}/V_+$ converges to f_{m,ω_k} in the topology of $\mathfrak{T}_M^\infty$ as $\varepsilon \to 0$.*

Proof. Let $v \in \{\omega_i ; i \in \mathbb{N}_0\}$, and set $v_\varepsilon(t) := e^{-\varepsilon t} v(t)$, $t \in]0,+\infty[$. Then from Propositions 5.11 and 5.36 we conclude that for every $m \in \mathbb{C}(-\infty,0)$ f_{m,v_ε} converges to $f_{m,v}$ in the topology of $\mathfrak{T}_M^\infty$ as $\varepsilon \to 0$. This implies the assertion in view of Theorem 5.37.(ix) and Theorem 5.42. ■

(f) The Locally Convex Spaces $\mathfrak{Q}_m(E_{\mathfrak{G}})$

Let $\mathfrak{G}$ satisfy the assumptions of Remark 2.67.(ii). In this section we are going to describe the spaces $Q_m(C_0^\infty(X)_{\mathfrak{G}})$ and $Q_m(\mathcal{S}(V)_{\mathfrak{G}})$ which are defined according to

Notation 5.44. For any subspace E of $C^{\infty,s}(X)$ we write $Q_m(E) := \{Q_m f ; f \in E\}$ and $E_{\mathfrak{G}} := \{f_{\mathfrak{G}} ; f \in E\}$.

In particular, for these spaces we shall introduce natural locally convex topologies required in Chapter 7. We begin by noting a few generalities on (almost) quasihomogeneous C^∞ functions on X. Recall that we are still assuming (5.17).

Proposition 5.45. *Let $R \in C^{\infty,s}(X)$. Then*

(i) the following conditions are equivalent:

(a) R is almost quasihomogeneous of degree m;

(b) $R = Q_m R$;

(c) $R = Q_m f$ for some $f \in C^{\infty,s}(X)$;

(ii) if one (and hence each) of the conditions of (i) is valid then R is almost quasihomogeneous of order not larger than

$$(5.65) \qquad N(m) := \begin{cases} 0 & \text{if } m \notin \mathfrak{A}(M) \\ \max\{N_\alpha ;\ \alpha \in \mathfrak{A}_m^+\} & \text{if } m \in \mathfrak{A}(M) \end{cases}$$

where N_α is defined by the condition that x^α is almost quasihomogeneous of order N_α (see Remark 1.30); in fact, R vanishes if $m \notin \mathfrak{A}(M)$.

Proof. The implication "*(b)⇒(c)*" is trivial. Since by Remark 1.30 the functions x^α, $\alpha \in \mathfrak{A}_m^+$, are almost quasihomogeneous of degree m and of order N_α the implication "*(c)⇒(a)*" and the assertion *(ii)* are an immediate consequence of the definition of $Q_m f$. By freezing the variables in ker M one reduces the proof of "*(a)⇒(b)*" to the case "$\sigma = \sigma_+$". Since in this case (5.17) means that X = V the assertion then follows from Propositions 1.58 and 1.34.(iii). ■

To formulate an assertion about the support of $Q_m f$ we require

Lemma 5.46. *If L is a closed subset of X then the sets $L \setminus X_+$ $(= L \cap M_O(L))$ and $M_O^{-1}(L \setminus X_+)$ are closed subsets of X.*

Proof. Since X_+ is open $L \setminus X_+$ is closed in X. If $(x_k)_{k\in\mathbb{N}}$ is a sequence in $M_O^{-1}(L \setminus X_+)$ converging to some point $x \in X$ as $k \to \infty$ then the sequence $(M_O x_k)$ converges to $M_O x$ as $k \to \infty$. Since by (5.17) $M_O x$ belongs to X and since $M_O x_k \in L$ it follows that $M_O x \in L \setminus X_+$, i.e. $x \in M_O^{-1}(L \setminus X_+)$. Note that one could do without assuming (5.17) if $L \setminus X_+$ were closed in $M_O(V)$. ■

Corollary 5.47. *Supp $Q_m f \subset M_0^{-1}((\mathrm{supp}\, f)\setminus X_+)$ for every $f \in C^{\infty,s}(X)$.*

Proof. If $x \notin M_0^{-1}((\mathrm{supp}\, f)\setminus X_+)$ then $M_0 x \notin \mathrm{supp}\, f$ so that $f^{(\alpha)}(M_0 x) = 0$ for every $\alpha \in \mathfrak{A}^+$. Consequently, the assertion follows from Lemma 5.46. ∎

Differentiation and multiplication by (almost) quasihomogeneous functions as well as taking $\mathfrak{G}$-averages commutes with the operation Q_m:

Proposition 5.48. *Let $f \in C^{\infty,s}(X)$. Then*

(i) $\partial^\alpha(Q_m f) = Q_{m-\alpha_M}(\partial^\alpha f)$ for every $\alpha \in \mathfrak{A}^{\infty,s}$;

(ii) $q\, Q_m f = Q_{m+\ell}(qf)$ for every $\ell \in \mathbb{C}$ and for every function $q: X \longrightarrow \mathbb{C}$ which is almost quasihomogeneous of degree ℓ and which is a $C^{\infty,s}$ function on a neighbourhood of $M_0^{-1}(\mathrm{supp}\, f)$;

(iii) $P_0(x,\partial)[Q_m f] = Q_{m+\ell}(P_0(x,\partial) f)$ for every $\ell \in \mathbb{C}$ and every $C^{\infty,s}$ copolynomial function $P_0 : X \times V^* \longrightarrow \mathbb{C}$ of degree $\le (\infty, s)$ which is almost quasihomogeneous of degree ℓ and of type $M \times (-M)^*$;

(iv) $(Q_m f)\circ A = Q_m(f \circ A)$ for every $A \in L(V,V)$ commuting with M;

(v) $(Q_m f)_{\mathfrak{G}} = Q_m(f_{\mathfrak{G}})$; in particular, if f is $\mathfrak{G}$-invariant so is $Q_m f$.

Proof. *(i):* this is a special case of (iii); for a direct proof we write $\alpha = \beta + \gamma$ where $\beta := \alpha_+$ and $\gamma := \alpha_0$ (see Notation 5.15) and obtain by Proposition 1.28.(i)

$$\partial^\alpha(Q_m f)(x) = \sum_{\eta \in \mathfrak{A}_m^+,\ \eta \ge \beta} \frac{1}{(\eta-\beta)!}\, x^{\eta-\beta}\, (\partial^{\gamma+\eta} f)(M_0 x)\,.$$

Since $\gamma + \eta = (\eta-\beta)+\alpha$ and $\{\eta - \beta\,;\ \eta \in \mathfrak{A}_m^+,\ \eta \ge \beta\} = \mathfrak{A}_{m-\alpha_M}^+$ the assertion follows.

(ii): again, this is a special case of (iii); for a direct proof we derive from the Leibniz formula that

$$Q_{m+\ell}(qf)(x) = \sum_{\gamma \in \mathfrak{A}_{m+\ell}^+}\ \sum_{\substack{\alpha,\beta \le \gamma \\ \alpha+\beta=\gamma}} \frac{1}{\alpha!}(\partial^\alpha f)(M_0 x)\, x^\alpha\, \frac{1}{\beta!}(\partial^\beta q)(M_0 x)\, x^\beta\,.$$

By (i) we have $\partial^\beta q = \partial^\beta(Q_\ell q) = Q_{\ell-\beta_M}(\partial^\beta q)$ near $M_0^{-1}(\mathrm{supp}\, f)$. This vanishes on $M_0(X)$ when $\beta_M \ne \ell$. Hence the assertion follows since by $(\alpha,\beta) \mapsto (\alpha+\beta,\alpha,\beta)$ a bijection from $\mathfrak{A}_m^+ \times \mathfrak{A}_\ell^+$ onto the set $\{(\gamma,\alpha,\beta)\,;\ \gamma \in \mathfrak{A}_{m+\ell}^+,\ \alpha,\beta \le \gamma,\ \alpha+\beta=\gamma,\ \beta_M = \ell\}$ is defined and since $q = Q_\ell q$.

(iii): Instead of performing a direct computation based on (i) and (ii) we employ Proposition 5.27 in order to reduce everything to the rules of computation already established. Indeed, supposing that $w = \chi_{]0,1]}$ and $\varepsilon = 0$ we obtain from Proposition 5.27 and (5.41) that $Q_m f = f - ((\partial_M - m) f)_{m,w}$ so that

$$P_0(x,\partial)[Q_m f] = P_0(x,\partial) f - P_0(x,\partial)((\partial_M - m) f)_{m,w} .$$

Setting $P_j := (\partial_{M\times(-M)^*} - \ell)^j P_0$ we deduce from Proposition 5.19 that

$$P_0(x,\partial)((\partial_M - m) f)_{m,w} = \sum_{j=0}^{\infty} (-1)^j [P_j(x,\partial)((\partial_M - m) f)]_{m+\ell, w_j} .$$

By Lemma 2.34 we have

$$P_j(x,\partial)((\partial_M - m) f) = (\partial_M - m - \ell)(P_j(x,\partial) f) - P_{j+1}(x,\partial) f .$$

Finally, applying Proposition 5.27 to $(w_j, (P_j(x,\partial) f, m+\ell)$ instead of (w,f,m) we obtain in view of (5.41)

$$[(\partial_M - m - \ell)(P_j(x,\partial) f)]_{m+\ell,w_j} + (P_j(x,\partial) f)_{m+\ell,w_{j-1}} =$$

$$= -\delta_{j0} Q_m(P_j(x,\partial) f) + w_j(1)\, P_j(x,\partial) f .$$

Since the right-hand side is equal to $\delta_{j0}[-Q_m(P_0(x,\partial) f) + P_0(x,\partial) f]$ one arrives at the desired formula by putting everything together.

(iv): Here, again, instead of performing a direct computation we apply Proposition 5.27 (to $w = \chi_{]0,1]}$, i.e. $k = 0$, $b = 1$, and $\varepsilon = 0$) twice and take Remark 2.67.(i) and Proposition 5.21.(i) into account to obtain

$$(Q_m f) \circ A = f \circ M_1 \circ A - ((\partial_M - m) f_{m,w}) \circ A =$$

$$= (f \circ A) \circ M_1 - (\partial_M - m)(f \circ A)_{m,w} = Q_m(f \circ A) .$$

(v): this follows from (iv) and the definition of $f_{\mathfrak{G}}$ (or from Propositions 5.27 and 5.21.(ii) and Remark 2.67.(ii)). ∎

Now we are going to deal with the case $E = C_0^\infty(X)$ first.

Notation 5.49. Let K be a G-invariant compact subset of V. By $Q_m^\infty(K;\mathfrak{G})$ we denote the space of all $\mathfrak{G}$-invariant functions $R \in C^\infty(V)$ with support contained in $M_0^{-1}(K \cap M_0(K))$ which are almost quasihomogeneous of degree m.

Lemma 5.50. *(i) Equipped with the semi-norms*

$$R \mapsto \Pi_\beta(R) := \sup\{\,|R^{(\alpha+\beta)}(x)|;\ x\in V,\ \alpha\in\mathfrak{A}_m^+\},\qquad \beta\in\mathfrak{A}^o,$$

$Q_m^\infty(K;\mathfrak{G})$ is a Fréchet space.

(ii) Let U be any compact neighbourhood of $K\cap M_o(K)$ in V, and set $L := U\cap M_o^{-1}(K\cap M_o(K))$; then

$$Q_m(C_o^\infty(K)_{\mathfrak{G}}) \subset Q_m^\infty(K;\mathfrak{G}) \subset Q_m(C_o^\infty(L)_{\mathfrak{G}}).$$

Proof. *(i)*: We first observe that $\Pi_\beta(R)$ which in view of $\partial^\alpha R = (\partial^\alpha R)\circ M_o$ is equal to

$$\sup\{\,|R^{(\alpha+\beta)}(x)|;\ x\in K\cap M_o(K),\ \alpha\in\mathfrak{A}_m^+\}$$

is finite for every $\beta\in\mathfrak{A}^o$. If $(R_j)_{j\in\mathbb{N}}$ is a Cauchy sequence in $Q_m^\infty(K;\mathfrak{G})$ then for every $\alpha\in\mathfrak{A}_m^+$ the sequence $(\partial^\alpha R_j)_{j\in\mathbb{N}}$ converges to a C^∞ function $R_\alpha : V\longrightarrow\mathbb{C}$ with support contained in $M_o^{-1}(K\cap M_o(K))$ such that $R_\alpha\circ M_o = R_\alpha$. Consequently, by

$$R(x) := \sum_{\alpha\in\mathfrak{A}_m^+}\frac{1}{\alpha!}R_\alpha(x)\,x^\alpha$$

the desired limit R of (R_j) is defined.

(ii): In view of Proposition 5.48.(v) and Corollary 5.47 the inclusion on the left-hand side is obvious. To prove the other inclusion we choose $\chi\in C_o^\infty(U)$ equal to 1 near $K\cap M_o(K)$. Then for every $R\in Q_m^\infty(K;\mathfrak{G})$ the Leibniz formula and Proposition 1.28.(i) show that $R = Q_m(\chi R)$. Again by Proposition 5.48.(v) we conclude that $R = Q_m[(\chi R)_{\mathfrak{G}}]$. ∎

For the following proposition we drop the assumption (5.17).

Proposition 5.51. *Via the identity*

$$Q_m(C_o^\infty(X)_{\mathfrak{G}}) = \operatorname*{ind}_{K\subset X} Q_m^\infty(K;\mathfrak{G})$$

the space $Q_m(C_o^\infty(X)_{\mathfrak{G}})$ is equipped with the structure of a strict (LF)-space (see, for example, § 23.5.3,4 in Floret-Wloka [51]); as a quotient of $C_o^\infty(X)$ it is a nuclear $(\overline{S})$-space.

Proof. The first assertion is an immediate consequence of Lemma 5.50. To see that $F := Q_m(C_o^\infty(X)_{\mathfrak{G}})$ is a quotient of $C_o^\infty(X)$ we define $P : C_o^\infty(X)\longrightarrow F$ by

$\varphi \mapsto Q_m \varphi_{\mathfrak{G}}$ and observe that P is linear, continuous and surjective. Since $C_0^\infty(X)$ is a strict (LF)-space, as well, the open mapping theorem (see e.g. § 24.4.1 in Floret-Wloka [5]) shows that P is a homomorphism. Since F is complete (see § 24.3.1 in [5]) the last assertion follows from § 27.2.4 and § 27.1.13 in [5]. ∎

Now we come to the case $E = \mathcal{S}(V)$.

Proposition 5.52. *(i)* $Q_m(\mathcal{S}(V)_{\mathfrak{G}})$ *is equal to the space of* $\mathfrak{G}$*-invariant* C^∞ *functions* $R: V \longrightarrow \mathbb{C}$ *which are almost quasihomogeneous of degree m such that*

$$\Pi_\ell(R) := \sup\{\,|R^{(\beta+\alpha)}(x)|\,(1+|M_o x|)^\ell;\ x \in V,\ \alpha \in \mathfrak{A}_m^+,\ \beta \in \mathfrak{A}^o,\ |\beta| \le \ell\}$$

is finite for every $\ell \in \mathbb{N}_o$.

(ii) The functions Π_ℓ *defined above are semi-norms providing* $Q_m(\mathcal{S}(V)_{\mathfrak{G}})$ *with the structure of a nuclear Fréchet-Schwartz space.*

For the proof and for other purposes the following lemma is required.

Lemma 5.53. *(i) If F and H are closed subsets of V such that*

(5.66) $\quad \operatorname{dist}(F, H) > 0$

then there exists a function $\chi \in C^\infty(V)$ *equal to 1 near F with support contained in* $V \setminus H$ *such that* $0 \le \chi \le 1$ *and all derivatives of* χ *are bounded.*

(ii) If J and K are disjoint closed subintervals of $[0, +\infty[$ *then (5.66) is satisfied for the choices (a)* $F = \Omega_J$ *and* $H = \Omega_K$ *(see Notation 5.10) and (b)* $F = \dot{\varkappa}^{-1}(J)$ *and* $H = \dot{\varkappa}^{-1}(K)$ *where* $\varkappa := \varkappa^+$ *(see Proposition 1.70).*

Proof. *(i):* see for example Corollary 1.4.11 in Hörmander [11].

(ii): Of course, we may assume that J and K are non-empty and that J or K, say K, is compact. Then $P_+(\Omega_K)$ is compact, as well. Since it is disjoint from the closed subset $P_+(\Omega_J)$ of V it follows that $\operatorname{dist}(P_+(\Omega_K), P_+(\Omega_J))$ is positive. Since $|x-y| \ge |P_+(x) - P_+(y)|$ the assertion is proved for the choice *(a)*. For the proof of *(b)* we first deduce from (1.76) and the defining equality in the proof of Proposition 1.63, "(b)⇒(a)", that $\varkappa \circ P_+ = \varkappa$ so that by continuity we have $\dot{\varkappa} \circ P_+ = \dot{\varkappa}$. Hence in view of (1.77) it follows that $P_+(\dot{\varkappa}^{-1}(K))$ is compact and

$P_+(\dot{x}^{-1}(J))$ is closed, and a similar argument as above leads to the desired result for the choice (b). ■

Proof of Proposition 5.52. (i): If $R = Q_m \varphi$ for some $\varphi \in \mathcal{S}(V)$ then in view of Propositions 5.45.(i) and 1.28.(i) it is obvious that $\Pi_\ell(R)$ is finite for every $\ell \in \mathbb{N}_0$. Conversely, let $R \in C^\infty(V)$ be almost quasihomogeneous of degree m such that $\Pi_\ell(R)$ is finite for every $\ell \in \mathbb{N}_0$. By Lemma 5.53 we choose a function $\chi \in C^\infty(V)$ equal to 1 near $\Omega_{\{0\}} = M_0(V)$ with support contained in $\Omega_{[0,1[}$ such that all the derivatives of χ are bounded. Then χR belongs to $\mathcal{S}(V)$, and we have $Q_m(\chi R) = R$. Since by Proposition 5.48.(v) we have $Q_m[(\chi R)_{\mathfrak{G}}] = R$ it follows that $R \in Q_m(\mathcal{S}(V)_{\mathfrak{G}})$.

(ii): If $(R_j)_{j \in \mathbb{N}}$ is a Cauchy sequence in $F := Q_m(\mathcal{S}(V)_{\mathfrak{G}})$ then it converges (with respect to the semi-norms Π_ℓ) to a $\mathfrak{G}$-invariant C^∞ function $R: V \longrightarrow \mathbb{C}$ such that $Q_m R = R$, i.e. R is almost quasihomogeneous of degree m and hence, by assertion (i), belongs to F. Note that the map $P: \mathcal{S}(V) \longrightarrow F$, $\varphi \mapsto Q_m \varphi_{\mathfrak{G}}$, is linear, continuous and surjective, hence a homomorphism by Banach's open mapping theorem. Consequently, F is a quotient of the nuclear Fréchet-Schwartz space $\mathcal{S}(V)$, hence a nuclear Fréchet-Schwartz space itself by § 27.2.4 and § 27.1.13 in Floret-Wloka [5]. ■

(g) The Locally Convex Spaces $\mathfrak{A}^\infty_{m,k}(E_{\mathfrak{G}})$

In this section we drop the assumption (5.17), fix $k \in \mathbb{N}_0$, and introduce natural locally convex topologies (required in Chapter 7) for the spaces $\mathfrak{A}^\infty_{m,k}(C_0^\infty(X)_{\mathfrak{G}})$ and $\mathfrak{A}^\infty_{m,k}(\mathcal{S}(V)_{\mathfrak{G}})$ which are defined according to

Notation 5.54. If E is a subspace of $\mathcal{S}(V)$ we write $\mathfrak{A}^\infty_{m,k}(E) := \{\varphi_{m,\omega_k};\ \varphi \in E\}$ and $\mathfrak{A}^\infty_m(E) := \mathfrak{A}^\infty_{m,0}(E)$.

To deal with the case $E = C_0^\infty(X)$ first we fix a G-invariant compact subset K of V and introduce

Notation 5.55. By $\mathfrak{A}^{\infty}_{m,k}(K;\mathfrak{G})$ we denote the space of all almost quasihomogeneous $\mathfrak{G}$-invariant C^{∞} functions $f: V_{+} \longrightarrow \mathbb{C}$ with support contained in $\overline{K_M}$ such that $(\partial_M - m)^{k+1} f$ extends to a C^{∞} function $R: V \longrightarrow \mathbb{C}$ belonging to $Q^{\infty}_{m}(K;\mathfrak{G})$ (see Notation 5.49). Note that the pairs (f,R) satisfy

$$(5.67) \qquad t^{-m} f \circ M_t = \sum_{i=0}^{k} \omega_i(t)\,(\partial_M - m)^i f + \sum_{i \in \mathbb{N}} \omega_{k+i}(t)\,(\partial_M - m)^{i-1} R|_{V_+}$$

for every $t \in]0,+\infty[$.

To handle the support of these functions f we require

Remark 5.56. *There is a compact subset L of V_+ such that*

$$(5.68) \qquad \overline{K_M} \cap V_+ = L_M .$$

Proof. The subset $L := \overline{K_M} \cap \Omega_{\{1\}}$ of V_+ is closed in V and in view of Remark 1.8 satisfies (5.68). It is bounded since by Lemma 5.9 it is contained in the set

$$\{x \in V;\ |P_+(x)| = 1,\ |M_o x| \le \max\{|M_o k|;\ k \in K\}\}. \quad \blacksquare$$

Lemma 5.57. *(i) Suppose that ν is chosen as in Proposition 5.41. Then, equipped with the semi-norms defined by*

$$f \mapsto \pi_\ell(f) := \sup\{|f^{(\alpha)}(x)|\,(1+|P_+(x)|)^{-\nu};\ x \in \Omega_{[1/\ell,+\infty[},\ |\alpha| \le \ell\}, \qquad \ell \in \mathbb{N},$$

$\mathfrak{A}^{\infty}_{m,k}(K;\mathfrak{G})$ becomes a Fréchet space.

(ii) Let L be any compact subset of V_+ satisfying (5.68), and let U be any open neighbourhood of $(K \cap M_o(K)) \cup L$; then we have

$$\mathfrak{A}^{\infty}_{m,k}(C^{\infty}_{o}(K)_{\mathfrak{G}}) \subset \mathfrak{A}^{\infty}_{m,k}(K;\mathfrak{G}) \subset \mathfrak{A}^{\infty}_{m,k}(C^{\infty}_{o}(U)_{\mathfrak{G}}).$$

For the proof the following lemma is required.

Lemma 5.58. *$\overline{L} \cap X \subset X_+$ for every $(M,]0,1])$-bounded subset L of X_+.*

Proof. We fix $x \in X \setminus X_+$ and choose $\varepsilon > 0$ such that the closed polydisc $\overline{P}(x,\varepsilon) := \{y \in V;\ |P_+ y| \le \varepsilon,\ |M_o(y-x)| \le \varepsilon\}$ is contained in X. Then $K := \{y \in \overline{P}(x,\varepsilon);\ |P_+ y| = \varepsilon\}$ is a compact subset of X_+. In view of the last assertion in Proposition 1.70 and in view of Remark 1.8.(i) we have $K_{M,1/]0,1]} = \overline{P}(x,\varepsilon) \cap X_+$. If L is an $(M,]0,1])$-bounded subset of X_+ it follows by Proposition 3.22 that $\overline{L} \cap (\overline{P}(x,\varepsilon) \cap X_+)$ is compact. Since $L \subset X_+$ this implies that x does not belong to $\overline{L}$. ■

Proof of Lemma 5.57. (ii): The inclusion on the left-hand side is obvious in view of Proposition 2.59.(i) and of the assertions (ii), (iii), and (viii) of Theorem 5.37.

To verify the other inclusion we set $\varepsilon := \mathrm{dist}(\partial U, K \cap M_O(K))/2$ and choose a finite subset N of $K \cap M_O(K)$ such that the set $K \cap M_O(K)$ is covered by the polydiscs $P(x,\varepsilon) := \{y \in V;\ |M_O(y-x)|, |P_+(y)| < \varepsilon\}$, $x \in N$. Then $W := \bigcup_{x \in N} \overline{P}(x,\varepsilon)$ is a compact neighbourhood of $K \cap M_O(K)$ contained in U such $M_O(W) \subset W$ and $M_O^{-1}(W) \subset W_M$. In particular, by Lemma 5.9 W_M is closed in V.

Now we fix $f \in \mathfrak{A}^{\infty}_{m,k}(K;\mathfrak{G})$, and let R be as in Notation 5.55. By Lemma 5.50.(ii) we choose $\phi \in C_O^{\infty}(W)$ such that $Q_m \phi = R$. Since by Theorem 5.37.(iii) ϕ_{m,ω_k} is almost quasihomogeneous of degree m such that $(\partial_M - m)^{k+1} \phi_{m,\omega_k} = (-1)^{k+1} R|_{V_+}$ it follows that $q := f + (-1)^k \phi_{m,\omega_k}$ is almost quasihomogeneous of degree m and of order $\leq k$. Moreover, by Theorem 5.37.(ii) the support of ϕ_{m,ω_k} is contained in W_M. Consequently, in view of Proposition 2.31 we can apply Theorem 3.48 to $X = V_+$, $\Lambda = \{m\}$, $j_m = k$, $q_{m,j} := (-1)^{k-j} (\partial_M - m)^{k-j} q$, $Z = V_+ \cap (\overline{K_M \cup W_M})$, and $Y = (V_+ \setminus Z) \cup (U \cap V_+)$ (note that in view of (5.68) and $U_M \supset L_M \cup W_M$ we have $Y_M = V_+$, indeed). Hence we obtain a function $\psi \in C_M^{\infty}(V_+)$ such that $\psi_{m,\omega_k} = q$ and $\mathrm{supp}\, \psi \subset Z \cap Y \subset U$. Since we may assume that U is relatively compact in X and since $\mathrm{supp}\, \psi$ is an M-bounded subset of V_+ Lemma 5.58 and the last inclusion imply that $\mathrm{supp}\, \psi$ is a compact subset of V_+ so that by setting $\psi|_{V \setminus V_+} :\equiv 0$ one extends ψ to a C^{∞} function defined on the whole of V and again denoted by ψ. It follows that $\varphi := \psi - (-1)^k \phi$ belongs to $C_O^{\infty}(U)$. Since f equals φ_{m,ω_k} and is $\mathfrak{G}$-invariant an application of Theorem 5.37.(viii) and Proposition 2.59 completes the proof of (ii).

(i): That the semi-norms π_k are well-defined follows in view of the assertion (ii) from Proposition 5.41.(ii). For the proof of completeness we first observe that the assumption (1.14) implies that $Mx = P_+ Mx = MP_+ x$, $x \in V$, so that by means of (1.32) one obtains

$$(5.69) \qquad \partial_M f(x) = \sum_{\substack{\alpha \in \mathfrak{A}^+ \\ |\alpha| = 1}} (MP_+ x)^{\alpha} \partial^{\alpha} f(x), \qquad x \in X,\ f \in C^{1,0}(X)$$

Moreover, we observe that the seminorms Π_β defined in Lemma 5.50 satisfy

$$\Pi_\beta(R) = \sup\{\,|R^{(\alpha+\beta)}(x)|\,;\ \alpha\in\mathfrak{A}^+_m,\ x\in\Omega_{\{1\}}\}\,, \qquad R\in Q^{\infty}_m(K;\mathfrak{G})\,.$$

Since (5.69) shows that the coefficients in the differential operator ∂_M are functions of P_+x, only, we conclude that for every $\beta\in\mathfrak{A}^0$ there are constants $C_\beta>0$ and $\ell_\beta\in\mathbb{N}$ satisfying

$$\Pi_\beta(R)\le C_\beta\,\pi_{\ell_\beta}(f)\,, \qquad f\in\mathfrak{A}^{\infty}_{m,k}(K;\mathfrak{G}),\ R=(\partial_M-m)^{k+1}f\,.$$

Now, let $(f_j)_{j\in\mathbb{N}}$ be a Cauchy sequence in $\mathfrak{A}^{\infty}_{m,k}(K;\mathfrak{G})$, and denote by R_j the extension of $(\partial_M-m)^{k+1}f_j$ to V. Then (R_j) is a Cauchy sequence in the space $Q^{\infty}_m(K;\mathfrak{G})$. By Lemma 5.50.(i) it converges to some $R\in Q^{\infty}_m(K;\mathfrak{G})$. Let f be the limit of the sequence (f_j) in the space $C^{\infty}(V_+)$. It then follows that supp f is contained in $\overline{K_M}$. Moreover, (5.67) is satisfied since it is valid with (f,R) replaced by (f_j,R_j). In particular, $(\partial_M-m)^{k+1}f=R|_{V_+}$. We conclude that f is the desired limit of (f_j) in the topology of $\mathfrak{A}^{\infty}_{m,k}(K;\mathfrak{G})$. ∎

Proposition 5.59. *Via the identity*

$$\mathfrak{A}^{\infty}_{m,k}(C^{\infty}_0(X)_{\mathfrak{G}}) = \operatorname*{ind}_{K\subset X}\ \mathfrak{A}^{\infty}_{m,k}(K;\mathfrak{G})$$

$\mathfrak{A}^{\infty}_{m,k}(C^{\infty}_0(X)_{\mathfrak{G}})$ *carries the structure of a strict (LF)-space (see e.g. § 23.5.3,4 in Floret-Wloka [5]); as a quotient of* $C^{\infty}_0(X)$ *it is a nuclear* $(\overline{S})$*-space.*

Proof. The first assertion is an immediate consequence of Lemma 5.57. To see that $F:=\mathfrak{A}^{\infty}_{m,k}(C^{\infty}_0(X)_{\mathfrak{G}})$ is a quotient of $C^{\infty}_0(X)$ we define $P: C^{\infty}_0(X)\longrightarrow F$ by $\varphi\mapsto(\varphi_{\mathfrak{G}})_{m,\omega_k}$ and observe that P is linear, continuous and surjective, hence open by the open mapping theorem (see e.g. § 24.4.1 in [5]). The rest of the assertion follows as in the proof of Proposition 5.51. ∎

Since we dropped the assumption (5.17) a remark is in order concerning the support of the functions φ_m and $Q_m\varphi$ if $\varphi\in C^{\infty}_0(X)$. Recall that the domain of definition of φ_m (resp. $Q_m\varphi$) is V_+ (resp. V).

Remark 5.60. *For every* $\varphi\in C^{\infty}_0(X)$ *we have*

(i) $\operatorname{supp}\varphi_{m,\omega_k}\subset X_+$, *and* *(ii)* $\operatorname{supp} Q_m\varphi\subset M_0^{-1}(X)\subset X$.

Proof. We first observe that

(5.70) $M_0^{-1}(X) \subset X$.

Indeed, let $x \in V$ be such that $y := M_0 x \in X$. Since $M_t x$ converges to y as $t \to 0$ and since X is open there is $t \in]0, +\infty[$ such that $M_t x \in X$, and since X is quasihomogeneous this implies that $x \in X$.

(i): By Theorem 5.37.(ii) the support of φ_m lies in the set $\overline{(\operatorname{supp}\varphi) \cap V_+}$ which is contained in X by Lemma 5.9 and (5.70).

(ii): By Corollary 5.47 the support of $Q_m \varphi$ lies in $M_0^{-1}((\operatorname{supp}\varphi) \setminus X_+)$ which is contained in X by (5.70). ■

Finally, we come to the case $E = \mathcal{S}(V)$.

Proposition 5.61. *(i) $\mathfrak{A}^{\infty}_{m,k}(\mathcal{S}(V)_{\mathfrak{G}})$ is equal to the space of all almost quasihomogeneous $\mathfrak{G}$-invariant C^{∞} functions $f: V_+ \longrightarrow \mathbb{C}$ such that $(\partial_M - m)^{k+1} f$ extends to a function $R \in Q_m(\mathcal{S}(V)_{\mathfrak{G}})$ and such that*

$$\pi_\ell(f) := \sup\{ |f^{(\alpha)}(x)| \, (1 + |P_+(x)|)^{-\nu} (1 + |M_0 x|)^{\ell} ; \, x \in \Omega_{[1/\ell, +\infty[} , \, |\alpha| \le \ell \}$$

is finite for every $\ell \in \mathbb{N}_0$ where ν is any constant chosen as in Proposition 5.41.

(ii) The functions π_ℓ defined above are semi-norms providing $\mathfrak{A}^{\infty}_{m,k}(\mathcal{S}(V)_{\mathfrak{G}})$ with the structure of a nuclear Fréchet-Schwartz space.

For the proof the following supplement to Theorem 3.48 is required.

Lemma 5.62. *Suppose that $X = V_+$, let Λ and $(j_m)_{m \in \Lambda}$ be as in Theorem 3.48 and for arbitrary $m \in \Lambda$ and $j \in \mathbb{N}_{j_m} \cup \{0\}$ let $q_{m,j} \in C^{\infty}(X)$ be given such that the condition (b) of Theorem 3.48 is valid and such that*

(5.71) $\sup\{ |q^{(\alpha)}(x)| \, (1 + |M_0(x)|)^{\ell} ; \, x \in \Omega_{\{1\}} \}$ *is finite*

for arbitrary $q \in \mathfrak{Q} := \{ q_{m,j} ; \, m \in \Lambda , \, j \in \mathbb{N}_{j_m} \cup \{0\} \}$, $\alpha \in \mathfrak{A}$, and $\ell \in \mathbb{N}$. Then for any non empty open subinterval I of $]0, +\infty[$ there exists a function $f \in \mathcal{S}(V)$ satisfying the condition (a) of Theorem 3.48 for $Y = \varkappa^{-1}(I)$ where $\varkappa$ is the function $\varkappa_+$ defined in Proposition 1.70.

Proof. Without loss of generality we may assume that I is relatively compact in $]0,+\infty[$. It then follows from (1.77) that $\varkappa^{-1}(I) \subset \Omega_J$ for some compact subinterval J of $]0,+\infty[$. Now, setting $Y = \varkappa^{-1}(I)$ we proceed as in Step 1 of the proof of Theorem 3.48. In particular, we obtain f by (3.43). In order to see that f belongs to $\mathcal{S}(V)$ we observe that in view of (1.76) and the definition of $\varkappa^{+}$ we have

(5.72) $$\text{(a)}\ \varkappa(x) = \varkappa(P_{+}x) \quad \text{and} \quad \text{(b)}\ \rho_{\varkappa}(x) = M_0(x) + \rho_{\varkappa}(P_{+}x), \qquad x \in X.$$

Since $P_{+}(\Omega_J)$ is a compact subset of V_{+} this implies that all the derivatives of $\varkappa$ and $\rho_{\varkappa}$ are bounded on Y. Hence it follows from (3.43) that for every $\alpha \in \mathfrak{A}$ there is a constant C_{α} such that

$$|f^{(\alpha)}(x)| \le C_{\alpha} \max\{\, |q^{(\beta)}(\rho_{\varkappa}(x))|\,;\ |\beta| \le |\alpha|,\ q \in \mathfrak{Q}\}, \qquad x \in Y.$$

Combining this with the assumption on $\mathfrak{Q}$ we obtain for arbitrary $\alpha \in \mathfrak{A}$ and $\ell \in \mathbb{N}$ a constant $C_{\alpha.\ell}$ such that

$$|f^{(\alpha)}(x)| \le C_{\alpha,\ell}(1+|M_0(x)|)^{-\ell}, \qquad x \in Y.$$

Since by the compactness of $P_{+}(\Omega_J)$, again, we have $1+|M_0(x)| \ge \text{const}\,(1+|x|)$, $x \in Y$, we conclude that f belongs to $\mathcal{S}(V)$, indeed. ■

Proof of Proposition 5.61. *(i)*: If $f = (\varphi_{\mathfrak{G}})_{m,\omega_k}$ for some $\varphi \in \mathcal{S}(V)$ then Theorem 5.37.(iii) and Proposition 5.41.(ii) show that f has the desired properties.

Conversely, let f be a function having these properties. Then we can find $\phi \in \mathcal{S}(V)$ such that $R := Q_m \phi_{\mathfrak{G}}$ extends $(\partial_M - m)^{k+1} f$. Consequently, by Theorem 5.37.(iii) $q := f + (-1)^k \phi_{m,\omega_k}$ is almost quasihomogeneous of degree m and of order $\le k$. We set $\Lambda = \{m\}$ and $j_m = k$. Since the functions $q_{m,j} := (-1)^{k-j}(\partial_M - m)^{k-j} q$ satisfy the estimates in the assumption of Lemma 5.62 and since in view of Proposition 2.31 the condition (b) of Theorem 3.48 is valid we obtain a function $\psi \in \mathcal{S}(V)$ such that supp ψ is an M-bounded subset of V_{+} and such that $\psi_{m,\omega_k} = q$, i.e. $f = \varphi_{m,\omega_k}$ where $\varphi := \psi - (-1)^k \phi \in \mathcal{S}(V)$. Since f is $\mathfrak{G}$-invariant an application of Theorem 5.37.(viii) and Propositions 2.59.(ii) and 2.62 completes the proof of (i).

(ii): That the semi-norms π_{ℓ} are well-defined follows in view of the assertion (i) from Proposition 5.41.(ii). We observe that for every $R \in Q_m(\mathcal{S}(V)_{\mathfrak{G}})$ the semi-norms $\Pi_{\ell}(R)$ defined in Proposition 5.52.(i) satisfy

$$\Pi_{\ell}(R) = \sup\{\, |R^{(\alpha+\beta)}(x)|\,(1+|M_0(x)|)^{\ell}\,;\ x \in \Omega_{\{1\}},\ \alpha \in \mathfrak{A}^{+}_{m},\ \beta \in \mathfrak{A}^{0},\ |\beta| \le \ell\}.$$

In view of (5.69) it follows that for every $\ell \in \mathbb{N}$ we find constants C_ℓ and $j_\ell \in \mathbb{N}$ such that

$$\Pi_\ell(R) \leq C_\ell \, \pi_{j_\ell}(f) , \qquad f \in \mathfrak{A}^\infty_{m,k}(\mathcal{S}(V)_{\mathfrak{G}}), \; R = (\partial_M - m)^{k+1} f .$$

Now, let $(f_j)_{j \in \mathbb{N}}$ be a Cauchy sequence in the space $F := \mathfrak{A}^\infty_{m,k}(\mathcal{S}(V)_{\mathfrak{G}})$, and choose $R_j \in Q_m(\mathcal{S}(V)_{\mathfrak{G}})$ such that R_j extends $(\partial_M - m)^{k+1} f_j$. Then (R_j) is a Cauchy sequence in $Q_m(\mathcal{S}(V)_{\mathfrak{G}})$. By Proposition 5.52 it converges to a limit R in this space. Since (5.67) is valid with (f,R) replaced by (f_j, R_j) we conclude that (5.67) remains valid for (f,R) if we denote by f the limit of (f_j) in the space $C^\infty(V_+)$. Hence f is almost quasihomogeneous of degree m satisfying the equation $R|_{V_+} = (\partial_M - m)^{k+1} f$. It follows that f is the limit of (f_j) in the topology of F.

To prove the other assertions we observe that the mapping $P: \mathcal{S}(V) \longrightarrow F$ defined by $\varphi \mapsto (\varphi_{\mathfrak{G}})_{m,\omega_k}$ is linear, continuous and surjective, hence open by Banach's open mapping theorem. The rest of the proof is done as in the proof of Proposition 5.52. ∎

Chapter VI

Constructing (Almost) Quasihomogeneous Distributions by Taking Quasihomogeneous Averages. The Case: (1.14) holds

In this chapter we continue to assume that (1.14) holds. Instead of (5.17) we only suppose that $X \setminus X_+ \neq \emptyset$. Our aim is to carry over the theory of Chapter 4. For very special weight functions w the distributions $u_{m,w}$ will be defined by (4.3), as before. This time one has to employ the functions $\varphi_{m,w}$, $\varphi \in C_0^\infty(X)$, introduced in Chapter 5. Moreover, the support of u has to be a *weakly (M,I)-bounded subset of X* in the sense of Definition 6.1 below. If m does not belong to the set $(-\mathfrak{A}(M)-\mu)$ the results are parallel to those of Chapter 4. However, if m lies in the exceptional set $(-\mathfrak{A}(M)-\mu)$ then complications occur which are mainly due to the fact that the quasihomogeneity order of the functions φ_{m,ω_k} may exceed k, the support of its $(k+1)^{th}$ order deficiency being contained in $X \setminus X_+$. So, in particular, in general u_m will only be almost quasihomogeneous of degree m, the support of its deficiency being contained in $X \setminus X_+$. In order later to be able to apply the Fourier transform we develop the theory for temperate distributions, as well.

In section (a) the weakly (M,I)-bounded subsets of X are introduced and studied. In section (b) for every $m \in \mathbb{C}$ the distributions $u_{m,w}$ are defined, and their basic properties are established. In section (c) the almost quasihomogeneous distributions on X with support contained in $X \setminus X_+$ are determined and described in a way which serves the purposes of the following sections. In section (d) those almost quasihomogeneous distributions T such that $\operatorname{supp}(\partial_M - m)^{k+1}T$ is contained in $X \setminus X_+$ are described in terms of quasihomogeneous averages of distributions with weakly M-bounded support. For these averages in section (f) a duality bracket is introduced required in Chapter 7. Section (e) deals with the solvability of the equation $(\partial_M - m)S = T$ for arbitrary $T \in \mathcal{D}'(X)$.

(a) Weakly (M,I)-bounded Subsets of X

Let I be a closed subinterval of $]0,+\infty[$. And let L be a subset of X.

Definition 6.1. L is called a *weakly (M,I)-bounded subset of X* if for every compact subset K of X the set $\overline{\overline{L} \cap X \cap K_{M,1/I}}$ is a compact subset of $X_{1/I}$ where

$$(6.1) \qquad X_I := \begin{cases} X & \text{if } I \text{ is bounded} \\ X_+ & \text{if } I \text{ is unbounded} \end{cases} .$$

If $I =]0,+\infty[$ we also say *weakly M-bounded* instead of *weakly (M,I)-bounded.*

Obviously, every compact subset of $X_{1/I}$ is a weakly (M,I)-bounded subset of X. Moreover, if I is compact then every subset of X is weakly (M,I)-bounded so that in this case the notions of weakly (M,I)-boundedness and (M,I)-boundedness coincide. The same turns out to be true if I is merely bounded. This follows from part A of the proposition below characterizing weakly (M,I)-bounded subsets of X. For its formulation we require

Notation 6.2. By $\tilde{X}$ we denote the set of all interior points of $X \cup \ker M$.

Note that $\tilde{X}$ is the set of all interior points of $X \cup M_0(X)$ and that

$$(6.2) \qquad \text{(a)} \ (\tilde{X})_+ = X_+, \qquad \text{and} \qquad \text{(b)} \ (X_+)^\sim = \tilde{X}.$$

Moreover, $\tilde{X}$ is a quasihomogeneous open subset of V. Note that there are two other important quasihomogeneous open subsets of V related to X, namely, the set $X^o = M_0^{-1}(X)$ already introduced in Notation 1.59 and the set $M_0^{-1}(M_0(X))$ Obviously, we have

$$(6.3) \qquad X^o \subset X \subset \tilde{X} \subset M_0^{-1}(M_0(X)).$$

The set X^0 will play an important role later in this chapter. The set $M_0^{-1}(M_0(X))$ already appeared in Chapter 5, though implicitly, in that the condition (5.17) amounts to equality being valid everywhere in (6.3).

Proposition 6.3.A. *Suppose that $I =]0, b]$ for some $b \in]0, +\infty[$. Then*

(i) the following conditions are equivalent:

(a) L is a weakly (M, I)-bounded subset of X;

(b) for some (resp. every) $Y \in \{X, X_+, \tilde{X}\}$ L is an (M, I)-bounded subset of Y;

(c) $\overline{L \cap K_{M,1/I}} \subset X_+$ for every compact subset K of Y for some (resp. every) $Y \in \{X, X_+, \tilde{X}\}$;

(ii) each of the conditions of (i) implies that

(d) $\bar{L} \cap \tilde{X} \subset X_+$;

the converse is valid provided that (5.17) holds.

However, if I is an unbounded proper subinterval of $]0, +\infty[$ then the description of the weakly (M,I)-bounded subsets of X looks different:

Proposition 6.3.B. *Suppose that $I = [a, +\infty[$ for some $a \in]0, +\infty[$. Then L is a weakly (M, I)-bounded subset of X if and only if $L \cap X_+$ is an (M, I)-bounded subset of X_+.*

So, if I is an unbounded proper subinterval of $]0, +\infty[$ then weakly (M,I)-bounded subsets of X may contain points of $X \setminus X_+$; and since by Remark 3.17 these points are not (M,I)-bounded in X one sees that not every weakly (M,I)-bounded subset of X is (M,I)-bounded.

Combining Propositions 6.3.A and 6.3.B one obtains

Proposition 6.3.C. *L is a weakly M-bounded subset of X if and only if L is an M-bounded subset of X_+.* ∎

Note that if L is an (M,I)-bounded subset of X then $L \cap X_+$ is an (M,I)-bounded subset of X_+. Hence, from Proposition 6.3 one obtains

Corollary 6.4. *If L is an (M,I)-bounded subset of X it is a weakly (M,I)-bounded subset of X.* ■

Another consequence of Proposition 6.3 is

Corollary 6.5. *Let Y be a quasihomogeneous open subset of V containing X. If L' is a weakly (M,I)-bounded subset of Y then $L' \cap X$ is a weakly (M,I)-bounded subset of X.*

Proof. This follows from Proposition 6.3 in view of the fact that if $L' \cap Y_I$ is an (M,I)-bounded subset of Y_I then $L' \cap X_I$ is an (M,I)-bounded subset of X_I. ■

For the proof of Proposition 6.3.A we require

Lemma 6.6. *Suppose that $0 \in \overline{I}$. Then for every compact subset K of $\tilde{X}$ there is a compact subset H of X_+ such that*

$$X_+ \cap K_{M,1/I} \subset H_{M,1/I}. \tag{6.4}$$

Proof. We fix $R \in]0,+\infty[$ such that $K \subset \Omega_{[0,R]}$. Then $H := \overline{K_{M,]0,1]}} \cap \Omega_{\{R\}}$ is compact for it is closed in V and - by Lemma 5.9 - contained in the bounded subset $\{x \in V; |P_+ x| = R, |M_0 x| \le \max\{|k|; k \in K\}\}$. Moreover, by Lemma 5.9 and (5.70 H lies in X_+. Now let $x \in X_+ \cap K_{M,1/I}$. We fix $k \in K$ and $t \in I$ such that $x = M_t k$ Then $P_+(k) \neq 0$, and by the choice of R and by the last property of Proposition 1.7 we find $s \in]0,1]$ such that $M_{1/s} k \in \Omega_{\{R\}}$. Since $st \in I$, as well, it follows tha $x = M_{st} M_{1/s} k \in H_{M,1/I}$. ■

Proof of Proposition 6.3.A. *(a) ⇒ (c) for Y = X:* this is a direct consequence c Definition 6.1 since under the present assumption on I we have $X_{1/I} = X_+$ an since for every compact subset K of X the set $\overline{L \cap X \cap K_{M,1/I}}$ contains $L \cap K_{M,1/}$ and hence the closure of the last set if it is closed itself.

(c) ∨ ((d) ∧ (5.17)) ⇒ (b): Let K be a compact subset of Y. If $J := \{t \in I; M_t(K) \cap L \neq \emptyset$ is not empty we choose a sequence $(t_j)_{j \in \mathbb{N}}$ in J converging to $s := \inf J \in [0,b$

We fix a sequence (k_j) in K such that $\ell_j := M_{t_j} k_j \in L$ for every $j \in \mathbb{N}$. By choosing subsequences if necessary we achieve that (k_j) converges to some $x \in K$. By continuity (see Corollary 1.9) it follows that (ℓ_j) converges to $M_s x \in \overline{L \cap K_{M,1/I}}$. Now, if (c) holds then $M_s x$ does not belong to $\ker M$ so that $s > 0$. On the other hand, if (5.17) is valid then the assumption $s = 0$ implies that $M_s x$ belongs to $\overline{L} \cap M_0(K) \subset X \setminus X_+$ so that (d) is violated. This establishes (b).

(b) for $Y = \tilde{X}$ $\Rightarrow$ (b) for $Y = X$ $\Rightarrow$ (b) for $Y = X_+$: this is trivial since the points of $\ker M$ are not (M,I)-bounded in $\tilde{X}$ or X.

(b) for $Y = X_+$ $\Rightarrow$ (d) $\wedge$ (c) for $Y = \tilde{X}$: In (b) we may assume that $I =]0,1]$. Hence, applying Lemma 5.58 to $\tilde{X}$ instead of X and taking (6.2.a) into account we deduce (d).

To derive (c) we fix a compact subset K of $\tilde{X}$. By Lemma 6.6 we find a compact subset H of X_+ such that (6.4) holds. Consequently, the set $L \cap K_{M,1/I}$ is contained in $\overline{L} \cap H_{M,1/I}$. Since by Proposition 3.22 the last set is a compact subset of X_+ the condition (c) follows for $Y = \tilde{X}$.

(b) $\wedge$ (d) $\Rightarrow$ (a): Let K be a compact subset of X. Then it follows by Proposition 3.22 from (b) for $Y = X$ that $\overline{L} \cap K_{M,1/I}$ is a compact subset of X. Hence in view of (d) it suffices to show that

$$\overline{K_{M,1/I}} \setminus K_{M,1/I} \subset M_0(X). \tag{6.5}$$

Since $1/(1/I) = I$ is bounded this follows from Lemma 5.9. ∎

Proof of Proposition 6.3.B. *"$\Rightarrow$"*: Let K be a compact subset of X_+. Since 1/I is bounded and since $K \cap M_0(K) = \emptyset$ it follows by Lemma 5.9 that $K_{M,1/I}$ is a closed subset of V. Consequently, $\overline{L} \cap X \cap \overline{K_{M,1/I}} = (\overline{L} \cap X_+) \cap K_{M,1/I}$, and the assumption on L implies that the set on the right-hand side is a compact subset of X_+. Hence Proposition 3.22 shows (note that (3.18.b) is satisfied for X_+ instead of X in view of Remark 1.67) that $\overline{L} \cap X_+$ is an (M,I)-bounded subset of X_+.

"$\Leftarrow$": Let K be a compact subset of X, and let $(\ell_m)_{m \in \mathbb{N}}$ be a sequence in $\overline{L} \cap X \cap \overline{K_{M,1/I}}$. We have to show that a subsequence converges to some $\ell \in X$. To this end we fix sequences (k_m) in K and (t_m) in I such that $(\ell_m - M_{t_m} k_m)$ converges to 0 as $m \to \infty$. By choosing subsequences if necessary we achieve that (k_m) converges to some $x \in K$. We fix $\varepsilon > 0$ such that the closed polydisc

$$\overline{P}(x,\varepsilon) := \{ y \in V ;\ |P_+(y-x)| \le \varepsilon ,\ |M_0(y-x)| \le \varepsilon \}$$

is contained in X. We may suppose that for some $N \in \mathbb{N}$ we have

$$(6.6) \qquad M_{1/a}\ell_m \notin \overline{P}(x,\varepsilon) \ \text{ if } m \ge N$$

for otherwise the desired subsequence of (ℓ_m) is found by the compactness of $M_a(\overline{P}(x,\varepsilon)) \subset X$.

Next we convince ourselves that it suffices to find a compact subset $\hat{K}$ of X_+ and sequences $(s_m)_{m \ge N}$ in I and $(y_m)_{m \ge N}$ in $\hat{K}$ such that

$$(6.7) \qquad \ell_m = M_{s_m} y_m \quad \text{ if } m \ge N .$$

In fact, this implies that $s_m \in \{ t \in I ;\ M_t(\hat{K}) \cap \overline{L} \cap X_+ \ne \emptyset \}$ so that by the assumption on L and by Remark 3.19 a subsequence of (s_m) converges to some $s \in I$ as $m \to \infty$; and since $\hat{K}$ is compact, by choosing subsequences for a second time we achieve that (y_m) converges to some $y \in \hat{K}$ so that by continuity and in view of (6.7) a subsequence of (ℓ_m) converges to $M_s y \in \hat{K}_{M,1/I} \subset X_+$ as $m \to \infty$.

To find $\hat{K}$, s_m, and y_m we first observe that since $(1/t_m)$ is a bounded sequence it follows by continuity (see Corollary 1.9) that $M_{1/t_m}(\ell_m - M_{t_m} k_m)$ converges to 0 as $m \to \infty$ so that

$$\lim_{m \to \infty} M_{1/t_m} \ell_m = x .$$

If $x \in X_+$ then for sufficiently large N the requirements above are satisfied by $s_m := t_m$, $y_m := M_{1/t_m}\ell_m$, and $\hat{K} := \{ y_m ;\ m \ge N \} \cup \{x\}$.

In case $x \notin X_+$ then - taking $M_0(M_{1/t_m}\ell_m) = M_0(M_{1/a}\ell_m)$ into account - we can choose N so large that $|M_0(M_{1/a}\ell_m - x)| < \varepsilon$ for $m \ge N$. By combining this with (6.6) and the assumption "$x \notin X_+$" we conclude that $|P_+(M_{1/a}\ell_m)| > \varepsilon$ for every $m \ge N$. In view of the last property in Proposition 1.70 we can therefore choose $s_m \in [a,+\infty[$ such that $y_m := M_{1/s_m}\ell_m$ lies in $\hat{K} := \{ z \in \overline{P}(x,\varepsilon) ;\ |P_+ z| = \varepsilon \}$ Since $\hat{K}$ is a compact subset of X_+ the requirements above are satisfied, again. ∎

From Proposition 6.3 we are going to deduce that Lemma 3.29 carries over to weakly (M,I)-bounded subsets of X. To this end (and for other purposes) we once and for all fix a continuous function $\varkappa : X_+ \longrightarrow]0,+\infty[$ which is quasihomogeneous of degree 1. This is possible by Corollary 3.36 and Theorem 3.39. Recall that by Proposition 1.61 $\varkappa$ extends to a continuous quasihomogeneous function $\dot{\varkappa}: X \longrightarrow [0,+\infty[$ such that $\dot{\varkappa}^{-1}(0) = X \setminus X_+$.

Lemma 6.7. *Let $\varkappa$ be as above. Then L is a weakly (M,I)-bounded subset of X if and only if*

$$(6.8) \qquad \dot{\varkappa}(L \cap K_M) \cap \overline{I} \text{ is a relatively compact subset of } I$$

for every compact subset K of X_I (resp. X_+).

Proof. If I is an unbounded proper subinterval of $]0,+\infty[$ then $X_I = X_+$, and one obtains the assertion by combining Lemma 3.29 (applied to X_+ instead of X) and Proposition 6.3.B.

Suppose now that I is bounded, i.e. $X_I = X$. Since for every $x \in X \setminus X_+$ we have $\{x\}_M = \{x\}$ and $\dot{\varkappa}(x) = 0 \in \overline{I}$ the condition (6.8) for $K = \{x\}$ shows that L is contained in X_+. Since by Proposition 6.3.A.(ii) the same is satisfied if L is a weakly (M,I)-bounded subset of X we may assume this to be valid. But then (6.8) becomes (3.19)' so that by Lemma 3.29 the condition (6.8) holds for every compact subset K of X_+ if and only if L is an (M,I)-bounded subset of X_+. By Proposition 6.3.A.(i) this means that L is a weakly (M,I)-bounded subset of X. Hence, to complete the proof we have to show that if (6.8) is valid for every compact subset K of X_+ it is so for every compact subset K of X, as well.

To prove this we choose such a compact subset K of X and by Remark 5.56 find a compact subset J of V_+ such that $\overline{K_M} \cap V_+ = J_M$. By Lemma 5.9 and by (5.70) J is a subset of X_+. Since we assumed that $L \subset X_+$ the set $\dot{\varkappa}(L \cap K_M) \cap \overline{I}$ is contained in $\varkappa(L \cap J_M) \cap I$. From this the desired conclusion follows.

Finally, if $I =]0,+\infty[$ then one obtains the assertion by combining the cases already dealt with above. ■

As a corollary of Lemma 6.7 the following variant of Remark 5.56 is obtained.

Remark 6.8. *For every compact subset K of X the set $L := \overline{K_M} \cap S^{\varkappa}$ is a compact subset of X_+ satisfying*

$$(6.9) \qquad \overline{K_M} \cap X_+ = L_M .$$

Proof. L is compact since by Lemma 6.7 $S^{\varkappa}$ is a weakly M-bounded subset of X. The condition (6.9) is satisfied since $(S^{\varkappa})_M = X_+$. ■

The following lemma is required for the proof of the assertion on supp $u_{m,w}$ in Proposition 6.15 below.

Lemma 6.9. *Let L be a weakly (M,I)-bounded subset of X. Then we have*

$$\text{(i)}\quad \overline{L_{M,I}} \cap X = (\overline{L} \cap X)_{M,I} \cup \begin{cases} \emptyset & \text{if } I \text{ is bounded} \\ X \cap M_0(\overline{L}) & \text{if } I \text{ is unbounded} \end{cases};$$

(ii) if I is unbounded then $X \cap M_0(\overline{L})$ *is a closed subset of X and equal to* $X \cap M_0(\overline{L} \cap X)$;

(iii) if K is a compact subset of X such that $K \cap \overline{L_{M,I}}$ *is empty then so is* $\overline{L} \cap X \cap \overline{K_{M,1/I}}$.

Proof. (i): Let $(t_j)_{j\in\mathbb{N}}$ be a sequence in I and (ℓ_j) be one in L such that $y_j := M_{1/t_j}\ell_j$ converges to some point $x \in X$. We set $K := \{y_j;\ j\in\mathbb{N}\} \cup \{x\}$. Then $\Lambda := \overline{L} \cap X \cap \overline{K_{M,1/I}}$ is a compact subset of $X_{1/I}$. Since the ℓ_j belong to Λ, by choosing subsequences if necessary we achieve that (ℓ_j) converges to some $z \in \overline{L} \cap X_{1/I}$ and (t_j) to some $s \in \overline{I}$. By continuity we obtain

$$(6.10)\qquad M_s x = \lim_{j\to\infty} M_{t_j} y_j = \lim_{j\to\infty} \ell_j = z \quad \text{if } s < +\infty .$$

If $s = 0$ then $1/I$ is unbounded, but then $X_{1/I} = X_+$ so that $z \in X_+$ which contradicts (6.10). Consequently, $s > 0$, and if $s < +\infty$ then (6.10) implies that $x = M_{1/s} z$ belongs to $(\overline{L} \cap X)_{M,I}$. Finally, if $s = +\infty$ then I is unbounded, and it follows by continuity that

$$x = \lim_{j\to\infty} y_j = M_0 z \in M_0(\overline{L} \cap X_{1/I}) .$$

Hence the proof of the inclusion "$\subset$" is complete. Note that we have even shown

$$(6.11)\qquad \overline{L_{M,I}} \cap X \subset (\overline{L} \cap X)_{M,I} \cup (X \cap M_0(\overline{L} \cap X)) .$$

To prove the inclusion "$\supset$" we first observe that by continuity $\overline{L_{M,I}} \cap X$ contains $(\overline{L} \cap X)_{M,I}$. Hence we may suppose that I is unbounded. Let $x \in X \cap M_0(\overline{L})$. We choose $y \in \overline{L}$ such that $x = M_0 y$ and a sequence $(\ell_j)_{j\in\mathbb{N}}$ in L converging to y as $j\to\infty$. Then $(M_0\ell_j)$ converges to x. Since $M_{1/t}\ell_j$ tends to $M_0\ell_j$ as $t\to+\infty$ and since I is unbounded we find a sequence (t_j) in I such that $x = \lim_{j\to\infty} M_{1/t_j}\ell_j$. This means that x belongs to $\overline{L_{M,I}}$.

(ii): We note first that $(\overline{L}\cap X)_{M,I}\setminus X_+ = (\overline{L}\cap X)\setminus X_+ \subset X\cap M_0(\overline{L}\cap X)$. Consequently, since $X\setminus X_+$ is a closed subset of X one obtains the assertion by intersecting both sides of (i), resp. (6.11) with $X\setminus X_+$.

(iii): It suffices to deal with the cases 1.° $I =]0,b]$ for some $b\in]0,+\infty[$ and 2.° $I = [a,+\infty[$ for some $a\in]0,+\infty[$.

If case 1 holds then by (i) we have $\overline{L_{M,I}}\cap X = (\overline{L}\cap X)_{M,I}$ so that the assumption on K implies that $\overline{L}\cap X\cap K_{M,1/I} = \emptyset$. In view of (6.5) and Proposition 6.3.A.(ii) the assertion follows.

If Case 2 holds we fix $\ell\in\overline{L}\cap X\cap\overline{K_{M,1/I}}$ and choose sequences $(t_j)_{j\in\mathbb{N}}$ in I and $(k_j)_{j\in\mathbb{N}}$ in K such that $M_{t_j}k_j$ converges to ℓ as $j\to\infty$. By choosing subsequences we achieve that $(1/t_j)$ converges to some $s\in[0,1/a]$ and (k_j) to some $x\in K$ as $j\to\infty$. By continuity it follows that

$$x = \lim_{j\to\infty} M_{1/t_j}M_{t_j}k_j = M_s\ell \in (\overline{L}\cap X)_{M,I}\cup M_0(\overline{L}\cap X).$$

By the assertion (i) this implies that $x\in K\cap\overline{L_{M,I}}$. Hence the contraposition of the desired implication is proved. ∎

The next assertion is the analogue of Lemma 3.26.

Lemma 6.10. *Suppose that $I\neq]0,+\infty[$. If L is a weakly (M,I)-bounded subset of X so is $L_{M,I}$.*

Proof. If $I =]0,b]$ for some $b\in]0,+\infty[$ then by applying Proposition 6.3.A twice we deduce the assertion from Lemma 3.26.

If $I = [a,+\infty[$ for some $a\in]0,+\infty[$ then by Proposition 6.3.B and Lemma 3.26 the set $L_{M,I}\cap X_+$ – being equal to $(L\cap X_+)_{M,I}$ – is an (M,I)-bounded subset of X_+ so that the assertion follows by another application of Proposition 6.3.B. ∎

The following analogue of Lemma 3.27 is required for the proof of Theorem 6.37 below.

Lemma 6.11. *Suppose that I is non-compact. Let J be a closed subinterval of $]0,+\infty[$*

such that $I \cap J$ is compact. If L is a weakly (M,I)-bounded subset of X then for every compact subset K of X the set $\overline{L_{M,I}} \cap \overline{K_{M,J}} \cap X$ is compact. Note that if L is compact or if J is bounded then this set is equal to $\overline{L_{M,I}} \cap \overline{K_{M,J}}$.

Proof. If J is compact then $K_{M,J}$ is a compact subset of X. Hence we suppose that J is non-compact. Then the assumptions on I and J mean that I and 1/J differ by a relatively compact subset of $]0,+\infty[$. Consequently, since by Lemma 6.10 $L_{M,I}$ is a weakly (M,I)-bounded subset of X it follows that $L_{M,I}$ is a weakly $(M,1/J)$-bounded subset of X. Replacing I by 1/J in Definition 6.1 we obtain the first assertion.

For the proof of the second assertion we first observe that in case J is bounded Lemma 5.9 and (5.70) imply that $\overline{K_{M,J}} \subset K_{M,J} \cup M_0^{-1}(K) \subset X$. On the other hand, if J is unbounded then I is bounded, so that if, in addition, L is compact the same argument shows that $\overline{L_{M,I}}$ is contained in X. ∎

Finally, we note the following consequence of Proposition 6.3.A.

Remark 6.12. *Every $\psi \in C^r_{]0,1]}(X_+)$ uniquely extends to a C^r function $\dot{\psi}$ on X with weakly (M,I)-bounded support.*

Proof. Let $x \in X \setminus X_+$. Since $L := \operatorname{supp}\psi$ is an $(M,]0,1])$-bounded subset of X_+ Proposition 6.3.A shows that ψ vanishes on $K(x,\varepsilon) \cap X_+$ for some $\varepsilon \in]0,+\infty[$. Hence by $\dot{\psi}|_{X \setminus X_+} :\equiv 0$ the desired extension $\dot{\psi}$ of ψ is defined. ∎

Below we shall identify the functions $\psi \in C^r_{]0,1]}(X_+)$ with their extensions to X.

(b) The Distributions $u_{m,w}$ and Their Basic Properties

Notation 6.13. By $\mathcal{D}'_I(X)$ we denote the space of distributions $u \in \mathcal{D}'(X)$ such that supp u is a weakly (M,I)-bounded subset of X. If $I =]0,+\infty[$ we also write $\mathcal{D}'_M(X)$.

We first observe that the assertion of Remark 6.12 carries over to distributions

Remark 6.14. *Suppose that $1/I$ is unbounded. Then every distribution $u \in \mathcal{D}'_I(X_+)$ (uniquely) extends to a distribution $\dot{u} \in \mathcal{D}'_I(X)$ which vanishes on a neighbourhood of $X \setminus X_+$. In this way $\mathcal{D}'_I(X)$ is canonically identified with $\mathcal{D}'_I(X_+)$.*

Proof. If $u \in \mathcal{D}'_I(X_+)$ then by Proposition 6.3.A the distance of every point of $X \setminus X_+$ to supp u is positive. Consequently, the desired extension $\dot{u}$ exists, and its support is equal to supp u. Hence Proposition 6.3.A implies that $\dot{u}$ belongs to $\mathcal{D}'_I(X)$. ∎

We now fix $k \in \mathbb{N}_0$ and suppose that $w:]0,+\infty[\longrightarrow \mathbb{R}$ is given by

$$w(t) = \chi_I(t)\,\omega_k(t), \qquad t \in]0,+\infty[. \tag{6.12}$$

Then by (1.65) we have

$$v(t) := w(1/t) = (-1)^k \chi_{1/I}(t)\,\omega_k(t), \qquad t \in]0,+\infty[. \tag{6.12'}$$

Proposition 6.15. *Let $u \in \mathcal{D}'_I(X)$. Then by (4.3) a distribution $u_{m,w} \in \mathcal{D}'(X)$ is well-defined. Its support is contained in $\overline{(\operatorname{supp} u)_{M,I}}$ (compare Lemma 6.9).*

Moreover, by $\mathfrak{M}_{u,w}(m) := u_{m,w}$, $m \in \mathbb{C} \setminus (-\mathfrak{A}(M)-\mu)$, a meromorphic function $\mathfrak{M}_{u,w}: \mathbb{C} \longrightarrow \mathcal{D}'(X)$ is defined, its poles lying in the set $(-\mathfrak{A}(M)-\mu)$; for every $m \in \mathbb{C}$ we have $a_0(m; \mathfrak{M}_{u,w}) = u_{m,w}$. If I is bounded then $\mathfrak{M}_{u,w}$ is holomorphic.

Proof. We set $L := \operatorname{supp} u$ and fix $\varphi \in C_0^\infty(X)$. Since by Remark 5.7, Proposition 5.16.(ii), and Theorem 5.37.(ii) we have

$$\operatorname{supp} \varphi_{m^*,v} \subset \overline{(\operatorname{supp} \varphi)_{M,1/I}} \cap X_{1/I} \tag{6.13}$$

and since by Definitions 6.1 and 6.13 the intersection of the right-hand side with L is a compact subset of $X_{1/I}$ we see that the function $\varphi_{m^*,v}$ belongs to the space $D_u(X_{1/I})$, i.e. Remark 4.1 (applied to $Y = X_{1/I}$) shows that the right-hand side of the equation in (4.3) is well-defined.

To prove that the linear functional $u_{m,w}$ defined by (4.3) is continuous we fix a compact subset K of X. Then by Definitions 6.1 and 6.13 the set $H := L \cap \overline{K_{M,1/I}}$ is a compact subset of $X_{1/I}$. Hence we can fix a compact neighbourhood W of H in $X_{1/I}$ and choose $\chi \in C_0^\infty(W)$ equal to 1 near H. Consequently, in view of (6.13) one concludes by Remark 4.1 that for $\varphi \in C_0^\infty(K)$ the definition of $u_{m,w}$ amounts to (4.5). Since by Propositions 5.11, 5.36, and 5.41.(ii), by (3.5), and by Remark 5.40 the map $C_0^\infty(K) \longrightarrow C_0^\infty(W)$, $\varphi \mapsto \chi \varphi_{m^*,v}$, is well-defined, linear and continuous it follows that $u_{m,w}$ is linear and continuous on $C_0^\infty(K)$, indeed. The assertion about $\operatorname{supp} u_{m,w}$ is an immediate consequence of Lemma 6.9.(iii) and Remark 4.1.

The assertions about $\mathfrak{M}_{u,w}$ follow by the continuity of the restriction of u to $C_0^\infty(W)$ from the corresponding assertions on $\mathfrak{M}_{\varphi,v}$ (see Propositions 5.11 and 5.16 and Theorem 5.37.(iv)) and because

$$(6.14) \qquad \langle a_j(m;\mathfrak{M}_{u,w}), \varphi \rangle = (-1)^j \langle u, a_j(m^*;\mathfrak{M}_{\varphi,v}) \rangle, \qquad j \in \mathbb{Z}. \ \blacksquare$$

Proposition 6.16. *For every $u \in \mathcal{D}_I'(X)$ the assertions of Proposition 4.4 as well as the condition (4.10) remain valid.*

Proof. We first observe that $\partial^\alpha u$, $\alpha \in \mathfrak{A}$, and $q_i u$ belong to $\mathcal{D}_I'(X)$, indeed. Let $A \in GL(V,V)$ commute with M. To verify that $u \circ A$ belongs to $\mathcal{D}_I'(A^{-1}(X))$ we set $L := \operatorname{supp} u$ and observe that $\operatorname{supp} u \circ A = A^{-1}(L)$ and that for every compact subset H of $A^{-1}(X)$ the set $K := A(H)$ is a compact subset of X satisfying

$$\overline{A^{-1}(L)} \cap A^{-1}(X) \cap \overline{H_{M,1/I}} = A^{-1}(\overline{L} \cap X \cap \overline{K_{M,1/I}}).$$

Finally, to see that $u_{\mathfrak{G}}$ belongs to $\mathcal{D}_I'(X)$ we observe - using Lemma 2.58.(i) - that for every compact subset K of X we have

$$\overline{L_G} \cap X \cap \overline{K_{M,1/I}} \subset (\overline{L} \cap X \cap \overline{(K_G)_{M,1/I}})_G$$

and that the right-hand side is compact by Lemma 2.58.(ii).

As for the formulas, in view of Lemma 5.4, Propositions 5.16, 5.19, and 5.21, Corollary 5.20, Remarks 5.3, 5.5, and 5.6, and Theorem 5.37 the proofs of Proposition 4.4 carry over. ■

The following lemma shows to what extent the equality (4.2)' remains valid in the present context.

Lemma 6.17. *Let $u \in \mathcal{D}_I'(X)$, and let $r \in \mathbb{N}_0 \cup \{\infty\}$ and $f \in C^r(X_I)$ such that $u|_{X_I} = T_f$. Then $f_{m,w}$ is a well-defined C^r function on X_I satisfying*

$$u_{m,w}|_{X_I} = T_{f_{m,w}}.$$

Proof. By Proposition 6.3 the set $L := \operatorname{supp} f$ is an (M,I)-bounded subset of X_I. Hence the first part of the assertion follows from Proposition 3.3. If I is compact or if $X_I = X_+$ then the second part of the assertion is valid by (4.2)'. Hence we may suppose that $I =]0,b]$ for some $b \in]0,+\infty[$; in particular, we have $X_I = X$. By Proposition 6.3.A L is then contained in X_+; by Lemma 6.9 it follows that $L_{M,I}$ is a closed subset of X which is contained in X_+. Hence we can fix a function $\chi \in C^\infty(X)$ equal to 1 on $L_{M,I}$ with support contained in X_+. Consequently, if $\varphi \in C_0^\infty(X)$ then $\chi\varphi \in C_0^\infty(X_+)$ so that by Propositions 3.10 and 3.22 and Lemma 3.11 it follows that

$$(6.15) \qquad \int_{X_+} f_{m,w}(x)\,(\chi\varphi)(x)\,dx = \int_{X_+} f(x)\,(\chi\varphi)_{m^*,v}(x)\,dx\,.$$

Since by Proposition 3.10 we have $\operatorname{supp} f_{m,w} \subset L_{M,I}$ the left-hand side of (6.15) coincides with $\int_X f_{m,w}(x)\,\varphi(x)\,dx$. On the other hand, since $\chi \equiv 1$ on $L_{M,I}$ and since by Proposition 5.11 $\varphi_{m^*,v}$ is defined by the integral formula (3.1)' it follows that $(\chi\varphi)_{m^*,v}(x) = \varphi_{m^*,v}(x)$ for every $x \in L$. Inserting this into the right-hand formula (3.1)' it follows that $(\chi\varphi)_{m^*,v}(x) = \varphi_{m^*,v}(x)$ for every $x \in L$. Inserting this into the right-hand side of (6.15) completes the proof. ■

We close this section by formulating conditions under which $u_{m,w}$ is temperate. The first step is to carry over Remark 4.1 to temperate distributions.

Notation 6.18. Let F be any subset of V, and let Y be an open subset of V.

(i) For any $\varepsilon > 0$ we set $F_\varepsilon := \{x \in V;\ \operatorname{dist}(x,F) < \varepsilon\}$;

(ii) by $\mathcal{S}(Y)$ we denote the space of all $\varphi \in \mathcal{S}(V)$ such that $(\operatorname{supp}\varphi)_\varepsilon \subset Y$ for some $\varepsilon > 0$;

(iii) by $\mathcal{S}(F;Y)$ we denote the space of all C^∞ functions $\varphi : Y \longrightarrow \mathbb{C}$ such that for some $\varepsilon > 0$ we have

$$(6.16.a) \qquad (F \cap \operatorname{supp}\varphi)_\varepsilon \subset Y$$

and

$$(6.16.b) \qquad \sup\{(1+|x|)^j\,|\varphi^{(\alpha)}(x)|\,;\ x \in (F \cap \operatorname{supp}\varphi)_\varepsilon\} < +\infty\,, \qquad j \in \mathbb{N}_0,\ \alpha \in \mathbb{N}_0^n.$$

One observes that by $\varphi \mapsto \varphi|_Y$ a linear isomorphism of $\mathcal{S}(Y)$ onto $\mathcal{S}(V;Y)$ is defined. In this way $\mathcal{S}(Y)$ will be identified with the subspace $\mathcal{S}(V;Y)$ of $\mathcal{S}(F;Y)$.

Lemma 6.19. *Let $S \in \mathcal{S}'(V)$. Then for any open subset Y of X there exists a unique linear functional $\tilde{S} : \mathcal{S}(\operatorname{supp} S\,;Y) \longrightarrow \mathbb{C}$ extending the restriction of S to $\mathcal{S}(Y)$ and vanishing on the subspace $\{\varphi \in C^{\infty}(Y);\ \operatorname{supp}\varphi \cap \operatorname{supp} S = \emptyset\}$. $\tilde{S}$ has the following properties:*

(i) $\langle (\partial^{\alpha} S)^{\sim}, \varphi \rangle = \langle \tilde{S}, (-\partial)^{\alpha}\varphi \rangle, \qquad \varphi \in \mathcal{S}(\operatorname{supp} S\,;Y),\ \ \alpha \in \mathbb{N}_o^n;$

(ii) $\langle f\tilde{S}, \varphi \rangle = \langle \tilde{S}, f\varphi \rangle, \qquad \varphi \in \mathcal{S}(\operatorname{supp} S\,;Y),\ f \in \mathcal{O}_{\mathcal{M}}(V).$

Below we shall omit the superscript "$\sim$" and any reference to Y, considering S as a functional on the union of all the spaces $\mathcal{S}(\operatorname{supp} S;Y)$.

<u>*Proof.*</u> Let $\varphi \in \mathcal{S}(\operatorname{supp} S\,;Y)$. We choose $\varepsilon > 0$ such that (6.16) is satisfied for $F := \operatorname{supp} S$. We write $L := F \cap \operatorname{supp}\varphi$. To prove the uniqueness part we choose $\eta \in \,]0,\varepsilon[$ and - by Lemma 5.53.(i) - find a function $\chi \in C^{\infty}(V)$ equal to 1 on L_{η} with support contained in L_{ε} such that all the derivatives of χ are bounded. Then by the Leibniz rule we derive from (6.16.b) that $\chi\varphi \in \mathcal{S}(Y)$. Since $\operatorname{supp}(1-\chi)$ is contained in $V \setminus L_{\eta}$ so that $\operatorname{supp}(1-\chi)\varphi \cap \operatorname{supp} S = \emptyset$ it follows that

$$(6.17) \qquad \langle \tilde{S}, \varphi \rangle = \langle S, \chi\varphi \rangle .$$

To prove the existence we define $\tilde{S}$ by (6.17) and have to verify that the definition does not depend on the choice of ε, η and χ. If ε', η' and χ' are other choices then $\operatorname{supp}(\chi-\chi')$ is contained in $L_{\gamma} \setminus \overset{\circ}{L}_{\delta}$ where $\gamma := \max\{\varepsilon, \varepsilon'\}$ and $\delta := \min\{\eta, \eta'\}$ so that by (6.16.b) the function $(\chi-\chi')\varphi$ belongs to $\mathcal{S}(V)$. Since $\operatorname{supp} S$ does not intersect $\operatorname{supp}(\chi-\chi')\varphi$ it follows that the right-hand side of (6.17) is equal to $\langle S, \chi'\varphi \rangle$, indeed.

The assertion on $\partial^{\alpha}\tilde{S}$ follows by the Leibniz rule since $L_{\eta} \cap \operatorname{supp}\partial^{\beta}\chi = \emptyset$ if $\beta \neq 0$. The last assertion is obvious. ∎

As we shall show below, if u is a distribution belonging to $\mathcal{S}'(V) \cap \mathcal{D}'_I(V)$ a sufficient condition for $u_{m,w}$ to be temperate is that its support belong to a special class of weakly (M,I)-bounded subsets of V. These are described by the following remark which is a simple consequence of the estimates (1.77).

Remark 6.20. *Let L be a subset of V. Then the following conditions are equivalent:*

(a) $L \subset \Omega_J$ (see Notation 5.10) for some closed subinterval J of $]0,+\infty[$ such that $J \cap \bar{I}$ is a compact subset of $]0,+\infty[$;

(b) $\dot{\varkappa}(L) \cap \bar{I}$ is a relatively compact subset of $]0,+\infty[$ if $\varkappa = \varkappa_+$. ■

Definition 6.21. (i) A subset L of V is called *(M,I)-temperate* if one (and hence each) of the conditions of Remark 6.20 is satisfied; if $I =]0,+\infty[$ we also say *M-temperate* instead of (M,I)-temperate.

(ii) By $\mathcal{S}'_I(V)$ we denote the space of all temperate distributions on V with (M,I)-temperate support; if $I =]0,+\infty[$ then we also write $\mathcal{S}'_M(V)$.

Note that it follows from Lemma 6.7 that (M,I)-temperate subsets of V are weakly (M,I)-bounded in V. In particular, $\mathcal{S}'_I(V)$ is contained in $\mathcal{S}'(V) \cap \mathcal{D}'_I(V)$.

Proposition 6.22. *(i) If L is an (M,I)-temperate subset of V then $\varphi_{m^*,v}$ belongs to the space $\mathcal{S}(L;V_{1/I})$ for every $\varphi \in \mathcal{S}(V)$.*

(ii) Let $u \in \mathcal{S}'_I(V)$. Then $u_{m,w}$ is temperate; more precisely, for every $\varphi \in \mathcal{S}(V)$ the right-hand side of the equation in (4.3) is well-defined, defining a continuous linear functional on $\mathcal{S}(V)$, denoted by $u_{m,w}$, as well. Moreover, $\mathfrak{M}_{u,w}$ (see Proposition 6.15) is a meromorphic function with values in $\mathcal{S}'(V)$.

Proof. *Case 1:* If I is compact then V is (M,I)-temperate, itself, and $\varphi \mapsto \varphi_{m^*,1/I}$ defines a continuous linear map from $\mathcal{S}(V)$ into itself, and the assertions follow.

Cases 2 and 3: Suppose that $I =]0,b]$ (resp. $[b,+\infty[$) for some $c \in]0,+\infty[$. For the proof of (i) one deduces from Remark 6.20 that L is contained in the set $\Omega_{[c,+\infty[}$ (resp. $\Omega_{[0,c]}$) for some $c \in]0,+\infty[$. Consequently, making use of Lemma 5.53.(ii) one finds $\varepsilon, d \in]0,+\infty[$ such that L_ε is contained in $\Omega_{[d,+\infty[}$ (resp. $\Omega_{[0,d]}$). Hence, for $F = L$ and $Y = V_{1/I}$ the condition (6.16.a) is trivially valid for arbitrary $\varphi \in C^\infty(X_{1/I})$, and by Proposition 5.11.(i) (resp. Proposition 5.36) the condition (6.16.b) holds with φ replaced by $\varphi_{m^*,v}$ for every $\varphi \in \mathcal{S}(V)$.

For the proof of (ii) we observe that by the assumption on supp u Lemma 5.53 provides us with a function $\chi \in C^\infty(V)$ equal to 1 on $(\operatorname{supp} u)_\varepsilon$ for some $\varepsilon > 0$

with support contained in the set $\Omega_{[d,+\infty[}$ (resp. $\Omega_{[0,d]}$) for some $d \in]0,+\infty[$ such that the derivatives of χ are bounded. Consequently, by Proposition 5.11.(i) (resp. Remark 5.35 and Proposition 5.36) the map defined by $\varphi \mapsto \chi\varphi_{m^*,v}$ is a continuous linear operator of $\mathcal{S}(V)$ into itself. In view of (6.17) the assertions follow.

Case 4: If $I =]0,+\infty[$ the proof is analogous or is obtained as a combination of the cases 2 and 3. ■

(c) Describing the Almost Quasihomogeneous Distributions on X with Support Contained in $X \setminus X_+$

To introduce a method for constructing distributions having the properties spelt out in the title of the present section we require

Lemma 6.23. *If K is a compact subset of X then $M_o^{-1}(K \cap M_o(K))$ is a closed subset of X which is contained in $L_{M,]0,1]}$ for some compact subset L of X^o.*

Proof. Since $K \cap M_o(K)$ is compact we find $\varepsilon > 0$ and a finite subset F of $K \cap M_o(K)$ such that

$$K \cap M_o(K) \subset L := \bigcup_{x \in F} \overline{P}(x,\varepsilon) \subset X$$

where the closed polydiscs $\overline{P}(x,\varepsilon)$ are defined as in the text preceding (6.6). Since in view of the last assertion in Proposition 1.70 we have

$$\overline{P}(x,\varepsilon)_{M,]0,1]} = \{y \in V ; |M_o(y-x)| \leq \varepsilon\} = M_o^{-1}(\overline{P}(x,\varepsilon))$$

we conclude that $M_o^{-1}(K \cap M_o(K)) \subset L_{M,]0,1]} \subset X$. By Lemma 5.46 the set $M_o^{-1}(K \cap M_o(K))$ is even closed in V. ■

Proposition 6.24. *Let $u \in \mathcal{D}'_{[1,+\infty[}(X)$. Then by*

(6.18) $$\langle Q'_m u, \varphi \rangle := \langle u, Q_{m^*} \varphi \rangle, \qquad \varphi \in C_o^\infty(X)$$

a distribution $Q'_m u \in \mathcal{D}'(X)$ is well-defined having the following properties:

(i) $Q'_m u$ *is almost quasihomogeneous of degree* m *and of order* $\leq N(m^*)$ *(for the definition of* $N(m^*)$ *see Proposition 5.45);*

(ii) *the support of* $Q'_m u$ *is contained in* $M_O(\operatorname{supp} u) \cap X$, *the latter being a closed subset of* X *which is contained in* $X \backslash X_+$;

(iii) *if* $X = V$ *and if* u *is temperate with* $(M, [1, +\infty[)$*-temperate support then* $Q'_m u$ *is temperate, as well, and (6.18) holds for every* $\varphi \in \mathcal{S}(V)$;

(iv)
$$Q'_m u = \sum_{\alpha \in \mathfrak{U}^+_{m^*}} \frac{1}{\alpha!} (-\partial)^\alpha \widehat{x^\alpha u}$$

where $\widehat{u}$ *is defined by*

$$\text{(6.19)} \qquad \widehat{u} := Q'_{-\mu} u, \qquad \text{i.e.} \qquad \langle \widehat{u}, \varphi \rangle = \langle u, \varphi \circ M_O \rangle, \qquad \varphi \in C_O^\infty(X).$$

Note that in case $X = V$ and $\ker M = \{0\}$ the equation in (iv) reads as

(iv)'
$$Q'_m u = \sum_{\alpha_M = m^*} \frac{1}{\alpha!} \langle u, x^\alpha \rangle (-\partial)^\alpha \delta_o$$

where $\delta_o \in \mathcal{D}'(V)$ is the Dirac distribution at 0.

<u>Proof.</u> Let K be a compact subset of X. Then by Lemma 6.23 and Definition 6.1 the set $Z := M_O^{-1}(K \cap M_O(K)) \cap \operatorname{supp} u$ is a compact subset of X. We fix a function $\chi \in C_O^\infty(X)$ equal to 1 near Z. Clearly, the map $C_O^\infty(X) \longrightarrow C_O^\infty(\operatorname{supp} \chi)$, $\varphi \mapsto \chi Q_{m^*} \varphi$, is well-defined and continuous. By Corollary 5.47 for every $\varphi \in C_O^\infty(K)$ the support of $Q_{m^*} \varphi$ is contained in $M_O^{-1}(K \cap M_O(K))$ so that $Q_{m^*} \varphi$ is equal to $\chi Q_{m^*} \varphi$ near $\operatorname{supp} u$. In view of Remark 4.1 it follows that $Q'_m u$ is a well-defined distribution on X satisfying $\langle Q'_m u, \varphi \rangle = \langle u, \chi Q_{m^*} \varphi \rangle$ for every $\varphi \in C_O^\infty(K)$.

<u>*(i)*</u>: Let $\ell := N(m^*)$. Then by Propositions 5.48.(iv) and 5.45 we deduce for every $\varphi \in C_O^\infty(X)$ that

$$\langle (Q'_m u) \circ M_t, \varphi \rangle = t^{-\mu} \langle Q'_m u, \varphi \circ M_{1/t} \rangle = t^{-\mu} \langle u, (Q_{m^*} \varphi) \circ M_{1/t} \rangle =$$
$$= t^{-\mu} (1/t)^{m^*} \sum_{i=0}^{\ell} \omega_i(1/t) \langle u, (\partial_M - m^*)^i Q_{m^*} \varphi \rangle$$

so that in view of (1.65) and (2.5) we obtain

$$(Q'_m u) \circ M_t = t^m \sum_{i=0}^{\ell} \omega_i(t)\, Q'_m((\partial_M - m)^i u).$$

<u>*(ii)*</u>: We first note that the condition "$K \cap M_O(\operatorname{supp} u) = \emptyset$" implies that $Z = \emptyset$

so that by Remark 4.1 we have $\langle Q'_m u, \varphi \rangle = 0$ for every $\varphi \in C_0^\infty(K)$. Since by Lemma 6.9.(ii) $M_0(\operatorname{supp} u) \cap X$ is a closed subset of X the assertion is proved.

(iii): Since by the assumption supp u is $(M, [1, +\infty[)$-temperate it follows from Proposition 5.52 that $Q_m * \varphi$ belongs to $\mathcal{S}(\operatorname{supp} u; V)$ for every $\varphi \in \mathcal{S}(V)$. More precisely, let $R \in]0, +\infty[$ be such that $\operatorname{supp} u \subset \Omega_{]0,R]}$. By Lemma 5.53 we find a function $\chi \in C^\infty(V)$ equal to 1 on $\Omega_{[0,R+1]}$ with support contained in $\Omega_{[0,R+2]}$ such that all the derivatives of χ are bounded. Then it follows by the Leibniz rule and by Proposition 5.52 that the map $\mathcal{S}(V) \longrightarrow \mathcal{S}(V)$, $\varphi \mapsto \chi\, Q_m * \varphi$, is well-defined and continuous. In view of Lemma 6.19 and (6.17) this proves the assertion.

(iv): is immediately deduced from the definitions. ■

We can now write down a convenient formula for the negative Laurent coefficients of the meromorphic functions $\mathfrak{M}_{u,w}$:

Remark 6.25. *Suppose that I is unbounded. Let $\mathfrak{M}_{u,w}$ be the meromorphic function defined in Proposition 6.15. Then for every $m \in (-\mathfrak{A}(M)-\mu)$ we have*

$$a_{-i}(m; \mathfrak{M}_{u,w}) = \begin{cases} 0 & \text{if } i \in \mathbb{N}_k \\ \binom{i-1}{k} (\partial_M - m)^{i-k-1} Q'_m u & \text{if } i \in \mathbb{N} \setminus \mathbb{N}_k \end{cases}.$$

Proof. One has to combine (6.14) and (6.12)' with Proposition 5.25 and Theorem 5.37.(iv), respectively, and with Proposition 5.48.(iii) and (2.5). ■

As for the computational properties of $Q'_m u$, making use of Corollary 5.47 (in order to cope with the fact that (5.17) may not be valid) one immediately derives the following result from Proposition 5.48.

Proposition 6.26. *Let $u \in \mathcal{D}'_{[1,+\infty[}(X)$. Then we have*

(i) $\partial^\alpha Q'_m u = Q'_{m-\langle\alpha,p\rangle}(\partial^\alpha u)$ *for every* $\alpha \in \mathfrak{A}$;

(ii) $q\, Q'_m u = Q'_{m+\ell}(qu)$ *for every* $\ell \in \mathbb{C}$ *and every* $q \in C^\infty(X)$ *which is almost quasihomogeneous of degree* ℓ;

(iii) $P(x,\partial)[Q'_m u] = Q'_{m+\ell}(P(x,\partial)u)$ *for every* $\ell \in \mathbb{C}$ *and every* C^∞ *copolynomial function* $P: X \times V^* \longrightarrow \mathbb{C}$ *which is almost quasihomogeneous of degree* ℓ *and of type* $M \times (-M)^*$;

Proposition 6.29. *Under the preceding conventions suppose that V_0 is non-trivial.*

(i) For every $R \in \mathcal{D}'(X'')$ the distribution

$$\delta_{V_1} \otimes R \tag{6.23}$$

- where δ_{V_1} denotes the Dirac distribution on V_1 - extends to a unique distribution on X with support contained in $X \setminus X_+$ which is quasihomogeneous of degree $-\mu$.

(ii) The restriction to X^O of every distribution $d \in \mathcal{D}'(X)$ with support contained in $X \setminus X_+$ which is quasihomogeneous of degree $-\mu$ is of the form (6.23) for some $R \in \mathcal{D}'(X'')$. This distribution R is uniquely determined; it will be denoted by $\breve{d}$; and it is given by

$$\langle \breve{d}, \varphi_2 \rangle = \langle d, e_1 \otimes \varphi_2 \rangle, \qquad \varphi_2 \in C_0^\infty(X''), \tag{6.24}$$

where $e_1 : V_1 \longrightarrow \mathbb{R}$ be the constant function $x' \mapsto 1$.

Proof. *(i)*: Denoting the distribution (6.23) by T we observe that the support of T, being equal to $\{0\} \times \operatorname{supp} R$, is a closed subset of X. Hence, Remark 4.1 shows that T extends to a distribution $d \in \mathcal{D}'(X)$ with the same support as T. It is immediately seen that T is quasihomogeneous of degree $-\mu$ so that $(\partial_M + \mu)T = 0$. Since d vanishes on X_+ so does $(\partial_M + \mu)d$. Hence, $(\partial_M + \mu)d = 0$, i.e. d is quasihomogeneous of degree $-\mu$, as well.

(ii). First of all, with the help of Remark 4.1 one deduces that the right-hand side of (6.24) defines a distribution $\breve{d}$ on X''. To compute it we fix $\varphi_1 \in C_0^\infty(V_1)$ and $\varphi_2 \in C_0^\infty(X'')$. Then $(\varphi_1 \otimes \varphi_2) \circ M_0 = \varphi_1(0)\, e_1 \otimes \varphi_2$. Consequently, since by Proposition 6.27 the distributions d and $\hat{d}$ coincide we conclude that

$$\langle d, \varphi_1 \otimes \varphi_2 \rangle = \varphi_1(0) \langle d, e_1 \otimes \varphi_2 \rangle = \langle \delta_{V_1}, \varphi_1 \rangle \langle \breve{d}, \varphi_2 \rangle.$$

This means that (6.23) is valid for $R = \breve{d}$, indeed. ■

If $d = \hat{u}$ for some $u \in \mathcal{D}'_{[1,+\infty[}(X)$ then properties of u carry over to $\breve{d}$:

Remark 6.30. *For any $u \in \mathcal{D}'_{[1,+\infty[}(X)$ the distribution*

$$\breve{u} := (Q'_{-\mu} u)^{\smile} \qquad (\text{i.e. } \hat{u}|_{X^o} = \delta_{V_1} \otimes \breve{u}) \tag{6.25}$$

(which is well-defined in view of Proposition 6.24) has the following properties:

(i) *if* $s \in \mathbb{N}_0 \cup \{\infty\}$ *and if* u *is induced by a* $C^{o,s}$ *function then* $\check{u}$ *is induced by a* C^s *function;*

(ii) *if* $X = V$ *and if* $u \in \mathcal{S}'_{[1,+\infty[}(V)$ *then* $\check{u}$ *belongs to* $\mathcal{S}'(V_0)$.

<u>*Proof*</u>. <u>*(i)*</u>: we let $f \in C^{o,s}(X)$ be such that $u = T_f$ and fix a compact subset K'' of X''. Then $K := \{x' \in V_1 ;\ |x'| \leq 1\} \times K''$ is a compact subset of X satisfying $K_{M,]0,1]} = V_1 \times K''$. Since $\operatorname{supp} f$ is a weakly $(M,[1,+\infty[)$-bounded subset of X it follows that the set $\{x' \in V_1 ;\ (x',x'') \in \operatorname{supp} f$ for some $x'' \in K''\}$ is compact. Consequently, by

$$g(x'') := \int_{V_1} f(x',x'')\, dx'$$

a C^s function $g : X'' \longrightarrow \mathbb{C}$ is defined satisfying $T_g = \check{u}$.

For the proof of <u>*(ii)*</u> we observe that the assumption on $\operatorname{supp} u$ means that $\operatorname{supp} u$ is contained in $K' \times V_0$ for some compact subset K' of V_1. We choose $\chi_1 \in C_0^\infty(V_1)$ equal to 1 on a neighbourhood of K'. Then the map $\mathcal{S}(V_1) \longrightarrow \mathcal{S}(V)$, $\varphi_2 \mapsto \chi_1 \otimes \varphi_2$, is well-defined, linear, and continuous so that the assertion follows from the equality $\langle \check{u}, \varphi_2 \rangle = \langle u, \chi_1 \otimes \varphi_2 \rangle$. ■

In Proposition 6.27 we saw that the distributions on X with support contained in $X \setminus X_+$ which are almost quasihomogeneous of degree m are precisely the distributions of the form $Q'_m u$ where $u \in \mathcal{D}'_{[1,+\infty[}(X)$. For some of the proofs in sections (d) and (e) it is crucial to know that u can be chosen to belong to $\mathcal{D}'_M(X)$. As we shall see in a moment, this is not always possible except when $\tilde{X} = X$.

Lemma 6.31. *Let* $d \in \mathcal{D}'(X)$ *be almost quasihomogeneous of degree* m *with support contained in* $X \setminus X_+$. *Then there exists* $u \in \mathcal{D}'_{[1,+\infty[}(X)$ *with support contained in* X_+ *satisfying the equation* $d = Q'_m u$ *and having the following properties:*

(i) *if* $0 \in \bar{I}$ *then* $\overline{(\operatorname{supp} u) \cap K_{M,1/I}} \cap \ker M \subset \tilde{X} \setminus X$ *for every compact subset* K *of* $\tilde{X}$;

(ii) *if* $s \in \mathbb{N}_0 \cup \{\infty\}$ *and if* $d = Q'_m T_f$ *for some* $f \in C^{o,s}(X)$ *then* u *is induced by a* $C^{\infty,s}$ *function on* X;

(iii) *if* $X = V$ *and if* $d \in \mathcal{S}'(V)$ *then* u *belongs to* $\mathcal{S}'_M(V)$, *more precisely,* u *is temperate with support contained in* $\Omega_{[1,2]}$.

Proof. In view of Propositions 6.27 and 6.26.(i) it suffices to prove the case $m=-\mu$. Note that by Corollary 6.28 d is then quasihomogeneous of degree $-\mu$. By $\varkappa$ we denote the function $\varkappa_+ : V_+ \longrightarrow]0,+\infty[$ given in Proposition 1.70 for $X=V$.

Case 1: $\ker M = \{0\}$. Then $X=V$ and $X\setminus X_+ = \{0\}$ so that by Corollary 2.20 we have $d = c\delta$ for some $c\in\mathbb{C}$. We choose $\psi\in C_0^\infty(K(0,2)\setminus\overline{K}(0,1))$ such that $\hat{\psi}(0)=1$ and set $u := cT_\psi \in \mathcal{E}'(\dot{V})\subset\mathcal{S}'(V)$. For every $\varphi\in C_0^\infty(V)$ we have

$$\langle \overset{\frown}{u}, \varphi\rangle = \langle u, \varphi\circ M_0\rangle = \varphi(0)\, c\,\hat{\psi}(0) = \langle d, \varphi\rangle .$$

Case 2: $\ker M \neq \{0\}$. Using the notation introduced in (6.22) we first suppose that $X=V$, choose $\psi\in C_0^\infty(V_1)$ with support contained in $K'(0,2)\setminus\overline{K'}(0,1)$ satisfying $\hat{\psi}(0)=1$, and set $u := T_\psi\otimes \overset{\smile}{d}$ where $\overset{\smile}{d}$ is defined according to (6.24). Then by Remark 6.20 the set $\dot{\varkappa}(\operatorname{supp} u)$ is a compact subset of $]0,+\infty[$. Hence by Lemma 6.7 $\operatorname{supp} u$ is a weakly M-bounded subset of V. For arbitrary $\varphi_1\in C_0^\infty(V_1)$ and $\varphi_2\in C_0^\infty(V_0)$ we have by (6.24) and (6.25) that

$$\langle \overset{\frown}{u}, \varphi_1\otimes\varphi_2\rangle = \langle T_\psi, \varphi_1(0)\, e_1\rangle \langle \overset{\smile}{d}, \varphi_2\rangle = \hat{\psi}(0)\,\langle \delta_{V_1}\otimes \overset{\smile}{d}, \varphi_1\otimes\varphi_2\rangle = \langle \overset{\frown}{d}, \varphi_1\otimes\varphi_2\rangle .$$

Since by Proposition 6.27 we have $\overset{\frown}{d} = d$ this means that $\overset{\frown}{u} = d$. If the assumption of *(ii)* is valid then by Remark 6.30.(i) there is a function $g\in C^s(V_0)$ such that $\overset{\smile}{d} = T_g$ so that $u = T_{\psi\otimes g}$. Under the assumption of *(iii)* we know from Remark 6.30 that the distribution $\overset{\smile}{d}$ belongs to $\mathcal{S}'(V_0)$ so that $u\in\mathcal{S}'(V)$.

To prove the general case we suppose that $X\neq V$ and define a function

$$\delta : \ker M \longrightarrow]0,+\infty[,\ x\mapsto \operatorname{dist}(\rho_\varkappa(V_+\cap M_0^{-1}(x)), V\setminus X).$$

Since by (5.72.b) and (1.76) we have

$$(6.26)\qquad \delta(x) = \operatorname{dist}(x+S, V\setminus X) \quad\text{where } S := \{v\in G_M(\sigma_+)\,;\ |v| = 1\}$$

we deduce that $|\delta(x)-\delta(x')| \le |x-x'|$ for arbitrary $x, x'\in\ker M$, i.e. δ is continuous. Next we are going to verify a property of δ which is crucial for the proof, namely, that for every $x\in\ker M$ we have

$$(6.27)\qquad \delta(x) > 0 \iff x\in\tilde{X}.$$

Indeed, supposing first that $x\notin\tilde{X}$ one finds a sequence of points x_m in the complement of $X\cup\ker M$ converging to x as $m\to\infty$. Then one can choose $t_m\in]0,+\infty[$ such that $y_m := M_{t_m}x_m$ lies in $S^\varkappa$. Since y_m belongs to $\rho_\varkappa(V_+\cap M_0^{-1}(M_0x_m))$ as well as to $V\setminus X$ one concludes

$$\delta(M_o x_m) \le \operatorname{dist}(y_m, V\setminus X) = 0$$

so that by the continuity of M_o and δ one obtains $\delta(x)=0$. Conversely, supposing that $x\in\tilde{X}$ one finds $\varepsilon>0$ such that the set

$$\{v\in V_+ ;\ |P_+v|<\varepsilon,\ |M_ov-x|<\varepsilon\}$$

and hence its quasihomogeneous hull $\{v\in V_+ ;\ |M_ov-x|<\varepsilon\}$ are contained in X. Since, in particular, this is true about its compact subset $x+S$ it follows by (6.26) that $\delta(x)>0$. This completes the proof of (6.27).

Now, since δ is continuous and since (6.27) tells us that, in particular, δ is positive on $X\cap\ker M$ we can find sequences $(z_k)_{k\in\mathbb{N}}$ in $\ker M$ and (η_k) in $]0,+\infty[$ such that the balls $K(z_k,\eta_k)$, $k\in\mathbb{N}$, cover $X\cap\ker M$, such that the balls $K(z_k,2\eta_k)$, $k\in\mathbb{N}$, are contained in X^o forming a locally finite family, and such that

$$\eta_k \le \delta(z_k) \le 2\delta(x), \qquad k\in\mathbb{N},\ x\in K(z_k,2\eta_k)\cap\ker M.$$

Next, for every $k\in\mathbb{N}$ we fix a point $y_k'\in S^{\varkappa}\cap M_o^{-1}(z_k)$ such that

$$\operatorname{dist}(y_k', V\setminus X) \le 2\delta(z_k).$$

Moreover, we choose a decreasing sequence $(t_k)_{k\in\mathbb{N}}$ in $]0,1]$ in such a way that the points $y_k := M_{t_k}y_k'$, $k\in\mathbb{N}$, have the following property: the sequence of numbers $|P_+y_k|$ is strictly decreasing and converges to zero as $k\to\infty$. Then we can find a decreasing sequence (ε_k) in $]0,+\infty[$ converging to zero and satisfying the estimates $|P_+y_k|-2\varepsilon_k \ge |P_+y_{k+1}|+2\varepsilon_{k+1}$, $k\in\mathbb{N}$. We set

$$L_k := \{x\in V ;\ |P_+(x-y_k)|\le\varepsilon_k,\ |M_ox-z_k|\le\eta_k\}, \qquad k\in\mathbb{N}.$$

Since $\rho_\varkappa$ is continuous, by making the ε_k smaller we can achieve that

$$|\rho_\varkappa(P_+x)-\rho_\varkappa(P_+y_k)| \le \delta(z_k), \qquad x\in L_k,\ k\in\mathbb{N}.$$

Referring to the notation introduced in (6.22) we can write $L_k=\overline{U_k}\times\overline{W_k}$ where the U_k are relatively compact open subsets of V_1 such that the $\overline{U_k}$, $k\in\mathbb{N}$, are pairwise disjoint subsets of $V_1\setminus\{0\}$ and where the W_k form an open covering $\mathfrak{W}$ of X'' such that the $\overline{W_k}$ constitute a locally finite family of compact subsets of X''. For every $k\in\mathbb{N}$ we fix a function $\psi_k\in C_o^\infty(U_k)$ such that $\hat{\psi}_k(0)=1$. Finally we choose a partition of unity $(\chi_k)_{k\in\mathbb{N}}$ on X'' subordinated to the covering $\mathfrak{W}$ and define

$$u_k := T_{\psi_k}\otimes(\chi_k\breve{d}).$$

Then $u_k \in \mathcal{E}'(V)$ with support contained in L_k. Since there exists a locally finite sequence of pairwise disjoint relatively compact open subsets Y_k of X such that $Y_k \supset L_k$ for every $k \in \mathbb{N}$, we conclude that

$$u := \sum_{k\in\mathbb{N}} u_k|_X$$

is a well-defined distribution on X with support contained in the closed subset

$$L := \bigcup_{k\in\mathbb{N}} L_k$$

of X. In order to see that u belongs to $\mathcal{D}'_{[1,+\infty[}(X)$ we have to verify that L is a weakly $(M,[1,+\infty[)$-bounded subset of X. In view of Proposition 6.3.B it suffices to show that L is an $(M,[1,+\infty[)$-bounded subset of X_+. So let K be a compact subset of X_+. Then $c := \inf \varkappa(K)$ is positive. Since $|P_+ y_k| + \varepsilon_k$ converges to 0 as $k\to\infty$, since the restriction of $\dot{\varkappa}$ to $G_M(\sigma_+)$ is continuous at 0, and since (5.72.a) holds we can choose $N \in \mathbb{N}$ so large that $\sup \varkappa(L_k) < c$ for every $k > N$. Consequently, for every $t \in [1,+\infty[$ we have

$$M_t(K) \cap L = M_t(K) \cap H \qquad \text{where} \qquad H := \bigcup_{k=1}^{N} L_k .$$

Since H is a compact, hence an M-bounded subset of X_+ it follows that the set $\{t \in [1,+\infty[\,;\ M_t(K) \cap L \neq \emptyset\}$ is a relatively compact subset of $]0,+\infty[$, indeed.

To see that $\overset{\frown}{u} = d$ we fix $\varphi \in C_0^\infty(X)$ and - by applying (6.24) first to $\overset{\smile}{u}_k$ instead of d and then to d itself - observe that

$$\langle \overset{\smile}{u}_k, \varphi(0,\cdot)\rangle = \hat{\psi}_k(0)\langle \overset{\smile}{d}, \chi_k\,\varphi(0,\cdot)\rangle = \langle d, (e_1 \otimes \chi_k)\,\varphi\circ M_0\rangle .$$

Consequently, making use of Remark 4.1 and of the fact that $\overset{\frown}{u}_k$ extends $\delta_{V_1} \otimes \overset{\smile}{u}_k$ we obtain

$$\langle \overset{\frown}{u}, \varphi\rangle = \sum_{k=1}^{+\infty} \langle \overset{\frown}{u}_k, \varphi\rangle = \sum_{k=1}^{+\infty} \langle \overset{\smile}{u}_k, \varphi(0,\cdot)\rangle = \sum_{k=1}^{+\infty} \langle d, (e_1 \otimes \chi_k)\,\varphi\circ M_0\rangle .$$

Since there are open neighbourhoods Y_k of L_k in X and Z_k of W_k in X" such that $(Y_k)_{k\in\mathbb{N}}$ (resp. $(Z_k)_{k\in\mathbb{N}}$) is locally finite in X (resp. X"), actually, the sums are finite. And since

$$\sum_{k=1}^{+\infty} e_1 \otimes \chi_k \equiv 1 \quad \text{on a neighbourhood of } \operatorname{supp} \varphi\circ M_0$$

the right-hand side of the preceding equation is equal to $\langle \overset{\frown}{d}, \varphi\rangle$ which coincides with $\langle d, \varphi\rangle$ by Proposition 6.27.(i).

For the proof of _(i)_ we first observe that by Lemma 6.6 it suffices to verify the inclusion in (i) for every compact subset K of X_+, only. To this end we fix $x \in L \cap K_{M,1/I}$ and choose $t \in I$, $z \in K$, and $k \in \mathbb{N}$ such that $x = M_t z$ and $x \in L_k$. Then $\rho_\varkappa(x) = \rho_\varkappa(z) \in \rho_\varkappa(K)$ so that by (5.72.b), by the equalities $\rho_\varkappa(y_k) = y'_k$ and $M_0(y_k) = z_k$, and by the estimates above we obtain

$$r := \operatorname{dist}(\rho_\varkappa(K), V\setminus X) \le \operatorname{dist}(\rho_\varkappa(x), V\setminus X) \le$$

$$\le |\rho_\varkappa(P_+x) - \rho_\varkappa(P_+y_k)| + |M_0(x) - M_0(y_k)| + \operatorname{dist}(\rho_\varkappa(y_k), V\setminus X) \le$$

$$\le \delta(z_k) + \eta_k + 2\delta(z_k) \le 4\delta(z_k) \le 8\delta(M_0x)$$

where the last inequality is valid since $M_0(x) \in \overline{K}(z_k, \eta_k)$. By the continuity of M_0 and δ this implies that $\delta \ge r/8$ on $\ker M \cap \overline{L \cap K_{M,1/I}}$. Since $\rho_\varkappa(K)$ is a compact subset of X it follows that $r > 0$. Consequently, it follows from (6.27) that $\ker M \cap \overline{L \cap K_{M,1/I}}$ is contained in $\tilde{X}$. Since $\overline{L} \cap (X\setminus X_+) = \emptyset$ the proof of (i) is complete.

If the assumption of *(ii)* is valid then Remark 6.30.(i) implies that $\breve{d}$ is induced by a C^s function so that the u_k and hence u are induced by $C^{\infty,s}$ functions. ■

In view of Proposition 6.3.A the condition (i) in Lemma 6.31 does not quite mean that the support of u is a weakly $(M,]0,1])$-bounded subset of X. In fact, it is not always possible to choose u with weakly M-bounded support:

Corollary 6.32. *Let d be as in Lemma 6.31. Then distributions u having the properties asserted in Lemma 6.31 can be chosen to belong to $\mathcal{D}'_M(X)$ if and only if d extends to an almost quasihomogeneous distribution on $\tilde{X}$.*

Proof. "$\Rightarrow$": In view of (6.2.a) Remark 6.14 says that u extends to a distribution $\tilde{u} \in \mathcal{D}'_M(\tilde{X})$ with the same support as u. Since $Q'_m\tilde{u}$ extends $Q'_m u$ the assertion follows by Proposition 6.24.(i).

"$\Leftarrow$": we may assume that $X = \tilde{X}$. Then in view of Proposition 6.3.A the condition (i) of Lemma 6.31, actually, means that supp u is an $(M,]0,1])$-bounded subset of X. Hence, Lemma 6.31 gives, in fact, the desired distribution $u \in \mathcal{D}'_M(X)$. ■

As a first application of Lemma 6.31 we obtain

Proposition 6.33. *Let $\ell \in \mathbb{C}$, and let $q \in C^\infty(X)$ be almost quasihomogeneous of degree ℓ such that*

$$(6.28) \qquad q^{-1}(0) \subset X \setminus X_+ .$$

Then for every distribution $d \in \mathcal{D}'(X)$ with support contained in $X \setminus X_+$ which is almost quasihomogeneous of degree m the equation $qc = d$ has a solution $c \in \mathcal{D}'(X)$ with support contained in $X \setminus X_+$ such that c is almost quasihomogeneous of degree $m - \ell$.

If in the case $X = V$ q is a polynomial function and d is temperate then c can be chosen to be temperate, as well.

Proof. By Lemma 6.31 we find $u \in \mathcal{D}'_{[1,+\infty[}(X)$ such that $Q'_m u = d$ and such that $\operatorname{supp} u$ is contained in X_+. Hence, it follows from (6.28) that $v := \frac{1}{q} u$ is a well-defined distribution on X belonging to $\mathcal{D}'_{[1,+\infty[}(X)$. Making use of Propositions 6.26.(ii) and 6.24 we deduce that $c := Q'_{m-\ell} v$ is the desired solution.

Under the assumptions of the supplementary assertion we first observe that by Lemma 6.31.(iii) we may assume that u is temperate and that its support is contained in $\Omega_{[1,2]}$. In view of Proposition 6.24.(iii) it suffices to show that v is temperate. By the Leibniz rule this follows from an estimate of the form

$$(6.29) \qquad |q(x)| \geq C\,(1 + |x|)^{-N}, \qquad x \in \Omega_{[1/2,3]},$$

where C and N are positive constants. By Hörmander [8] there are positive constants C', ε, and N such that

$$|q(x)| \geq C'\,(1 + |x|)^{-N} \operatorname{dist}(x, q^{-1}(0))^{\varepsilon}, \qquad x \in V.$$

Since by (6.28) we have $\operatorname{dist}(x, q^{-1}(0)) \geq |P_+(x)|$ this implies (6.29), indeed. ■

(d) Characterizing (Almost) Quasihomogeneous Distributions on X in Terms of Quasihomogeneous Averages

Notation 6.34. For every $u \in \mathcal{D}'_I(X)$ we denote u_{m,χ_I} by $u_{m,I}$; if $I =]0,+\infty[$ we also write u_m.

Proposition 6.35. *Let $u \in \mathcal{D}'_M(X)$. Then u_{m,ω_k} is almost quasihomogeneous of degree m such that*

$$(6.30) \qquad (\partial_M - m)^i\, u_{m,\omega_k} = \begin{cases} (-1)^i\, u_{m,\omega_{k-i}} & \text{if } i \in \mathbb{N}_k \\ (-1)^k\, (\partial_M - m)^{i-k-1}\, Q'_m u & \text{if } i \in \mathbb{N} \setminus \mathbb{N}_k \end{cases}.$$

In particular, u_m is almost quasihomogeneous of degree m with deficiency $Q'_m u$, and hence it is quasihomogeneous of degree m if and only if $m \notin (-\mathfrak{A}(M) - \mu)$ or $(x^\alpha u)^\frown = 0$ for every $\alpha \in \mathfrak{A}^+_{m^*}$.

Proof. Let $\varphi \in C_0^\infty(X)$. The assertions (vii) and (iii) of Theorem 5.37 and (1.65) imply that

$$(6.31) \qquad (\varphi \circ M_{1/t})_{m^*,\omega_k} = \varphi_{m^*,\omega_k} \circ M_{1/t} =$$

$$= t^{-m^*} \Big(\sum_{i=0}^{k} \omega_i(t)\, \varphi_{m^*,\omega_{k-i}} + \sum_{i=0}^{\infty} (-1)^i\, \omega_{k+1+i}(t)\, (\partial_M - m^*)^i\, Q_{m^*} \varphi\big|_{X_+} \Big)$$

for every $t \in]0,+\infty[$. Making use of (6.12)' and (2.5) repeatedly, one deduces the first part of the assertion. The second part is a consequence of Proposition 6.27. ■

Example 6.36. *Let $y \in V_+$, and let δ_y be the Dirac distribution at y. Then for every polynomial function $Q : V^* \longrightarrow \mathbb{C}$ the distribution $(Q(-\partial)\,\delta_y)_m$ is almost quasihomogeneous of degree m with deficiency*

$$(6.32) \qquad \sum_{\alpha \in \mathfrak{A}^+_{m^*}} \frac{1}{\alpha!} \sum_{\beta \in \mathfrak{A}^0} \frac{1}{\beta!}\, Q^{(\beta)}(\partial_y)\, y^\alpha\, (-\partial)^{\alpha+\beta}\, \delta_{M_0 y} .$$

If Q is almost quasihomogeneous of degree m^* and of type M^* then this is equal to $Q(-\partial)\,\delta_{M_0 y}$.

Proof. For arbitrary $\alpha \in \mathfrak{A}^+_{m^*}$ and $\varphi \in C_0^\infty(V)$ we deduce by the Leibniz formula in the form of (1.54) that

$$\langle (-\partial)^\alpha [(x^\alpha Q(-\partial)\delta_y)^{\frown}], \varphi \rangle = [Q(\partial)(x^\alpha(\partial^\alpha \varphi)\circ M_0)](y) =$$

$$= \sum_{\beta \in \mathfrak{A}} \frac{1}{\beta!} Q^{(\beta)}(\partial_y) y^\alpha \partial^\beta((\partial^\alpha \varphi)\circ M_0)(y) .$$

If $\beta \notin \mathfrak{A}^0$ then $\partial^\beta((\partial^\alpha\varphi)\circ M_0)$ vanishes identically. Consequently, the summation is over $\mathfrak{A}^0$, and the first part of the assertion follows by Propositions 6.35 and 6.24.(iv). The second part is a consequence of Propositions 6.35 and 6.26.(iii). ■

We now formulate the main theorem of this section, the analogue of Theorem 4.8.

Theorem 6.37. *Let $T \in \mathcal{D}'(X)$ and $k \in \mathbb{N}_0$.*

(i) Then the following conditions are equivalent:

(a) T is almost quasihomogeneous of degree m such that the support of its $(k+1)^{th}$ order deficiency $(\partial_M - m)^{k+1} T$ is contained in $X \setminus X_+$;

(b) $\langle T, \varphi \rangle = 0$ for every $\varphi \in C_0^\infty(X)$ satisfying $\varphi_{m^*,\omega_k} \equiv 0$;

(c) there exists a distribution $u \in \mathcal{D}'_M(X)$ and a distribution $d \in \mathcal{D}'(X)$ satisfying the condition (a) of Proposition 6.27.(i) such that

(6.33) $$T = u_{m,\omega_k} + d .$$

(ii) If $\tilde{X} = X$ then the distribution u in condition (c) of (i) can be chosen in such a way that (6.33) is valid for $d = 0$.

(iii) If $X = V$ and if T is temperate and satisfies the conditions of (i) then the condition (b) is valid for every $\varphi \in \mathcal{S}(V)$, and the distribution u in condition (c) of (i) can be chosen to be temperate, as well, with M-temperate support, in such a way that (6.33) is valid for $d = 0$.

(iv) The distributions u and d in condition (c) of (i) can be chosen to be $\mathfrak{G}$-invariant provided that T is $\mathfrak{G}$-invariant and that $\mathfrak{G}$ satisfies the assumptions of Remark 2.67.(ii).

The proof follows the lines of the proof of Theorem 4.8 except that in case $m \in (-\mathfrak{A}(M) - \mu)$ complications arise; in particular, the direct proof of the implication "(b)⇒(a)" does not seem to carry over. We begin with the

Proof of "(a)⇒(b)". Let $\varphi \in \mathcal{S}(V)$ satisfy $\varphi_{m^*,\omega_k} \equiv 0$. Then $\phi_k := \varphi_{m^*,\chi_{]0,1]}\omega_k}$ is equal to $-\varphi_{m^*,\chi_{[1,+\infty[}\omega_k}$ on V_+. Applying Proposition 5.11 to the derivatives of the latter function and taking Lemma 5.4 into account we deduce that

$$(6.34) \qquad \sup\{\,|\phi_k^{(\alpha)}(x)|\,(1+|x|)^N\,;\ x \in \Omega\} < +\infty\,, \qquad N \in \mathbb{N},\ \alpha \in \mathfrak{A},$$

is valid for $\Omega = \Omega_{[1,+\infty[}$. Similarly, applying Proposition 5.13 and Corollaries 5.29 and 5.33 to $b=1$ and

$$W(x) = (1+|M_0(x)|)^{-N}\,(1+|P_+(x)|)^{-N}$$

we conclude that the condition (6.34) holds for $\Omega = \Omega_{[0,1]}$, as well. Consequently, ϕ_k belongs to $\mathcal{S}(V)$. Moreover, if φ belongs to $C_0^\infty(X)$ then so does ϕ_k. In fact, since by Remark 3.25.(i) $\operatorname{supp}\varphi$ is a weakly $(M,[1,+\infty[)$-bounded subset of X it follows by Lemma 6.11 that $\overline{(\operatorname{supp}\varphi)_{M,[1,+\infty[}} \cap \overline{(\operatorname{supp}\varphi)_{M,]0,1]}}$ is a compact subset of X so that in view of Remark 5.7 and Proposition 5.16.(ii) the support of ϕ_k is a compact subset of X.

As in the proof of Proposition 6.27 we now derive from Proposition 5.27 that the equation (6.21) is valid for $j=k$. Since both φ and ϕ_k and hence ${}^t(\partial_M - m)^{k+1}\phi_k$ belong to $\mathcal{S}(V)$ this implies, in particular, that $Q_{m^*}\varphi \equiv 0$. Assuming that φ belongs to $C_0^\infty(X)$ in case $T \in \mathcal{D}'(X)$ we deduce that

$$\langle T, \varphi\rangle = -\langle d, \phi_k\rangle$$

where $d := (\partial_M - m)^{k+1}T$. Since by the assumption on T Proposition 6.27 shows that $d = Q'_m(d)$ and since by Corollary 5.31 we have $Q_{m^*}\phi_k \equiv 0$ we conclude that $\langle d, \phi_k\rangle = 0$ and hence $\langle T, \varphi\rangle = 0$, as desired. ■

For the proof of the implication "(b)⇒(c)" we are going to derive a variant of Proposition 4.13. To this end we fix a sequence $(\psi_i)_{0\le i\le k}$ in $C_M^\infty(X)$ satisfying (4.18) for $\mathfrak{J} = \{0\} \cup \mathbb{N}_{k+N+1}$ and (4.19) where the constants $c_{k,i}$ are defined by (4.15) and where

$$(6.35) \qquad N := N(m^*) \qquad \text{- see (5.65).}$$

That this is possible - even for $\mathfrak{J} = \mathbb{N}_0$ - follows by Lemma 4.12.(ii) and Remark 6.12. When working in the context of temperate distributions it is necessary to choose the functions ψ_i more carefully. This requires

Lemma 6.38. *Suppose that $\varkappa = \varkappa_+$ (see Proposition 1.70). Let I be a compact subset of $]0,+\infty[$, and let J be a relatively compact open neighbourhood of I in $]0,+\infty[$. Then there exists a C^∞ function $\psi : V \longrightarrow \mathbb{R}$ equal to 1 on $\varkappa^{-1}(I)$ satisfying*

$$\psi_0 \equiv 1 \tag{6.36}$$

and having the following properties:

(i) $\operatorname{supp}\psi \subset \varkappa^{-1}(J)$;

(ii) *all the derivatives of ψ are bounded;*

(iii) *for every $j \in \mathbb{N}_0$ all the derivatives of ψ_{0,ω_j} are bounded on every M-temperate subset of V;*

(iv) *if $I \subset]r, er[$ for some $r \in]0,+\infty[$ then $\psi \geq 0$.*

Proof. We choose $a, b, c, d \in]0,+\infty[$ such that $b > d > c > a$ and $[a,b] \subset J \setminus I$. By Lemma 5.53 we find non-negative functions $\chi, \varphi \in C^\infty(V)$ with values in $[0,1]$ such that all their derivatives are bounded, such that χ (resp. φ) is equal to 1 on $\varkappa^{-1}([c,d])$ (resp. $\varkappa^{-1}(I)$) and such that the support of χ (resp. φ) is contained in $\varkappa^{-1}(]a,b[)$ (resp. $\varkappa^{-1}(J \setminus [a,b])$). Since $t\varkappa(x) = \varkappa(M_t x) \in [c,d]$ if and only if $t \in [c/\varkappa(x), d/\varkappa(x)]$ it follows in view of Theorem 5.37.(i) that

$$\chi_0(x) \geq \int_{c/\varkappa(x)}^{d/\varkappa(x)} \frac{dt}{t} = \log\frac{d}{c} > 0\,, \qquad x \in V_+\,. \tag{6.37}$$

Consequently, by $\psi := \varphi + (1-\varphi_0)\chi/\chi_0$ a C^∞ function on V is well-defined. Obviously, it is equal to 1 on $\varkappa^{-1}(I)$, and its support is contained in $\varkappa^{-1}(J)$. Since $1-\varphi_0$ and χ_0 are quasihomogeneous of degree 0 it follows from Remark 5.3 that $\psi_0 = \varphi_0 + (1-\varphi_0) \equiv 1$. To prove (ii) we choose $R > \rho > 0$ such that $J =]\rho, R[$ so that $\operatorname{supp}\varphi \subset \varkappa^{-1}(]\rho, R[)$. Moreover, we fix $\ell \in \mathfrak{A}(M)$ and a polynomial function $P : V^* \longrightarrow \mathbb{C}$ which is almost quasihomogeneous of degree ℓ and of type M^*. For every $i \in \mathbb{N}$ we set $P_i := (\partial_{M^*} - \ell)^i P$. Applying Proposition 3.4 we deduce that

$$|(P(\partial)\varphi_0)(x)| \leq \sum_{i\in\mathbb{N}_0} \left| \int_{\rho/\varkappa(x)}^{R/\varkappa(x)} t^\ell (P_i(\partial)\varphi)(M_t x)\,\omega_i(t)\,\frac{dt}{t} \right| \leq \tag{6.38}$$

$$\leq \sum_{i\in\mathbb{N}_0} \int_{\rho/b}^{R/a} t^{\operatorname{Re}\ell - 1}\,|\omega_i(t)|\,dt\,\|P_i(\partial)\varphi\|_{L_\infty}$$

and

$$|(P(\partial)\chi_0)(x)| \leq \sum_{i\in\mathbb{N}_0} \int_{a/b}^{b/a} t^{\mathrm{Re}\,\ell-1}\,|\omega_i(t)|\,dt\;\|P_i(\partial)\chi\|_{L_\infty}$$

for arbitrary $x \in \varkappa^{-1}([a,b])$. Combining the last estimate with (6.37) shows that the derivatives of $1/\chi_0$ are bounded on $\operatorname{supp}\chi$, and (ii) follows by (6.38).

The property *(iii)* is a consequence of the following estimate which - in view of Proposition 3.4 and (1.38) - is valid for arbitrary $\varepsilon > 0$ and $x \in \varkappa^{-1}([\varepsilon, 1/\varepsilon])$:

$$|P(\partial)\psi_{0,\omega_j}(x)| \leq \sum_{i\in\mathbb{N}_0} \binom{j+i}{i} \left| \int_{\rho/\varkappa(x)}^{R/\varkappa(x)} t^{\ell}\,(P_i(\partial)\psi)(M_t x)\,\omega_{j+i}(t)\,\frac{dt}{t} \right| \leq$$

$$\leq \sum_{i\in\mathbb{N}_0} \binom{j+i}{i} \int_{\rho\varepsilon}^{R/\varepsilon} t^{\mathrm{Re}\,\ell-1}\,|\omega_{j+i}(t)|\,dt\;\|P_i(\partial)\psi\|_{L_\infty}.$$

Finally, if the assumption of *(iv)* holds then in the proof of (ii) above we can choose ρ and R such that $R = e\rho$. Consequently, making use of (6.38) for $P \equiv 1$ we obtain

$$\varphi_0(x) \leq \int_{\rho/\varkappa(x)}^{R/\varkappa(x)} \frac{dt}{t} = 1\,.$$

Since $\chi_0 > 0$ this shows that $\psi \geq 0$. ∎

By Lemma 4.12.(ii) we obtain

Corollary 6.39. *Under the assumptions of Lemma 6.38 the functions ψ_i, $0 \leq i \leq k$, can be chosen so as to have the properties (i), (ii), and (iii) of Lemma 6.38.* ∎

Besides the properties spelt out in Lemma 4.12 the constants $c_{k,i}$ possess other important properties required for the proof of Theorem 6.37.

Lemma 6.40. *For every $j \in \mathbb{N}_0$ we have*

$$\sum_{i=0}^{k} \binom{j+k-i}{j} c_{k,i} = \begin{cases} (-1)^k \delta_{0j} & \text{if } j \leq k \\ \binom{j-1}{k} & \text{if } j > k \end{cases}. \tag{6.39}$$

Proof. Inserting the defining equations (4.15), interchanging the order of summation and substituting $i = i - I$ we see that the left-hand side of (6.39) is equal to

$$\sum_{I=0}^{k}\sum_{i=I}^{k} (-1)^i \binom{j+k-i}{j}\binom{k-I}{k-i} = \sum_{I=0}^{k} (-1)^I \Big[\sum_{i=0}^{k-I} (-1)^i \binom{j+k-I-i}{j}\binom{k-I}{i}\Big].$$

Applying Lemma 1.76 to $(j,j,k-I)$ instead of (j,k,ℓ) we see that the sum in square brackets vanishes in case $j<k-I$ and equals $\binom{j}{j+I-k}$ in case $j\geq k-I$. Hence the the left-hand side of (6.39) is equal to

$$\sum_{\substack{I=0\\ I\geq k-j}}^{k} (-1)^I \binom{j}{k-I} = \sum_{i=0}^{\min(j,k)} (-1)^{k-i} \binom{j}{i}.$$

If $j\leq k$ this equals $(-1)^k(1-1)^j = (-1)^k \delta_{0j}$, and the assertion follows in this case. If, on the other hand $j>k$ then by (4.20) this equals $\binom{j-1}{k}$, as desired. ∎

Now, let T be a distribution satisfying the condition (b) of Theorem 6.37.(i). Then applying Proposition 4.13 to $(T|_{X_+}, X_+, k)$ instead of (T,X,N) we observe that the support of the distribution

$$d := T - \sum_{i=0}^{k} (\psi_i (\partial_M - m)^i T)_{m,\omega_k} \tag{6.40}$$

is contained in $X\backslash X_+$. However, at this stage it is not clear that d is, in fact, almost quasihomogeneous of degree m. In order to verify this we require another auxilliary function: we fix a C^∞ function $\chi_\psi : X \longrightarrow \mathbb{C}$ such that $\operatorname{supp}\chi_\psi$ is a weakly $(M,[1,+\infty[)$-bounded subset of X and such that we have

$$(\chi_\psi|_{X^o})_{o,\omega_{k+j}} = (\psi_o)_{o,\omega_{2k+1+j}}|_{(X^o)_+}, \qquad j\in\mathfrak{I}, \tag{6.41}$$

for $\mathfrak{I}=\{0\}\cup\mathbb{N}_N$ with N being defined by (6.35) (recall that $X^o := M_o^{-1}(X)$). Note that in view of Propositions 6.3.B and 3.3 the assumption on $\operatorname{supp}\chi_\psi$ implies that $(\chi_\psi)_{o,\omega_j}\chi_{[1,+\infty[}$ is well-defined on X_+ by (3.1)' whereas for $f := \chi_\psi|_{X^o}$ the function $f_{o,\omega_j}\chi_{]0,1]}$ is well-defined on X^o by the results of § 5.(b) so that f_{o,ω_j} is well-defined on $(X^o)_+$ by (5.60) with $w=\omega_j$. In fact, f satisfies the assumptions of Theorem 5.37 with X replaced by X^o. By way of Lemma 6.47 below we shall verify the existence of functions χ_ψ having the above properties and, at the same time, shed some light on the meaning of the condition (6.41). At present, in the following lemma we note a very easy way of obtaining the desired data by first prescribing χ_ψ and then constructing the functions ψ_i depending on χ_ψ.

Lemma 6.41. *Let $\chi : V \longrightarrow \mathbb{C}$ be any C^∞ function such that $\operatorname{supp}\chi$ is a weakly $(M,[1,+\infty[)$-bounded subset of V and such that $\chi\equiv 1$ on a neighbourhood of $\ker M$. Then the functions*

(6.42) $$\psi_i := c_{k,i}\,(-\partial_M)^{k+1-i}\chi\,, \qquad 0 \le i \le k\,,$$

belong to $C^\infty_M(V)$ *and satisfy* (4.18) *for* $\mathfrak{J} = \mathbb{N}_0$, (4.19), *and* (6.41) *for* $X = V$, $\chi_\psi = \chi$, *and* $\mathfrak{J} = \mathbb{N}_0$.

Proof. Since $\chi \equiv 1$ near $V \setminus V_+$ one deduces with the help of Proposition 6.3.A.(ii) that $\operatorname{supp}\psi_i$ is a weakly $(M,]0,1])$-bounded subset of V. By the assumption on $\operatorname{supp}\chi$ this means that $\operatorname{supp}\psi_i$ is a weakly M-bounded subset of V, i.e. $\psi_i \in C^\infty_M(V)$. It follows from (3.7) and (5.63) that $(-\partial_M\chi)_O = (-\partial_M)\chi_O = \chi \circ M_O|_{V_+} \equiv 1$. Hence, the function $\psi = -\partial_M\chi|_{V_+}$ satisfies the assumptions of Lemma 4.12.(ii) so that (4.18) for $\mathfrak{J} = \mathbb{N}_0$ and (4.19) hold. Moreover, the conditions (3.7) and (5.63), again, imply that

$$(\psi_O)_{O,\omega_{k+1+j}} = (-\partial_M)^{k+1}\chi_{O,\omega_{k+1+j}} = \chi_{O,\omega_j}\,, \qquad j \in \mathbb{N}_0\,. \quad ■$$

Functions ψ_i and χ defined on X satisfying (4.18), (4.19), and (6.41) are obtained as restrictions to X of the corresponding functions in Lemma 6.41. This is obvious in view of Corollary 6.5.

We now come to the description of the distribution d defined by (6.40).

Theorem 6.42. *Let* $T \in \mathcal{D}'(X)$ *and* $k \in \mathbb{N}_0$ *be such that the condition (b) of Theorem 6.37.(i) is satisfied. Then*

(6.43) $$T = \sum_{i=0}^{k} (\psi_i\,(\partial_M - m)^i T)_{m,\omega_k} + Q'_m(\chi_\psi T)\,.$$

Note that in case the functions ψ_i *are given as in Lemma 4.12.(ii) then*

(6.44) $$\sum_{i=0}^{k} (\psi_i\,(\partial_M - m)^i T)_{m,\omega_k} = \sum_{i=0}^{k} (-1)^i (\psi\,(\partial_M - m)^i T)_{m,\omega_i}\,.$$

Proof. We define N by (6.35). Let $\varphi \in C^\infty_O(X)$. Then by applying the formula (3.8) to $(\varphi_{m^*,\omega_k}, \chi|_{X_+}, m^*, 0, \omega_q)$ instead of (q, f, ℓ, m, w) and making use of the equalit[y] $(\chi|_{X_+})_{O,w} = \chi_{O,w}$ we obtain in view of (5.63) for arbitrary $\chi \in C^\infty_M(X)$ and $q \in \mathbb{N}$ that

(6.45) $$(\chi\varphi_{m^*,\omega_k})_{m^*,\omega_q} = \sum_{j=0}^{k} (-1)^j \chi_{O,\omega_q\omega_j}\,\varphi_{m^*,\omega_{k-j}} - $$
$$- (-1)^k \sum_{j=0}^{N} \chi_{O,\omega_q\omega_{k+1+j}}\,(\partial_M - m^*)^j Q_{m^*}\varphi|_X$$

Since for every $i \in \{0\} \cup \mathbb{N}_k$ $\operatorname{supp} \psi_i$ is a weakly M-bounded subset of X it follows in view of Theorem 5.37.(ii) that $\psi_i \varphi_{m^*,\omega_k}$ extends to a function in $C_0^\infty(X_+)$, again denoted by $\psi_i \varphi_{m^*,\omega_k}$, so that by

$$\phi := \sum_{i=0}^{k} (-1)^{k+i} (\partial_M - m^*)^i (\psi_i \varphi_{m^*,\omega_k})$$

a function in $C_0^\infty(X_+)$ is well-defined. By (3.7) and (5.63) and by (6.45) it follows that

$$\phi_{m^*,\omega_k} = \sum_{i=0}^{k} (-1)^k (\psi_i \varphi_{m^*,\omega_k})_{m^*,\omega_{k-i}} = A - B$$

where

$$A := \sum_{i=0}^{k} (-1)^k \sum_{j=0}^{k} (-1)^j (\psi_i)_{0,\omega_{k-i}\omega_j} \varphi_{m^*,\omega_{k-j}}$$

and

$$B := \sum_{i=0}^{k} \sum_{j=0}^{N} (\psi_i)_{0,\omega_{k-i}\omega_{k+1+j}} (\partial_M - m^*)^j Q_{m^*} \varphi\big|_{X_+}.$$

By (1.38) and (4.18) and by changing the order of summation and substituting $J = k-j$ we obtain

$$A = \sum_{i=0}^{k} \sum_{j=0}^{k} (-1)^{k+j} \binom{k-i+j}{j} c_{k,i} (\psi_0)_{0,\omega_{k+j}} \varphi_{m^*,\omega_{k-j}} =$$

$$= \sum_{J=0}^{k} (-1)^J \Big[\sum_{i=0}^{k} \binom{k-J+k-i}{k-J} c_{k,i} \Big] (\psi_0)_{0,\omega_{2k-J}} \varphi_{m^*,\omega_J}.$$

Since by (6.39) the sum in square brackets is equal to $(-1)^k \delta_{0k-J}$ it follows by (4.19) that $A = \varphi_{m^*,\omega_k}$. Moreover, again by (1.38) and (4.18) and by (6.39) we conclude that

$$B = \sum_{j=0}^{N} \Big[\sum_{i=0}^{k} \binom{k+1+j+k-i}{k+1+j} c_{k,i} \Big] (\psi_0)_{0,\omega_{2k+1+j}} (\partial_M - m^*)^j Q_{m^*} \varphi\big|_{X_+} =$$

$$= \sum_{j=0}^{N} \binom{k+j}{k} (\psi_0)_{0,\omega_{2k+1+j}} (\partial_M - m^*)^j Q_{m^*} \varphi\big|_{X_+}.$$

Since by Corollary 5.47 the support of $Q_{m^*}\varphi$ is a closed subset of V being contained in X^0 we deduce that $\chi_\psi Q_{m^*}\varphi$ extends to a C^∞ function $h: V \longrightarrow \mathbb{C}$ with (unchanged) support contained in X^0. Since by Lemma 6.23 $\operatorname{supp} Q_{m^*}\varphi$ is contained in $L_{M,]0,1]}$ for some compact subset L of X^0 and since $\operatorname{supp} \chi_\psi$ is a weakly $(M,[1,+\infty[)$-bounded subset of X it follows that $\operatorname{supp} \chi_\psi \cap \operatorname{supp} Q_{m^*}\varphi$ is a compact subset of X, i.e. h belongs to $C_0^\infty(X^0)$. Finally, we deduce from (3.8)' applied to $q = Q_{m^*}\varphi\big|_{X^0}$), (1.38), and (6.41) that the equality

$$h_{m^*,\omega_k} = \sum_{j=0}^{N} (\chi_\psi|_{X^0})_{0,\omega_k\omega_j} (\partial_M - m^*)^j Q_{m^*}\varphi =$$

$$= \sum_{j=0}^{N} \binom{k+j}{k} (\psi_0)_{0,\omega_{2k+1+j}} (\partial_M - m^*)^j Q_{m^*}\varphi .$$

is valid on $(X^0)_+$. Since by Remark 5.60 the support of h_{m^*,ω_k} is contained in $(X^0)_+$, as well, equality holds on the whole of V_+. Consequently, in view of what was proved above we see that the functions ϕ_{m^*,ω_k} and $\varphi_{m^*,\omega_k} - h_{m^*,\omega_k}$ coincide on X_+. Since by Remark 5.60.(i) both functions vanish on $V_+ \setminus X_+$ we conclude that the function $\Phi := \varphi - \phi - h$ satisfies $\Phi_{m^*,\omega_k} \equiv 0$. Hence, by the condition (b) of Theorem 6.37.(i) we conclude that $\langle T, \Phi\rangle = 0$, i.e.

$$\langle T,\varphi\rangle = \langle T,\phi\rangle + \langle T,h\rangle =$$

$$= \sum_{i=0}^{k} (-1)^k \langle (\partial_M - m)^i T, \psi_i\, \varphi_{m^*,\omega_k}\rangle + \langle \chi_\psi T, Q_{m^*}\varphi\rangle,$$

and (6.43) follows.

For an alternative proof of (6.43) in case $\chi_\psi \equiv 1$ on a neighbourhood of $X \cap M_0(X)$ (compare Lemma 6.47 below) one proceeds as follows: Denoting the distribution $\sum_{i=0}^{k} (\psi_i (\partial_M - m)^i T)_{m,\omega_k}$ by S one deduces from Proposition 4.13 that the support of $T - S$ is contained in $X \setminus X_+$ so that the assumption on χ_ψ and Proposition 6.27 imply that $T - S = Q'_m(\chi_\psi T - \chi_\psi S)$. Consequently, to prove (6.43) one has to show that $Q'_m(\chi_\psi S) = 0$. This can be carried out by way of similar computations as above.

For the proof of (6.44) one easily checks that the computations done in the *End of the proof of Theorem 4.8* and in the derivation of (4.14)' remain valid in the present context. ■

As our final preparation for the end of the proof of Theorem 6.37 we derive the following formula which is a variant of (6.43) in case T is of the form $T = u_{m,\omega_k}$.

Remark 6.43. *For every $u \in \mathcal{D}'_M(X)$ the distribution u_{m,ω_k} is equal to*

$$\sum_{i=0}^{k} (\psi_i (\partial_M - m)^i u_{m,\omega_k})_{m,\omega_k} + (-1)^k \sum_{j=0}^{\infty} (-1)^j \binom{k+j}{k} (\partial_M - m)^j Q'_m((\psi_0)_{0,\omega_{2k+j+1}} u)$$

Proof. As before, we define N by (6.35). Let $\varphi \in C_0^\infty(X)$. Then from (6.12) and (6.45) we deduce for arbitrary $\chi \in C_M^\infty(X)$ and $q \in \mathbb{N}_0$ that

$$\langle (\chi u_{m,\omega_q})_{m,\omega_k}, \varphi \rangle = (-1)^{k+q} \langle u, (\chi \varphi_{m^*,\omega_k})_{m^*,\omega_q} \rangle =$$

$$= \sum_{j=0}^{k} (-1)^{k+q+j} \langle u, \chi_{0,\omega_q\omega_j} \varphi_{m^*,\omega_{k-j}} \rangle -$$

$$- \sum_{j=0}^{N} (-1)^q \langle u, \chi_{0,\omega_q\omega_{k+1+j}} (\partial_M - m^*)^j Q_{m^*} \varphi\big|_{X_+} \rangle .$$

By (1.38) and (2.5) this implies that

$$(6.46) \qquad (\chi u_{m,\omega_q})_{m,\omega_k} = \sum_{j=0}^{k} (-1)^q \binom{q+j}{q} (\chi_{0,\omega_{q+j}} u)_{m,\omega_{k-j}} -$$

$$- \sum_{j=0}^{N} (-1)^{q+j} \binom{q+k+1+j}{q} (\partial_M - m)^j Q'_m(\chi_{0,\omega_{q+k+1+j}} u) .$$

Inserting the equations $(\partial_M - m)^i u_{m,\omega_k} = (-1)^i u_{m,\omega_{k-i}}$, $0 \le i \le k$, and applying (6.46) to $(\psi_i, k-i)$ instead of (χ, q) we obtain that

$$\sum_{i=0}^{k} (\psi_i (\partial_M - m)^i u_{m,\omega_k})_{m,\omega_k} = A - B$$

where

$$A := \sum_{i=0}^{k} (-1)^i \sum_{j=0}^{k} (-1)^{k-i} \binom{k-i+j}{k-i} ((\psi_i)_{0,\omega_{k-i+j}} u)_{m,\omega_{k-j}}$$

and

$$B := \sum_{i=0}^{k} (-1)^i \sum_{j=0}^{N} (-1)^{k-i+j} \binom{2k-i+1+j}{k-i} (\partial_M - m)^j Q'_m((\psi_i)_{0,\omega_{2k-i+1+j}} u) .$$

Inserting (4.18) we deduce that

$$A = (-1)^k \sum_{j=0}^{k} \Big[\sum_{i=0}^{k} \binom{j+k-i}{j} c_{k,i} \Big] ((\psi_0)_{0,\omega_{k+j}} u)_{m,\omega_{k-j}} .$$

By (6.39) and (4.19) this is equal to u_{m,ω_k}.

Moreover, inserting (4.18) and interchanging the order of summation we obtain

$$B = (-1)^k \sum_{j=0}^{N} (-1)^j (\partial_M - m)^j Q'_m\Big(\Big[\sum_{i=0}^{k} \binom{(k+j+1)+k-i}{k+j+1} c_{k,i} \Big] (\psi_0)_{0,\omega_{2k+1+j}} u \Big) .$$

Since by (6.39) the sum in square brackets is equal to $\binom{k+j}{k}$ the proof is complete.

For an alternative proof one can apply Theorem 6.42 to $T = u_{m,\omega_k}$ and - making use of (6.41) - verify the equality

$$Q'_m(\chi_\psi u_{m,\omega_k}) = (-1)^k \sum_{j=0}^{N} (-1)^j \binom{k+j}{k} (\partial_M - m)^j Q'_m((\psi_0)_{0,\omega_{2k+1+j}} u) . \quad ■$$

As a consequence of the case "k=0" of Remark 6.43 we deduce the following step in the proof of Theorem 6.37.

Lemma 6.44. *Let $d \in \mathcal{D}'(X)$ be almost quasihomogeneous of degree m with support contained in $X \setminus X_+$. Then*

(i) $d = u_m$ for some $u \in \mathcal{D}'_M(X)$ if and only if d extends to a distribution $\tilde{d} \in \mathcal{D}'(\tilde{X})$ which is almost quasihomogeneous of degree m, as well;

(ii) if X = V and if T is temperate then in (i) one can choose u to be temperate, as well, with M-temperate support.

Proof. *(i), ⇒:* In view of (6.2.a) Remark 6.14 shows that u extends to a distribution $\tilde{u} \in \mathcal{D}'_M(\tilde{X})$. Since $(\tilde{u})_m$ extends u_m it follows that it extends d. Since by Proposition 6.35 it is almost quasihomogeneous of degree m the implication "⇒" is proved.

(i), ⇐: Obviously, we may suppose that $\tilde{X} = X$. Then by Lemma 6.31 we can choose $v \in \mathcal{D}'_M(X)$ such that $d = Q'_m v$. By Remark 6.14 and by Proposition 3.42 we find an M-bounded open subset L of X_+ such that $L \supset \operatorname{supp} v$. Moreover, applying Lemma 3.45 to $I := [e, +\infty[$ we obtain another M-bounded open subset Y of X_+ having the properties listed in the conclusion of Lemma 3.45. By Theorem 3.39 and Remark 6.12 we fix a non-negative C^∞ function ψ on X with support contained in Y such that (6.36) holds. Then from (3.1)' and (3.37) we deduce that $\psi_{0,\omega_1} \geq \psi_0 \equiv 1$ on L. Consequently, by induction we can find a sequence of functions $\chi_\ell \in C^\infty(X)$, $\ell \in \mathbb{N}_0$, such that

$$\sum_{\ell=0}^{J} (-1)^{J+\ell} \psi_{0,\omega_{J-\ell+1}} \chi_\ell \equiv \delta_{0J} \quad \text{on } L, \qquad J \in \mathbb{N}_0. \tag{6.47}$$

Finally, we observe that by Corollary 6.28 the number N defined by (6.35) has the property that $(\partial_M - m)^{N+1} c = 0$ for every distribution $c \in \mathcal{D}'(X)$ with support contained in $X \setminus X_+$ which is almost quasihomogeneous of degree m. Employing Remark 6.43 with (u, ψ_0, k) replaced by $(\chi_\ell v, \psi, 0)$ and applying $(\partial_M - m)^\ell$ we obtain in view of (4.10) (see Proposition 6.16) that

$$[(\partial_M - m)^\ell (\chi_\ell v - \psi(\chi_\ell v)_m)]_m = \sum_{j=0}^{N-\ell} (-1)^j (\partial_M - m)^{j+\ell} Q'_m(\psi_{0,\omega_{j+1}} \chi_\ell v)$$

Summing over ℓ, substituting $J = j + \ell$, and changing the order of summation we see that the distribution

$$(6.48) \qquad u := \sum_{\ell=0}^{N} (\partial_M - m)^\ell \, (\chi_\ell v - \psi(\chi_\ell v)_m)$$

satisfies the equation

$$u_m = \sum_{\ell=0}^{N} \sum_{J=\ell}^{N} (-1)^{J+\ell} (\partial_M - m)^J \, Q'_m(\psi_{0,\omega_{J-\ell+1}} \chi_\ell v) =$$

$$= \sum_{J=0}^{N} (\partial_M - m)^J \, Q'_m\Big(\Big[\sum_{\ell=0}^{J} (-1)^{J+\ell} \, \psi_{0,\omega_{J-\ell+1}} \chi_\ell\Big] v\Big) .$$

By (6.47) this is equal to $Q'_m v$, as desired.

(ii). We now suppose that $X = V$, that d is temperate, and that $\varkappa = \varkappa_+$. Then by Lemma 6.31.(iii) the distribution v can be chosen to be temperate with support contained in $\Omega_{[1,2]}$ so that we may suppose that $L = \Omega_{]1/2,3[}$. By (1.77) we have $\varkappa \leq 3^d$ on L for a suitable constant $d > 0$. We set $a := e\,3^d$ and $Y := \varkappa^{-1}(]a, a+1[)$. Then $\{t \in]0,+\infty[\,;\, M_t(L) \cap Y \neq \emptyset\} \subset [e,+\infty[$. Hence we choose ψ as in Lemma 6.38 for $J =]a, a+1[$. Then, as before, we deduce that $\psi_{0,\omega_1} \geq 1$ on L so that the derivatives of $1/\psi_{0,\omega_1}$ are bounded on L, as well. Consequently, the functions χ_ℓ satisfying (6.47) can be chosen in such a way that their derivatives are bounded on L. It follows by Proposition 6.22 that the distribution u defined in (6.48) is temperate. Since its support is contained in $L \cup \varkappa^{-1}([a, a+1])$ we see that u has the desired properties. ■

End of the Proof of Theorem 6.37. (b)⇒(c): By Theorem 6.42 and Proposition 6.24 the distribution d defined by (6.40) is almost quasihomogeneous of degree m, its support being contained in $X \setminus X_+$. Consequently, the equation (6.33) is valid for

$$(6.49) \qquad u = \sum_{i=0}^{k} \psi_i \, (\partial_M - m)^i \, T .$$

Since the implication *"(c)⇒(a)"* follows from Proposition 6.35 the proof of *(i)* is complete.

(ii). If $\tilde{X} = X$ then by Lemma 6.44 we find $v \in \mathcal{D}'_M(X)$ such that $d = v_m$. Since by (4.10) and (6.30) we have $v_m = ((-\partial_M + m)^k v)_{m,\omega_k}$ the equation $T = u_{m,\omega_k}$ holds for

$$(6.49)' \qquad u = \Big[\sum_{i=0}^{k} \psi_i \, (\partial_M - m)^i \, T\Big] + (-\partial_M + m)^k \, v .$$

(iii). If T is temperate then in view of Corollary 6.39 we can achieve that the distributions $\psi_i \, (\partial_M - m)^i \, T$, $0 \leq i \leq k$, belong to $\mathcal{S}'_M(V)$. Hence, by Proposition 6.22

the distribution d defined by (6.40) is temperate, as well. Then by Lemma 6.44 the distribution v can be chosen to belong to $\mathcal{S}'_M(V)$ so that the distribution u defined by (6.49)' lies in $\mathcal{S}'_M(V)$, as well.

(iv): Suppose that (6.33) is valid and that T is $\mathfrak{G}$-invariant. Since by Propositions 6.16.(v) $u_{\mathfrak{G}}$ belongs to $\mathcal{D}'_M(X)$ and satisfies the equation $(u_{m,\omega_k})_{\mathfrak{G}} = (u_{\mathfrak{G}})_{m,\omega_k}$ it follows by Proposition 2.64.(iii) that $T = T_{\mathfrak{G}} = (u_{\mathfrak{G}})_{m,\omega_k} + d_{\mathfrak{G}}$. Here $u_{\mathfrak{G}}$ is $\mathfrak{G}$-invariant by Proposition 2.64. Moreover, by the same Proposition $d_{\mathfrak{G}}$ is $\mathfrak{G}$-invariant with support contained in $X \setminus X_+$ (note that ker M is G-invariant), and by Remark 2.67.(ii) $d_{\mathfrak{G}}$ is almost quasihomogeneous of degree m. Note that d = 0 implies $d_{\mathfrak{G}} = 0$.

Finally, note that for arbitrary $b \geq \varepsilon > 0$ there are $b' > \varepsilon' > 0$ such that $(\Omega_{[\varepsilon,b]})_G$ is contained on $\Omega_{[\varepsilon',b']}$. Hence, in view of Proposition 2.64.(ii) if u belongs to $\mathcal{S}'_M(V)$ so does $u_{\mathfrak{G}}$. ∎

We end this section with a few supplements to Theorem 6.42. For the first two supposing that X = V we ask under which conditions the constituting parts of the formula (6.43) are temperate distributions. It turns out that under special hypotheses on M every almost quasihomogeneous distribution on V is temperate. This is the contents of the following consequence of Theorem 6.42 (or rather Proposition 4.13) and Corollary 6.39.

Theorem 6.45. *Suppose that ker M = {0}. Then every distribution $T \in \mathcal{D}'(V)$ satisfying one (and hence each) of the conditions of Theorem 6.37.(i) is temperate.*

Proof. In view of Remarks 6.14 and 3.23 the assumption on M implies that $\mathcal{D}'_M(V) = \mathcal{E}'(\dot{V}) = \mathcal{S}'_M(V)$. Hence Proposition 6.22 implies that the distribution $(\psi_i (\partial_M - m)^i T)_{m,\omega_k}$ are temperate. Since, moreover, the support of the distributio[n] d defined by (6.40) is contained in $V \setminus V_+ = \{0\}$, i.e. $d \in \mathcal{E}'(V)$ it follows fro[m] (6.40) that T is temperate, as well, indeed. ∎

However, in case X = V and ker M ≠ {0} then, of course, T is not necessarily temperate. If it is so by assumption, then each summand on the right-hand side [of] (6.43) should belong to $\mathcal{S}'(V)$, as well. Establishing this requires special choice[s] of ψ and χ_ψ which are possible by Lemmata 5.53 and 6.41:

Remark 6.46. *Suppose that $X = V$, that the functions ψ_i, $0 \le i \le k$, have bounded derivatives and M-temperate support, and that χ_ψ has bounded derivatives and $(M, [1, +\infty[)$-temperate support. Let T be a distribution on V satisfying one (and hence each) of the conditions of Theorem 6.37.(i). If T is temperate then so are $(\psi_i T)_{m,\omega_k}$, $(x^\alpha \chi_\psi T)^\frown$ and $(x^\alpha \psi_i T)^\frown$ for $0 \le i \le k$ and $\alpha \in \mathfrak{A}^+_{m*}$.*

Proof. This is a consequence of Propositions 6.22 and 6.24.(iii) & (iv). ■

To prepare for the final result of this section we come to the precise lemma about the existence and the properties of the functions χ_ψ, alluded to in the text preceding Lemma 6.41.

Lemma 6.47. *Let $\chi : X \longrightarrow \mathbb{C}$ be any C^∞ function such that $\operatorname{supp} \chi$ is a weakly $(M, [1, +\infty[)$-bounded subset of X, and let $h \in \mathbb{N}_0$. Then the following assertions hold.*

(i) The following conditions are equivalent:

(a) there exists a function $\chi_\psi \in C^\infty(X)$ satisfying (6.41) for $\mathfrak{I} = \{0\} \cup \mathbb{N}_h$ such that the support of $\chi - \chi_\psi$ is a weakly $(M, [1, +\infty[)$-bounded subset of X and, at the same time, an M-bounded subset of $(X^o)_+$;

(b) $\chi|_{M_o(X) \cap X} \equiv 1$.

(ii) The following assertions are equivalent:

(a) there exists a function $\chi_\psi \in C^\infty(X)$ satisfying (6.41) for $\mathfrak{I} = \{0\} \cup \mathbb{N}_h$ such that the support of $\chi - \chi_\psi$ is a weakly M-bounded subset of X;

(b) for some (resp. every) $\tilde{\psi} \in C^\infty_M(X)$ the function $\chi_\psi := \chi - \tilde{\psi}$ satisfies (6.41) with $\mathfrak{I} = \{0\} \cup \mathbb{N}_h$ for a suitable choice of the functions ψ_i;

(c) $(\chi|_{X^o})_{o,\omega_{k+h}}$ extends to a C^∞ function $g : X_+ \longrightarrow \mathbb{C}$ such that $(-\partial_M)^{k+h+1} g \equiv 1$.

Note that the existence of functions χ satisfying the condition (c) in (ii) is established by the assertion (i). Of course, if (5.17) holds, i.e. if $X = X^o$ then (c) is trivially satisfied, and the formulation of Lemma 6.47 becomes much simpler.

Proof. Throughout the proof we set $W := (X^o)_+$ and $f := \chi|_{X^o}$.

(i): "$\Rightarrow$". From Theorem 5.37.(iii) we deduce that the function $(\chi_\psi|_{X^o})_{0,\omega_k}$ is almost quasihomogeneous of degree 0 such that

$$(-\partial_M)^{k+1}(\chi_\psi|_{X^o})_{0,\omega_k} = Q_0(\chi_\psi|_{X^o})|_W = \chi_\psi \circ M_0|_W .$$

Since by Proposition 3.13 the function $(\psi_0)_{0,\omega_{2k+1}}$ is almost quasihomogeneous of degree 0 satisfying

$$(-\partial_M)^{k+1}(\psi_0)_{0,\omega_{2k+1}} = (\psi_0)_{0,\omega_k}$$

the condition (b) is a consequence of (6.41) for $j=0$ and (4.19).

"$\Leftarrow$". We define $q : W \longrightarrow \mathbb{C}$ by

$$q := f_{0,\omega_{k+h}} - (\psi_0)_{0,\omega_{2k+1+h}}|_W .$$

By (5.63) and (3.11) we have $(-\partial_M)^{k+h+1}q = \chi \circ M_0|_W - (\psi_0)_{0,\omega_k}|_W$. By the conditions (b) and (4.19) we conclude that the latter function vanishes identically, i.e. q is almost quasihomogeneous of degree 0 and of order $\le j_0 := k+h$. In other words, the condition (b) of Theorem 3.48 is satisfied for $\Lambda = \{0\}$ and $q_{0,j} = (-\partial_M)^{j_0-j}q$ with X replaced by W.

Before going to apply Theorem 3.48 we have to construct a suitable subset Y of X. To this end we fix $a \in]0,+\infty[$. Then

$$L_a := \{x \in X ;\ \dot{\varkappa}(x) \le a \min\{1, \mathrm{dist}(M_0(x), V \setminus X)\}\}$$

is a closed subset of X contained in X^o. By Lemma 6.7 L_a is a weakly $(M,[1,+\infty[)$-bounded subset of X. Moreover, L_a is a neighbourhood of $M_0(X) \cap X$; more precisely, if $\varepsilon \in]0,a[$ then $\overset{\circ}{L}_a$ is an open neighbourhood of L_ε. Note that $(\overset{\circ}{L}_a \setminus L_\varepsilon)_M = W$. Consequently, we can apply Theorem 3.48 to the data described above and to $Y := \overset{\circ}{L}_a \setminus L_\varepsilon$ and obtain a function $\Psi \in C^\infty_M(W)$ with support contained in $\overset{\circ}{L}_a \setminus L_\varepsilon$ such that $\Psi_{0,\omega_j} = q_{0,j}$ for $0 \le j \le j_0$.

Since $X \cap \overline{(L_a \setminus L_\varepsilon)} \subset W$ the support of Ψ is a closed subset of X. Hence the extension $\tilde{\psi}$ of Ψ to X defined by $\tilde{\psi}|_{X \setminus W} :\equiv 0$ is a C^∞ function with the same support as Ψ. In particular, supp $\tilde{\psi}$ is a weakly $(M,[1,+\infty[)$-bounded subset of X and an M-bounded subset of W. Consequently, the function $\chi_\psi := \chi - \tilde{\psi}$ has the desired properties.

(ii): *(a)* $\Rightarrow$ *(c)*. If $\tilde{\psi} := \chi - \chi_\psi$ belongs to $C^\infty_M(X)$ then by Remark 6.12 and Proposition 3.13 $\tilde{\psi}_0$ is quasihomogeneous of degree 0, and in view of (6.41) $f_{0,\omega_{k+h}}$ is

extended by the function $g := (\tilde{\psi})_{0,\omega_{k+h}} + (\psi_0)_{0,\omega_{2k+1+h}}$ which by (3.11) and (4.19) satisfies the equation $(-\partial_M)^{k+h+1} g \equiv 1$.

<u>(c)⇒(b)</u>. Let $g \in C^\infty(X_+)$ be an extension of $f_{0,\omega_{k+h}}$ having the properties asserted in (c). Then for any $\tilde{\psi} \in C^\infty_M(X)$ the function $q := g - (\tilde{\psi})_{0,\omega_{k+h}}$ belongs to $C^\infty(X_+)$ and - by the assumptions on g and by (3.11) - is almost quasihomogeneous of degree 0 and of order $\leq k+h+1$ satisfying $(-\partial_M)^{k+h+1} q \equiv 1$. Hence, by Theorem 3.48 and Remark 6.12 one finds a function $\psi \in C^\infty_M(X)$ such that $\psi_{0,\omega_{k+h+1}} = q$. In particular, by (5.63) we have $\psi_0 \equiv 1$. Defining ψ_i by (4.16) we see from Lemma 4.12.(ii) that (4.18) for $\mathfrak{J} = \mathbb{N}_0$ and (4.19) are valid. Moreover, since $\psi_0 = (-\partial_M)^k \psi$ the condition (5.63) shows that $q = (\psi_0)_{0,\omega_{2k+h+1}}$. Finally, setting $\chi_\psi := \chi - \tilde{\psi}$ we derive from (5.63) that

$$(6.50) \qquad (\chi_\psi|_{X^0})_{0,\omega_{k+j}} = (-\partial_M)^{h-j}\left(f_{0,\omega_{k+h}} - (\tilde{\psi})_{0,\omega_{k+h}}|_W\right), \qquad j \in \{0\} \cup \mathbb{N}_h,$$

and since the term in large brackets is equal to $q|_W$ we deduce (6.41) for $\mathfrak{J} = \{0\} \cup \mathbb{N}_h$ by another application of (5.63).

<u>(b)⇒(c)</u>. If $\tilde{\psi} \in C^\infty_M(X_+)$ is such that $\chi_\psi := \chi - \tilde{\psi}$ satisfies (6.41) for $\mathfrak{J} = \{0\} \cup \mathbb{N}_h$ then in view of (5.63) and (4.19) $(\tilde{\psi})_{0,\omega_{k+h}} + (\psi_0)_{0,\omega_{2k+1+h}}$ is the desired extension of $f_{0,\omega_{k+h}}$.

<u>(c)⇒(a)</u>. Let $g \in C^\infty(X_+)$ be an extension of $f_{0,\omega_{k+h}}$ having the properties asserted in (c). Then by the assumptions on g and by (5.63) and (4.19) the function $q := g - (\psi_0)_{0,\omega_{2k+1+h}}$ is almost quasihomogeneous of degree 0 and of order $\leq k+h$. Hence, by Theorem 3.48 and Remark 6.12 we obtain a function $\tilde{\psi} \in C^\infty_M(X)$ such that $(\tilde{\psi})_{0,\omega_{k+h}} = q$. Setting $\chi_\psi := \chi - \tilde{\psi}$ we see that the term in large brackets in equation (6.50) is equal to $(\psi_0)_{0,\omega_{2k+1+h}}|_W$. Hence the condition (6.41) for $\mathfrak{J} = \{0\} \cup \mathbb{N}_h$ follows from (6.50) by means of (5.63). ∎

How does the sum on the right-hand side of (6.43) depend on the choice of the functions ψ_i? In view of (6.43) and (6.41) it only depends on $(\psi_0)_{0,\omega_{2k+N+1}}$. Is it always possible to choose ψ_i in such a way that

$$(6.43)' \qquad T = \sum_{i=0}^{k} (\psi_i (\partial_M - m)^i T)_{m,\omega_k}\ ?$$

Of course, if $T \neq 0$ then a necessary condition is that $\operatorname{supp} T$ is not contained in $X \setminus X_+$ for otherwise we have $\psi_i (\partial_M - m)^i T = 0$.

Proposition 6.48. *Suppose that $m^* \in \mathfrak{U}(M)$. Let $T \in \mathcal{D}'(X)$ satisfy one (and hence each) of the conditions of Theorem 6.37.(i). Then the equation (6.43)' is valid for some choice of the functions ψ_i if and only if for some (resp. every) $\chi \in C^\infty(X)$ satisfying the assumption and the condition (c) of Lemma 6.47.(ii) there exists $\tilde{\psi} \in C^\infty_M(X)$ such that*

$$(6.51) \qquad Q'_m(\chi T) = Q'_m(\tilde{\psi} T).$$

Proof. "$\Rightarrow$": By Lemma 6.47.(ii) we can choose χ_ψ in such a way that $\tilde{\psi} := \chi - \chi_\psi$ belongs to $C^\infty_M(X)$. If (6.43)' is valid then (6.43) shows that $Q'_m(\chi_\psi T) = 0$, i.e. (6.51) holds.

"$\Leftarrow$": By Lemma 6.47.(ii).(b) we can choose ψ_i in such a way that $\chi_\psi := \chi - \tilde{\psi}$ satisfies (6.41) for $\mathfrak{J} = \{0\} \cup \mathbb{N}_N$. Since (6.51) means that $Q'_m(\chi_\psi T) = 0$ the assertion follows by (6.43), again. ■

(e) Solving the Equation $(\partial_M - m)S = T$

Theorem 6.49. *Suppose that*

$$(6.52) \qquad \text{(a) } m \notin (-\mathfrak{U}(M) - \mu), \qquad \text{or} \qquad \text{(b) } \tilde{X} = X.$$

Then for every $T \in \mathcal{D}'(X)$ there exists a solution $S \in \mathcal{D}'(X)$ of the equation

$$(6.53) \qquad (\partial_M - m)S = T$$

having the following properties:

(i) let $r \in \mathbb{N}_0 \cup \{\infty\}$; if T is induced by a C^r function then so is $S|_{X_+}$;

(ii) suppose that $X = V$; if T is temperate so is S;

(iii) let $\mathfrak{G}$ satisfy the assumptions of Remark 2.67.(ii); if T is $\mathfrak{G}$-invariant so is S.

The main step in the proof is summarized in the following lemma. We fix a function $\chi \in C^\infty(X)$ such that

$$(6.54) \qquad \text{(a) } \operatorname{supp}\chi \subset \dot{x}^{-1}([0,a[), \qquad \text{(b) } \chi \equiv 1 \text{ on } \dot{x}^{-1}([0,\varepsilon])$$

for some $a > \varepsilon > 0$. Then the support of $1-\chi$ is contained in $\varkappa^{-1}([\varepsilon,+\infty[)$, and in view of Lemma 6.7 it makes sense to define $\tilde{T}_m$ by (4.24). In Chapter 7 we require the more general distributions

$$(6.55) \qquad \tilde{T}_{m,k} := (-1)^k \left(((1-\chi)T)_{m,\omega_{k,0}} - (\chi T)_{m,\omega_{k,\infty}} \right), \qquad k \in \mathbb{N}_0,$$

where $\omega_{k,0} := \omega_k \chi_{]0,1]}$ and $\omega_{k,\infty} := \omega_k \chi_{[1,+\infty[}$. Note that $\tilde{T}_{m,0} = \tilde{T}_m$.

Lemma 6.50. *For every $T \in \mathcal{D}'(X)$ by (6.55) a sequence of distribution $\tilde{T}_{m,k} \in \mathcal{D}'(X)$, $k \in \mathbb{N}_0$, is well-defined having the following properties:*

(i) $(\partial_M - m)\tilde{T}_{m,k+1} = \tilde{T}_{m,k}$;

(ii) $(\partial_M - m)^{k+1}\tilde{T}_{m,k} = (\partial_M - m)\tilde{T}_m = T - Q'_m(\chi T)$;

(iii) let $r \in \mathbb{N}_0 \cup \{\infty\}$; if $T/_{X_+}$ is induced by a C^r function so is $\tilde{T}_{m,k}/_{X_+}$;

(iv) if $q \in C^\infty(X)$ is quasihomogeneous of degree $\ell \in \mathbb{C}$ then $q\tilde{T}_{m,k} = (\widetilde{qT})_{m+\ell,k}$;

(v) suppose that $X = V$ and $\varkappa = \varkappa_+$; if T is temperate so is $\tilde{T}_{m,k}$.

Proof. *(i)*: Let $\varphi \in C_0^\infty(X)$. By (4.10) and Proposition 5.27 we have

$$((\partial_M - m^*)\varphi)_{m^*,\omega_{k+1,0}} = -\varphi_{m^*,\omega_{k,0}}$$

Since $\varphi_{m^*,\omega_{k,\infty}} = \varphi_{m^*,\omega_k} - \varphi_{m^*,\omega_{k,0}}$ on X_+, the preceding equation combined with (4.10) and (5.63) leads to

$$((\partial_M - m^*)\varphi)_{m^*,\omega_{k+1,\infty}} = -\varphi_{m^*,\omega_{k,\infty}}.$$

Consequently, since $\omega_{k,0}(1/t) = (-1)^k \omega_{k,\infty}(t)$, $t \in]0,+\infty[$, we obtain by (2.5) that

$$\langle (\partial_M - m)\tilde{T}_{m,k+1}, \varphi \rangle = -\langle \tilde{T}_{m,k+1}, (\partial_M - m^*)\varphi \rangle =$$

$$= -\langle (1-\chi)T, ((\partial_M - m^*)\varphi)_{m^*,\omega_{k+1,\infty}} \rangle + \langle \chi T, ((\partial_M - m^*)\varphi)_{m^*,\omega_{k+1,0}} \rangle$$

$$= \langle (1-\chi)T, \varphi_{m^*,\omega_{k,\infty}} \rangle - \langle \chi T, \varphi_{m^*,\omega_{k,0}} \rangle = \langle \tilde{T}_{m,k}, \varphi \rangle.$$

(ii): The first equality sign follows from (i) by induction. For the proof of the second equality we fix $\varphi \in C_0^\infty(X)$. By (4.10) and Proposition 5.27 we have

$$((\partial_M - m^*)\varphi)_{m^*,]0,1]} = \varphi - Q_{m^*}\varphi.$$

Since $\varphi_{m^*,[1,+\infty[} = \varphi_{m^*} - \varphi_{m^*,]0,1]}$ on X_+, the preceding equation combined with (4.10) and (5.63) leads to

$$((\partial_M - m^*)\varphi)_{m^*,[1,+\infty[} = -\varphi|_{X_+}.$$

Hence we obtain by (2.5) that

$$\langle(\partial_M - m)\tilde{T}_m, \varphi\rangle = -\langle\tilde{T}_m, (\partial_M - m^*)\varphi\rangle =$$

$$= -\langle(1-\chi)T, ((\partial_M - m^*)\varphi)_{m^*,[1,+\infty[}\rangle + \langle\chi T, ((\partial_M - m^*)\varphi)_{m^*,]0,1]}\rangle =$$

$$= \langle(1-\chi)T, \varphi\rangle + \langle\chi T, \varphi - Q_{m^*}\varphi\rangle = \langle T, \varphi\rangle - \langle\chi T, Q_{m^*}\varphi\rangle.$$

(iii): this follows from Lemma 6.17.

(iv): this is a consequence of Proposition 6.16.(ii).

(v): see Proposition 6.22. ■

Proof of Theorem 6.49. If $m \notin (-\mathfrak{A}(M) - \mu)$ then Lemma 6.50 and Corollary 6.28 show that $S := \tilde{T}_m$ is a solution satisfying (i) and (ii). Hence we suppose that $m \in (-\mathfrak{A}(M) - \mu)$, i.e. $m^* \in \mathfrak{A}(M)$, and $\tilde{X} = X$. Then by Lemma 6.31 we can choose $w \in \mathcal{D}'_M(X)$ such that

(6.56) $$Q'_m(\chi T) = Q'_m w.$$

By Proposition 6.35 w_m is almost quasihomogeneous of degree m such that $(\partial_M - m)w_m = Q'_m w$. In view of (6.56) Lemma 6.50 implies that $S := \tilde{T}_m + w_m$ is the desired solution of (6.53).

If T is induced by a C^r function so is w by Lemma 6.31, and by Lemma 6.17 this remains valid for w_m. Hence the property *(i)* follows by Lemma 6.50.(iii).

If $X = V$ and if T is temperate then by Lemma 6.31.(iii) w is temperate, as well, with M-temperate support. Consequently, by Proposition 6.22 w_m is temperate, and property *(ii)* results from Lemma 6.50.(v).

Finally, we observe that if S is a solution of (6.53) satisfying (i) and (ii) and if T is $\mathfrak{G}$-invariant then by Propositions 2.64 and 2.59 and by (2.36) and (2.39) $S_{\mathfrak{G}}$ is a $\mathfrak{G}$-invariant solution of (6.53) satisfying (i) and (ii), as well. ■

Corollary 6.51. *Suppose that (6.52) holds. Then for every $d \in \mathcal{D}'(X)$ which is almost quasihomogeneous of degree m there exists a distribution $T \in \mathcal{D}'(X)$ having the properties (i) - (iii) of Theorem 6.49 and being almost quasihomogeneous of degree m with deficiency d.* ■

Is the assumption (6.52) necessary? To examine this we let $S \in \mathfrak{D}'(X)$ be a solution of the equation (6.53), assuming that $m \in (-\mathfrak{A}(M) - \mu)$ and that T is almost quasihomogeneous of degree m with support contained in $X \setminus X_+$. Then S is almost quasihomogeneous of degree m. Hence by Theorem 6.37 we find $u \in \mathfrak{D}'_M(X)$ and a distribution $d \in \mathfrak{D}'(X)$ satisfying the condition (a) of Proposition 6.27.(i) such that $S = u_m + d$. Inserting this into (6.53) and taking (4.10) into account we obtain $(\partial_M - m)^k T = v_m$ where $k := \mathrm{ord}_M(d)$ and $v := (\partial_M - m)^{k+1} u \in \mathfrak{D}'_M(X)$. Since by Remark 6.14 v extends to a distribution $\tilde{v} \in \mathfrak{D}'_M(\tilde{X})$ it follows that $(\tilde{v})_m$ extends $(\partial_M - m)^k T$. Since $k \le N(m^*)$ and since $N(m^*) = 0$ in case $m = \mu$ or M is semi-simple we have proved the assertion (ii) of the following remark the first part of which is a direct consequence of Theorem 6.49.

Remark 6.52. *Let $d \in \mathfrak{D}'(X)$ be a distributions satisfying the condition (a) of Proposition 6.27.(i).*

(i) If d extends to a distribution on $\tilde{X}$ which is almost quasihomogeneous of degree m then the equation (6.53) has a solution $S \in \mathfrak{D}'(X)$ for $T = d$.

(ii) The converse of (i) is valid provided that $m = \mu$ or M is semi-simple. ■

Theorem 6.49 also leads to the following improvement of Theorem 6.45.

Theorem 6.53. *Suppose that $\ker M = \{0\}$. Let $T \in \mathfrak{D}'(V)$. If $(\partial_M - m)\, T$ is temperate then T is temperate itself. In particular, if $T|_{V \setminus \{0\}}$ is almost quasihomogeneous then T is temperate.*

Proof. By Theorem 6.49 one finds a temperate distribution $w \in \mathcal{S}'(V)$ such that $(\partial_M - m)\, w = (\partial_M - m)\, T$. By Proposition 2.11 this means that $T - w$ is quasihomogeneous of degree m, hence temperate by Theorem 6.45, and the first assertion follows.

If $T|_{V \setminus \{0\}}$ is almost quasihomogeneous of degree m then for some $N \in \mathbb{N}$ the support of $S := (\partial_M - m)^N T$ is contained in the set $\{0\}$, i.e. $S \in \mathcal{E}'(V) \subset \mathcal{S}'(V)$. Thus it follows from the first assertion by induction that T is temperate. ■

(f) Duality Brackets for the Spaces $Q_m(E_{\mathfrak{G}})$ and $\mathfrak{A}^{\infty}_{m,k}(E_{\mathfrak{G}})$

Let $\mathfrak{G}$ satisfy the assumptions of Remark 2.67.(ii). Let E be one of the spaces $\mathcal{D}(X)$ or $\mathcal{S}(V)$. The condition (b) in Proposition 6.27.(i) gives rise to a duality bracket for $Q_m(E_{\mathfrak{G}})$. We employ the following

Notation 6.54. (i) If F is a subset of $\mathcal{D}'(X)$ we write $F_{\mathfrak{G}} := \{T_{\mathfrak{G}} ; T \in F\}$.

(ii) By $Q'_m(E)$ we denote the set of all distributions of the form $Q'_m u$ where $u \in \mathcal{D}'_{[1,+\infty[}(X)$ and where in case $E = \mathcal{S}(V)$ u is temperate with $(M,[1,+\infty[)$-temperate support.

Remark 6.55. *$Q'_m(E)_{\mathfrak{G}}$ is equal to the space of all $\mathfrak{G}$-invariant distributions $d \in E'$ with support contained in $X \setminus X_+$ which are almost quasihomogeneous of degree m. It is a weakly closed subspace of E'.*

Proof. The first assertion follows by Propositions 6.24, 6.27.(i) and 2.64 and Remark 2.67.(ii). The second assertion is an immediate consequence of Proposition 2.31 and Definition 2.56. ∎

Proposition 6.56. *By $Q'_m(E)_{\mathfrak{G}} \times Q_{m^*}(E_{\mathfrak{G}'}) \longrightarrow \mathbb{C}$, $(d,R) \mapsto \langle d,R \rangle$, a duality braket is well-defined exhibiting $Q'_m(E)_{\mathfrak{G}}$ as the (topological) dual of the space $Q_{m^*}(E_{\mathfrak{G}'})$ if the latter is equipped with the locally convex topology described in Propositions 5.51 and 5.52 respectively.*

Proof. Let $d \in Q'_m(E)_{\mathfrak{G}}$ and $\varphi \in E$. Recall that the condition (b) of Proposition 6.27.(i) means that

$$(6.57) \qquad \langle d, Q_{m^*} \varphi_{\mathfrak{G}'} \rangle = \langle d, \varphi \rangle .$$

If the restriction of d to $Q_{m^*}(E_{\mathfrak{G}'})$ vanishes identically then (6.57) implies that $d = 0$. On the other hand, suppose that $\langle \cdot, Q_{m^*} \varphi_{\mathfrak{G}'} \rangle \equiv 0$. Since for arbitrary $y \in X \setminus X_+$ and $\alpha \in \mathfrak{A}^+_{m^*}$ the distribution $d = Q'_m((-\partial)^{\alpha} \delta_y)_{\mathfrak{G}}$ belongs to $Q'_m(E)_{\mathfrak{G}}$ (where δ_y denotes the Dirac distribution at y) it then follows by (6.57) and by Propositions 5.48.(v) and 5.45 that

$$0 = \langle (-\partial)^{\alpha}\delta_y , Q_{m*}\big((Q_{m*}\varphi_{\mathfrak{G}'})_{\mathfrak{G}'}\big)\rangle = \partial^{\alpha}(Q_{m*}\varphi_{\mathfrak{G}'})(y) .$$

Hence we conclude (in the case $E = \mathfrak{D}(X)$ using Remark 5.60 and (5.70)) that $Q_{m*}\varphi_{\mathfrak{G}'} \equiv 0$.

To prove the last assertion we recall from the proofs of Propositions 5.51 and 5.52, respectively, that the map $P: E \longrightarrow Q_{m*}(E_{\mathfrak{G}'})$, defined by $\varphi \mapsto Q_{m*}\varphi_{\mathfrak{G}'}$, is a surjective continuous linear homomorphism. Consequently, every $d \in Q'_m(E)_{\mathfrak{G}}$ induces a continuous linear functional on $Q_{m*}(E_{\mathfrak{G}'})$, and, conversely, every continuous linear functional ν on $Q_{m*}(E_{\mathfrak{G}'})$ is induced by the distribution $d := \nu \circ P$. ∎

Now we come to the spaces $\mathfrak{A}^{\infty}_{m,k}(E_{\mathfrak{G}})$.

Notation 6.57. We set $\mathfrak{A}'_{m,k}(E) := \{T \in E' ;\ (\partial_M - m)^{k+1}T \in Q'_m(E)\}$.

Remark 6.58. *(i) $\mathfrak{A}'_{m,k}(E)_{\mathfrak{G}}$ (see Notation 6.54.(i)) consists of all $\mathfrak{G}$-invariant distributions $T \in E'$ satisfying the condition (a) of Theorem 6.37.(i).*

(ii) If (6.52) holds then $\mathfrak{A}'_{m,k}(E)_{\mathfrak{G}}$ is equal to the space of all distributions of the form $(u_{\mathfrak{G}})_{m,\omega_k}$ where $u \in \mathfrak{D}'_M(X)$ and where in case $E = \mathcal{S}'(V)$ u is temperate with M-temperate support.

(iii) $\mathfrak{A}'_{m,k}(E)_{\mathfrak{G}}$ is a weakly closed subspace of E'.

<u>*Proof.*</u> <u>*(i)*</u>: This is an immediate consequence of Propositions 2.64 and 6.27 and of (2.39).

<u>*(ii)*</u>: One obtains this by combining Theorem 6.37 and Proposition 6.16.(iv).

<u>*(iii)*</u>: This is deduced from Definition 2.63, from the weak continuity of the operator $\partial_M - m$, and from the second assertion of Remark 6.55. ∎

The condition (b) of Theorem 6.37.(i) leads to a canonical duality bracket for the space $\mathfrak{A}'_{m,k}(E)_{\mathfrak{G}}$:

Proposition 6.59. *By*

(6.58) $\qquad \langle T, (\varphi_{\mathfrak{G}'})_{m^*,\omega_k}\rangle_M := \langle T, \varphi\rangle$

a duality bracket $\mathfrak{A}'_{m,k}(E)_{\mathfrak{G}} \times \mathfrak{A}^{\infty}_{m^*,k}(E_{\mathfrak{G}'}) \longrightarrow \mathbb{C}$ is well-defined exhibiting the *space $\mathfrak{A}'_{m,k}(E)_{\mathfrak{G}}$ as the topological dual of the space $\mathfrak{A}^{\infty}_{m^*,k}(E_{\mathfrak{G}'})$ if the latter is equipped with the locally convex topology described in Propositions 5.59 and 5.61, respectively.*

Proof. If $T \in E'$ is $\mathfrak{G}$-invariant then by Proposition 2.64.(iii) we have $\langle T, \varphi \rangle = \langle T, \varphi_{\mathfrak{G}'} \rangle$. Hence the condition (b) of Theorem 6.37.(i) (see also Theorem 6.37.(iii) in case $E = \mathcal{S}(V)$) shows that $\langle \cdot, \cdot\cdot \rangle_M$ is well-defined. If $\langle T, \cdot \rangle_M \equiv 0$ then $T = 0$ by (6.58). If, on the other hand, $\langle \cdot, (\varphi_{\mathfrak{G}'})_{m^*,\omega_k} \rangle_M \equiv 0$ then it follows by Theorem 5.37.(viii) that for every $y \in X_+$ we have

$$0 = (-1)^k \langle ((\delta_y)_{\mathfrak{G}})_{m,\omega_k}, (\varphi_{\mathfrak{G}'})_{m^*,\omega_k} \rangle_M = (-1)^k \langle ((\delta_y)_{\mathfrak{G}})_{m,\omega_k}, \varphi \rangle =$$
$$= \langle \delta_y, (\varphi_{\mathfrak{G}'})_{m^*,\omega_k} \rangle = (\varphi_{\mathfrak{G}'})_{m^*,\omega_k}(y) .$$

In view of Remark 5.60.(i) this implies that $(\varphi_{\mathfrak{G}'})_{m^*,\omega_k} \equiv 0$. Consequently, the bracket $\langle \cdot, \cdot\cdot \rangle_M$ is non-degenerate.

For the proof of the last part we recall from the proofs of Propositions 5.59 and 5.61, respectively, that the map $P : E \longrightarrow \mathfrak{A}^{\infty}_{m^*,k}(E_{\mathfrak{G}'})$ defined by $\varphi \mapsto (\varphi_{\mathfrak{G}'})_{m^*,\omega_k}$ is a surjective continuous linear homomorphism. Consequently, for every distribution $T \in \mathfrak{A}'_{m,k}(E)_{\mathfrak{G}}$ the map $\langle T, \cdot \rangle_M$ is a continuous linear functional on the space $\mathfrak{A}^{\infty}_{m^*,k}(E_{\mathfrak{G}'})$, and, conversely, every continuous linear functional ν on $\mathfrak{A}^{\infty}_{m^*,k}(E_{\mathfrak{G}'})$ defines a distribution $T := \nu \circ P \in E'$ which by Theorem 6.37.(i) and Propositions 2.64 and 2.59 belongs to $\mathfrak{A}'_{m,k}(E)_{\mathfrak{G}}$ and obviously satisfies $\langle T, \cdot \rangle_M = \nu$. ∎

Chapter VII

Solvability of Quasihomogeneous Multiplication Equations and Partial Differential Equations

As in the preceding chapters, we continue to assume that (1.14) holds and that $X \setminus X_+ \neq \emptyset$. In sections (a), (b), and (c) we are going to solve quasihomogeneous multiplication equations in the set of (almost) quasihomogeneous distributions on X. The main results are Theorems 7.3, 7.5, and 7.14 in section (a), Theorem 7.31 and Corollary 7.32 in section (b), and Theorem 7.35 and Corollary 7.36 in section (c).

If $X = V$ then via the Fourier transform the results can be restated as assertions on the solvability of partial differential equations with constant coefficients in the space of (almost) quasihomogeneous temperate distributions. This is done in section (d). Here the main results are Theorems 7.44, 7.45, and 7.51 and Supplements 7.53 and 7.54.

It turns out that for exceptional m one does not always have solvability. In fact, for these m solvability is characterized completely in terms of the function defining the operator, only. This is illustrated by way of several examples given in section (e).

Finally, in section (f), for the heat and for the Schrödinger operator we determine those fundamental solutions which are (almost) quasihomogeneous of type $(2,1,\ldots,1)$ and which are invariant under orthogonal transformations of the space variables.

For the whole chapter, in order to handle additional invariance properties we fix a compact subgroup G of GL(V) satisfying the assumptions of Remark 2.67, choose two continuous homomorphisms $\sigma, \tau : G \longrightarrow \dot{\mathbb{C}}$ as in section 2.(f), and set $\mathfrak{G} := (G, \sigma)$ and $\mathfrak{H} := (G, \tau)$.

(a) Quasihomogeneous Multiplication Operators

For this and the following two sections we fix a number $\ell \in \mathbb{C}$ and a C^∞ function $q : X \longrightarrow \mathbb{C}$ which is quasihomogeneous of degree ℓ and which is $(G,\tau/\sigma)$-invariant. The results we are going to derive are valid for the spaces $\mathcal{D}'(X)$ and $\mathcal{S}'(V)$. To unify the treatment we denote by E either $\mathcal{D}(X)$ or - if $X = V$ - $\mathcal{S}(V)$; in the latter case we assume that q belongs to the space $\mathcal{O}_{\mathfrak{M}}(V)$ of continuous multiplication operators on $\mathcal{S}'(V)$.

Our aim is for every $\mathfrak{H}$-invariant distribution $T \in E'$ which is (almost) quasihomogeneous of degree m to find a $\mathfrak{G}$-invariant distribution $S \in E'$ which is (almost) quasihomogeneous of degree $m - \ell$ and solves the equation

(7.1) $\qquad qS = T.$

If $k \in \mathbb{N}$ and if d denotes the kth order deficiency of T then Corollary 2.36.(ii) implies that the kth order deficiency c of S satisfies the equation

(7.2) $\qquad qc = d.$

Note that by Proposition 2.64.(iii) and Remark 2.67.(ii) d (resp. c) is $\mathfrak{H}$-invariant (resp. $\mathfrak{G}$-invariant). Thus we are lead to the aim of solving (7.1) in the following more precise way: given a $\mathfrak{G}$-invariant solution c of (7.2) which is almost quasi-homogeneous of degree $m - \ell$ we would like to find a solution S of (7.1) such that its kth order deficiency is equal to the prescribed distribution c.

The solvability results given below are based on the solvability of (7.1) in the whole space E'; more precisely, throughout sections (a), (b) and (c) we assume that

(7.3) $\qquad qE' = E'$;

(7.4) $\qquad$ the ideal generated by $q|_{X_+}$ in $C^\infty(X_+)$ is closed;

(7.5) $\qquad$ if $E = \mathcal{S}(V)$ then every $f \in C^\infty(V)$ such that $qf \in \mathcal{S}(V)$ belongs to $\mathcal{S}(V)$.

Note that (7.3) implies that

(7.6) $\qquad \operatorname{supp} q = X$, i.e. $X \setminus q^{-1}(0)$ is dense in X.

Since by Proposition 5.45 the restriction of q to the non-empty set X^o is a polynomial with respect to the variables in $G_M(\sigma_+)$, this implies, in particular, that ℓ belongs to $\mathfrak{A}(M)$.

Remark 7.1. *(i) If q is real analytic then the condition (7.3) for $E = \mathcal{D}(X)$ and the condition (7.4) are valid by the Lojasiewicz division theorem in [14] (see also Remarque VI.1.9 in Tougeron [17]).*

(ii) If q is a polynomial function and if $E = \mathcal{S}(V)$ then the condition (7.3) holds by Hörmander's division theorem in [8] while the condition (7.5) is an immediate consequence of the estimate (4.3) in Hörmander [8]. ■

We first deal with the equation (7.2) for almost quasihomogeneous distributions with support contained in $X \setminus X_+$.

Proposition 7.2. *Let $d \in E'$ be $\mathfrak{H}$-invariant and almost quasihomogeneous of degree m. If $\operatorname{supp} d \subset X \setminus X_+$ then the equation (7.2) has a $\mathfrak{G}$-invariant solution $c \in E'$ which is almost quasihomogeneous of degree $m - \ell$ such that $\operatorname{supp} c \subset X \setminus X_+$.*

In other words, multiplication by q defines a surjective operator from the space $Q'_{m-\ell}(E)_{\mathfrak{G}}$ onto $Q'_m(E)_{\mathfrak{H}}$ (see Notation 6.54).

<u>Proof</u>. By (7.3) we find $v \in E'$ such that $qv = d$. If $E = \mathcal{D}(X)$ then multiplying by a cut-off function we achieve that the support of v is contained in $\dot{x}^{-1}([0,1[)$ so that by Lemma 6.7 v belongs to $\mathcal{D}'_{[1,+\infty[}(X)$. Hence by Propositions 6.26.(ii) and 6.27.(i).(b) and by Remark 2.67.(ii) and Proposition 2.65.(ii) the distribution $c := Q'_{m-\ell}(v)_{\mathfrak{G}}$ is the desired solution of (7.2).

If $E = \mathcal{S}(V)$ then multiplying by a cut-off function chosen according to Lemma 5.53 we achieve that $\operatorname{supp} v \subset \Omega_{[0,1]}$. Then the solution c of (7.2) defined above is temperate by Proposition 6.24.(iii). ■

Next we come to the equation (7.1) for $T \in \mathfrak{A}'_{m,k}(E)_{\mathfrak{H}}$ (see Notations 6.57 and 6.54).

Theorem 7.3. *Let $T \in E'$ be $\mathfrak{H}$-invariant and almost quasihomogeneous of degree m. Then the equation (7.1) has a $\mathfrak{G}$-invariant solution $S \in E'$ which is almost quasihomo-*

geneous of degree $m-\ell$. More precisely, for every $k \in \mathbb{N}_0$ the multiplication by q induces a surjective operator from the space $\mathfrak{A}'_{m-\ell,k}(E)_{\mathfrak{G}}$ onto $\mathfrak{A}'_{m,k}(E)_{\mathfrak{H}}$.

Proof. Let $k \in \mathbb{N}_0$. By Theorem 6.37 we find $\mathfrak{H}$-invariant distributions $u \in \mathfrak{D}'_M(X) \cap E'$ and $d \in E'$ satisfying $T = u_{m,\omega_k} + d$ such that $\operatorname{supp} d \subset X \setminus X_+$ and such that in case $E = \mathcal{S}(V)$ $\operatorname{supp} u$ is M-temperate. An application of Proposition 7.2 reduces the proof to the case $d = 0$. By (7.3) we find $v \in E'$ such that $qv = u$. If $E = \mathfrak{D}(X)$ then multiplying by a suitable cut-off function we achieve that v belongs to $\mathfrak{D}'_M(X)$, as well. Hence, in view of Propositions 6.16.(ii) and 2.65.(ii) the distribution $S := (v_{m-\ell,\omega_k})_{\mathfrak{G}}$ is the desired solution. If $E = \mathcal{S}(V)$ then by employing a cut-off function given by Lemma 5.53 we achieve that $\operatorname{supp} v$ is M-temperate, as well. It then follows by Propositions 6.22 and 2.64.(ii) that the solution S defined above is temperate. ■

Note that in view of Corollary 6.28 in case $m^* \notin \mathfrak{A}(M)$ the space $Q'_m(E)$ is trivial. Consequently, in case $m^* + \ell \notin \mathfrak{A}(M)$, i.e. $m - \ell \notin (-\mathfrak{A}(M) - \mu)$, the assertion of Theorem 7.3 means that the following solvability condition holds for every $k \in \mathbb{N}_0$:

$(7.7)_k$ for every $\mathfrak{H}$-invariant $T \in E'$ which is almost quasihomogeneous of degree m and of order $\leq k$ the equation (7.1) has a ($\mathfrak{G}$-invariant) solution $S \in E'$ which is almost quasihomogeneous of degree $m-\ell$ and of order $\leq k$.

In the exceptional case "$m^* + \ell \in \mathfrak{A}(M)$", fixing an almost quasihomogeneous distribution T we give a first answer to the question under which conditions the equation (7.1) has a solution S which is almost quasihomogeneous of order not exceeding the order of T.

Proposition 7.4. *Suppose that $m \in (-\mathfrak{A}(M) - \mu + \ell)$ and that $X = \tilde{X}$. Let $k \in \mathbb{N}$ be such that $k > N := N(m^* + \ell)$ (see (5.65)), and let $T \in E'$ be $\mathfrak{H}$-invariant and almost quasihomogeneous of degree m and of order $\leq k$.*

(i) Then the following conditions are equivalent:

(a) the equation (7.1) has a ($\mathfrak{G}$-invariant) solution $S \in E'$ which is almost quasihomogeneous of degree $m-\ell$ and of order $\leq k$;

(b) for some (resp. every) $j \in \mathbb{N}_{k-N}$ the equation (7.2) with $d = (\partial_M - m)^j T$ has a ($\mathfrak{G}$-invariant) solution $c \in E'$ which is almost quasihomogeneous of degree $m-\ell$ and of order $\leq k-j$.

(ii) If c is a solution as in (b) then in (a) S can be chosen in such a way that $(\partial_M - m + \ell)^j S$ *and c coincide on* X_+.

Proof. "$\Rightarrow$": Corollary 2.36.(ii) shows that $c := (\partial_M - m + \ell)^j S$ is the desired solution of (7.2).

"$\Leftarrow$": By Corollary 6.51 we find a distribution $R \in E'$ such that $c = (\partial_M - m + \ell)^j R$. Since then, in particular, R is almost quasihomogeneous of degree $m - \ell$ and of order $\leq k$ it follows by Corollary 2.36.(ii), again, that qR is almost quasihomogeneous of degree m and of order $\leq k$ such that $(\partial_M - m)^j (qR) = qc = d$. Hence $T - qR$ is almost quasihomogeneous of degree m and of order $\leq j-1$. Therefore by Theorem 7.3 we find a solution $U \in E'$ of the equation $qU = T - qR$ which is almost quasihomogeneous of degree $m - \ell$ such that the support of $a := (\partial_M - m + \ell)^j U$ is contained in $X \setminus X_+$. Since by the definition of N we have $(\partial_M - m + \ell)^{N+j+1} U = 0$ the assumption on j implies that U is almost quasihomogeneous of degree $m - \ell$ and of order $\leq k$. Hence $S := R + U$ is the desired solution of (7.1) (note that $(\partial_M - m + \ell)^j S$ is equal to $c + a$ which may be different from c!). ■

For the exceptional case "$m^* + \ell \in \mathfrak{A}(M)$" the preceding results do not answer the question whether for every $T \in E'_{\mathfrak{H}}$ which is quasihomogeneous of degree m the equation (7.1) has a solution $S \in E'$ which is quasihomogeneous of degree $m - \ell$.

Theorem 7.5. *Suppose that*

(7.8) $\qquad$ (a) $m \notin (-\mathfrak{A}(M) - \mu + \ell)$, $\quad$ *or* $\quad$ (b) $X = \tilde{X}$.

Then the following conditions are equivalent:

(a) the solvability condition $(7.7)_k$ *is valid for some* $k \in \mathbb{N}_0$;

(a)' the condition $(7.7)_k$ *is valid for every* $k \in \mathbb{N}_0$; *in particular, for every* $\mathfrak{H}$*-invariant* $T \in E'$ *which is quasihomogeneous of degree m the equation (7.1) has a* ($\mathfrak{G}$*-invariant) solution* $S \in E'$ *which is quasihomogeneous of degree* $m - \ell$;

(b) for some (resp. every) $k \in \mathbb{N}_0$ *the following condition holds:*

$(7.9)_k$ $\qquad$ *for every pair* $(T, c) \in E' \times E'$ *such that T (resp. c) is* $\mathfrak{H}$*-invariant (resp.* $\mathfrak{G}$*-invariant) satisfying the equation* $qc = (\partial_M - m)^{k+1} T$ *there exists a (*$\mathfrak{G}$*-invariant) solution* $S \in E'$ *of the equation (7.1) satisfying* $(\partial_M - m + \ell)^{k+1} S = c$.

(c) for some (resp. every) $k \in \mathbb{N}_0$ the following condition holds:

$(7.10)_k$ *for every $\mathfrak{G}$-invariant distribution $c \in E'$ with support contained in $X \setminus X_+$ which solves the equation $qc = 0$ and is almost quasihomogeneous of degree $m - \ell$ there exists a $\mathfrak{G}$-invariant solution $S \in E'$ of the equation $qS = 0$ such that $(\partial_M - m + \ell)^{k+1} S = c$.*

The proof is easily reduced to Theorems 7.3 and 6.49. The reduction step is completely elementary and is formulated as

Lemma 7.6. *Let V_1, V_2, W_1, and W_2 be complex vector spaces, and for every $j \in \{1, 2\}$ let $A_j : V_j \longrightarrow W_j$ be a linear map. Moreover, let $B : V_1 \longrightarrow V_2$ and $C : W_1 \longrightarrow W_2$ be linear maps such that*

$$(7.11) \qquad C \circ A_1 = A_2 \circ B .$$

(i) We define linear maps $\beta : V_1 \longrightarrow V_2 \times W_1$ by $v_1 \mapsto (B(v_1), A_1(v_1))$ and $\gamma : V_2 \times W_1 \longrightarrow W_2$ by $(v_2, w_1) \mapsto C(w_1) - A_2(v_2)$. Then $\operatorname{im}\beta \subset \ker\gamma$, and by

$$\Phi : (\ker A_2)/B(\ker A_1) \longrightarrow \ker\gamma/\operatorname{im}\beta , \qquad v_2 + B(\ker A_1) \mapsto (v_2, 0) + \operatorname{im}\beta ,$$

a linear monomorphism is defined which is surjective if A_1 is surjective.

(ii) Let U_1 be a subspace of V_1 such that

$$(7.12) \qquad (a) \quad U_1 \supset \ker A_1 , \qquad \text{and} \qquad (b) \quad B(U_1) \supset \ker A_2 .$$

Then by

$$\Psi : (A_1(U_1) \cap \ker C)/A_1(U_1 \cap \ker B) \longrightarrow (\ker A_2)/B(\ker A_1),$$
$$A_1(u_1) + A_1(U_1 \cap \ker B) \mapsto B(u_1) + B(\ker A_1),$$

a linear isomorphism is well-defined.

Proof. (i): The condition (7.11) means that $\gamma \circ \beta \equiv 0$, i.e. $\operatorname{im}\beta \subset \ker\gamma$.

If $v_2 \in \ker A_2$ then $\gamma(v_2, 0) = -A_2(v_2) = 0$; and if, in addition, $v_2 = B(v_1)$ for some $v_1 \in \ker A_1$ then $(v_2, 0) = \beta(v_1) \in \operatorname{im}\beta$. Hence Φ is well-defined.

If $\Phi(v_2 + B(\ker A_1)) = \operatorname{im}\beta$, i.e. $(v_2, 0) = (B(v_1), A_1(v_1))$ for some $v_1 \in V_1$, then $v_2 \in B(\ker A_1)$. This shows that Φ is injective.

Let $(v_2, w_1) \in \ker\gamma$, i.e. $C(w_1) = A_2(v_2)$. If A_1 is surjective we find $v_1 \in V_1$ such that $A_1(v_1) = w_1$. It follows by (7.11) that $A_2(v_2) = C(A_1(v_1)) = A_2(B(v_1))$, i.e.

$v_2 - B(v_1) \in \ker A_2$. Since $(B(v_1), w_1) = \beta(v_1) \in \operatorname{im}\beta$ we conclude that the image of $v_2 - B(v_1) + B(\ker A_1)$ under Φ is equal to $(v_2, w_1) + \operatorname{im}\beta$. Hence Φ is surjective provided A_1 is so.

(ii): If $u_1 \in \mathfrak{U}_1$ is such that $A_1(u_1) \in \ker C$ then (7.11) implies that $B(u_1) \in \ker A_2$; if, in addition, $A_1(u_1) = A_1(u_1')$ for some $u_1' \in \mathfrak{U}_1 \cap \ker B$ then $u_1 - u_1' \in \ker A_1$ and $B(u_1) = B(u_1 - u_1') \in B(\ker A_1)$, i.e. Ψ is well-defined.

If $B(u_1) \in B(\ker A_1)$ then $B(u_1) = B(u_1')$ for some $u_1' \in \ker A_1$, i.e. $u_1 - u_1' \in \ker B$, and $A_1(u_1) = A_1(u_1 - u_1') \in A_1(\ker B)$. Since by (7.12.a) $u_1 - u_1'$ belongs to $\mathfrak{U}_1$ this shows that Ψ is injective.

If $v_2 \in \ker A_2$ then by (7.12.b) we find $u_1 \in \mathfrak{U}_1$ such that $B(u_1) = v_2$. By (7.11) it follows that $C(A_1(u_1)) = A_2(B(u_1)) = 0$, i.e. $A_1(u_1) \in A_1(\mathfrak{U}_1) \cap \ker C$. Consequently, Ψ is surjective. ■

Proof of Theorem 7.5. Let $k \in \mathbb{N}_0$. We are going to apply Lemma 7.6 to $V_1 = W_1 = E'_{\mathfrak{G}}$, $V_2 = W_2 = E'_{\mathfrak{H}}$, $A_1 := (\partial_M - m + \ell)^{k+1}$, $A_2 := (\partial_M - m)^{k+1}$, $\mathfrak{U}_1 := \mathfrak{A}'_{m-\ell,k}(E)_{\mathfrak{G}}$, and $B = C :=$ multiplication by q. Then A_1 (as well as A_2) is surjective by Theorem 6.49; the condition (7.11) follows from Lemma 2.34; (7.12.a) is valid by Proposition 2.31.(i); and (7.12.b) holds by Theorem 7.3 and Proposition 2.31.(i). Note that by Corollary 6.51 $A_1(\mathfrak{U}_1)$ is equal to $Q'_{m-\ell}(E)_{\mathfrak{G}}$.

Now, in view of Proposition 2.31.(i) the condition $(7.7)_k$ can be rephrased as "$\ker A_2 = B(\ker A_1)$". By Lemma 7.6.(i) this is equivalent to the condition "$\ker\gamma = \operatorname{im}\beta$" which amounts to $(7.9)_k$.

Moreover, by Lemma 7.6.(ii) the condition $(7.7)_k$ is equivalent to the condition "$A_1(\mathfrak{U}_1) \cap \ker C = A_1(\mathfrak{U}_1 \cap \ker B)$" which reads as $(7.10)_k$.

Since by Corollary 2.36.(ii) we have $q(\partial_M - m + \ell)^k S = (\partial_M - m)^k (qS)$ we see that $(7.10)_k$ implies $(7.10)_0$. Hence, in order to complete the proof we have to verify the converse implication. Indeed, suppose that $(7.10)_0$ is valid, and let c be given having the properties listed in $(7.10)_k$. By $(7.10)_0$ we find $S_0 \in E'$ such that $qS_0 = 0$ and $(\partial_M - m + \ell)S_0 = c$. Applying $(7.9)_0$ to $(0, S_{j-1})$ instead of (T, c) we recursively construct a sequence of distributions $S_j \in E'$ such that $qS_j = 0$ and $(\partial_M - m + \ell) S_j = S_{j-1}$, $j \in \mathbb{N}$. It follows that $S := S_k$ has the properties required in $(7.10)_k$. ■

From now on for the rest of the present section and for sections (b) and (c) we suppose that (7.8) holds. Actually, most of the following results are trivial in case (7.8.a) is valid but a few are not. In order to examine the validity of the conditions of Theorem 7.5 we fix $k \in \mathbb{N}_0$ and rephrase the condition $(7.10)_k$ in the form

$$(7.10)'_k \qquad \ker C_m = A_{m-\ell}(\ker B_m)$$

employing the following

Notation 7.7. (i) By $A_m : \mathfrak{A}'_{m,k}(E)_{\mathfrak{G}} \longrightarrow Q'_m(E)_{\mathfrak{G}}$ we denote the map defined by $T \mapsto (\partial_M - m)^{k+1} T$;

(ii) by $B_m : \mathfrak{A}'_{m-\ell,k}(E)_{\mathfrak{G}} \longrightarrow \mathfrak{A}'_{m,k}(E)_{\mathfrak{H}}$ we denote the map defined by $T \mapsto qT$;

(iii) by $C_m : Q'_{m-\ell}(E)_{\mathfrak{G}} \longrightarrow Q'_m(E)_{\mathfrak{H}}$ we denote the map defined by $c \mapsto qc$.

We observe that C_m and hence the left-hand side of $(7.10)'_k$ do not depend on k. In order to see how the right-hand side of $(7.10)'_k$ depends on k we write $(A_{m,k}, B_{m,k})$ instead of (A_m, B_m) in the following

Remark 7.8. *(i)* $W_k := A_{m-\ell,k}(\ker B_{m,k}) \supset W_{k+1}$ *for every* $k \in \mathbb{N}_0$.

(ii) If $k_0 \in \mathbb{N}_0$ *is such that* $W_{k_0+1} = W_{k_0}$ *then* $W_k = W_{k_0}$ *for every* $k \geq k_0$.

Moreover, for $N := N(\ker C_m) := \max\{\operatorname{ord}_M(c);\ c \in Q'_{m-\ell}(E)_{\mathfrak{G}},\ qc = 0\}$ *the following assertions hold:*

(iii) $N \leq N(m^* + \ell)$;

(iv) $\mathfrak{C}_{i-1} \supsetneq \mathfrak{C}_i := (\partial_M - m + \ell)^i (\ker C_m)$ *for every* $i \in \mathbb{N}_{N+1}$;

(v) $W_k = W_N + \mathfrak{C}_{k+1}$ *for every* $k \in \mathbb{N}_0$*; in particular,* $W_k = W_N$ *for every* $k \geq N$.

Proof. *(iii):* In view of $\ker C_m \subset Q'_{m-\ell}(E)$ this is a consequence of Corollary 6.28.

(iv): If $c \in \ker C_m$ then by Corollary 2.36.(ii) we have $q(\partial_M - m + \ell)c = (\partial_M - m)(qc) = 0$ so that $\mathfrak{C}_i \subset \mathfrak{C}_{i-1}$ for every $i \in \mathbb{N}$. On the other hand, we fix $i \in \mathbb{N}_{N+1}$ and choose $c \in \ker C_m$ such that $(\partial_M - m + \ell)^{i-1} c \in \mathfrak{C}_i$, i.e. $(\partial_M - m + \ell)^{i-1} c = (\partial_M - m + \ell)^i b$ for some $b \in \ker C_m$. Applying $(\partial_M - m + \ell)^{N+1-i}$ to both sides of this equation we deduce that $(\partial_M - m + \ell)^N c = 0$. Hence, $\operatorname{ord}_M(c) \leq N-1$. In other words, if $c \in \ker C_m$ is such that $\operatorname{ord}_M(c) = N$ then $(\partial_M - m + \ell)^{i-1} c$ does not belong to $\mathfrak{C}_i$. Since by the definition of N there exists $c \in \ker C_m$ such that $\operatorname{ord}_M(c) = N$ the proof of (iv) is complete.

For the proof of the other assertions we fix $k \in \mathbb{N}_0$ and $R \in W_k$ and choose $S \in \ker B_{m,k}$ such that $R = (\partial_M - m + \ell)^{k+1} S$. Note that $S \in \mathfrak{A}'_{m-\ell,k}(E)_{\mathfrak{S}}$ and $qS = 0$.

(i): Since $T := (\partial_M - m + \ell) S$ belongs to $\mathfrak{A}'_{m-\ell,k-1}(E)_{\mathfrak{S}}$ and - by Corollary 2.36.(ii) - satisfies $qT = (\partial_M - m)(qS) = 0$ we conclude that $R = (\partial_M - m + \ell)^k T$ belongs to W_{k-1}, indeed.

For the next steps we fix, in addition, $\varkappa$ and χ as in § 6.(e) supposing that in case $E = \mathcal{S}(V)$ $\varkappa$ be equal to $\varkappa_+$. Moreover, we choose $\nu \in \{0, \max\{N-k, 0\}\}$. By Proposition 2.65.(ii) and Lemma 6.50.(iv) and by (2.39), Lemma 6.50.(ii) and Proposition 2.64.(iii) the distribution $T := (\tilde{S}_{m-\ell,\nu})_{\mathfrak{S}}$ satisfies the equations

$$(7.13) \qquad \text{(a)} \quad qT = ((\widetilde{qS})_{m,\nu})_{\mathfrak{H}} = 0, \qquad \text{and} \qquad \text{(b)} \quad (\partial_M - m + \ell)^{\nu+1} T = S - c,$$

where $c := Q'_{m-\ell}(\chi S)_{\mathfrak{S}}$. It follows from (7.13.b) and Lemma 6.50.(v) that T lies in $\mathfrak{A}'_{m-\ell,k+\nu+1}(E)_{\mathfrak{S}}$. Moreover, we note that by the assertions (v) and (ii) of Proposition 6.26 we have $qc = Q'_m(\chi q S)_{\mathfrak{H}} = 0$.

(v): Now, suppose that $k \geq N$. Then $\nu = 0$ and $(\partial_M - m + \ell)^{k+1} c = 0$ so that by (7.13.b)

$$(\partial_M - m + \ell)^{k+2} T = (\partial_M - m + \ell)^{k+1} S = R.$$

In view of (7.13.a) this shows that R belongs to W_{k+1}. Hence in view of (i) we have proved that $W_{k+1} = W_k$ for every $k \geq N$. From this it follows by induction that $W_k = W_N$ for every $k \geq N$.

On the other hand, in case $k < N$ we deduce from (7.13.b) for $\nu = N-k$ and from (7.13.a) that

$$R = (\partial_M - m + \ell)^{k+1} S = (\partial_M - m + \ell)^{N+1} T + (\partial_M - m + \ell)^{k+1} c \in W_N + \mathfrak{C}_{k+1}.$$

Hence, in this case we have proved that $W_k \subset W_N + \mathfrak{C}_{k+1}$. Since in view of the inclusions "$\ker C_m \subset \ker B_m$" and (i) the inverse inclusion is obvious the proof of (v) is complete.

(ii): We suppose that $k \in \mathbb{N}_N$ and that $W_k = W_{k-1}$. It suffices to show that $W_{k+1} = W_k$, i.e. we have to show that the distribution $R \in W_k$ fixed above belongs to W_{k+1}. Now, since in case $E = \mathcal{S}(V)$ by Proposition 6.24.(iii) and by the choice of χ the distribution *c* defined above belongs to $\mathcal{S}'(V)$, Proposition 6.26.(iii) shows that $c \in \mathfrak{A}'_{m+\ell,k-1}(E)_{\mathfrak{S}}$. Consequently, $(\partial_M - m + \ell)^k c$ belongs to W_{k-1} and hence, by assumption, to W_k, i.e. we find $U \in \mathfrak{A}'_{m-\ell,k}(E)_{\mathfrak{S}}$ such that $(\partial_M - m + \ell)^k c = (\partial_M - m + \ell)^{k+1} U$ and $qU = 0$. It follows from (7.13.b) for $\nu = 0$ that

$$R = (\partial_M - m + \ell)^{k+1} S = (\partial_M - m + \ell)^{k+2} T + (\partial_M - m + \ell)^{k+1} c = (\partial_M - m + \ell)^{k+2}(T + U).$$

In view of (7.13.a) this shows that $R \in W_{k+1}$, as desired. ■

Let us note a few consequences of Remark 7.8. First of all, Remark 7.8 leads to an alternative proof of the fact that the conditions $(7.10)'_k$ are equivalent for all $k \in \mathbb{N}_0$. Moreover, if in a concrete situation one sets out to verify the validity of $(7.10)'_k$ one should work with k as small as possible. On the other hand, if the aim is to show that $(7.10)'_k$ is false it is most promising to deal with the case $k \geq N(m^* + \ell)$. However, I cannot produce examples having the property that the constant $N(\ker C_m)$ is positive or that $W_k \neq W_{k-1}$ for some (resp. every) $k \in \mathbb{N}_N$.

Next we are going to see that the conditions of Theorem 7.5 are not always valid. The first step is to see that the condition $(7.10)'_k$ becomes simpler if (6.28) holds:

Remark 7.9. *Suppose that $q^{-1}(0) \cap X_+ = \emptyset$. Then*

(i) M is semi-simple provided that $\ell \neq 0$;

(ii) $\ker B_m = \ker C_m$;

(iii) $A_{m-\ell}(\ker B_m)$ is trivial.

Proof. *(i)*: Let Z be any M-irreducible (and hence, in particular, M-cyclic) subspace Z of $G_M(\sigma_+)$. By the assumption "$X^o \neq \emptyset$" one finds $x_o \in \ker M$ such that $G_M(\sigma_+) + x_o$ is contained in X (see also (6.22)). Then the polynomial function $q_Z : Z \longrightarrow \mathbb{C}$, $z \mapsto q(z + x_o)$, is quasihomogeneous of degree ℓ and of type M_Z. If $\dim' Z > 1$ then Proposition 1.44 implies that q_Z has a non-trivial zero. Since this contradicts (6.28) it follows that $\dim' Z = 1$. This means that M is semi-simple.

(ii). If $\ell = 0$ then q has no zeros at all so that $\ker B_m = \{0\}$. If $\ell \neq 0$ then by (i) we have $N(m^* + \ell) = 0$. And since the assumption on q implies that $\operatorname{supp} S \subset X \setminus X_+$ for every $S \in \ker B_m$ the assertion follows from Proposition 6.27.

(iii). In view of Corollary 6.28 this follows from (i) and (ii). ■

As a first information about the left-hand side of $(7.10)'_k$ we note

Remark 7.10. *Suppose that $G = \{Id_V\}$. Then $\dim \ker C_m \geq |\mathfrak{A}^+_{m^*+\ell}| - |\mathfrak{A}^+_{m^*}|$. If $\ker M = \{0\}$ then equality holds.*

Proof. We set $A := \mathfrak{A}^{+}_{m^*+\ell}$ and fix $y \in X \setminus X_+$. From the Leibniz rule and from the fact that $q^{(\gamma)}(y) = 0$ if $\gamma \in \mathfrak{A}^+ \setminus \mathfrak{A}^+_\ell$ it follows for every family $c = (c_\alpha)_{\alpha \in A}$ of complex numbers c_α that the distribution

$$S_c := \sum_{\alpha \in A} c_\alpha (-\partial)^\alpha \delta_y$$

satisfies the equation $q S_c = 0$ if and only if c is a solution of the system of linear equations

$$\sum_{\alpha \in A, \alpha \geq \beta} c_\alpha \binom{\alpha}{\beta} q^{(\alpha-\beta)}(y) = 0, \qquad \beta \in B := \mathfrak{A}^+_{m^*}.$$

Since the dimension of the solution space of this system is not smaller than $|A| - |B|$ the first assertion follows.

If $\ker M = \{0\}$ then the spaces $Q_j'(\mathcal{S}(V))$ are finite-dimensional so that Proposition 7.2 implies that $\dim \ker C_m = d_{m-\ell} - d_m$ where for $j \in \{m-\ell, m\}$ $d_j := \dim Q_j'(\mathcal{S}(V))$ is equal to $|\mathfrak{A}^+_{j^*}|$. ■

Proposition 7.11. *Suppose that $q^{-1}(0) \cap X_+ = \emptyset$ and $G = \{Id_V\}$. Then each of the conditions of Theorem 7.5 is violated if and only if*

(7.14) (a) $m^* + \ell \in \mathfrak{A}(M)$; (b) $\ell \neq 0$; and (c) $m^* \notin \mathfrak{A}(M)$ or $\dim' G_M(\sigma_+) \geq 2$.

Proof. In view of what was said in the text following the proof of Theorem 7.3 we may assume that (7.14.a) holds. In view of Remark 7.9.(iii) the condition $(7.10)_k'$ is valid if and only if $\ker C_m = \{0\}$. If $\ell = 0$ then the assumption (6.28) implies that $q^{-1}(0) = \emptyset$ so that C_m is injective, indeed. Hence we suppose that (7.14.b) is valid, as well.

If (7.14.c) holds it suffices to show that $|\mathfrak{A}^+_{m^*+\ell}| > |\mathfrak{A}^+_{m^*}|$ for then Remark 7.10 yields that $\ker C_m \neq \{0\}$. Indeed, if $m^* \notin \mathfrak{A}(M)$ then the desired estimate is trivial in view of (7.14.a). And if $\dim' G_M(\sigma_+) \geq 2$ then (6.28) and (7.14.b) imply that $|\mathfrak{A}^+_\ell| \geq 2$, i.e. we can fix $\beta, \gamma \in \mathfrak{A}^+_\ell$ such that $\beta \neq \gamma$ so that in case m^* belongs to $\mathfrak{A}(M)$ the injective maps $\mathfrak{A}^+_{m^*} \longrightarrow \mathfrak{A}^+_{m^*+\ell}$ defined by $\alpha \mapsto \alpha + \beta$ and $\alpha \mapsto \alpha + \gamma$, respectively, have distinct images, and the desired estimate follows again.

Finally, we suppose that (7.14.c) is false. Then by Remark 1.40 $|\mathfrak{A}^+_{m^*+\ell}| = 1 = |\mathfrak{A}^+_{m^*}|$ so that in case $\dim' V = 1$ Remark 7.10 directly implies that $\ker C_m = \{0\}$. This leaves us with the case $\dim' G_M(\sigma_+) = 1 < \dim' V$. Writing $V = \mathbb{K} \times V_0$ where $\mathbb{K} \times \{0\} = G_M(\sigma_+)$ and $\{0\} \times V_0 = \ker M$ we see that X^0 is of the form $X^0 = \mathbb{K} \times X'$ where X' is a non-empty open subset of V_O. By Proposition 5.45 we find $g \in C^\infty(X')$ such that $q(x) = x_1^\nu g(x')$, $x = (x_1, x') \in X^O$, where ν is the unique element of $\mathfrak{A}_\ell^+$. Now let $d \in Q'_{m-\ell}(\mathfrak{D}(X))$. Then by Propositions 6.27.(i) and 6.29 there exists a distribution $u \in \mathfrak{D}'(X')$ such that $d|_{X^O} = [(-\partial)^{\rho+\nu}\delta] \otimes u$ where δ denotes the Dirac distribution on $\mathbb{K}$ and where ρ is the unique element of $\mathfrak{A}_m^+$. If $d \in \ker C_m$ then

$$(x_1^\nu(-\partial)^{\rho+\nu}\delta) \otimes (gu) = 0.$$

Since by the Leibniz rule we have

$$x_1^\nu(-\partial)^{\rho+\nu}\delta = \frac{(\rho+\nu)!}{\rho!}(-\partial)^\rho\delta \neq 0$$

it follows that $gu = 0$. Since by (6.28) g has no zeros this implies that $u = 0$, i.e. $d|_{X^O} = 0$. Since by Proposition 6.24.(ii) the support of d is contained in X^O it follows that $d = 0$. Hence $\ker C_m = \{0\}$. ∎

In order to analyze the condition $(7.10)'_k$ further we are now going to employ the duality brackets of section 6.(f) and make use of the locally convex properties of $\mathfrak{A}^\infty_{m,k}(E_{\mathfrak{G}})$. In particular, we are going to describe the polar set of $A_{m-\ell}(\ker B_m)$ and of $\ker C_m$. To this end we first compute the transposes of the operators A_m, B_m, and C_m (introduced in Notation 7.7) with respect to the duality brackets defined in section 6.(f).

A simple technical remark is in order: if $\varphi \in C_0^\infty(X)$ then by Remark 5.60 we have $\operatorname{supp} Q_m\varphi \subset X$ and $\operatorname{supp} \varphi_{m,\omega_k} \subset X_+$; although, actually, the domain of definition of $Q_m\varphi$ (resp. φ_{m,ω_k}) is the whole of V (resp. V_+) we may consider it as a function on X (resp. X_+); in other words, we shall not always distinguish between $Q_m\varphi$ (resp. φ_{m,ω_k}) and $(Q_m\varphi)|_X$ (resp. $\varphi_{m,\omega_k}|_{X_+}$). Moreover, if q is extended to V by $q|_{V\setminus X} :\equiv 0$ then the equations involving q appearing below can also be read as equations for functions on V (resp. V_+). This has to be taken into account, in particular, when Proposition 5.48.(ii) or Theorem 5.37.(vi) are applied.

Remark 7.12. *(i)* ${}^tA_m : Q_{m*}(E_{\mathfrak{G}'}) \longrightarrow \mathfrak{A}^\infty_{m*,k}(E_{\mathfrak{G}'})$ *is well-defined and given by* $R \mapsto R/_{X_+}$;

(ii) ${}^tB_m : \mathfrak{A}^\infty_{m*,k}(E_{\mathfrak{H}'}) \longrightarrow \mathfrak{A}^\infty_{m*+\ell,k}(E_{\mathfrak{G}'})$ *is well-defined and given by* $f \mapsto q/_{X_+}f$;

(iii) ${}^tC_m : Q_{m*}(E_{\mathfrak{H}'}) \longrightarrow Q_{m*+\ell}(E_{\mathfrak{G}'})$ *is well-defined and given by* $R \mapsto qR$.

Proof. *(i):* We let $R \in Q_{m^*}(E_{\mathfrak{G}'})$ and fix $\varphi \in E$ such that $R = Q_{m^*}(\varphi_{\mathfrak{G}'})$. Then by (5.63), (4.10), and (2.39) we have

$$(7.15) \qquad R|_{X_+} = (-1)^{k+1}(((\partial_M - m^*)^{k+1}\varphi)_{\mathfrak{G}'})_{m^*,\omega_k}.$$

In particular, we see that $R|_{X_+}$ belongs to $\mathfrak{A}^{\infty}_{m^*,k}(E_{\mathfrak{G}'})$. By (6.57), by (2.5), by (6.58), and by (7.15) it follows for every $T \in \mathfrak{A}'_{m,k}(E)_{\mathfrak{G}}$ that

$$\langle A_m(T), R\rangle = \langle (\partial_M - m)^{k+1} T, Q_{m^*}(\varphi_{\mathfrak{G}'})\rangle =$$

$$= \langle (\partial_M - m)^{k+1} T, \varphi\rangle = (-1)^{k+1}\langle T, (\partial_M - m^*)^{k+1}\varphi\rangle =$$

$$= (-1)^{k+1}\langle T, (((\partial_M - m^*)^{k+1}\varphi)_{\mathfrak{G}'})_{m^*,\omega_k}\rangle_M = \langle T, R|_{X_+}\rangle_M .$$

(ii): For arbitrary $T \in \mathfrak{A}'_{m-\ell,k}(E)_{\mathfrak{G}}$ and $\varphi \in E$ we deduce by (6.58) and by Propositions 2.61.(ii) and Theorem 5.37.(vi) that

$$\langle B_m(T), (\varphi_{\mathfrak{H}'})_{m^*,\omega_k}\rangle_M = \langle qT, \varphi\rangle = \langle T, q\varphi\rangle =$$

$$= \langle T, ((q\varphi)_{\mathfrak{G}'})_{m^*+\ell,\omega_k}\rangle_M = \langle T, q|_{X_+}(\varphi_{\mathfrak{H}'})_{m^*,\omega_k}\rangle_M .$$

(iii): For arbitrary $d \in Q'_{m-\ell}(E)_{\mathfrak{G}}$ and $\varphi \in E_{\mathfrak{H}'}$ we conclude by (6.57) and by Proposition 5.48.(ii) that

$$\langle C_m(d), Q_{m^*}(\varphi)\rangle = \langle qd, \varphi\rangle = \langle d, q\varphi\rangle =$$

$$= \langle d, Q_{m^*+\ell}(q\varphi)\rangle = \langle d, q\,Q_{m^*}(\varphi)\rangle . \quad \blacksquare$$

The relevant topological properties of the operators introduced in Notation 7.7 are stated in

Proposition 7.13. *The operators A_m, B_m, and C_m are (surjective) weak homomorphisms, the images of their transposes being closed.*

Proof. We denote by $u : U \longrightarrow W$ any one of the operators ${}^tA_m$, ${}^tB_m$, or ${}^tC_m$. Since X_+ is dense in X and in view of (7.6) u is injective. By Theorems 6.49 and 7.3 and Proposition 7.2, respectively, its transpose ${}^tu$ is surjective.

The case "$E = \mathcal{S}(V)$": Here by Propositions 5.52 and 5.61 the spaces U and W are Fréchet spaces; and the assertion follows by Proposition VI.2.2 in de Wilde [4].

The case "$E = \mathcal{D}(X)$": Here by Propositions 5.51 and 5.59 U is a complete nuclear Schwartz space, and W is a countable strict inductive limit of Fréchet spaces and hence barrelled. By combining Propositions IV.3.3 and VI.3.8 in de Wilde [4] we deduce that u is relatively open. Consequently, since $\mathcal{U}$ is complete so is im u. In particular, im u is closed in W. The bipolar theorem then implies that $\operatorname{im} u = (\ker {}^t u)^o$. Applying Proposition VI.1.4 in de Wilde [4] to ${}^t u$ we conclude that ${}^t u$ is a weak homomorphism. ∎

As a first consequence of Proposition 7.13 we note that

(7.16) $\quad$ (a) $\ker D_m = (\operatorname{im} {}^t D_m)^o$, and (b) $(\ker D_m)^o = \operatorname{im} {}^t D_m$, $\quad D \in \{A, B, C\}$.

Indeed, the equation (a) is obvious, and the equation (b) follows from (a) by the bipolar theorem since by Proposition 7.13 $\operatorname{im} {}^t D_m$ is closed in the range space of D. In particular, since $(A_{m-\ell}(\ker B_m))^o = {}^t A_{m-\ell}^{-1}((\ker B_m)^o)$ the condition (7.16.b) for D = B leads to

(7.17) $\quad (A_{m-\ell}(\ker B_m))^o = {}^t A_{m-\ell}^{-1}(\operatorname{im} {}^t B_m)$.

From these equations we obtain the following dual description of $(7.10)_k$.

Theorem 7.14. *The conditions of Theorem 7.5 are valid if and only if for some (resp. every) $k \in \mathbb{N}_0$ the following conditions hold:*

(7.18) $\quad$ every $f \in \mathfrak{A}^{\infty}_{m^*,k}(E_{\mathfrak{H}'})$ such that $q|_{X_+} f$ extends to an element of $Q_{m^*+\ell}(E_{\mathfrak{G}'})$ extends to a function in $Q_{m^*}(E_{\mathfrak{H}'})$,

and

(7.19) $\quad A_{m-\ell}(\ker B_m)$ is weakly closed in $Q'_{m-\ell}(E)_{\mathfrak{G}}$.

<u>*Proof*</u>. Since $\ker C_m$ is weakly closed in $Q'_{m-\ell}(E)_{\mathfrak{G}}$ it follows by the bipolar theorem that $(7.10)'_k$ is valid if and only if (7.19) holds and

(7.18)" $\quad (A_{m-\ell}(\ker B_m))^o = (\ker C_m)^o$.

By (7.17) and (7.16.b) the last condition amounts to

(7.18)' $\quad {}^t A_{m-\ell}^{-1}(\operatorname{im} {}^t B_m) = \operatorname{im} {}^t C_m$.

Since by (7.6) the map $C^{\infty}(X_+) \longrightarrow C^{\infty}(X_+)$, $f \mapsto q|_{X_+} f$, is injective it follows by Remark 7.12 that (7.18) is equivalent to (7.18)'. ∎

We are now going to illustrate the usefulness of condition (7.18), beginning with a rather simple example.

Remark 7.15. *If $m^* = -\ell$ then the condition (7.18) is valid provided that $\ker M = \{0\}$ and $q^{-1}(0) \cap \dot{V} \neq \emptyset$.*

Proof. The assumptions on m^* and M imply that $Q_{m^*+\ell}(E)$ consists of all constant functions. Consequently, if $q|_{X_+} f$ extends to an element of $Q_{m^*+\ell}(E)$ then by the assumption on q it vanishes identically so that by (7.6) f does so, as well. ■

For dealing with another special case where the condition (7.18) can easily be checked we have to show that the condition (7.18) can be formulated in an apparently slightly weaker form:

Remark 7.16. *Let $R \in Q_{m^*+\ell}(E_{\mathfrak{H}'})$ and $h \in C^{\infty}(X_+)$ such that $R/_Y = q/_Y h$ where $Y := (X^o)_+$. If h extends to a C^{∞} function on X^o then, in fact, it extends to a function $P \in Q_{m^*}(E_{\mathfrak{H}'})$ satisfying $R = qP$.*

Proof. It suffices to find $P \in Q_{m^*}(E_{\mathfrak{H}'})$ such that $R = qP$; for it then follows by continuity from (7.6) that the functions h and P coincide on X (in case $E = \mathcal{D}(X)$ one has to take into account that the support of R and hence that of P and h are contained in X^o).

If $E = \mathcal{D}(X)$ then by Corollary 5.47 we choose a G-invariant compact subset K of $X \cap M_0(X)$ such that $\operatorname{supp} R \subset M_0^{-1}(K)$. Moreover, we fix a G-invariant function $\chi \in C_0^{\infty}(X)$ equal to 1 on a neighbourhood U of K. Then by the assumption on h it is obvious that χh extends to a function in $E_{\mathfrak{H}'}$ - again denoted by χh -, and by Propositions 5.48.(ii) and 5.45 we conclude that

$$q\, Q_{m^*}(\chi h) = Q_{m^*+\ell}(q \chi h) = Q_{m^*+\ell}(\chi R) = Q_{m^*+\ell}(R) = R .$$

If $X = V$ (hence $X^o = V$) and $E = \mathcal{S}(V)$ then by Lemma 5.53 we choose a G-invariant C^{∞} function $\chi : V \longrightarrow \mathbb{C}$ equal to 1 near $V \setminus V_+$ with support contained in $\Omega_{[0,r]}$ for some $r > 0$ such that all its derivatives are bounded. Since χR and hence $q(\chi h)$ belong to $\mathcal{S}(V)$ it follows by (7.5) that χh belongs to $\mathcal{S}(V)$, as well. As above we conclude that $q\, Q_{m^*}(\chi h) = R$. Since in view of the assertions (ii) and (v) of

Proposition 5.48 we may assume that h is $\mathfrak{H}'$-invariant we see that $P := Q_{m^*}(\chi h)$ is the desired function. ∎

The proof of the following remark shows that in case q is sufficiently simple the condition (7.18) can sometimes be checked directly for arbitrary m.

Remark 7.17. *(i) If there exists $v \in G_M(\sigma_+) \setminus \{0\}$ such that the directional derivative $\partial_v q$ of q vanishes identically on X^o then the condition (7.18) of Theorem 7.14 is always satisfied.*

(ii) If $\mathrm{Re}\,\ell < \lambda_{max} := \max\{\mathrm{Re}\,\lambda;\ \lambda \in \sigma_+\}$ then the hypothesis of assertion (i) is automatically satisfied.

Proof. (i): By choosing a suitable basis we identify V with $\mathbb{R} \times W$ in such a way that $\mathbb{R} \times \{0\}$ is identified with $\mathbb{R}v$ and W with $C \oplus \ker M$ where C is a complement of $\mathbb{R}v$ in $G_M(\sigma_+)$. We set $X' := \{x' \in W;\ (0, x') \in X^o\}$. Then a look at (6.22) shows that $\mathbb{R} \times X' = X^o$.

Now, let $R \in Q_{m^*+\ell}(E_{\mathfrak{G}'})$ be such that $R|_{X_+} = q|_{X_+} h$ for some $h \in C^\infty(X_+)$. By the definition of $Q_{m^*+\ell}\,\varphi$ we find $d \in \mathbb{N}$ such that for every $y \in \ker M$ the function $G_M(\sigma_+) \longrightarrow \mathbb{C}$, $z \mapsto R(z+y)$, is a polynomial function of degree $\le d$. In particular, fixing $x' \in X'$ we see that $R(\cdot, x')$ is a polynomial function of degree not larger than d. Now, by the assumption on q we have $q(\cdot, x') \equiv q(0, x')$. Since in case $q(0,x') \ne 0$ the function $h(\cdot, x')$ coincides with $R(\cdot, x')/q(0,x')$ on $]0,+\infty[$ we deduce that $h(\cdot, x')|_{]0,+\infty[}$ is a polynomial function of degree not larger than d and hence equal to $P(\cdot, x')|_{]0,+\infty[}$ where $P: X^o \longrightarrow \mathbb{C}$ is the C^∞ function defined by

$$P(x_1, x') := \sum_{j=0}^{d} \frac{1}{j!} (\partial_1^j h)(1, x')\,(x_1 - 1)^j, \qquad x_1 \in \mathbb{R},\ x' \in X'.$$

It follows that the polynomial functions $R(\cdot, x')$ and $(qP)(\cdot, x')$ coincide on $]0,+\infty[$ and hence on the whole of $\mathbb{R}$. Since this is trivially valid if $q(\cdot, x') \equiv 0$ we have proved that $R = qP$ on X^o.

If $X = V$ then $X^o = V$, and the proof is complete. If $X \ne V$ then $E = \mathfrak{D}(X)$, and by Lemma 5.50.(ii) we see that, in fact, R belongs to $Q_{m^*+\ell}(\mathfrak{D}(X^o)_{\mathfrak{G}'})$. Applying Remark 7.16 to $E = \mathfrak{D}(X^o)$ we deduce that P extends to a function H belonging to $Q_{m^*}(\mathfrak{D}(X^o)_{\mathfrak{H}'}) \subset Q_{m^*}(E_{\mathfrak{H}'})$. Since the supports of R and H are contained in X^o it follows that $R = qH$ on X.

(ii): We choose $\alpha \in \mathfrak{A}^+$ such that $|\alpha| = 1$ and $\mathrm{Re}\,\alpha_M > \mathrm{Re}\,\ell$ and observe that $\partial^\alpha q$ is almost quasihomogeneous of degree $\ell - \alpha_M$. Since $\mathrm{Re}(\ell - \alpha_M) < 0$ Lemma 1.60 implies that $\partial^\alpha q$ vanishes on X^o. ■

Now we are going to have a look at the condition (7.19) in Theorem 7.14. If $q^{-1}(0) \cap X_+ = \emptyset$ then by Remark 7.9 the condition (7.19) is satisfied for trivial reasons. It is also valid if $\ker M = \{0\}$ for then $Q'_{m-\ell}(E)$ and hence all of its subspaces are finite-dimensional and hence closed. Whether (7.19) is valid for general q and M I do not know. There is, however, an elementary reformulation of it for which we make use of

Notation 7.18. We set $\mathfrak{A}\mathfrak{Q}'_{m,k}(E) := \mathfrak{A}\mathfrak{Q}'_{m,k}(X) \cap E'$ (see Notation 4.28); i.e. the space $\mathfrak{A}\mathfrak{Q}'_{m,k}(E)$ consists of all distributions $T \in E'$ which are almost quasihomogeneous of degree m and of order $\leq k$.

Lemma 7.19. *The condition (7.19) holds if and only if the space* $q\,\mathfrak{A}\mathfrak{Q}'_{m-\ell,k}(E)_{\mathfrak{G}}$ *is weakly closed in* $\mathfrak{A}'_{m,k}(E)_{\mathfrak{H}}$.

Proof. In view of Proposition 7.13 this is a special case of Lemma 7.20 below. ■

Lemma 7.20. *Let F, G, and H be locally convex vector spaces, let* $A : F \longrightarrow G$ *be a surjective continuous linear homomorphism, and let* $B : F \longrightarrow H$ *be a continuous linear map. If* $B(\ker A)$ *is closed in H then* $A(\ker B)$ *is closed in G.*

Proof. Let $g \in G$ be in the closure of $A(\ker B)$. Since A is surjective we find $f \in F$ such that $A(f) = g$. It suffices to show that $B(f)$ is in the closure of $B(\ker A)$. For then by the assumption we find $f' \in \ker A$ such that $B(f) = B(f')$ so that $f - f' \in \ker B$ and $g = A(f) = A(f - f') \in A(\ker B)$.

So let W be an open neighbourhood of $B(f)$ in H. Since B is continuous the set $U := B^{-1}(W)$ is an open neighbourhood of f in F. Since A is open the set $A(U)$ is an open neighbourhood of g in G, hence by the assumption on g contains an element of the form $A(f'')$ where $f'' \in \ker B$. Choosing $f''' \in U$ such that $A(f''') = A(f'')$ we conclude that $f' := f''' - f'' \in \ker A$ and $B(f') = B(f''') \in B(U) \subset W$. ■

(b) Reformulating (7.18); the Test Space $\mathfrak{Z}_{m,k}(q)$

In this section we are now going to analyze the condition (7.18) further. To this end we introduce a canonical subspace of $\ker B_m$ consisting of almost quasihomogeneous distributions whose $(1+k)^{th}$ order deficiencies are easily computed.

Notation 7.21. (i) If $f \in C^\infty(X)$ we define $\mathfrak{B}(f)$ as the set of all pairs (y,Q) consisting of a point $y \in X$ and a polynomial function $Q: V^* \longrightarrow \mathbb{C}$ satisfying

$$(7.20) \qquad f\,Q(-\partial)\,\delta_y = 0$$

where δ_y denotes the Dirac distribution at y. Since by the Leibniz rule in the form of (1.54) we have

$$\langle f\,Q(-\partial)\,\delta_y, \varphi\rangle = \sum_{\alpha \in \mathfrak{A}} \frac{1}{\alpha!}\left(Q^{(\alpha)}(\partial)\,f\right)(y)\,\varphi^{(\alpha)}(y)\,, \qquad \varphi \in C_0^\infty(X),$$

this amounts to

$$(7.20)' \qquad \left(Q^{(\alpha)}(\partial)\,f\right)(y) = 0\,, \qquad \alpha \in \mathfrak{A}.$$

(ii) By $\mathfrak{Z}_{m,k}(q)$ we denote the subspace of $\mathfrak{A}'_{m,k}(E)$ generated by the distributions $(Q(-\partial)\,\delta_y)_{m,\omega_k}$, $(y,Q) \in \mathfrak{B}(q|_{X_+})$.

Remark 7.22. *(i)* $\mathfrak{Z}_{m-\ell,k}(q)_{\mathfrak{S}} \subset \ker B_m$.

(ii) The space $(\partial_M - m)^{k+1}\mathfrak{Z}_{m,k}(q)$, *consisting of the* $(1+k)^{th}$ *order deficiencies of the elements of* $\mathfrak{Z}_{m,k}(q)$, *is generated by the distributions of the form (6.32) where* $(y,Q) \in \mathfrak{B}(q|_{X_+})$; *it does not depend on the choice of* k.

Proof. *(i)*: this follows from (7.20) since by Propositions 6.16.(ii) and 2.65.(ii) we have $q\left((Q(-\partial)\,\delta_y)_{\mathfrak{S}}\right)_{m-\ell,\omega_k} = \left((q\,Q(-\partial)\,\delta_y)_{\mathfrak{S}}\right)_{m,\omega_k}$.

(ii): see (6.30) and Example 6.36. ∎

That the $(1+k)^{th}$ order deficiencies of the distributions in $\mathfrak{Z}_{m-\ell,k}(q)_{\mathfrak{S}}$ play a decisive role when one wants to decide whether or not the condition $(7.10)_k$ holds becomes clear by

Lemma 7.23. *If $k \geq N(m^*+\ell)$ then the weak closures of the subspaces $A_{m-\ell}(\ker B_m)$ and $A_{m-\ell}(\mathfrak{Z}_{m-\ell,k}(q)_{\mathfrak{S}})$ in the space E' are equal.*

For the proof of this lemma, employing the duality brackets of section 6.(f), again, by the bipolar theorem we have to show that the polar sets of both subspaces are equal. To prepare this we are going to compute the polar set of $\mathfrak{Z}_{m,k}(q)_{\mathfrak{S}}$ first. This requires the following reformulation of a special case of Whitney's spectral theorem on the divisibility of C^∞ functions.

Lemma 7.24. *Let Ω be an open subset of V, let $f \in C^\infty(\Omega)$ be such that the ideal $\mathfrak{J}$ generated by f in $C^\infty(\Omega)$ is closed, and let $g \in C^\infty(\Omega)$. Then the equation $g = fh$ has a solution $h \in C^\infty(\Omega)$ if and only if $\mathfrak{V}(g) \supset \mathfrak{V}(f)$.*

Proof. "$\Rightarrow$": this is an immediate consequence of the Leibniz formula.
"$\Leftarrow$": Choosing real coordinates we may assume that $V = \mathbb{R}^n$. Moreover, as the definition of $\mathfrak{V}(f)$ does not depend on M, we may assume that $M = \mathrm{Id}_V$ so that $\mathfrak{A} = \mathbb{N}_0^n$. We note first that the formulation of Whitney's spectral theorem in Tougeron [17] (see Corollaire V.1.6 there) is valid for complex valued functions, as well: one reduces the proof to the real valued case by introducing the $C^\infty(\Omega;\mathbb{R})$ submodule of $C^\infty(\Omega;\mathbb{R})^2$ generated by $(\mathrm{Re}\, f, \mathrm{Im}\, f)$ and $(-\mathrm{Im}\, f, \mathrm{Re}\, f)$. Consequently, since $\mathfrak{J}$ is closed in $C^\infty(\Omega)$, in order to verify that g belongs to $\mathfrak{J}$ one has to check that for every $x \in X$ the *Taylor series* $T_x g$ of g at x belongs to the set $T_x\mathfrak{J} := \{T_x h;\ h \in \mathfrak{J}\}$. If $f(x) \neq 0$ then this is the case for trivial reasons. Hence we may suppose that $f(x) = 0$.

By Borel's theorem (see Remarque IV.3.5 in [17]) it suffices to find a formal power series $d = \sum_{\alpha\in\mathfrak{A}} d_\alpha Z^\alpha$ such that $T_x g = d\, T_x f$. This equation can be written as the infinite system of linear equations

$$L_\alpha(d) := \sum_{\gamma\leq\alpha} f_{\alpha-\gamma} d_\gamma = \frac{1}{\alpha!} g^{(\alpha)}(x) =: g_\alpha\,, \quad \alpha\in\mathfrak{A}, \qquad \text{where} \quad f_\alpha := \frac{1}{\alpha!} f^{(\alpha)}(x).$$

By Lemma 6.3.7 in Hörmander [9] there exists a solution $(d_\alpha)_{\alpha\in\mathfrak{A}}$ if for every family $(c_\alpha)_{\alpha\in\mathfrak{A}}$ such that c_α is non-zero for finitely many α, only, the condition

$$\text{(7.21.A)} \qquad \sum_{\alpha\in\mathfrak{A}} c_\alpha L_\alpha \equiv 0$$

implies that

$$\text{(7.21.B)} \qquad \sum_{\alpha\in\mathfrak{A}} c_\alpha g_\alpha = 0\,.$$

By the definition of L_α the condition (7.21.A) amounts to

$$\text{(7.21.A)'} \qquad \sum_{\beta\in\mathfrak{A}} c_{\beta+\gamma} f_\beta = 0 , \qquad \gamma\in\mathfrak{A} .$$

Now, if we set

$$Q(Z) := \sum_{\alpha\in\mathfrak{A}} \frac{1}{\alpha!} c_\alpha Z^\alpha$$

then a simple computation shows that for arbitrary $h\in C^\infty(\Omega)$ and $\gamma\in\mathfrak{A}$ we have

$$\text{(7.22)} \qquad (Q^{(\gamma)}(\partial) h)(x) = \sum_{\beta\in\mathfrak{A}} c_{\beta+\gamma} h_\beta .$$

Hence the condition (7.21.A)' means that $(x,Q)\in\mathfrak{B}(f)$. By the assumption on g it follows that $(Q(\partial)g)(x) = 0$. This is identical with (7.21.B) as a look at (7.22) - applied to $(h,\gamma) = (g,0)$ - shows. ■

Now we come to the description of the polar set of $\mathfrak{Z}_{m-\ell,k}(q)_{\mathfrak{G}}$.

Lemma 7.25. *The polar set* $(\mathfrak{Z}_{m-\ell,k}(q)_{\mathfrak{G}})^o$ *consists of all* $g\in\mathfrak{A}^\infty_{m^*+\ell,k}(E_{\mathfrak{G}'})$ *such that for some* $h\in C^\infty(X_+)$ *we have*

$$\text{(7.23)} \qquad g = q|_{X_+} h .$$

Proof. We fix a $\mathfrak{G}'$-invariant function $\varphi\in E$ such that $g = \varphi_{m^*+\ell,\omega_k}$. That g belongs to $(\mathfrak{Z}_{m-\ell,k}(q)_{\mathfrak{G}})^o$ means that for every $(y,Q)\in\mathfrak{B}(q|_{X_+})$ we have

$$\begin{aligned} 0 &= (-1)^k \langle ((Q(-\partial)\delta_y)_{\mathfrak{G}})_{m-\ell,\omega_k} , g\rangle_M = \\ &= (-1)^k \langle ((Q(-\partial)\delta_y)_{\mathfrak{G}})_{m-\ell,\omega_k} , \varphi\rangle = \\ &= \langle Q(-\partial)\delta_y , (\varphi_{m^*+\ell,\omega_k})_{\mathfrak{G}'}\rangle = (Q(\partial)g)(y) . \end{aligned}$$

In view of (7.4) and Lemma 7.24 this is equivalent to the existence of a function $h\in C^\infty(X_+)$ satisfying (7.23). ■

Note that by (7.6) any function h satisfying (7.23) is uniquely determined by that property. What other properties does it have? In particular, when does it belong to $\mathfrak{A}^\infty_{m^*,k}(E_{\mathfrak{H}'})$? To find an answer, making use of Lemma 2.34 we deduce from (7.23) that

$$\text{(7.23)'} \qquad (\partial_M - m^* - \ell)^i g = q|_{X_+} (\partial_M - m^*)^i h , \qquad i\in\mathbb{N} .$$

Since g is almost quasihomogeneous of degree $m^*+\ell$ and since $X_+\setminus q^{-1}(0)$ is dense in X_+ this implies first of all that h is almost quasihomogeneous of degree m^* and of order $\leq \mathrm{ord}_M(g)$. Moreover, since the $(1+k)^{th}$ order deficiencies of the elements of $\mathfrak{A}^{\infty}_{m^*,k}(E_{\mathfrak{H}'})$ are restrictions of functions from $Q_{m^*}(E_{\mathfrak{H}'})$ we deduce from (7.23)' that a necessary condition for h to belong to $\mathfrak{A}^{\infty}_{m^*,k}(E_{\mathfrak{H}'})$ is that $(\partial_M-m^*)^{k+1}h$ extends to a C^∞ function P on X, belonging to $Q_{m^*}(E_{\mathfrak{H}'})$. Denoting by $R\in Q_{m^*+\ell}(E_{\mathfrak{G}'})$ the extension of $(\partial_M-m^*-\ell)^{k+1}g$ to X we can rephrase this by stating that the equation $R=qP$ be valid. This condition is also sufficient:

Lemma 7.26. *Let $g\in\mathfrak{A}^{\infty}_{m^*+\ell,k}(E_{\mathfrak{G}'})$. Then the equation (7.23) is valid for some $h\in\mathfrak{A}^{\infty}_{m^*,k}(E_{\mathfrak{H}'})$ if and only if g belongs to $(\mathfrak{Z}_{m-\ell,k}(q)_{\mathfrak{G}})^{o}$ and*

(7.24) $$(\partial_M-m^*-\ell)^{k+1}g=(qP)|_{X_+} \text{ for some } P\in Q_{m^*}(E_{\mathfrak{H}'}).$$

Note that the functions described in Lemma 7.26 constitute the image of ${}^tB_m$.

<u>*Proof.*</u> <u>"$\Rightarrow$"</u>: see the text preceding the statement of the lemma.

<u>"$\Leftarrow$"</u>: First of all we fix a $\mathfrak{G}'$-invariant function $\varphi\in E$ such that $g=\varphi_{m^*+\ell,\omega_k}$. Moreover, we choose $\phi\in E_{\mathfrak{H}'}$ such that (7.24) is valid for $P=(-1)^{k+1}Q_{m^*}(\phi)$. Since X_+ is dense in X it follows from (5.63) that

(7.24)' $$Q_{m^*+\ell}(\varphi)=q\,Q_{m^*}(\phi).$$

By Lemma 7.25 we find $h\in C^\infty(X_+)$ such that (7.23) holds. Since $X_+\setminus q^{-1}(0)$ is dense in X_+ we observe that $\mathrm{supp}\,h\subset\mathrm{supp}\,g$. We fix functions $\psi_0,\ldots,\psi_k\in C^\infty_M(X)$ satisfying (4.18) for $\mathfrak{J}=\mathbb{N}_0$ and (4.19). If $E=\mathcal{S}(V)$ then in view of Corollary 6.39 these functions can be chosen to have the properties (i) - (iii) of Lemma 6.38. We claim that for every $i\in\{0\}\cup\mathbb{N}_k$ the functions

(7.25) $$\psi_i\,(\partial_M-m^*)^i h \text{ and } \psi_i\,(\partial_M-m^*)^i\phi_{m^*,\omega_k} \text{ belong to } E.$$

Indeed, we first suppose that $E=\mathfrak{D}(X)$. Then the set $L:=\mathrm{supp}\,\varphi$ is compact, and by Remark 6.8 we can choose a compact subset K of X_+ such that $\overline{L_M}\cap X_+=K_M$. Since by Theorem 5.37.(ii) the support of g and hence that of h and each of its derivatives is contained in this set and since by Proposition 3.22 $K_M\cap\mathrm{supp}\,\psi_i$ is a compact subset of X_+ we conclude that the support of $\psi_i\,(\partial_M-m^*)^i h$, is so, as well. That the same is valid for the support of $\psi_i\,(\partial_M-m^*)^i\phi_{m^*,\omega_k}$ follows similarly.

Supposing now that $E = \mathscr{S}(V)$ we deduce from Proposition 5.41 and Remark 5.40 that $\psi_i (\partial_M - m^* - \ell)^i g$ and $\psi_i (\partial_M - m^*)^i \phi_{m^*,\omega_k}$ belong to $\mathscr{S}(V)$. In view of (7.23)' we deduce from (7.5) that $\psi_i (\partial_M - m^*)^i h$ belongs to $\mathscr{S}(V)$, as well. Hence the proof of (7.25) is complete.

Now, applying Proposition 4.13 to $(g, m^* + \ell)$ instead of (T,m) and taking (4.2)' and (5.63) into account we deduce

$$g = \sum_{i=0}^{k} (\psi_i (\partial_M - m^* - \ell)^i g)_{m^* + \ell, \omega_k} + \sum_{j=0}^{\infty} \binom{k+j}{k} (\psi_0)_{0,\omega_{2k+1+j}} (\partial_M - m^* - \ell)^j (Q_{m^* + \ell} \varphi)\big|_{X_+}. \tag{7.26}$$

In view of (7.23)' and (7.25) it follows by (3.9) that

$$(\psi_i (\partial_M - m^* - \ell)^i g)_{m^* + \ell, \omega_k} = q|_{X_+} (\psi_i (\partial_M - m^*)^i h)_{m^*, \omega_k}.$$

Since by Lemma 2.34, again, we derive from (7.24)' that

$$(\partial_M - m^* - \ell)^j Q_{m^* + \ell} \varphi = q\, (\partial_M - m^*)^j Q_{m^*} \phi$$

we conclude that $g = q|_{X_+} f$ where

$$f := \sum_{i=0}^{k} (\psi_i (\partial_M - m^*)^i h)_{m^*, \omega_k} + \sum_{j=0}^{\infty} \binom{k+j}{k} (\psi_0)_{0,\omega_{2k+1+j}} (\partial_M - m^*)^j Q_{m^*} \phi\big|_{X_+}.$$

Similarly as above, applying Proposition 4.13 to $(\phi_{m^*,\omega_k}, m^*)$ instead of (T,m) and taking (4.2)' and (5.63) into account we see that the second sum is equal to

$$\phi_{m^*,\omega_k} - \sum_{i=0}^{k} (\psi_i (\partial_M - m^*)^i \phi_{m^*,\omega_k})_{m^*,\omega_k}.$$

Consequently, $f = \chi_{m^*,\omega_k}$ where

$$\chi := \phi + \sum_{i=0}^{k} \psi_i (\partial_M - m^*)^i (h - \phi_{m^*,\omega_k}).$$

Note that by (7.25) χ belongs to E. Since by Propositions 3.7.(ii), 2.61.(ii), and 2.59.(ii), the equation $g = q|_{X_+} f$ remains valid if χ is replaced by $\chi_{\mathfrak{H}}$, the proof is complete. ∎

Note that by Remark 7.16 the condition (7.24) in Lemma 7.26 is slightly weaker than it looks.

Making use of Lemma 7.26 we now give a description of the polar sets of the subspaces $A_{m-\ell}(\mathfrak{Z}_{m-\ell,k}(q)_{\mathfrak{G}})$ and $A_{m-\ell}(\ker B_m)$. To this end we introduce

Notation 7.27. (i) $\mathfrak{R}_m(q;E_{\mathfrak{G}'}) := \{R \in Q_m(E_{\mathfrak{G}'})\,;\ R|_{X_+} = q|_{X_+}h$ for some $h \in C^\infty(X_+)\}$;

(ii) $\mathfrak{R}_{m,i}(q;E_{\mathfrak{G}'}) := \{R \in \mathfrak{R}_m(q;E_{\mathfrak{G}'})\,;\ (\partial_M - m)^i R = qP$ for some $P \in Q_{m-\ell}(E_{\mathfrak{H}'})\}$ for every $i \in \mathbb{N}_0$.

We also write $\mathfrak{R}_m$ (resp. $\mathfrak{R}_{m,i}$) instead of $\mathfrak{R}_m(q;E_{\mathfrak{G}'})$ (resp. $\mathfrak{R}_{m,i}(q;E_{\mathfrak{G}'})$).

In view of the conditions (v) and (vi) the conditions (i)-(iii) of the following Lemma correspond to the conditions (i), (ii), and (v) of Remark 7.8.

Lemma 7.28. *(i)* $\mathfrak{R}_{m,i} \subset \mathfrak{R}_{m,i+1}$ *for every* $i \in \mathbb{N}_0$.

(ii) If $i_o \in \mathbb{N}_0$ *is such that* $\mathfrak{R}_{m,i_o} = \mathfrak{R}_{m,i_o+1}$ *then* $\mathfrak{R}_{m,i} = \mathfrak{R}_{m,i_o}$ *for every* $i \geq i_o$.

(iii) $\mathfrak{R}_{m,i} = \mathfrak{R}_m$ *for every* $i \geq 1+N(m)$.

(iv) $(A_{m-\ell}(\mathfrak{Z}_{m-\ell,k}(q)_{\mathfrak{G}}))^o = \mathfrak{R}_{m^*+\ell}(q;E_{\mathfrak{G}'})$.

(v) $(A_{m-\ell}(\ker B_m))^o = \mathfrak{R}_{m^*+\ell,k+1}(q;E_{\mathfrak{G}'})$.

(vi) $(\ker C_m)^o = \mathfrak{R}_{m^*+\ell,0}(q;E_{\mathfrak{G}'})$.

(vii) More generally, for every $i \in \mathbb{N}$ *we have*

$$(\mathcal{C}_i)^o = \{R \in Q_{m^*+\ell}(E_{\mathfrak{G}'})\,;\ (\partial_M - m^* - \ell)^i R = qP \text{ for some } P \in Q_{m^*}(E_{\mathfrak{H}'})\}$$

where $\mathcal{C}_i$ *denotes the subspace of* $Q'_{m-\ell}(E)_{\mathfrak{G}}$ *introduced in Remark 7.8.(iv).*

Proof. *(i)*: Let $R \in \mathfrak{R}_{m,i}$ and choose $P \in Q_{m-\ell}(E_{\mathfrak{H}'})$ such that $(\partial_M - m)^i R = qP$. It then follows by Proposition 2.36.(ii) that $(\partial_M - m)^{i+1} R = q(\partial_M - m + \ell)P$. Since by Proposition 5.48.(iii) - applied to $\partial_M - m + \ell$ instead of $P_0(x,\partial)$ - and by (2.39) the function $(\partial_M - m + \ell)P$ belongs to $Q_{m-\ell}(E_{\mathfrak{H}'})$ we conclude that $R \in \mathfrak{R}_{m,i+1}$.

(ii): Let $i \in \mathbb{N}$ be such that $\mathfrak{R}_{m,i} = \mathfrak{R}_{m,i-1}$. It suffices to verify the eqality $\mathfrak{R}_{m,i+1} = \mathfrak{R}_{m,i}$. Hence we fix $R \in \mathfrak{R}_m$ such that $(\partial_M - m)^{i+1} R = qP$ for some $P \in Q_{m-\ell}(E_{\mathfrak{H}'})$. Since by Lemma 2.34 $(\partial_M - m)R$ belongs to $\mathfrak{R}_m$, as well, this means that $(\partial_M - m)R$ belongs to $\mathfrak{R}_{m,i}$ and hence, by the assumption, to $\mathfrak{R}_{m,i-1}$. This means that $(\partial_M - m)^{i-1}(\partial_M - m)R = q\tilde{P}$ for some $\tilde{P} \in Q_{m-\ell}(E_{\mathfrak{H}'})$, i.e. $R \in \mathfrak{R}_{m,i}$, as desired.

(iii): this is clear since $(\partial_M - m)^{i+1} R = 0$ for arbitrary $R \in Q_m(E_{\mathfrak{G}'})$ and $i \geq N(m)$.

(iv): In view of $(A_{m-\ell}(\mathfrak{Z}_{m-\ell,k}(q)_{\mathfrak{G}}))^o = {}^tA^{-1}_{m-\ell}\big((\mathfrak{Z}_{m-\ell,k}(q)_{\mathfrak{G}})^o\big)$ this is a consequence of Remark 7.12.(i) and Lemma 7.25.

(v): in view of (7.17) and Remark 7.12.(i) the assertion is a consequence of Lemma 7.26 and of the fact that X_+ is dense in X.

(vi): Since in view of Remark 7.12.(iii) the subspace $\operatorname{im}{}^t C_m$ equals $\mathfrak{R}_{m^*+\ell,0}(q;E_{\mathfrak{S}'})$ the assertion is a consequence of (7.16.b).

(vii): Let $R \in Q_{m^*+\ell}(E_{\mathfrak{S}'})$. By (2.5) R belongs to the polar set $(\mathfrak{C}_i)^o$ if and only if $(\partial_M - m^* - \ell)^i R \in (\ker C_m)^o$. Hence the assertion follows from (vi). ■

Proof of Lemma 7.23. By combining the assertions (iv), (iii), and (v) of Lemma 7.28 one obtains

(7.27) $$(A_{m-\ell}(\mathfrak{Z}_{m-\ell,k}(q)_{\mathfrak{S}}))^o = (A_{m-\ell}(\ker B_m))^o \qquad \text{if } k \geq N(m^*+\ell).$$

Hence the assertion follows by the bipolar theorem. ■

The following proposition is the desired description of the condition (7.18).

Proposition 7.29. *Each of the following conditions is equivalent to (7.18):*

(a) the weak closure of $A_{m-\ell}(\ker B_m)$ in $Q'_{m-\ell}(E)_{\mathfrak{S}}$ is equal to $\ker C_m$;

(b) the weak closure of $A_{m-\ell}(\mathfrak{Z}_{m-\ell,k}(q)_{\mathfrak{S}})$ in $Q'_{m-\ell}(E)_{\mathfrak{S}}$ equals $\ker C_m$;

(c) the weak closure of $\mathfrak{Z}_{m-\ell,k}(q)_{\mathfrak{S}}$ in $\mathfrak{A}'_{m-\ell}(E)_{\mathfrak{S}}$ is equal to $\ker B_m$;

(d) $\mathfrak{R}_{m^*+\ell,0}(q;E_{\mathfrak{S}'}) = \mathfrak{R}_{m^*+\ell}(q;E_{\mathfrak{S}'})$, i.e.: for every $R \in Q_{m^*+\ell}(E_{\mathfrak{S}'})$ such that $R/_{X_+} = q/_{X_+}h$ for some $h \in C^\infty(X_+)$ there exists a function $P \in Q_{m^*}(E_{\mathfrak{S}'})$ (resp. $C^\infty(X)$) satisfying $R = qP$.

Proof. *(7.18)⟺(a)*: This follows by the proof of Theorem 7.14.

(a)⟺(d): In view of the assertions (v) and (vi) of Lemma 7.28 and since the inclusion "$A_{m-\ell}(\ker B_m) \subset \ker C_m$" is always valid the bipolar theorem shows that the condition (a) is equivalent to

(a)' $$\mathfrak{R}_{m^*+\ell,k+1}(q;E_{\mathfrak{S}'}) = \mathfrak{R}_{m^*+\ell,0}(q;E_{\mathfrak{S}'}).$$

By the assertions (i), (ii), and (iii) of Lemma 7.28 this is equivalent to (d).

(b)⟺(d): In view of the assertions (vi) and (iv) of Lemma 7.28 and since the inclusion "$A_{m-\ell}(\mathfrak{Z}_{m-\ell,k}(q)_{\mathfrak{S}}) \subset \ker C_m$" always holds the condition (d) can be rephrased as

(d)' $$(A_{m-\ell}(\mathfrak{Z}_{m-\ell,k}(q)_{\mathfrak{S}}))^o = (\ker C_m)^o.$$

The bipolar theorem, again, shows that (b) is equivalent to (d)'.

To prove the equivalence of (c) and (d) we first note that in view of (7.16.b) and Remark 7.22.(i) the condition (c) is equivalent to

(c)' $\quad (\mathfrak{Z}_{m-\ell,k}(q)_{\mathfrak{S}})^{o} \subset \operatorname{im}{}^{t}B_m$.

<u>*(c)'⇒(d)*</u>: Let R be as in condition (d), and choose a function $\varphi \in E_{\mathfrak{S}'}$ such that $R = Q_{m^*+\ell}(\varphi)$. It follows by Lemma 2.34 that the functions $(\partial_M - m^* - \ell)^j R|_{X_+}$, $j \in \mathbb{N}_0$, are divided by $q|_{X_+}$. Choosing functions $\psi_0, \ldots, \psi_k \in C_M^\infty(X)$ as in the proof of Lemma 7.26 and applying (7.26) to $g := \varphi_{m^*+\ell,\omega_k}$ we see that the function

$$\phi := \varphi - \sum_{i=0}^{k} \psi_i \, (\partial_M - m^* - \ell)^i g$$

has the property that $\phi_{m^*+\ell,\omega_k}$ is divisible by $q|_{X_+}$ and, consequently, by Lemma 7.25 belongs to $(\mathfrak{Z}_{m-\ell,k}(q)_{\mathfrak{S}})^{o}$ and hence, by (c)', to $\operatorname{im}{}^{t}B_m$. In view of Lemma 7.26, Remark 7.12.(ii), and (5.63), and since X_+ is dense in X this means that $Q_{m^*+\ell}(\phi) = qP$ for some $P \in Q_{m^*}(E_{\mathfrak{S}'})$. Since supp $\psi_i \subset X_+$ we have $Q_{m^*+\ell}(\phi) = Q_{m^*+\ell}(\varphi)$ so that $R = qP$, as desired.

<u>*(d)⇒(c)'*</u>: Let $g \in (\mathfrak{Z}_{m-\ell,k}(q)_{\mathfrak{S}})^{o}$. By Lemma 7.25 this means that g satisfies (7.23) for some $h \in C^\infty(X_+)$. Since then for $i = k+1$ the equation in (7.23)' is valid and since by (5.63) we have $(\partial_M - m^* - \ell)^{k+1} g = R|_{X_+}$ for some $R \in Q_{m^*+\ell}(E_{\mathfrak{S}'})$ it follows from (d) that $R = qP$ for some $P \in Q_{m^*}(E_{\mathfrak{S}'})$. Hence in view of Remark 7.12.(ii) Lemma 7.26 shows that $g \in \operatorname{im}{}^{t}B_m$.

Since by Remark 7.16 both versions of the "i.e. ..." part of condition (d) are equivalent the proof is complete. ∎

In passing we note

Remark 7.30. *Let M' be the semi-simple part of M, i.e. the unique semi-simple linear endomorphism of V in $\mathbb{R}[M]$ such that M − M' is nilpotent. Then the real-complex coordinates chosen for M in Conventions 1.24.A and 1.24.B.(i) are suitable for M', as well. Consequently, the spaces $Q_m(E)$ remain the same if M is replaced by M'. Hence, the condition (d) of Proposition 7.29 does not change if M is replaced by M'. Note that if $X^o = X$ then q is quasihomogeneous of type M', as well.* ∎

Putting Theorem 7.14, Proposition 7.29, and Lemma 7.19 together one can formulate the main theorem on the solvability of (7.1) in the set of almost quasihomogeneous distributions as follows.

Theorem 7.31. *One (and hence each) of the conditions of Theorem 7.5 is valid if and only if one (and hence each) of the conditions of Proposition 7.29 and one (and hence both) of the conditions of Lemma 7.19 hold.* ∎

In view of Remark 7.9 the special case that q satisfies (6.28) reads as follows (if $G = \{\mathrm{Id}_V\}$ see also Proposition 7.11 above):

Corollary 7.32. *Suppose that* $q^{-1}(0) \cap X_+ = \emptyset$. *Then the conditions of Lemma 7.19 are automatically satisfied, and the conditions of Theorem 7.5 are valid if and only if one (and hence each) of the following equivalent conditions holds:*

(a) C_m *is injective;*

(b) B_m *is injective;*

(c) for every $R \in Q_{m^*+\ell}(E_{\mathfrak{G}'})$ *there exists a function* $P \in Q_{m^*}(E_{\mathfrak{H}'})$ *satisfying* $R = qP$. ∎

The simplifications which occur in the formulation of Theorem 7.31 when $\ker M = \{0\}$ are discussed in section (d) below.

(c) Solvability of (7.1) for Individual T

In this section we fix $k \in \mathbb{N}_0$. If the conditions of Theorem 7.5 do not hold one would like to be able to decide for individual $T \in \mathfrak{A}\mathfrak{Q}'_{m,k}(E)_{\mathfrak{H}}$ whether or not the equation (7.1) has a solution $S \in \mathfrak{A}\mathfrak{Q}'_{m-\ell,k}(E)_{\mathfrak{G}}$. From the proof of Theorem 7.14 we are going to extract a necessary condition which is often also sufficient. It requires the introduction of a special set of test functions appearing in (7.18)':

Notation 7.33. By $\mathfrak{T}_{m,k}(q;E)$ we denote the set of all functions $f \in \mathfrak{A}^\infty_{m,k}(E)$ such that $q|_{X_+}f$ extends to a function $R \in Q_{m+\ell}(E)$. This extension R is unique and will be denoted by $e_q(f)$. If $E = \mathcal{S}(V)$ we also write $\mathfrak{T}_{m,k}(q)$.

Remark 7.34. *(i) Every* $f \in \mathfrak{T}_{m,k}(q;E)$ *is almost quasihomogeneous of degree m and of order* $\leq N(m+\ell)$.

(ii) $\mathfrak{T}_{m,k}(q;E)_{\mathfrak{H}'} = \mathfrak{T}_{m,k}(q;E_{\mathfrak{H}'}) \subset \mathfrak{T}_{m,k}(q;E)$.

(iii) $Q_m(E_{\mathfrak{H}'})/_{X_+} \subset \mathfrak{T}_{m,k}(q;E)_{\mathfrak{H}'}$.

(iv) *The condition (7.18) reads as* $\mathfrak{T}_{m*,k}(q;E)_{\mathfrak{H}'} = Q_{m*}(E_{\mathfrak{H}'})/_{X_+}$.

(v) *If* q *satisfies (6.28) and if* $\ell \neq 0$ *then* $\mathfrak{T}_{m,k}(q;E)_{\mathfrak{H}'} = \frac{1}{q} Q_{m+\ell}(E_{\mathfrak{G}'})/_{X_+}$.

(vi) e_q *maps the space* $\mathfrak{T}_{m*,k}(q;E)_{\mathfrak{H}'}$ *bijectively onto* $(A_{m-\ell}(\ker B_m))^o$; *in particular, if* $\ker M = \{0\}$ *then* $\mathfrak{T}_{m*,k}(q;E)_{\mathfrak{H}'}$ *is finite-dimensional with*

(7.28) $$\dim \mathfrak{T}_{m*,k}(q;E)_{\mathfrak{H}'} = \dim Q_{m*+\ell}(E_{\mathfrak{G}'}) - \dim A_{m-\ell}(\ker B_m).$$

(vii) If $X = V$ *then* $\mathfrak{T}_{m,k}(q;\mathfrak{D}(V))_{\mathfrak{H}'}$ *is dense in* $\mathfrak{T}_{m,k}(q)_{\mathfrak{H}'}$ *in the topology of the space* $\mathfrak{A}^\infty_{m,k}(\mathcal{S}(V))_{\mathfrak{H}'}$.

Proof. (i): Let $f \in \mathfrak{T}_{m,k}(q;E)$. Since by Lemma 2.34 we have

$$q|_{X_+}(\partial_M - m)^{N+1} f = (\partial_M - m - \ell)^{N+1}(q|_{X_+} f)$$

and since by (7.6) $X \setminus q^{-1}(0)$ is dense in X we conclude from Proposition 5.45 that $(\partial_M - m)^{N+1} f = 0$ for $N = N(m+\ell)$.

(ii): This follows from Propositions 5.48.(ii) and 2.61.(ii).

(iii) follows from (7.15) and Proposition 5.48.(ii). (iv): is clear.

(v): The inclusion "$\subset$" is trivial. For the proof of "$\supset$" we fix $R \in Q_{m+\ell}(E_{\mathfrak{G}'})$. Since by Remark 7.9.(i) M is semi-simple it follows from Corollary 6.28 that the C^∞ function $f := (R|_{X_+})/(q|_{X_+})$ is $\mathfrak{H}'$-invariant and quasihomogeneous of degree m. Now we fix $\psi \in C^\infty_M(X)$ satisfying $\psi_0 \equiv 1$. By Propositions 3.14 and 6.3.B we then have $f = (\psi f)_m$.

In case $E = \mathfrak{D}(X)$ it follows from Corollary 5.47, Lemma 6.23 and Remark 5.56 that supp f is contained in K_M for some compact subset K of X_+ so that the function ψf belongs to $\mathfrak{D}(X)$ by Proposition 3.22. In case $E = \mathcal{S}(V)$ we may assume that the function ψ has the properties (i)-(iii) of Lemma 6.38. Hence we deduce from Proposition 5.52 that $q\psi f = \psi R \in \mathcal{S}(V)$. Consequently, it follows from (7.5) that the function ψf belong to $\mathcal{S}(V)$, as well.

So in each case we conclude that f belongs to $\mathfrak{A}^\infty_m(E)$ and - since it is $\mathfrak{H}'$-invariant - to $\mathfrak{A}^\infty_m(E)_{\mathfrak{H}'}$. Since $\mathfrak{A}^\infty_m(E)_{\mathfrak{H}'} \subset \mathfrak{A}^\infty_{m,k}(E)_{\mathfrak{H}'}$ the proof of (v) is complete.

(vi): By (7.6) e_q is injective. In view of (7.17) the rest of the first part of the assertion is an immediate consequence of the assertions (i) and (ii) of Remark 7.12. In view of Proposition 6.56 the second part is obvious.

(vii): We fix a G-invariant function $\varphi \in C_0^\infty(V)$ with values in $[0,1]$ equal to 1 on the polycircle $P := \{x \in V;\ P_+(x), M_0(x) \in K(0,1)\}$ (take $\varphi = \phi_G$ where $\phi \in C_0^\infty(V)$ is equal to 1 on P_G) and for every $j \in \mathbb{N}$ define $\chi_j \in C^\infty(V)$ by $\chi_j(x) := \varphi(\frac{1}{j} M_0 x)$. Then χ_j is G-invariant, quasihomogeneous of degree 0 and equal to 1 on the set $L_j := \{x \in V;\ |M_0 x| \leq j\}$.

Now let $f \in \mathfrak{A}^\infty_{m,k}(\mathcal{S}(V)_{\mathfrak{H}'})$ be such that $q|_{X_+} f$ extends to some $R \in Q_{m+\ell}(\mathcal{S}(V)_{\mathfrak{G}'})$. Since $\operatorname{supp}\chi_j \subset M_0^{-1}(j \operatorname{supp}\varphi)$ it follows by Lemma 6.23 and Lemma 5.57.(ii) that $\chi_j f$ belongs to $\mathfrak{A}^\infty_{m,k}(\mathcal{D}(V)_{\mathfrak{G}'})$. By the Leibniz rule we obtain for every $\alpha \in \mathfrak{A}$ that

$$(f\chi_j - f)^{(\alpha)}(x) = f^{(\alpha)}(x)(\chi_j(x) - 1) + \sum_{\substack{\beta \in \mathfrak{A}^0 \setminus \{0\} \\ \beta \leq \alpha}} \binom{\alpha}{\beta} f^{(\alpha-\beta)}(x)\, j^{-|\beta|} \varphi^{(\beta)}(\tfrac{1}{j} M_0 x).$$

Since $\chi_j - 1$ has values in $[0,1]$ and vanishes on L_j and since the functions $\varphi^{(\beta)}$ are bounded it follows that $f\chi_j$ converges to f in the topology of $\mathfrak{A}^\infty_{m,k}(\mathcal{S}(V)_{\mathfrak{H}'})$ as $j \to \infty$ (see Proposition 5.61). Finally, since $\chi_j R$ is almost quasihomogeneous of degree $m+\ell$ with support contained in $M_0^{-1}(j \operatorname{supp}\varphi)$ it follows by Lemma 5.50.(ii) that $\chi_j R$ belongs to $Q_{m+\ell}(\mathcal{D}(V)_{\mathfrak{G}'})$. Hence $\chi_j f \in \mathfrak{T}_{m,k}(q;\mathcal{S}(V))_{\mathfrak{H}'}$. ■

We now come to a result which for individual distributions T gives a necessary and sufficient conditions for the equation (7.1) to have a solution S with prescribed $(1+k)^{th}$ order deficiency c.

Theorem 7.35. *Let $T \in \mathfrak{A}'_{m,k}(E)_{\mathfrak{H}}$ and $c \in Q'_{m-\ell}(E)_{\mathfrak{G}}$.*

(i) If the equation (7.1) has a $\mathfrak{G}$-invariant solution $S \in E'$ which is almost quasi-homogeneous of degree $m-\ell$ such that $(\partial_M - m + \ell)^{k+1} S = c$ then the following condition holds:

(7.29) $$\langle T, f\rangle_M = \langle c, e_q(f)\rangle, \qquad f \in \mathfrak{T}_{m^*,k}(q;E)_{\mathfrak{H}'}.$$

(ii) The converse implication is valid provided that (7.19) holds.

The special case c = 0 is formulated as

Corollary 7.36. *Let $T \in E'$ be $\mathfrak{H}$-invariant and almost quasihomogeneous of degree m and of order $\leq k$. Then the following holds.*

(i) If (7.1) has a $\mathfrak{G}$*-invariant solution* $S \in E'$ *which is almost quasihomogeneous of degree* $m-\ell$ *and of order* $\leq k$ *then* $\langle T, f \rangle_M = 0$ *for every* $f \in \mathfrak{X}_{m^*}(q;E)_{\mathfrak{H}'}$.

(ii) The converse implication is valid provided that (7.19) holds. ∎

<u>*Proof of Theorem 7.35*</u>. We claim: if $R \in E'$ is almost quasihomogeneous of degree $m-\ell$ such that $d := (\partial_M - m + \ell)^{k+1} R \in Q'_{m-\ell}(E)_{\mathfrak{G}}$ then

$$\langle d, e_q(f) \rangle = \langle qR, f \rangle_M , \qquad f \in \mathfrak{X}_{m^*,k}(q;E)_{\mathfrak{H}'} . \tag{7.30}$$

In fact, we choose $\varphi \in E_{\mathfrak{G}'}$ such that $e_q(f) = Q_{m^*+\ell}(\varphi)$; and since by Proposition 6.27 we have $Q'_{m-\ell}(d) = d$ we conclude by (6.18), (2.5), (6.58), and (2.39) that

$$\langle d, e_q(f) \rangle = \langle Q'_{m-\ell}(d), \varphi \rangle = \langle d, \varphi \rangle = \langle (\partial_M - m + \ell)^{k+1} R, \varphi \rangle =$$

$$= (-1)^{k+1} \langle R, (\partial_M - m^* - \ell)^{k+1} \varphi \rangle = (-1)^{k+1} \langle R, ((\partial_M - m^* - \ell)^{k+1} \varphi)_{m^*+\ell,\omega_k} \rangle_M .$$

Since by (3.7) and (5.63) we have

$$((\partial_M - m^* - \ell)^{k+1} \varphi)_{m^*+\ell,\omega_k} = (\partial_M - m^* - \ell)^{k+1} \varphi_{m^*+\ell,\omega_k} = (-1)^{k+1} \, e_q(f)|_{X_+}$$

it follows that $\langle d, e_q(f) \rangle = \langle R, e_q(f)|_{X_+} \rangle_M$. Now, choosing $\phi \in E_{\mathfrak{H}'}$ such that $f = \phi_{m^*,\omega_k}$ we deduce from Theorem 5.37.(vi) that $e_q(f)|_{X_+} = (q\phi)_{m^*+\ell,\omega_k}$ so that

$$\langle R, e_q(f)|_{X_+} \rangle_M = \langle R, q\phi \rangle = \langle qR, \phi \rangle = \langle qR, f \rangle_M .$$

This implies (7.30).

<u>*(i)*</u>: we apply (7.30) to (S,c) instead of (R,d).

<u>*(ii)*</u>: In a first step we prove the assertion for <u>*the special case* $c = 0$</u>. By Theorem 7.3 we choose a $\mathfrak{G}$-invariant solution $R \in E'$ of $qR = T$ which is almost quasihomogeneous of degree $m-\ell$ such that $d := (\partial_M - m + \ell)^{k+1} R \in Q'_{m-\ell}(E)$. By Lemma 7.6.(ii) - applied to the situation described in the proof of Theorem 7.5 - it suffices to show that d belongs to the space $H := A_{m-\ell}(\ker B_m)$. Since by the assumption of (ii) the subspace H is weakly closed in $Q'_{m-\ell}(E)_{\mathfrak{G}}$ the bipolar theorem and Remark 7.34.(vi) imply that d belongs to H if and only if it lies in the polar set of $e_q(\mathfrak{X}_{m^*,k}(q;E)_{\mathfrak{H}'})$. In view of (7.30) this means that $\langle T, f \rangle_M = 0$ for every $f \in \mathfrak{X}_{m^*,k}(q;E)_{\mathfrak{H}'}$. Since by (7.29) and the assumption $c = 0$ the latter condition is valid the proof of the first step is complete.

To prove <u>*the general case*</u>, by Corollary 6.51 we choose a distribution $R \in E'$ such that $(\partial_M - m + \ell)^{k+1} R = c$. By Proposition 2.64.(iii) and Remark 2.67.(ii) we may assume that R is $\mathfrak{G}$-invariant. Applying (7.30) to $d=c$ we deduce from (7.29) that $\langle T-qR, f\rangle_M = 0$ for every $f \in \mathfrak{T}_{m^*,k}(q;E)_{\mathfrak{H}'}$. Hence by the first step of the proof we find an almost quasihomogeneous $\mathfrak{G}$-invariant solution $S' \in E'$ of the equation $qS' = T-qR$ satisfying $(\partial_M - m + \ell)^{k+1} S' = 0$ so that $S := S'+R$ is the desired solution of (7.1). ∎

In case M is semi-simple when checking (7.29) one may take advantage of

Remark 7.37. *Suppose that M is semi-simple. Then the left-hand side of (7.29) is equal to*

(i) $\langle T, [(-\partial_M)^k \psi] f\rangle$ *where* $\psi \in C_M^\infty(X)$ *is any function satisfying* $\psi_O \equiv 1$ *and - in case* $E = \mathcal{S}(V)$ *- having the properties (i) - (iii) of Lemma 6.38;*

and - in case $E = \mathcal{D}(X)$ *- to*

(ii) $\langle [(\partial_M - m)^k T]|_{S^\varkappa}, f|_{S^\varkappa}\rangle$ *where* $\varkappa : X_+ \longrightarrow]0,+\infty[$ *is any* C^∞ *function which is quasihomogeneous of degree 1.*

If M is not semi-simple it is more complicated to compute the left-hand side of (7.29) - see, for example, (8.12) below.

Proof. In view of the assumption on M Remark 7.34.(i) shows that f is quasihomogeneous of degree m^*. Hence, by Theorem 3.48 and Lemma 5.62 we find $\varphi \in E_{\mathfrak{H}'}$ such that $\operatorname{supp}\varphi$ is an M-bounded subset of X_+ and such that $f = \varphi_{m^*,\omega_k}$. In view of (6.58) this implies that

$$\langle T, f\rangle_M = \langle T, \varphi\rangle = \langle T|_{X_+}, \varphi\rangle .$$

Consequently, defining functions ψ_i by (4.16) we deduce from Proposition 4.13, (2.5), and Corollary 2.36.(ii) that

$$\langle T, f\rangle_M = \sum_{i=0}^{k} \langle (\psi_i (\partial_M - m)^i T)_{m,\omega_k}, \varphi\rangle =$$

$$= \sum_{i=0}^{k} (-1)^k \langle \psi_i (\partial_M - m)^i T, f\rangle = (-1)^k \sum_{i=0}^{k} (-1)^i \langle T, (\partial_M - m^*)^i (\psi_i f)\rangle =$$

$$= (-1)^k \sum_{i=0}^{k} \langle T, [(-\partial_M)^i \psi_i] f\rangle = c_k \langle T, [(-\partial_M)^k \psi] f\rangle$$

where by an application of (6.39) to $j=0$ we see that $c_k := (-1)^k \sum_{i=0}^{k} c_{k,i} = 1$.

Hence, the first equality is proved. For the proof of the second equality we observe that by Theorem 4.29 we have

$$\sum_{i=0}^{k} \langle \psi_i (\partial_M - m)^i T, f \rangle = \sum_{i=0}^{k} (-1)^i \langle (\partial_M - m)^i T|_{S^\varkappa}, [(\partial_M - m^*)^{k-i} f]|_{S^\varkappa} \rangle$$

and that $(\partial_M - m^*)^{k-i} f = \delta_{i,k} f$. ■

If q satisfies (6.28) it is sometimes useful to choose $\varkappa$ as follows:

Remark 7.38. *Suppose that $\ell \neq 0$ and that $q^{-1}(0) \cap X_+ = \emptyset$. Then the restriction of $|q|^{1/\mathrm{Re}\,\ell}$ to X_+ is a C^∞ function $\varkappa_q : X_+ \longrightarrow]0, +\infty[$ which is quasihomogeneous of degree 1 such that $q\bar{q} \equiv 1$ on $S^{\varkappa_q}$.* ■

Example 7.39. *Suppose that $\ell \neq 0$, that q is non-negative and has no zeros in X_+, and that there exists $\alpha \in \mathfrak{A}^+_{m^*+\ell}$ such that each component of α is even.*

Then for every $T \in \mathcal{D}'(X)$ which is almost quasihomogeneous of degree m such that on X_+ its kth order deficiency is induced by a positive continuous function the equation $qS = T$ has no solution $S \in \mathcal{D}'(X)$ which is almost quasihomogeneous of degree $m - \ell$ and of order $\leq k$.

Proof. We assume that $E = \mathcal{D}(X)$. In view of the discussion concerning the equations (7.1) and (7.2) at the beginning of section (a) the proof is reduced to the case $k = 0$. Hence, we assume that $T|_{X_+} = T_g$ for a suitable continuous function $g : X_+ \longrightarrow]0, +\infty[$. We set $\varkappa := \varkappa_q$ (see Remark 7.38). Moreover, we fix a point x in $X \setminus X_+$ and choose ε, K, and L as in the proof of the implication "(a)' ⇒ (a)" of Proposition 1.91. Then we fix a non-negative function $\varphi \in C_0^\infty(X)$ equal to 1 on K. By (1.107) $\varphi \circ M_0$ is equal to 1 on L. Finally, by the assumption we choose $\alpha \in \mathfrak{A}^+_{m^*+\ell}$ such that each component of α is even and define a non-negative function $f : X_+ \longrightarrow \mathbb{C}$ by $f(x) := x^\alpha \varphi(M_0 x)/q(x)$. Since by our choice of $\varkappa$ we have $f(\vartheta) = \vartheta^\alpha$, $\vartheta \in L$, it follows that

$$\int_{S^\varkappa} g(\vartheta)\, f(\vartheta)\, d\tilde{\varkappa}(\vartheta) \geq \int_L g(\vartheta)\, \vartheta^\alpha\, d\tilde{\varkappa}(\vartheta)\,.$$

Here the right-hand side is positive: this follows by Proposition 1.86 since the set $Z := \{ M_t \vartheta \,;\, \vartheta \in L,\, t \in [1,2] \}$ is of positive Lebesgue measure so that $\int_Z g(x)\, x^\alpha dx$ is positive. Hence, in view of Remark 7.37.(ii) (the assumption of which is valid by Remark 7.9.i) and Corollary 7.36.(i) the proof will be complete once it is verified that f belongs to $\mathfrak{T}_{m^*}(q;E)$. But this follows from Remark 7.34.(v) since by Corollary 5.47 and Proposition 5.48.(ii) the function $x \mapsto x^\alpha \varphi \circ M_0$ belongs to $Q_{m^*+\ell}(E)$. ■

The following example illustrates in a less simple situation that it is sometimes possible to verify the condition (7.29) by direct computation.

Example 7.40. *Suppose that $V = \mathbb{R} \times \mathbb{R}^{n-1}$. Let $w \in \mathbb{C} \setminus i\mathbb{R}$, define*

$$(7.31) \qquad P_w(t,\xi) = iwt + |\xi|^2, \qquad (t,\xi) \in \mathbb{R} \times \mathbb{R}^{n-1},$$

and set $p := (2,1,\dots,1)$, $G := \{Id_V\}$, and $\sigma := \tau :\equiv 1$. Then for every $h \in \mathbb{N}$ $q := P_w^h$ satisfies the condition of Corollary 7.36 for $T \equiv 1$, $m = 0$, and $k = 0$. Consequently, the equation $qS = 1$ has a solution $S \in \mathcal{S}'(\mathbb{R}^n)$ which is quasihomogeneous of degree $-\ell$.

Proof. Indirectly this follows from Corollary 7.36 and the existence of quasihomogeneous fundamental solutions for q(D) (see the references in Ortner [15]). We sketch a direct computation.

Let $u, v \in \mathbb{R}$ be such that $iw = u + iv$. We assume that $|w| = 1$ and set $\varkappa := |P_w|^{1/2}$. Then $\varkappa(\tau,\xi) = 1$ if and only if $|\xi|^2 \le R := \frac{1}{v}$ and $\tau \in \{\tau_\pm(\xi)\}$ where

$$\tau_\pm(\xi) := -u|\xi|^2 \pm \sqrt{1 - v^2|\xi|^4}\,.$$

Moreover, if $\varkappa(\tau,\xi) = 1$ then

$$\partial_\tau \varkappa(\tau,\xi) = \frac{1}{2}(\tau + u|\xi|^2) = \pm\frac{1}{2}\sqrt{1 - v^2|\xi|^4}\,.$$

We choose $j \in \mathbb{N}_0$ and $\beta \in \mathbb{N}_0^{n-1}$ such that

$$(7.32) \qquad 2j + |\beta| = 2h - n - 1.$$

In view of Remark 7.34.(v), Remark 7.37.(ii) and Lemma 1.87 and since on $S^\varkappa$ we have $1/q = \overline{q}$ it suffices to show that

$$\int_{K(0,\sqrt{R})} \left(\overline{q(\tau_+(\xi),\xi)}\,\tau_+(\xi)^j + \overline{q(\tau_-(\xi),\xi)}\,\tau_-(\xi)^j \right) \xi^\beta \frac{2\,d\xi}{\sqrt{1 - v^2|\xi|^4}} = 0\,.$$

Introducing polar coordinates with respect to the space variables ξ one sees that the left-hand side is equal to

$$F_j(w) \int_{S^{n-2}} \vartheta^\beta \, d\vartheta$$

where

$$F_j(w) := \int_0^{\sqrt{R}} \Big((-i\overline{w}\sigma_+ + \rho^2)^h \sigma_+^j + (-i\overline{w}\sigma_- + \rho^2)^h \sigma_-^j \Big) \rho^{n-3+|\beta|} \frac{2\rho\, d\rho}{\sqrt{1-v^2\rho^4}}$$

with

$$\sigma_\pm = \sigma_\pm(\rho) := -u\rho^2 \pm \sqrt{1-v^2\rho^4} \; .$$

By substituting $\rho^2 = r/v$, using (7.32) and taking $-i\overline{w}(-\frac{u}{v}r \pm \tilde{r}) + \frac{r}{v} = \overline{w}(r \mp i\tilde{r})$ into account where

$$\tilde{r} := \sqrt{1-r^2}$$

we derive

$$F_j(w) = \overline{w}^h v^{j+1-h} G_j(\tfrac{u}{v})$$

where

$$G_j(z) := \int_0^1 \Big(\varphi(r)^h (-zr + \tilde{r})^j + \overline{\varphi(r)}^h (-zr - \tilde{r})^j \Big) r^{h-j-2} \frac{dr}{\tilde{r}}$$

with $\varphi(r) := r - i\tilde{r}$. Obviously, G_j is a polynomial of degree $\le j$ satisfying

$$G_j^{(s)} = (-1)^s \frac{j!}{(j-s)!} G_{j-s} \, , \qquad s \in \mathbb{N}_j \, .$$

Hence it suffices to prove that $G_j(0) = 0$. Observing that $r^{h-j-2} = \frac{d}{dr}\frac{r^{h-j-1}}{h-j-1}$ we do partial integration and in view of

$$(7.33) \qquad (\varphi^h \pm \overline{\varphi}^h)'(r) = \frac{hi}{\tilde{r}} (\varphi^h \mp \overline{\varphi}^h)(r)$$

obtain

$$G_j(0) = -\frac{ih}{h-j-1} G_{j-1}(0) + \frac{j-1}{h-j-1} G_{j-2}(0) \, , \qquad 1 \le j \le h-2 \, .$$

This reduces the proof to the verification of $G_0(0) = 0$. To this end we set

$$b_j := \int_0^1 (\varphi(r)^h + \overline{\varphi(r)}^h) \, r^j \frac{dr}{\tilde{r}} \, .$$

Doing partial integration first on the basis of

$$(r^2\tilde{r})^{-1} = -\frac{d}{dr}(\tilde{r}/r)$$

and then in the first of the resulting integrals on the basis of $r^{j+1} = \frac{d}{dr}\frac{r^{j+2}}{j+2}$ we obtain

$$b_{j+2} = -\frac{(j+1)(j+2)}{h^2-(j+2)^2} b_j \qquad \text{if } 0 \le j < h-2 \, .$$

If h is even then it follows from (7.33) that $b_0 = 0$ and hence by induction that $G_0(0) = b_{h-2} = 0$. If h is odd then first substituting $\rho = \tilde{r}$, i.e. $d\rho = (r/\tilde{r})\,dr$, and then performing partial integration based on

$$(\varphi^h - \overline{\varphi}^h)(\rho) = \frac{\tilde{\rho}}{hi}(\varphi^h + \overline{\varphi}^h)'(\rho)$$

(see (7.33)) we obtain

$$b_1 = i^h \int_0^1 (\varphi^h - \overline{\varphi}^h)(\rho)\, d\rho = \frac{1}{h}\, i^{h-1}\, b_1 .$$

Since by (7.32) and because of $n \geq 2$ we have $h \geq 2$ this shows that $b_1 = 0$ so that by induction, again, we deduce that $G_0(0) = b_{h-2} = 0$. ∎

In order to avoid repetitions we postpone the discussion of further examples after having reformulated Theorem 7.31 for the case $E = \mathcal{S}(V)$ by employing the Fourier transform in the following section.

(d) Quasihomogeneous Linear Partial Differential Equations with Constant Coefficients

In this and the following section we suppose that q is a non-trivial complex-valued polynomial function defined on V^* which is quasihomogeneous of degree $\ell \in \mathfrak{A}(M)$ and of type M^* and is $(G,\tau/\sigma)^*$-invariant (see Notation 2.60). Then for $E = \mathcal{S}(V^*)$ the conditions (7.3), (7.4) and (7.5) are valid (see Remark 7.1). Via the Fourier transform the results (for $E = \mathcal{S}(V^*)$ and for $(M^*, G^*, \sigma^*, \tau^*, m^*)$ instead of (M, G, σ, τ, m)) of the preceding sections turn into assertions on the solvability of the equation

(7.34) $\qquad q(D)\,S = T$

in the set of (almost) quasihomogeneous temperate distributions. For the kth order deficiencies d of T and c of S this involves the equation

(7.35) $\qquad q(D)\,c = d$

as is seen by Corollary 2.36.(i). We fix $k \in \mathbb{N}_0$ and define

Notation 7.41. (i) $\mathcal{Q}\mathcal{S}'_m(V) := \mathcal{F}^{-1}\left(Q'_{m^*}(\mathcal{S}(V^*))\right)$ where in Notation 6.54 M has to be replaced by M^*;

(ii) $\mathcal{AS}'_{m,k}(V) := \mathcal{F}^{-1}\left(\mathfrak{A}'_{m^*,k}(\mathcal{S}(V^*))\right)$ where in Notation 6.57 M has to be replaced by M^*.

Remark 7.42. *(i.1) If* $\ker M = \{0\}$ *then* $\mathcal{QS}'_m(V)_{\mathfrak{G}}$ *is equal to the space* $\mathfrak{P}_m(V)_{\mathfrak{G}}$ *of all* $\mathfrak{G}$*-invariant complex-valued polynomial functions on* V *which are almost quasihomogeneous of degree* m;

(i.2) if $\ker M \neq \{0\}$ *then making use of the notation introduced in (6.22) we have* $\mathcal{QS}'_m(V)_{\mathfrak{G}}$ *is equal to the space of all* $\mathfrak{G}$*-invariant distributions of the form*

$$\sum_{\alpha \in \mathfrak{A}^+_m} x^\alpha \otimes S_\alpha \qquad \text{where } S_\alpha \in \mathcal{S}'(V_0).$$

(ii) $\mathcal{AS}'_{m,k}(V)_{\mathfrak{G}}$ *is the space of all* $\mathfrak{G}$*-invariant temperate distributions on* V *which are almost quasihomogeneous of degree* m *with* $(1+k)^{th}$ *order deficiency belonging to* $\mathcal{QS}'_m(V)_{\mathfrak{G}}$.

<u>*Proof.*</u> Everything follows from Propositions 2.40, 6.24.(iv), 2.64.(ii), and 6.29. ∎

As an immediate consequence of Propositions 7.2, 2.64.(ii), and 2.40 one obtains

Proposition 7.43. *For every* $d \in \mathcal{QS}'_m(V)_{\mathfrak{H}}$ *the equation (7.35) has a solution* $c \in \mathcal{QS}'_{m+\ell}(V)_{\mathfrak{G}}$. ∎

Likewise, Theorem 7.3 transforms into

Theorem 7.44. *For any* $\mathfrak{H}$*-invariant distribution* $T \in \mathcal{AS}'_{m,k}(V)_{\mathfrak{H}}$ *the equation (7.34) has a* $\mathfrak{G}$*-invariant solution* $S \in \mathcal{AS}'_{m+\ell,k}(V)_{\mathfrak{G}}$. ∎

As a special case we obtain the assertion (i) of

Theorem 7.45. *Suppose that* $\tau \equiv 1$. *Then*

(i) $q(D)$ *has an almost quasihomogeneous (of degree* $\ell-\mu$*)* $\mathfrak{G}$*-invariant temperate fundamental solution whose deficiency belongs to* $\mathcal{QS}'_{\ell-\mu}(V)_{\mathfrak{G}}$; *note that such a fundamental solution is quasihomogeneous if* $\ell < \mu$.

(ii.1) if $q(D)$ *has a quasihomogeneous temperate fundamental solution then*

(7.36) $\quad \langle 1, f \rangle_M = 0 \quad$ *for every* $f \in \mathfrak{T}_{-\mu}(q; \mathcal{D}(V^*))_{t_{\mathfrak{H}}}$ *(see Notation 7.33)* *where* ${}^t\mathfrak{H} := (G^*, {}^t\tau)$ *with* ${}^t\tau(A) := \tau(A^*)$, *i.e.* ${}^t\tau = 1/\tau^*$.

(ii.2) the converse is valid provided that $q^{-1}(0) \cap (V^*)_+ = \emptyset$ *or* $\ker M = \{0\}$.

Note that by Remark 7.37 in case M is semi-simple the left-hand side of (7.36) is equal to $\int_{S^\varkappa} f(\vartheta)\, d\tilde{\varkappa}(\vartheta)$.

Proof. Since the Dirac distribution δ is G-invariant and quasihomogeneous of degree $-\mu$ the assertion (i) is a corollary of Theorem 7.44. The assertion (ii) follows from Corollary 7.36 and Remark 7.34.(vii). ■

We refrain from carrying over the full statement of Theorem 7.35 via the Fourier transform. Note that in case $m+\ell \notin \mathfrak{A}(M)$ the spaces $\mathfrak{Q}\mathcal{S}'_m(V)$ and $\mathfrak{Q}\mathcal{S}'_{m+\ell}(V)$ are trivial by Corollary 6.28 so that $\mathfrak{A}\mathcal{S}'_{m,k}(V) = \mathfrak{A}\mathfrak{Q}'_{m,k}(\mathcal{S}(V))$ (see Notation 7.18). Hence, in this case the assertion of Theorem 7.44 becomes

$(7.37)_k$ *for every* $\mathfrak{H}$*-invariant* $T \in \mathcal{S}'(V)$ *which is almost quasihomogeneous of degree m and of order* $\le k$ *the equation (7.34) has a* ($\mathfrak{G}$*-invariant) solution* $S \in \mathcal{S}'(V)$ *which is almost quasihomogeneous of degree* $m+\ell$ *and of order* $\le k$.

As for the exceptional case "$m+\ell \in \mathfrak{A}(M)$", in view of Lemma 2.21 Proposition 7.4 becomes

Proposition 7.46. *Suppose that* $m+\ell \in \mathfrak{A}(M)$ *and that* $k > N := N(m+\ell)$. *Let T be an* $\mathfrak{H}$*-invariant temperate distribution on V which is almost quasihomogeneous of degree m and of order* $\le k$.

(i) Then the following conditions are equivalent:

(a) the equation (7.34) has a ($\mathfrak{G}$*-invariant) temperate solution S which is almost quasihomogeneous of degree* $m+\ell$ *and of order* $\le k$;

(b) for some (resp. every) $j \in \mathbb{N}_{k-N}$ *the equation (7.35) for* $d = (\partial_M - m)^j T$ *has a* ($\mathfrak{G}$*-invariant) temperate solution c which is almost quasihomogeneous of degree* $m+\ell$ *and of order* $\le k-j$.

(ii) If c is such a solution as in (b) then in (a) S can be chosen in such a way that $(\partial_M - m - \ell)^j S$ *differs from c by an element of* $\mathfrak{Q}\mathcal{S}'_{m+\ell}(V)_{\mathfrak{G}}$. ■

In order to deal with the case "$m+\ell \in \mathfrak{A}(M)$" further we introduce

Notation 7.47. (i) $\mathfrak{N}_{m,k}(q) := \mathcal{F}^{-1}(\mathfrak{Z}_{m^*,k}(q))$ (see Notation 7.21).

(ii) $\mathfrak{D}_m(q) := (\partial_M - m)^{k+1}\mathfrak{N}_{m,k}(q) = \mathcal{F}^{-1}((\partial_{M^*} - m^*)^{k+1}\mathfrak{Z}_{m^*,k}(q))$; this space does not depend on k, indeed, as can be seen from Remark 7.22.(ii) or from Remark 7.48.(i) below.

It does not seem to be easy to give a fairly explicit description (not involving the Fourier transform) of the elements of the space $\mathfrak{N}_{m,k}(q)$ (compare section 8.(d) below). However, for the space $\mathfrak{D}_m(q)$ of its $(1+k)^{th}$ order deficiencies it is easy:

Remark 7.48. *(i) $\mathfrak{D}_m(q)$ is generated by the functions of the form*

$$(7.38)\qquad V \ni x \mapsto \sum_{\alpha \in \mathfrak{A}_m^+} \frac{1}{\alpha!} \sum_{\beta \in \mathfrak{A}^o} \frac{1}{\beta!}\, Q^{(\beta)}(\partial_\eta)\eta^\alpha (ix)^{\alpha+\beta} \exp\langle i(M^*)_o\eta, x\rangle$$

where $(\eta, Q) \in \mathfrak{B}(q|_{(V^*)_+})$ (see Notation 7.21.(i)).

(ii) $\mathfrak{D}_m(q)_{\mathfrak{G}}$ is a weakly dense subspace of $(\partial_M - m)^{k+1}\big(\mathfrak{A}\mathcal{S}'_{m,k}(V)_{\mathfrak{G}} \cap \ker q(D)\big)$.

Proof. *(i)*: this follows from Lemma 2.21 and Example 6.36.

(ii): this is a consequence of Lemma 2.21 and Lemma 7.23. ∎

Proposition 7.29 turns into

Proposition 7.49. *The following conditions are equivalent:*

(a) the weak closure of $\mathfrak{D}_{m+\ell}(q)_{\mathfrak{G}}$ in $\mathfrak{Q}\mathcal{S}'_{m+\ell}(V)_{\mathfrak{G}}$ is equal to the subspace $\mathfrak{Q}\mathcal{S}'_{m+\ell}(V)_{\mathfrak{G}} \cap \ker q(D)$;

(b) the weak closure of $\mathfrak{N}_{m+\ell,k}(q)_{\mathfrak{G}}$ in $\mathfrak{A}\mathcal{S}'_{m+\ell,k}(V)_{\mathfrak{G}}$ is equal to the subspace $\mathfrak{A}\mathcal{S}'_{m+\ell,k}(V)_{\mathfrak{G}} \cap \ker q(D)$;

(c) the condition (d) of Proposition 7.29 is valid with $(V, M, m^*, \mathfrak{G}', \mathfrak{H}')$ replaced by $(V^*, M^*, m, {}^t\mathfrak{G}, {}^t\mathfrak{H})$, i.e. $\mathfrak{T}_{m,k}(q)_{{}^t\mathfrak{H}}$ (see Notation 7.33) is equal to $Q_m(\mathcal{S}(V^*)_{{}^t\mathfrak{H}})|_{(V^*)_+}$ where ${}^t\mathfrak{H} := (G^*, {}^t\tau)$ with ${}^t\tau(A) := \tau(A^*)$. ∎

Lemma 7.19 becomes

Lemma 7.50. *The space* $q(D)\,\mathfrak{U}\mathfrak{Q}'_{m+\ell,k}(\mathcal{S}(V))_{\mathfrak{G}}$ *(see Notation 7.18) is weakly closed in* $\mathcal{S}'(V)$ *if and only if* $(\partial_M - m - \ell)^{k+1}\big(\mathfrak{U}\mathcal{S}'_{m+\ell,k}(V)_{\mathfrak{G}} \cap \ker q(D)\big)$ *is weakly closed in* $\mathfrak{Q}\mathcal{S}'_{m+\ell}(V)_{\mathfrak{G}}$. ∎

Finally, Theorem 7.31 transforms into our main solvability theorem for (7.34):

Theorem 7.51. *The following conditions are equivalent:*

(a) the solvability condition $(7.37)_k$ *is valid for some* $k \in \mathbb{N}_0$;

(a)' the condition $(7.37)_k$ *is valid for every* $k \in \mathbb{N}_0$; *in particular, for every* $\mathfrak{H}$*-invariant* $T \in \mathcal{S}'(V)$ *which is quasihomogeneous of degree m the equation (7.34) has a* ($\mathfrak{G}$*-invariant) solution* $S \in \mathcal{S}'(V)$ *which is quasihomogeneous of degree* $m+\ell$;

(b) for some (resp. every) $k \in \mathbb{N}_0$ *the following condition holds:*

$(7.39)_k$ *for every pair (T, c) of temperate distributions on V such that T (resp. c) is* $\mathfrak{H}$*-invariant (resp.* $\mathfrak{G}$*-invariant) satisfying* $q(D)c = (\partial_M - m)^{k+1}T$ *there exists a (*$\mathfrak{G}$*-invariant) temperate solution S of the equation (7.34) such that* $(\partial_M - m - \ell)^{k+1}S = c$;

(c) for some (resp. every) $k \in \mathbb{N}_0$ *the following condition holds:*

$(7.40)_k$ *for every solution* $c \in \mathfrak{Q}\mathcal{S}'_{m+\ell}(V)_{\mathfrak{G}}$ *of the equation* $q(D)c = 0$ *there exists a (*$\mathfrak{G}$*-invariant) solution* $S \in \mathcal{S}'(V)$ *of the equation* $q(D)S = 0$ *which is almost quasihomogeneous of degree* $m+\ell$ *with* $(1+k)^{th}$ *order deficiency c;*

(d) for some (resp. every) $k \in \mathbb{N}_0$ *one (and hence each) of the conditions of Proposition 7.49 and one (and hence both) of the conditions of Lemma 7.50 are valid.* ∎

There are two important general situations where the condition (d) of Theorem 7.51 can be simplified, namely if (6.28) holds or if $\ker M = \{0\}$. To handle the second one we require the following

Remark 7.52. *Suppose that* $\ker M = \{0\}$.

(i) Then the space $\mathfrak{P}_m(V)$ *of complex-valued polynomial functions on V which are almost quasihomogeneous of degree m is finite-dimensional, and*

$$(7.41)\qquad \dim \mathfrak{P}_m(V^*)_{t_{\mathfrak{H}}} = \dim \mathfrak{P}_m(V)_{\mathfrak{H}}\,.$$

(ii) $\dim\left(\mathfrak{P}_{m+\ell}(V)_{\mathfrak{G}} \cap \ker q(D)\right) = \dim \mathfrak{P}_{m+\ell}(V)_{\mathfrak{G}} - \dim \mathfrak{P}_m(V)_{\mathfrak{H}}\,.$

(iii) $\mathfrak{D}_{m+\ell}(q)$ *is the subspace of* $\mathfrak{P}_{m+\ell}(V)$ *generated by the polynomial functions*

$$(7.42)\qquad V \ni x \mapsto \sum_{\alpha_M = m+\ell} \frac{1}{\alpha!}\, Q(\partial_\eta)\,\eta^{\alpha}(ix)^{\alpha}, \qquad (\eta, Q) \in \mathfrak{B}(q|_{V^*\setminus\{0\}}).$$

(iv) For the subspace $\mathfrak{T}_{m,k}(q)$ *of* $\mathfrak{U}^{\infty}_{m,k}(\mathcal{S}(V^*))$ *introduced in Notation 7.33 we have*

$$(7.43)\qquad \dim \mathfrak{D}_{m+\ell}(q)_{\mathfrak{G}} \le \dim \mathfrak{P}_{m+\ell}(V)_{\mathfrak{G}} - \dim \mathfrak{T}_{m,k}(q)_{t_{\mathfrak{H}}}$$

with equality being valid provided that $k \ge N(m+\ell)$.

Proof. (i): the first assertion is clear. For the proof of (7.41) one observes that $\mathfrak{P}_m(V) = \mathfrak{Q}\mathcal{S}'_m(V)$ (by Remark 7.42) and $Q_m(\mathcal{S}(V^*)) = \mathfrak{P}_m(V^*)$ and that by the isomorphism property of the Fourier transform and by Propositions 2.64.(ii) and 6.56 one has: $\dim \mathfrak{Q}\mathcal{S}'_m(V)_{\mathfrak{H}} = \dim Q'_{m^*}(\mathcal{S}(V^*))_{\mathfrak{H}^*} = \dim Q_m(\mathcal{S}(V^*))_{t_{\mathfrak{H}}}$.

(ii): in view of Remark 7.42.(i) this is a consequence of Proposition 7.43.

(iii): is a special case of Remark 7.48.(i).

(iv): follows from (ii) - applied to $(m+\ell, \mathfrak{G})$ instead of $(\mathfrak{H}, m)$ - and from Propositions 2.64.(ii) and Lemma 7.23. ■

The following supplement to Theorem 7.51 deals with the case that (6.28) holds.

Supplement 7.53. *Suppose that* $q^{-1}(0) \cap (V^*)_+ = \emptyset$.

(i) Then the conditions of Lemma 7.50 are automatically satisfied, and the conditions of Proposition 7.49 amount to

$$(7.44)\qquad \mathfrak{E}_{\mathfrak{G}} \cap \ker q(D) = \{0\} \quad \text{for} \quad \mathfrak{E} = \mathfrak{Q}\mathcal{S}'_{m+\ell}(V) \text{ or } \mathfrak{E} = \mathfrak{U}\mathcal{S}'_{m+\ell,k}(V)\,;$$

resp.

$$(7.44)'\qquad Q_{m+\ell}(\mathcal{S}(V^*)_{t_{\mathfrak{G}}}) = q\, Q_m(\mathcal{S}(V^*)_{t_{\mathfrak{H}}})\,.$$

If $\ker M = \{0\}$ *then (7.44) is equivalent to*

$$(7.44)''\qquad \dim \mathfrak{P}_{m+\ell}(V)_{\mathfrak{G}} = \dim \mathfrak{P}_m(V)_{\mathfrak{H}}\,.$$

(ii) In particular, if $G = \{Id_V\}$ *then the conditions of Theorem 7.51 are violated if and only if* $m+\ell \in \mathfrak{U}(M)$, *if* $\ell \ne 0$, *and if* $\operatorname{Re} m < 0$ *or if* $\dim' G_M(\sigma_+) \ge 2$.

Proof. *(i)*: In view of the assertions (iii) and (iv) of Remark 7.34 the first part is a reformulation of Corollary 7.32 ; by Remark 7.42.(i) the second part follows from Remark 7.52.(ii).

(ii) is a consequence of Proposition 7.11. ■

Concerning the assertion (ii) of Supplement 7.53, it should be pointed out that if $\mathfrak{G}$-invariance is taken into account then the chances are better that there are positive results. This is illustrated by the first three examples in section (e) below.

Next we formulate the simplifications in the condition (d) of Theorem 7.51 which follow from the assumption $\ker M = \{0\}$.

Supplement 7.54. *Suppose that $\ker M = \{0\}$. Then the conditions of Lemma 7.50 are automatically satisfied, and the condition (a) of Proposition 7.49 becomes*

$$(7.45) \qquad \mathfrak{D}_{m+\ell}(q)_{\mathfrak{G}} = \mathfrak{P}_{m+\ell}(V)_{\mathfrak{G}} \cap \ker q(D)$$

which is equivalent to

$$(7.45)' \qquad \dim \mathfrak{D}_{m+\ell}(q)_{\mathfrak{G}} \geq \dim \mathfrak{P}_{m+\ell}(V)_{\mathfrak{G}} - \dim \mathfrak{P}_{m}(V)_{\mathfrak{H}},$$

the converse inequality being always valid. Another equivalent condition is

$$(7.46) \qquad \mathfrak{T}_{m,k}(q)_{t_{\mathfrak{H}}} = \mathfrak{P}_m(V^*)_{t_{\mathfrak{H}}}\Big/_{V^* \setminus \{0\}}$$

which is equivalent to

$$(7.46)' \qquad \dim \mathfrak{T}_{m,k}(q)_{t_{\mathfrak{H}}} \leq \dim \mathfrak{P}_m(V)_{\mathfrak{H}},$$

the converse inequality being always valid.

Proof. By Remark 7.42.(i) the assumption on M implies that $\mathfrak{Q}\mathcal{S}'_{m+\ell}(V)_{\mathfrak{G}}$ is equal to $\mathfrak{P}_{m+\ell}(V)$ and hence finite-dimensional so that its subspaces $\mathfrak{D}_{m+\ell}(q)_{\mathfrak{G}}$ and $(\partial_M - m - \ell)^{k+1}(\mathfrak{A}\mathcal{S}'_{m+\ell,k}(V)_{\mathfrak{G}} \cap \ker q(D))_{\mathfrak{G}}$ are so, as well. In particular, all these spaces are closed. Hence, the second condition of Lemma 7.50 holds, and the condition (a) (resp. (c)) of Proposition 7.49 is identical with (7.45) (resp. (7.46)). Moreover, the equivalence of (7.45) and (7.45)' (resp. of (7.46) and (7.46)') is a consequence of Remark 7.52.(ii) (resp. (7.41) and Remark 7.34.(ii)). ■

In passing we observe

Remark 7.55. *Suppose that* $\ker M = \{0\}$. *Let* M' *be (as in Remark 7.30) the semi-simple part of* M. *Then the spaces* $\mathfrak{D}_{m+\ell}(q)$ *and* $\mathfrak{P}_{m+\ell}(V)$ *remain the same if* M *is replaced by* M'. *Hence, in particular, the conditions of Theorem 7.51 are valid if and only if they are so when* M *is replaced by* M'. ■

As a first application of Supplement 7.54 we obtain

Proposition 7.56. *Suppose that* $\ker M = \{0\}$. *Then* q *satisfies the conditions of Theorem 7.51 for* $G = \{Id_V\}$ *provided that one of the following conditions holds:*

(a) $\partial_v q \equiv 0$ *for some* $v \in V^* \setminus \{0\}$;

(b) there exist two non-trivial M*-invariant subspaces* V_1 *and* V_2 *of* V *satisfying* $V = V_1 \oplus V_2$ *such that* q *is of the form* $q = q_1 \otimes q_2$ *where for every* $j \in \{1,2\}$ $q_j : V_j^* \longrightarrow \mathbb{C}$ *is a polynomial function which is quasihomogeneous of type* $(M_{V_j})^*$.

Proof. Under the assumption *(a)* it follows in view of Remark 7.34.(iv) from Remark 7.17.(i) that the condition (7.46) is satisfied. If under the assumption *(b)* $q_j \equiv 1$ for some $j \in \{1,2\}$ then (a) is valid, as well. Since $q_1 \otimes q_2 = (q_1 \otimes 1)(1 \otimes q_2)$ one obtains the general assertion from the special one by the following obvious remark on the surjectivity of composites of multiplication operators. ■

Remark 7.57. *For every* $j \in \{1,2\}$ *let* q_j *be a polynomial function on* V^* *which is quasihomogeneous of degree* ℓ_j *and of type* M^*. *Suppose that* $q = q_1 q_2$. *For simplicity we assume that* $G = \{Id_V\}$. *Then*

(i) q *satisfies the condition (a)' of Theorem 7.51 provided that* q_1 *does so for* ℓ_1 *instead of* ℓ *and* q_2 *does so with* (m, ℓ) *replaced by* $(m + \ell_1, \ell_2)$;

(ii) conversely, if q *satisfies the condition (a)' of Theorem 7.51 then* q_1 *and* q_2 *do so with* ℓ *replaced by* ℓ_1 *and* ℓ_2, *respectively.* ■

As a simple consequence of Proposition 7.56 we are going to derive

Proposition 7.58. *Suppose that* $M = Id_V$ *and* $\ell = 1$. *If* q *has a zero in* $V^* \setminus \{0\}$ *- which is always the case if* $n \geq 3$ *- then* q *satisfies each of the conditions of Theorem 7.51.*

Proof. By the assumption on q we find $v \in V^* \setminus \{0\}$ such that $q(v) = 0$. Since the

the assumptions on M and ℓ imply that q is linear we conclude that $\partial_v q \equiv 0$. Hence the assertion follows by Proposition 7.56. ∎

We close this section by observing that the condition (6.28) is not necessary for the non-solvability of (7.34) in the set of quasihomogeneous distributions. Namely, in view of Supplement 7.53.(ii) the contraposition of Remark 7.57.(ii) leads to

Remark 7.59. *Suppose that $\dim' G_M(\sigma_+) \geq 2$ and $m+\ell \in \mathfrak{A}(M)$. Let q_1 be a polynomial function on V^* which is quasihomogeneous of degree $\ell_1 \neq 0$. Suppose that $q_1^{-1}(0) \cap (V^*)_+ = \emptyset$. Then $q := q_1 q_2$ does not satisfy the condition (a)' of Theorem 7.51 for every polynomial function q_2 on V^* which is quasihomogeneous of degree $\ell - \ell_1$.* ∎

To see how under the assumptions of Remark 7.59 the condition (7.45) is violated we require

Lemma 7.60. *Let $g, h \in C^\infty(X)$ and $y \in X$, and let Q be a polynomial function on V^*. Then (i) $(y, Q) \in \mathfrak{B}(g) \Rightarrow (y, Q) \in \mathfrak{B}(gh)$; (ii) the converse is valid if $h(y) \neq 0$.*

Proof. *(i)* is an immediate consequence of the Leibniz formula and of the fact that $(y, Q^{(\alpha)}) \in \mathfrak{B}(g)$ for arbitrary (y, Q) and $\alpha \in \mathfrak{A}$.
(ii) is obtained by an application of (i) to $(gh, 1/h)$ instead of (g, h). ∎

Remark 7.61. *Suppose that $\ker M = \{0\}$ and $m+\ell \in \mathfrak{A}(M)$. Let q_1 and ℓ_1 be as in Remark 7.59, and let q_2 be a polynomial function on V^* which is quasihomogeneous of degree $\ell_2 := \ell - \ell_1$ such that q_2 satisfies the condition (a)' of Theorem 7.51 for $(\{Id_V\}, m+\ell_1)$ instead of (G, m). Then for $q := q_1 q_2$ we have*

$$\dim \mathfrak{P}_{m+\ell}(V) - \dim \mathfrak{P}_m(V) - \dim \mathfrak{D}_{m+\ell}(q) = \dim \mathfrak{P}_{m+\ell_1}(V) - \dim \mathfrak{P}_m(V).$$

Proof. By Lemma 7.60 we have $\mathfrak{D}_{m+\ell}(q) = \mathfrak{D}_{m+\ell}(q_2)$. Hence the assertion follows by Supplement 7.54 applied to $(q_2, m+\ell_1)$ instead of (q, m). ∎

Example 7.62. *Suppose that $M = Id_V$, $V = \mathbb{R}^2$, and $q(x,y) = (x - iy)(x - y)$, $(x,y) \in \mathbb{R}^2$. Then $\dim \mathfrak{D}_{m+2}(q) = 1 = \frac{1}{2}(\dim \mathfrak{P}_{m+2}(V) - \dim \mathfrak{P}_m(V))$ for every $m \in \mathbb{N}_0$.* ∎

(e) Examples

In this section we are going to treat several examples. The first one discusses the elementary case "dim' V = 1". For the sake of clarity we formulate the cases "V = ℝ" and "V = ℂ" separately.

Example 7.63$_{\mathbb{R}}$. *Suppose that $V = \mathbb{R}$ and $M = Id_{\mathbb{R}}$.*

(i) If $G = \{Id_{\mathbb{R}}\}$ then the conditions of Theorem 7.51 hold if and only if m is not an integer satisfying $-\ell \leq m \leq -1$.

(ii) Suppose that $G = \{\pm Id_{\mathbb{R}}\}$; then the assumptions on $\mathfrak{G}$, $\mathfrak{H}$, and q mean that $\sigma = \rho^{\ell}\tau$ where $\rho(\pm Id_{\mathbb{R}}) := \pm 1$; consequently, if m belongs to the exceptional set of assertion (i) then the conditions of Theorem 7.51 hold if and only if $\tau = \rho^{1-m}$.

Proof. *(i)*: The condition (7.44)" of Supplement 7.53 is violated if and only if $m + \ell \in \mathbb{N}_0$ but $m \notin \mathbb{N}_0$.

(ii): If $m \in -\mathbb{N}_{\ell}$ then the condition (7.44)" of Supplement 7.53 amounts to $\mathfrak{P}_{m+\ell}(\mathbb{R})_{\mathfrak{G}} = \{0\}$. Since $\mathfrak{P}_{m+\ell}(\mathbb{R}) = \mathbb{C}\, x^{m+\ell}$ this means that $\rho^{m+\ell} \neq \sigma$, i.e. $\rho^{m+\ell} = \rho\sigma = \rho\rho^{\ell}\tau$, i.e. $\tau = \rho^{-m+1}$. ∎

Note that under the assumptions of Example 7.63$_{\mathbb{R}}$ the Dirac distribution δ_o is homogeneous of degree $m = -1$; hence it is the assertion (ii) (applied to $\tau \equiv 1$ and $\sigma = \rho^{\ell}$) - and not (i)! - that implies the existence of a homogeneous fundamental solution (with parity $(-1)^{\ell}$). Of course, such a fundamental solution is explicitly given by $\frac{1}{2}(E + (-1)^{\ell}\check{E})$ where $E := \frac{1}{\ell!}x^{\ell-1}H$ and H denotes the Heaviside function.

More generally, suppose that under the assumptions of Example 7.63$_{\mathbb{R}}$ the number m belongs to the exceptional set $(-\mathbb{N}_{\ell})$. In order to describe the operator

$$O_{\ell,m} : \mathfrak{A}\mathfrak{Q}'_{m+\ell}(\mathbb{R}) \longrightarrow \mathfrak{A}\mathfrak{Q}'_m(\mathbb{R}),\ T \mapsto T^{(\ell)},$$

more precisely, we first recall from Proposition 2.13 that the argument and the target spaces are two-dimensional. Since $\ell + m + 1 \geq 1$ the preceding argument shows that there is a homogeneous distribution T such that $T^{(\ell+m+1)} = \delta$. Since $\delta_o^{(-m-1)}$ belongs to $\mathfrak{A}\mathfrak{Q}'_m(\mathbb{R})$ we conclude that $O_{\ell,m}(T) = \delta_o^{(-m-1)}$. Since by Example 7.63$_{\mathbb{R}}$

the operator O_ℓ is not surjective we conclude that $\operatorname{im} O_{\ell,m} = \mathbb{C}\,\delta_0^{(-m-1)}$. This, in turn, implies that the kernel of $O_{\ell,m}$ is 1-dimensional; indeed, it is spanned by the function $x^{m+\ell}$ - note that $0 \le m+\ell \le \ell-1$. Of course, in view of the equation

$$(\underline{x}^{-1})^{(-m-1)} = (-m-1)!\,(-1)^{-m-1}\,\underline{x}^{m}$$

the fact that the distribution $\underline{x}^m$ does not belong to $\operatorname{im} O_{\ell,m}$ is a direct consequence of the fact that every solution $u \in \mathfrak{D}'(\mathbb{R})$ of the equation $u' = \underline{x}^{-1}$ is of the form $u = T_{c+\log|\cdot|}$ for some constant $c \in \mathbb{C}$ (see, for example, Hörmander [11], (3.2.13), p.73).

Example 7.63$_{\mathbb{C}}$. *Suppose that $\dim_{\mathbb{C}} V = 1$, and let $\lambda \in \mathbb{C} \setminus \mathbb{R}$ be such that $M = \lambda\,Id_V$. Then there is a unique element $\alpha' \in \mathfrak{A}$ such that $q(x) = \text{const}\,x^{\alpha'}$, and the following assertions hold:*

(i) If $G = \{Id_V\}$ then the conditions of Theorem 7.51 are violated if and only if m belongs to $(\mathfrak{A}(M)-\ell) \setminus \mathfrak{A}(M)$.

(ii) Suppose that $G = \{\pm Id_V\}$; then the assumptions on $\mathfrak{G}$, $\mathfrak{H}$, and q mean that $\sigma = \rho^{|\alpha'|}\tau$ where $\rho(\pm Id_V) := \pm 1$; consequently, if m belongs to the exceptional set of assertion (i) then the conditions of Theorem 7.51 hold if and only if

$$(7.47)_{\mathbb{C}} \qquad \tau = \rho^{|\alpha| - |\alpha'| + 1}$$

where α is the unique element of $\mathfrak{A}$ satisfying $\alpha_M = m+\ell$; note that $(7.47)_{\mathbb{C}}$ means that $\tau = \rho$ in case $|\alpha|-|\alpha'|$ is even and $\tau \equiv 1$ in case $|\alpha|-|\alpha'|$ is odd.

Of course, under the assumptions of Example 7.63$_{\mathbb{C}}$ similar considerations are valid if G is an arbitrary subgroup of SO(V).

Proof. The first assertion is a consequence of Remark 1.40.(i). Hence, q satisfies (6.28), and the solvability of (7.34) is handled by Supplement 7.53.

(i): In fact, Remark 1.40 shows that $\dim \mathfrak{P}_m(\mathbb{C}) = 1$ for every $m \in \mathfrak{A}(M)$. Consequently, the condition (7.44)" is valid for $G = \{Id_V\}$ if and only if m and $m+\ell$ both belong to $\mathfrak{A}(M)$ or both do not belong to $\mathfrak{A}(M)$.

(ii): Suppose that m belongs to the exceptional set of assertion (i). Then it is by Remark 1.40 that we can fix a unique $\alpha \in \mathfrak{A}$ such that $\alpha_M = m+\ell$ so that $\mathfrak{P}_{m+\ell}(\mathbb{C}) = \mathbb{C}\,x^\alpha$. Moreover, the condition (7.44)" of Supplement 7.53 amounts to

$\mathfrak{P}_{m+\ell}(\mathbb{C})_{\mathfrak{G}} = \{0\}$. Since $x^\alpha \circ (-\mathrm{Id}_{\mathbb{C}}) = (-1)^{|\alpha|} x^\alpha$ this means that $\rho^{|\alpha|} \neq \sigma$, i.e. $\rho^{|\alpha|} = \rho\sigma = \rho\rho^{|\alpha'|}\tau$, i.e. $(7.47)_{\mathbb{C}}$ holds. ∎

Note that under the assumptions of Example 7.63$_{\mathbb{C}}$ the Dirac distribution δ_o is quasihomogeneous of degree $m = -\mu$ where here $\mu = 2\,\mathrm{Re}\,\lambda = \lambda + \bar{\lambda} = (1,1)_M$; hence, if $\alpha' \geq (1,1)$ then – as in the case $V = \mathbb{R}$ – the assertion (i) does not imply the existence of homogeneous fundamental solutions. But in contrast to the case $V = \mathbb{R}$ the same is true for the assertion (ii). In fact, choosing $\alpha = (\beta, \gamma)$ as in assertion (ii) and writing $\alpha' = (\beta', \gamma')$ we compute

$$\beta\lambda + \gamma\bar{\lambda} = \alpha_M = m + \ell = -(1,1)_M + (\alpha')_M = (\beta'-1)\lambda + (\gamma'-1)\bar{\lambda}$$

so that by Lemma 1.41 we have $\beta = \beta'-1$ and $\gamma = \gamma'-1$. Since this implies $|\alpha| - |\alpha'| = -2$ we conclude that for $\tau \equiv 1$ the condition $(7.47)_{\mathbb{C}}$ is always violated, indeed. Note that under the assumptions of Example 7.63$_{\mathbb{C}}$ the operator $q(D)$ in general does not have homogeneous fundamental solutions, indeed; for example, if $q = 4z\bar{z}$ then $q(D)$ is equal to the Laplacian Δ and has the fundamental solution $\frac{1}{2\pi}\log|\cdot|$ which is not homogeneous.

The Laplacian in n dimensions is the object of

Example 7.64. *Suppose that $V = \mathbb{R}^n$ and $M = \mathrm{Id}_V$. Let $h \in \mathbb{N}$ and $q(D) = \Delta^h$ where by Δ we denote the Laplacian. Then $\ell = 2h$. If $m \in \mathbb{Z}$ with $m \geq -2h$ then the condition (a)' of Theorem 7.51 is valid for $G = SO(V)$ and $\sigma = \tau \equiv 1$ if and only if $m \geq 0$ or if m is odd.*

In fact, it follows from Example 7.39 that if n is even and not larger than 2h then Δ^h has no homogeneous fundamental solution; indeed, as is well-known, in this case Δ^h has a fundamental solution of the form $c\,|\cdot|^{2h-n}\log|\cdot|$ which is almost homogeneous but not homogeneous so that it also follows by Proposition 2.48 that in this case Δ^h has no homogeneous fundamental solution.

<u>*Proof.*</u> It is easily seen that $\dim \mathfrak{P}_h(\mathbb{R}^n)_{SO(n)}$ equals 1 if $h \in 2\mathbb{N}_0$ and 0 otherwise. Hence one obtains the assertion via the condition (7.44)" of Supplement 7.53. ∎

Next we are going to have a look at the heat operator.

Example 7.65. *We define P_w, p, σ and τ as in Example 7.40. Let G be the group $SO(n-1)$ operating on the space variables ξ. Then $q = P_w$ satisfies the conditions of Theorem 7.51 if and only if $m \notin 2\mathbb{N}_0 - 2$.*

Proof. It is easily verified (see Proposition 7.73 below) that

$$\dim \mathfrak{P}_m(\mathbb{R}^n)_{SO(n-1)} = \begin{cases} 0 & \text{if } m \in 2\mathbb{N}_0+1 \\ \frac{m}{2}+1 & \text{if } m \in 2\mathbb{N}_0 \end{cases} .$$

Hence (7.44)" is valid if and only if $m \notin 2\mathbb{N}_0 - 2$, and the assertion follows by Supplement 7.53. ∎

Note that for every $h \in \mathbb{N}$ the assertion of Example 7.65 implies that the operator $P_w^h(D)$ has a quasihomogeneous (of type $p := (2,1,\ldots,1)$) fundamental solution provided that $n \geq 2h$ or n is even. In fact, unlike the situation in Example 7.64 the last assertion is valid without any restrictions on n or h as a look at the explicit form of the standard fundamental solution of $P_w^h(D)$ shows (see the references in Ortner [15]). A less explicit proof is provided by Example 7.40.

For two further special, but classical examples we are going to show that the conditions of Theorem 7.51 are satisfied. This time we do this by verifying the condition (7.45)'. Here the following lemma is required.

Lemma 7.66. *Suppose that $\ker M = \{0\}$. Let $m \in \mathfrak{A}(M)$, and let Υ be an open subset of $V \setminus \{0\}$. Then there is a subset Ξ of Υ such that the polynomial functions*

$$x \mapsto \sum_{\alpha_M = m} \frac{1}{\alpha!} \xi^\alpha x^\alpha , \qquad \xi \in \Xi ,$$

constitute a basis of the space $\mathfrak{P}_m(V)$ of all polynomial functions $V \longrightarrow \mathbb{C}$ which are almost quasihomogeneous of degree m. If M is semi-simple then one can choose Ξ to be a subset of $\Upsilon_M \cap S_V$.

Proof. We set $A := \{\alpha \in \mathfrak{A} ; \alpha_M = m\}$. By Propositions 1.34.(iii) and 1.28.(ii) the functions $\frac{1}{\alpha!} x^\alpha$, $\alpha \in A$, constitute a $\mathbb{C}$-basis of $\mathfrak{P}_m(V)$. Consequently, for proving the main part of the assertion it suffices to find a family $(\xi_\alpha)_{\alpha \in A}$ in Υ such that the $A \times A$-Matrix $\left((\xi_\alpha)^{\alpha'}\right)_{(\alpha,\alpha') \in A \times A}$ with complex entries has a non-zero determinant, i.e. is invertible. For then the functions

$$V \longrightarrow \mathbb{C}, \; x \mapsto \sum_{\alpha \in A} \frac{1}{\alpha!} (\xi_{\alpha'})^{\alpha} x^{\alpha}, \qquad \alpha' \in A,$$

constitute a $\mathbb{C}$-basis of $\mathfrak{P}_m(V)$, as well.

Now, for arbitrary $\alpha \in A$ and $j \in \mathbb{N}_n$ we denote by $Y_{\alpha,j}$ a transzendental variable. We suppose that all these variables are independent. We set

$$Y_\alpha := (Y_{\alpha,1}, \ldots, Y_{\alpha,n})$$

and define primitive monomials

$$Y_\alpha^{\alpha'} := \prod_{j=1}^{d} (Y_{\alpha,j})^{\beta_j} \prod_{j=1}^{c} (Y_{\alpha,j+d})^{\gamma_j} (Y_{\alpha,j+d+c})^{\delta_j}, \qquad \alpha' = (\beta,\gamma,\delta) \in A,$$

(compare Notation 1.25.A). By $\mathfrak{N}$ we denote the $A \times A$-Matrix

$$\left(Y_\alpha^{\alpha'}\right)_{(\alpha,\alpha') \in A \times A}.$$

Then $P := \det \mathfrak{N}$ is a polynomial in the variables $Y_{\alpha,j}$ with coefficients in $\mathbb{Z}$. Explicitly, we have

$$(7.48) \qquad P(Y) = \sum_{\sigma \in \Sigma(A)} \varepsilon(\sigma) \prod_{\alpha \in A} Y_\alpha^{\sigma(\alpha)}$$

where $\Sigma(A)$ denotes the set of permutations of A and ε the sign homomorphism given by any ordering of the elements of A. We observe that all the primitive monomials appearing on the right-hand side of (7.48) are different, and hence independent. Consequently, P is not the zero polynomial. This implies that P induces a non-trivial function $\tilde{P}: (\mathbb{C}^n)^A \longrightarrow \mathbb{C}$. Note that via the pseudo-real coordinates introduced in Notation 1.25.A one obtains a function on V^A which is denoted by $(\xi_\alpha)_{\alpha \in A} \mapsto P((\xi_\alpha))$. It follows that one can find a family $(\xi_\alpha)_{\alpha \in A}$ in Υ such that $P((\xi_\alpha)) \neq 0$. Indeed, denoting by $\tilde{\Upsilon}$ the subset of $\mathbb{C}^n$ consisting of all the n-tuples of pseudo-real coordinates of the points of Υ, we claim that the Taylor coefficients of $\tilde{P}$ at any point of the set $\tilde{\Upsilon}^A$ can be computed from the knowledge of the restriction of $\tilde{P}$ to $\tilde{\Upsilon}^A$. To see this one has to take into account that for any holomorphic function h defined on an open subset of $\mathbb{C}^2$ we have

$$(7.49) \qquad \text{(a)} \; \partial_{z_1} h = \tfrac{1}{2}(\partial_+ h - i\partial_- h), \quad \text{and} \quad \text{(a)} \; \partial_{z_2} h = \tfrac{1}{2}(\partial_+ h + i\partial_- h)$$

where for any z in the domain of definition of h by $\partial_+ h(z)$ ($\partial_- h(z)$) we denote the derivative at $t=0$ of the real function $t \mapsto h(z_1+t, z_2+t)$ (resp. $h(z_1+it, z_2-it)$). Since the family $(\xi_\alpha)_{\alpha \in A}$ has the desired properties the proof of the main part of the assertion is complete.

Suppose now that M is semi-simple. If $(t_\alpha)_{\alpha\in A}$ is any family in $]0,+\infty[$ then

$$P((M_{t_\alpha}\xi_\alpha)_{\alpha\in A}) = \det(t_\alpha^m \xi_\alpha^\beta)_{(\alpha,\beta)\in A\times A} = \Big(\prod_{\alpha\in A} t_\alpha\Big)^m P((\xi_\alpha)) \neq 0 .$$

Consequently, we can replace ξ_α by $M_{t_\alpha}\xi_\alpha$ and - in view of Proposition 1.70 - prescribe the modulus of ξ_α as we want. In particular, we can achieve that $\xi_\alpha \in S_V$. ■

The following example concerns the Schrödinger operator.

Example 7.67. *Let $u\in\dot{\mathbb{R}}$, and suppose that $q(t,\xi) = ut + |\xi|^2$, $(t,\xi)\in\mathbb{R}\times\mathbb{R}^{n-1}$. Then q is quasihomogeneous of type $p := (2,1,\dots,1)$ and of degree $\ell := 2$, and for every m the conditions of Theorem 7.51 are valid for $G = \{Id_V\}$. In particular, for every $h\in\mathbb{N}$ the operator $q^h(D)$ has an $O(n-1)$-invariant fundamental solution which is quasihomogeneous of degree $2h-n-1$ (Bernstein's fundamental solution has these properties - see [7]).*

Proof. For every $k\in\mathbb{N}_0$ we set $I_n(k) := \{\alpha\in\mathbb{N}_0^n ;\ |\alpha| = k\}$. We first observe that by

$$\Phi(\alpha) := \begin{cases} \alpha + e_1 & \text{if } \alpha\in\mathfrak{A}_m^+ \\ (0,\alpha) & \text{if } \alpha\in I_{n-1}(m+2)\end{cases}$$

a bijective map $\Phi : \mathfrak{A}_m^+ \cup I_{n-1}(m+2) \longrightarrow \mathfrak{A}_{m+2}^+$ is defined. Since $\dim\mathfrak{P}_m(\mathbb{R}^n) = |\mathfrak{A}_m^+|$ this implies

$$\text{(7.50)}\qquad \dim\mathfrak{P}_{m+2}(\mathbb{R}^n) - \dim\mathfrak{P}_m(\mathbb{R}^n) = |I_{n-1}(m+2)| .$$

We set $\rho := -\frac{1}{u}$, fix $\xi\in S^{n-2}$ and set $\eta := (\rho,\xi)$. Then $q(\eta) = 0$, and the polynomial function in (7.42) for $Q\equiv 1$ becomes

$$\mathbb{R}\times\mathbb{R}^{n-1} \ni x = (x_1,x') \mapsto d_\xi(x) := \sum_{j=0}^{+\infty} \frac{1}{j!}(i\rho x_1)^j R_{j,\xi}(x')$$

where

$$R_{j,\xi}(x') := \sum_{\beta\in I_{n-1}(m+2-2j)} \frac{1}{\beta!}\xi^\beta (ix')^\beta .$$

By Lemma 7.66 we find a family $(\xi_\delta)_{\delta\in I_{n-1}(m+2)}$ in S^{n-2} such that the polynomials R_{0,ξ_δ}, $\delta\in I_{n-1}(m+2)$, are linearly independent. It follows that the polynomials d_{ξ_δ}, $\delta\in I_{n-1}(m+2)$, are linearly independent. In view of (7.50) this implies the condition (7.45)' of Supplement 7.54. ■

The next example is about the wave operator.

Example 7.68. *Suppose that* $q(t,\xi)=(ct)^2-\langle\xi,\xi\rangle$, $(t,\xi)\in\mathbb{R}\times\mathbb{R}^{n-1}$, *for some positive constant* c. *Then* q *is homogeneous of degree* 2, *and for every* m *the conditions of Theorem 7.51 are valid for* $M=Id_{\mathbb{R}^n}$ *and* $G=\{Id_{\mathbb{R}^n}\}$. *In particular, for every* $h\in\mathbb{N}$ *the operator* $q^h(D)$ *has an* $O(n-1)$*-invariant homogeneous fundamental solution.*

Proof. We define $I_n(k)$ as in the proof of Example 7.67 and set $B:=I_{n-1}(m+2)$ and $\Gamma:=I_{n-1}(m+1)$. Since by

$$\Phi(\alpha):=\begin{cases}\alpha-2e_1 & \text{if } \alpha_1\geq 2\\ (\alpha_2,\ldots,\alpha_n) & \text{if } \alpha_1\in\{0,1\}\end{cases}$$

a bijection $\Phi: I_n(m+2)\longrightarrow I_n(m)\cup B\cup\Gamma$ is defined we obtain

$$|I_n(m+2)|-|I_n(m)|=|B|+|\Gamma|. \tag{7.51}$$

We now fix $\xi\in S^{n-2}$. Then $\eta_\pm:=(\pm 1/c,\xi)$ is a non-trivial real zero of q so that for $Q\equiv 1$ and $\eta=\eta_\pm$ the polynomial function in (7.42) becomes

$$\mathbb{R}\times\mathbb{R}^{n-1}\ni x=(x_1,x')\mapsto d_{\eta_\pm}(x):=\sum_{j=0}^{+\infty}\frac{1}{j!}(\pm i x_1/c)^j R_{j,\xi}(x')$$

where

$$R_{j,\xi}(x'):=\sum_{\beta\in I_{n-1}(m+2-j)}\frac{1}{\beta!}\xi^\beta(ix')^\beta.$$

If $m=-2$ then $d_{\eta_\pm}=R_{0,\xi}\equiv 1$, and since $\dim\mathfrak{P}_0(\mathbb{R}^n)=1$ the condition (7.45)' of Supplement 7.54 is satisfied.

Suppose now that $m\geq -1$. Then we set $Q_\xi^\pm:=\frac{1}{2}(d_{\eta_+}\pm d_{\eta_-})$ so that

$$Q_\xi^+(x)=\sum_{j\in 2\mathbb{N}_0}\frac{1}{j!}(ix_1/c)^j R_{j,\xi}(x') \quad\text{and}\quad Q_\xi^-(x)=\sum_{j\in 2\mathbb{N}_0+1}\frac{1}{j!}(ix_1/c)^j R_{j,\xi}(x').$$

By Lemma 7.66 we can choose families $(\xi_\beta)_{\beta\in B}$ and $(\zeta_\gamma)_{\gamma\in\Gamma}$ in S^{n-2} such that the polynomials R_{0,ξ_β}, $\beta\in B$, (resp. R_{1,ζ_γ}, $\gamma\in\Gamma$) are linearly independent over $\mathbb{C}$. It follows that the polynomials $Q_{\xi_\beta}^+$, $\beta\in B$, and $Q_{\zeta_\gamma}^-$, $\gamma\in\Gamma$, are linearly independent over $\mathbb{C}$. Since they belong to $\mathfrak{D}_{m+2}(q)$ we conclude that

$$\dim_{\mathbb{C}}\mathfrak{D}_{m+2}(q)\geq |B|+|\Gamma|. \tag{7.52}$$

In view of (7.51) this implies that the condition (7.45)' of Supplement 7.54 is satisfied. ∎

We are now going to discuss a few examples in case M is not semi-simple.

Proposition 7.69. *Suppose that V is M-cyclic and that $\dim' V = 3$ (resp. 2) in case the only eigenvalue λ of M is real (resp. non-real). Then the conditions of Theorem 7.51 are always valid for $G = \{Id_V\}$.*

Proof. The case $\lambda \in \mathbb{R}$. We set $d := \ell/\lambda$ and $\rho := m/\lambda$. We may suppose that $V^* = \mathbb{R}^3$ and that $M^* \cdot (x,y,z) = (\lambda x + y, \lambda y + z, \lambda z)$. By Proposition 1.46 one finds $c \in \mathbb{N}_0$ such that $c \le d/2$ and complex numbers $a_0, \dots, a_c$ such that $a_c \neq 0$ and

$$q(x,y,z) = z^d P((y^2 - 2xz)/z^2)$$

where P is the polynomial defined by $P := \sum_{j=0}^{c} a_j T^j$. We first observe that by Proposition 7.56 the polynomial function q_0 defined by $q_0(x,y,z) := a_c z^{d-2c}$ satisfies the conditions of Theorem 7.51. Hence we may suppose that $c > 0$. If $\nu_1, \dots, \nu_c \in \mathbb{C}$ are such that $P = a_c \prod_{j=1}^{c} (T - \nu_j)$ we conclude that $q = \prod_{j=0}^{c} q_j$ where

$$q_j(x,y,z) := (y^2 - 2xz) - \nu_j z^2, \qquad j \in \mathbb{N}.$$

In view of Remark 7.57 it suffices to prove the assertion for the q_j. In view of what was said above about q_0 this leaves us with the case $q(x,y,z) = y^2 - 2xz - \nu z^2$ for some $\nu \in \mathbb{C}$. In case ν is real q has many real zeros so that the method of the proof of Example 7.68 leads to a simple proof:

The subcase $\nu \in \mathbb{R}$. We first observe that the set $\Upsilon := \{(x,z) \in \mathbb{R}^2;\ 2xz + \nu z^2 > 0\}$ is non-empty. We fix $\zeta = (x,z) \in \Upsilon$ and – setting $y := \sqrt{2xz + \nu z^2}$ – observe that $\eta_\pm := (x, \pm y, z)$ is a non-trivial real zero of q. For $Q \equiv 1$ and $\eta = \eta_\pm$ the polynomial function in (7.42) becomes

$$R_\zeta^\pm := \sum_{j=0}^{\rho+2} \frac{1}{j!} (\pm y)^j S_{\zeta,j} (iY)^j$$

where

$$S_{\zeta,j} := \sum_{(h,k) \in I_2(\rho+2-j)} \frac{1}{h!} \frac{1}{k!} x^h z^k (iX)^h (iZ)^k .$$

with $I_2(\rho)$ being defined as in the proof of Example 7.67. If $\rho = -2$ then $R_\zeta^\pm = S_{\zeta,0} \equiv 1$, and since $\dim \mathfrak{P}_0(\mathbb{R}^3) = 1$ the condition (7.45)' of Supplement 7.54 is satisfied. Suppose now that $\rho \ge -1$. Then setting $Q_\zeta^\pm := \frac{1}{2}(R_\zeta^+ \pm R_\zeta^-)$ we obtain

$$Q_\zeta^+ = \sum_{j \in 2\mathbb{N}_0} \frac{1}{j!} y^j S_{\zeta,j} (iY)^j \quad \text{and} \quad Q_\zeta^- = \sum_{j \in 2\mathbb{N}_0+1} \frac{1}{j!} y^j S_{\zeta,j} (iY)^j .$$

Setting $B := I_2(\rho+2)$ and $\Gamma := I_2(\rho+1)$, by Lemma 7.66 we can choose families $(\zeta_\beta)_{\beta \in B}$ and $(\omega_\gamma)_{\gamma \in \Gamma}$ in Υ such that the polynomials $S_{\zeta_\beta,0}$, $\beta \in B$, (resp. $S_{\omega_\gamma,1}$, $\gamma \in \Gamma$,) are linearly independent. It follows that the polynomials $Q_{\zeta_\beta}^+$, $\beta \in B$, and $Q_{\zeta_\gamma}^-$,

$\gamma \in \Gamma$, are linearly independent. Since they belong to $\mathfrak{D}_{m+2}(q)$ we conclude that (7.52) is valid. In view of (7.51) (for $n=3$ and $m=\rho$) this implies that the condition (7.45)' is satisfied.

The subcase $\nu \notin \mathbb{R}$. In this case there are not so many real zeros: $q(x,y,z)=0$ if and only if $y=0=z$. Consequently, a proof via the the condition (7.45)' has to make use of the elements (η, Q) in $\mathfrak{B}(q)$ where Q is non-trivial. However, here we are going to proceed along different lines in that we verify the condition (c) of Proposition 7.49. Hence we fix $R \in \mathfrak{P}_{m+2}(V^*)$ such that $R|_\Omega = q|_\Omega H$ for some $H \in C^\infty(\Omega)$ where $\Omega := \mathbb{R}^3 \setminus \{0\}$. Now we fix $r \in \mathbb{R}\setminus\{0\}$ and observe that by

$$q_r(x,y) := q(x,y,ry) = (1-\nu r^2)\,y^2 - 2rxy$$

a polynomial function is defined which is induced by a polynomial of degree 2 with respect to y with coefficients in the polynomial ring $\mathbb{C}[x]$ with invertible leading coefficient. Consequently, since the function R_r can be considered as a polynomial with respect to y with coefficients in $\mathbb{C}[x]$ the Euclidean algorithm provides us with functions $T, U \in C^\infty(\mathbb{R}^3)$ which are polynomials with respect to the first two variables such that

$$R_r = q_r T(\cdot,\cdot\cdot,r) + U(\cdot,\cdot\cdot,r)$$

and such that the degree of the polynomial function $U(\cdot,\cdot\cdot,r)$ with respect to y is not larger than 1. In view of the assumption on H we deduce for every $x \in \mathbb{R}\setminus\{0\}$ that $\lim_{y\to 0} U(x,y,r)/q_r(x,y)$ exists. Since $\deg_y q_r = 2$ this implies that $U(x,\cdot,r) \equiv 0$ for $x \in \mathbb{R}\setminus\{0\}$. By continuity this means that $U \equiv 0$, i.e.

(7.53) $\qquad$ (a) $R_r = q_r T(\cdot,\cdot\cdot,r)$, $\quad$ and $\quad$ (b) $T(\cdot,\cdot\cdot,r) = H_r$ on $(\mathbb{R}\setminus\{0\})\times\mathbb{R}$.

In order to complete the proof we have to show that

(7.54) $\qquad$ by $(x,y,z) \mapsto T(x,y,z/y)$ a polynomial function $P: \mathbb{R}^3 \longrightarrow \mathbb{C}$ is well-defined.

For then it follows from (7.53) that $R = qP$, and the condition (c) of Proposition 7.49 is verified. In order to prove (7.54) we have to gain information about the coefficients

$$b_{j,k}(r) := \frac{1}{j!\,k!}\,(\partial_x^j \partial_y^k T)(0,0,r)$$

of the polynomial function $T(\cdot,\cdot\cdot,r)$. Since this is to be obtained out of (7.53.a) we compute

$$q_r(x,y)\,T(x,y,r) = \sum_{k\geq 2} (1-\nu r^2)\, b_{0,k-2}(r)\, y^k - \sum_{j\in\mathbb{N}} 2r\, b_{j-1,0}(r)\, x^j y \; +$$
$$+ \sum_{j\in\mathbb{N}} \sum_{k\geq 2} \Big((1-\nu r^2)\, b_{j,k-2} - 2r\, b_{j-1,k-1}(r) \Big)\, x^j y^k .$$

On the other hand, denoting the coefficient of $x^j y^k z^h$ in R by $a_{j,k,h}$ we obtain

$$R_r(x,y) = \sum_{j,k,h\in\mathbb{N}_0} a_{j,k,h}\, x^j y^{k+h} r^h = \sum_{j,k\in\mathbb{N}_0} A_{j,k}(r)\, x^j y^k$$

where

$$A_{j,k}(r) := \sum_{h=0}^{k} a_{j,k-h,h}\, r^h .$$

Inserting the preceding equations into (7.53.a) and comparing coefficients we arrive at the equations

$$\begin{aligned}
&A_{j,k}(r) = 0 && \text{if } k=0 \text{ or if } k=1 \text{ and } j=0,\\
&(1-\nu r^2)\, b_{0,k-2}(r) = A_{0,k}(r) && \text{if } k\geq 2,\\
&-2r\, b_{j-1,0}(r) = A_{j,1}(r) && \text{if } j\in\mathbb{N},\\
&(1-\nu r^2)\, b_{j,k-2} - 2r\, b_{j-1,k-1}(r) = A_{j,k}(r) && \text{if } j\in\mathbb{N} \text{ and } k\geq 2.
\end{aligned}$$

Observing that the elements of $\mathfrak{P}_{m+2}(V^*)$ are precisely the polynomial functions which are homogeneous of degree $\rho+2$ so that $a_{j,k-h,h} = 0$ and hence $A_{j,k} \equiv 0$ if $j+k \neq \rho+2$ we deduce that $b_{j,k} \equiv 0$ if $j+k\neq\rho$. It follows that the first of the preceding equations gives

$$\text{(7.55)} \qquad a_{\rho+2,0,0} = 0$$

and the other ones lead to

$$\text{(7.56)} \qquad \text{(a)}\;\; B_\rho(r) = (1-\nu r^2)^{-1} A_{\rho+2}(r), \qquad \text{(b)}\;\; B_0(r) = -\frac{1}{2r} A_1(r), \qquad \text{and}$$
$$\text{(c)}\;\; B_k(r) = \frac{1}{2r}\Big((1-\nu r^2)\, B_{k-1}(r) - A_{k+1}(r) \Big), \qquad k\in\mathbb{N}_\rho ,$$

where

$$B_k := b_{\rho-k,k} \quad\text{and}\quad A_k(r) := \sum_{h=0}^{k} a_{\rho+2-k,k-h,h}\, r^h , \qquad k\in\mathbb{N}_{\rho+2} .$$

We can now show by induction on k that, actually, the function B_k is a polynomial function of degree $\leq k$. Indeed, since the coefficients of a polynomial function on $\mathbb{R}^2$ are universal linear combinations of its derivatives at an arbitrarily chosen point of $(\mathbb{R}\setminus\{0\})\times\mathbb{R}$, say $(1,1)$, we conclude in view of (7.53.b) that the functions $b_{j,k}$ are linear combinations of the functions $\mathbb{R}\longrightarrow\mathbb{C}$, $r\mapsto r^{\gamma_3} H^{(\gamma)}(1,1,r)$, $\gamma\in\mathbb{N}_0^3$. In particular, the functions B_k are C^∞ on the whole of $\mathbb{R}$. Hence, in view of (7.56.b) we deduce that the zero order coefficient in A_1, $a_{\rho+1,1,0}$, vanishes so that $B_0 \equiv -a_{\rho+1,0,1}/2$. Since A_{k+1} is a polynomial function of degree $\leq k+1$ it follows by induction by a similar argument from (7.56.c) that B_k is a polynomial function of degree $\leq k$. Consequently, by $(y,z) \mapsto B_k(z/y)\, y^k$ a polynomial function on $\mathbb{R}^2$ is well-defined. This implies (7.54), as desired.

The case $\lambda \notin \mathbb{R}$. Here we may suppose that $V^* = \mathbb{C}^2$ and that

$$M^* \cdot (z_1, z_2) = (\lambda z_1 + z_2, \lambda z_2).$$

By Proposition 1.47 we find $c \in \mathbb{N}_0$ such that $c \leq \min\{m_R, m_I\}$ and complex numbers $a_0, \ldots, a_c$ such that $a_c \neq 0$ and

$$q(z) = (z_2)^{m_R} (\overline{z_2})^{m_I} P((z_1 \overline{z_2} - \overline{z_1} z_2) |z_2|^{-2})$$

where P is the polynomial defined by $P := \sum_{j=0}^{c} a_j T^j$. We first observe that in view of Proposition 7.56 and Remark 7.57 the polynomial function q_0 defined by

$$q_0(z) = a_c (z_2)^{m_R - c} (\overline{z_2})^{m_I - c}$$

satisfies the conditions of Theorem 7.51. Hence we may suppose that $c > 0$. If $\nu_1, \ldots, \nu_c \in \mathbb{C}$ are such that $P = a_c \prod_{j=1}^{c} (T - \nu_j)$ we conclude that $q = \prod_{j=0}^{c} q_j$ where

$$q_j(x,y,z) := (z_1 \overline{z_2} - \overline{z_1} z_2) - \nu_j z_2 \overline{z_2}, \qquad j \in \mathbb{N}.$$

In view of Remark 7.57 it suffices to prove the assertion for the q_j. In view of what was said above about q_0 this leaves us with the case $q(z) = (z_1 \overline{z_2} - \overline{z_1} z_2) + \nu z_2 \overline{z_2}$ for some $\nu \in \mathbb{C}$. Note that if $\nu \in \mathbb{R}$ then $q(z)$ vanishes if and only if $z_2 = 0$.

First of all we observe that $\ell = \lambda + \overline{\lambda} = 2 \operatorname{Re} \lambda$. We set $b := (m+\ell)_R$ and $c := (m+\ell)_I$. It follows from Remark 1.40 that

$$\dim \mathfrak{P}_m(V) = (m_R + 1)(m_I + 1), \qquad m \in \mathfrak{A}(M).$$

Consequently, if m and $m+\ell$ both belong to $\mathfrak{A}(M)$ then $b = m_R + 1$ and $c = m_I + 1$ so that

$$\dim \mathfrak{P}_{m+\ell}(V) - \dim \mathfrak{P}_m(V) = (b+1)(c+1) - bc = b + c + 1. \tag{7.57}$$

Moreover, if $m + \ell$ belongs to $\mathfrak{A}(M)$ but m does not then by Lemma 1.41 we have $b = 0$ or $c = 0$, and $\dim \mathfrak{P}_{m+\ell}(V) = b + c + 1$, and (7.57) remains valid.

For the computation of $\mathfrak{D}_{m+\ell}(q)$ we first observe that $q(z_1, 0) = 0$ for every $z_1 \in \mathbb{C}$. For every $j \in \mathbb{N}$ we set

$$Q_j := \sum_{i=1}^{j} a_{j,i} (Z_1)^{j-i} (Z_2 + \overline{Z_2})^i$$

where $a_{j,j} := 1$ and where $a_{j,1}, \dots, a_{j,j-1}$ are constants to be determined below in such a way that $((1,0), Q_j)$ belongs to $\mathfrak{B}(q)$. To this end we fix $\alpha = ((r,s),(t,u)) \in \mathfrak{A}$ and first observe that in case $|\alpha| \le j-3$ or $|\alpha| \ge j+1$ or $r \ge j$ or $t \ge 1$ we have $Q_j^{(\alpha)}(\partial)\, q \equiv 0$. Moreover, if $|\alpha| = j$ then $Q_j^{(\alpha)}$ is constant so that $Q_j^{(\alpha)}(\partial)\, q$ vanishes at $z = (1,0)$. Hence we suppose that $j > |\alpha| \ge j-2$ and $r \le j-1$ and $t = 0$. Now, observing that

$$(\partial_{z_2})^s (\partial_{\overline{z_2}})^u (Z_2 + \overline{Z_2})^i = \frac{i!}{(i-s-u)!} (Z_2 + \overline{Z_2})^{i-s-u} \qquad \text{if } s+u \le i$$

we deduce that

$$(Q_j)^{(\alpha)} = \sum_{\substack{i=1 \\ i \ge s+u}}^{j-r} a_{j,i} \frac{(j-i)!}{(j-i-r)!} \frac{i!}{(i-s-u)!} (Z_1)^{j-i-r} (Z_2 + \overline{Z_2})^{i-s-u}.$$

If $s+u = 0$ then $r = |\alpha| \ge j-2$ so that in case $r = j-1$ this equals

$$a_{j,1} (j-1)!\, (Z_2 + \overline{Z_2})$$

and in case $r = j-2$ is equal to

$$a_{j,1} (j-1)!\, Z_1 (Z_2 + \overline{Z_2}) + a_{j,2} (j-2)!\, (Z_2 + \overline{Z_2})^2.$$

In case $v := s+u \ge 1$ substituting $I = i - v$ we obtain

$$Q_j^{(\alpha)} = \sum_{I=0}^{j-|\alpha|} a_{j,I+v} \frac{(j-I-v)!}{(j-I-|\alpha|)!} \frac{(I+v)!}{I!} (Z_1)^{j-I-|\alpha|} (Z_2 + \overline{Z_2})^I.$$

If $|\alpha| = j-1$ then this equals

$$a_{j,v} (j-v)!\, v!\, Z_1 + a_{j,1+v} (j-1-v)!\, (v+1)!\, (Z_2 + \overline{Z_2})$$

and in case $|\alpha| = j-2$ this is equal to

$$a_{j,v} \tfrac{1}{2} (j-v)!\, v!\, (Z_1)^2 + a_{j,v+1} (j-1-v)!\, (v+1)!\, Z_1 (Z_2 + \overline{Z_2}) + \\ + a_{j,v+2} (j-2-v)!\, \tfrac{1}{2} (v+2)!\, (Z_2 + \overline{Z_2})^2$$

Taking the equations

$$(\partial_{z_2} + \partial_{\overline{z_2}})\, q = (z_1 - \overline{z_1}) + \nu (\overline{z_2} + z_2), \qquad (\partial_{z_2} + \partial_{\overline{z_2}})^2 q \equiv 2\nu,$$

$$\partial_{z_1} (\partial_{z_2} + \partial_{\overline{z_2}})\, q \equiv 1, \qquad \partial_{z_1} q = \overline{z_2}, \qquad (\partial_{z_1})^2 q \equiv 0,$$

into account we deduce that $[(Q_j)^{(\alpha)}(\partial)\, q](1,0) = 0$ provided that in case $j \ge 2$ the following conditions are valid:

$$a_{j,1} (j-1)! + a_{j,2} (j-2)!\, 2\nu = 0$$

and

$$a_{j,v+1}(j-1-v)!\,(v+1)! + a_{j,v+2}(j-2-v)!\,\tfrac{1}{2}(v+2)!\,2\nu = 0, \qquad v \in \mathbb{N}_{j-2}.$$

This leads to the following recursive definition of the $a_{j,i}$:

$$a_{j,i} := -\nu\,\frac{i+1}{j-i}\,a_{j,i+1}, \qquad i \in \mathbb{N}_{j-1}.$$

Since the recursion starts with $a_{j,j}=1$ we see that the $a_{j,i}$ are well-definied non-zero complex numbers. Consequently, setting $Q_0 :\equiv 1$ we conclude that the pairs $((1,0), Q_j)$, $j \in \mathbb{N}_0$, belong to $\mathfrak{B}(q)$.

Now, for every $j \in \mathbb{N}_0$ we denote the polynomial (7.42) for $Q = Q_j$ and $\eta = (1,0)$ by P_j. Our claim is that the polynomials P_j, $0 \le j \le b+c$, are linearly independent. In view of Supplement 7.54 and (7.57) this implies the assertion.

Choosing α as above, fixing $j \in \mathbb{N}_{b+c}$, and applying the binomial formula we compute

$$Q_j(\partial)\,\eta^\alpha = \sum_{\substack{i=1\\ i \ge j-r}}^{j} a_{j,i}\,\frac{r!}{(r-j+i)!}\,\eta_1^{r-j+i}\,\overline{\eta_1}^{t} \sum_{\substack{k=0\\ i-u \le k \le s}}^{i} \binom{i}{k} \frac{s!}{(s-k)!}\,\eta_2^{s-k}\,\frac{u!}{(u-i+k)!}\,\overline{\eta_2}^{u-i+k}.$$

Setting $\eta_2 = 0$ we see that the terms in the second sum vanish except when $s = k = i-u$. This means, in particular, that $i = v := s+u$. It follows that $[Q_j(\partial)\,\eta^\alpha](1,0)$ vanishes except when $\max\{1, j-r\} \le s+u \le j$ in which case it is equal to

$$a_{j,v}\,\frac{r!}{(r-j+v)!}\binom{v}{s}\,s!\,u!\,.$$

Note that if $\alpha_M = m+\ell$ then by Remark 1.40.(ii) we have $r = b-s$ and $t = c-u$. Hence

$$P_j = \sum_{(s,u)\in I_j} a_{j,s+u}\,\frac{1}{(b+u-j)!\,(c-u)!}\binom{s+u}{s}(iZ_1)^{b-s}(iZ_2)^{s}\,(\overline{iZ_1})^{c-u}\,(\overline{iZ_2})^{u}$$

where

$$I_j := \{(s,u) \in \mathbb{N}_0^2;\ s \le b,\ u \le c,\ \max\{1, j-b+s\} \le s+u \le j\} =$$

$$= \{(s,u) \in \mathbb{N}_0^2;\ s \le b,\ j-b \le u \le c,\ s+u \in \mathbb{N}_j\}.$$

Finally, it is immediately clear from (7.42) that

$$P_0 = \frac{1}{b!\,c!}\,(iZ_1)^b\,(\overline{iZ_1})^c.$$

Now since the coefficients $a_{j,j}$ are equal to 1 we observe that for every $j \in \mathbb{N}_{b+c}$ the monomials

$$(iZ_1)^{b+u-j}\,(iZ_2)^{j-u}\,(\overline{iZ_1})^{c-u}\,(\overline{iZ_2})^{u}, \qquad j-b \le u \le c,$$

appear in P_j but not in P_i for $0 \le i \le j-1$. Hence we conclude that $P_0, \dots, P_{b+c}$ are linearly independent, as desired. ∎

Note that in view of Remark 7.55 the polynomial functions q appearing in Proposition 7.69 satisfy the conditions of Theorem 7.51 if M is replaced by Id_V; in other words, for these q the equation (7.34) has a homogeneous solution S for every homogeneous distribution T on V.

Finally, I would like to point out that in case M is not semi-simple I cannot produce an example q such that the conditions of Theorem 7.51 are violated. On the other hand, the q's appearing in Proposition 7.69 seem to be too simple as to lend support to the conjecture that in case M is not semi-simple the conditions of Theorem 7.51 be always valid.

(f) The Invariant Fundamental Solutions of the Heat and of the Schrödinger Equation

In this section we suppose that $V = \mathbb{R} \times \mathbb{R}^n$ and that for some fixed $w = v - iu \in \mathbb{C} \setminus \{0\}$ q is of the form

(7.58) $$q(\tau,\xi) = i w \tau + |\xi|^2, \qquad (\tau,\xi) \in V^* = \mathbb{R} \times \mathbb{R}^n.$$

Moreover, in this section we set

$$M \cdot (t,x) := (2t,x), \text{ i.e. } M_s \cdot (t,x) = (s^2 t, sx), \qquad s > 0, \ (t,x) \in \mathbb{R} \times \mathbb{R}^n.$$

Obviously, q is quasihomogeneous (of degree 2). It is also invariant under orthogonal transformations of the space variables. We are going to describe the fundamental solutions of q(D) having the corresponding invariance properties. For the sake of brevity, in the present section we are going to use the following terminology:

Definition 7.70. Suppose that X is quasihomogeneous of type M and invariant under the action of the orthogonal group *O(n)* on the space variables x. Then a distribution $T \in \mathcal{D}'(X)$ is called *(almost) invariant of degree* $m \in \mathbb{C}$ if and only if it is (almost) quasihomogeneous of degree m and invariant under the action of O(n) on the x variables.

From the theory of partial differential equations we recall the standard method for constructing fundamental solutions of $q(D)$ with support in a half space. By a formal application of the partial Fourier transform with respect to the space variables x - here denoted by "$\sim$" - the equation $q(D)E = \delta_0$ becomes

$$(7.59) \qquad w\,\partial_t \tilde{E} + |\xi|^2 \tilde{E} = \delta_{\mathbb{R}} \otimes 1_{\mathbb{R}^n}$$

where $\delta_{\mathbb{R}}$ denotes the Dirac distribution on $\mathbb{R}^1$ at $t = 0$. Although in general the restriction of distributions to lines is not well-defined one looks upon (7.59) as an ordinary differential equation depending on ξ as a parameter. As is immediately checked, it has two distinguished solutions, namely

$$(7.60) \qquad t \mapsto \tfrac{\sigma}{w} H(\sigma t) \exp(-\tfrac{t}{w}|\xi|^2), \qquad \sigma \in \{+1, -1\},$$

where $H: \mathbb{R} \longrightarrow \mathbb{R}$ denotes the Heaviside function which is equal to 1 on $]0, +\infty[$ and vanishes on $]-\infty, 0]$. Since $\frac{t}{w} = \sigma t \frac{\sigma}{w}$, as a function of (t, ξ) the expression to the right of the arrow in (7.60) defines a temperate distribution if and only if $\sigma v |w|^{-2} = \operatorname{Re} \frac{\sigma}{w} \geq 0$, i.e. $\sigma v \geq 0$.

Notation 7.71. We fix $\sigma \in \{1, -1\}$ such that $\sigma = \operatorname{sign} v$ if $v \neq 0$, and denote by E the temperate distribution on $\mathbb{R} \times \mathbb{R}^n$ whose partial Fourier transform with respect to x is induced by the function (7.60). If $v = 0$ and $\sigma = \pm 1$ we also write $E^{\pm}$ instead of E.

It is then obvious that E is a fundamental solution of $q(D)$. In order to express E without recourse to the partial Fourier transform one makes use of

$$\mathcal{F} \exp(-a|\cdot|^2) = (\pi/a)^{n/2} \exp(-|\cdot|^2/4a), \qquad a \in \mathbb{C}\setminus\{0\},\ \operatorname{Im} a \geq 0,$$

where $(\cdot)^{1/2}$ denotes that branch of the complex root function defined on $\mathbb{C} \setminus]-\infty, 0]$ which is positive on $]0, +\infty[$. Introducing

$$(7.61) \qquad e(t,x) := (\sigma w)^{(n-2)/2} H(\sigma t)\, (4\pi\sigma t)^{-n/2} \exp(-\tfrac{w}{4t}|x|^2)$$

we arrive at

$$(7.62) \qquad \langle E, \varphi \rangle := \int_{-\infty}^{+\infty} \int_{\mathbb{R}^n} e(t,x)\, \varphi(t,x)\, dx\, dt, \qquad \varphi \in \mathcal{S}'(\mathbb{R} \times \mathbb{R}^n).$$

One observes that the support of E is contained in the half plane $\sigma[0, +\infty[\times \mathbb{R}^n$, its analytic singular support being equal to the hyperplane $\{0\} \times \mathbb{R}^n$. Note that a priori the order of integration cannot be changed. However, in case $v \neq 0$ the function e is C^∞ outside the origin, and the order of integration in (7.62) is arbitrary. In particular, q is hypoelliptic in this case.

Moreover, one observes that E is invariant of degree $-n$ (in the sense of Definition 7.70). The main object of the present section is to prove

Theorem 7.72. *(i) If $v \neq 0$ then E is the unique fundamental solution of $q(D)$ which is almost invariant.*

(ii) If $v = 0$ then a distribution $F \in \mathcal{D}'(\mathbb{R} \times \mathbb{R}^n)$ is an invariant fundamental solution of $q(D)$ if and only if it is of the form

$$(7.63) \qquad F = (1-z)E^+ + zE^- \qquad \text{for some } z \in \mathbb{C};$$

in particular, $E^\pm$ is the unique invariant fundamental solution with support contained in the half space $\pm[0,+\infty[\times \mathbb{R}^n$.

Proof. First of all we note that by Example 2.2 and Corollary 2.36.(i) every almost invariant fundamental solution of q(D) is so of degree $-n$. Since the difference T of two fundamental solutions is a solution of the homogeneous equation

$$(7.64) \qquad q(D)T = 0$$

we have to determine all solutions $T \in \mathcal{D}'(\mathbb{R} \times \mathbb{R}^n)$ of (7.64) which are almost invariant of degree $-n$. For the case $v \neq 0$ this is done in Theorem 7.75 below. In case $v = 0$ one has to employ Theorem 7.77'. It shows, in particular, that the space of invariant solutions of (7.64) is 1-dimensional so that it is spanned by $E^+ - E^-$, indeed. ∎

First of all we determine all almost invariant polynomial functions. Note that by Corollary 1.36 they are all invariant.

Proposition 7.73. *Let $Q: \mathbb{R} \times \mathbb{R}^n \longrightarrow \mathbb{C}$ be a non-constant polynomial function, and let $\ell \in \mathbb{C}$. Then Q is invariant of degree ℓ if and only if ℓ is an even natural number and there are complex numbers $a_0, \ldots, a_{\ell/2}$ such that*

$$(7.65) \qquad Q(\tau,\xi) = \sum_{j=0}^{\ell/2} a_j \tau^j |\xi|^{\ell-2j}, \qquad (\tau,\xi) \in \mathbb{R} \times \mathbb{R}^n.$$

Proof. Let e_1 be a unit vector in $\mathbb{R}^n$. For any fixed $x \in \mathbb{R}^n \backslash \{0\}$ we choose $S \in O(n)$ such that $Sx = s e_1$ where $s := |x|$. If Q is invariant of degree ℓ then

$$Q(t,x) = Q(s^2 t |x|^{-2}, se_1) = s^\ell P(t/|x|^2), \qquad t \in \mathbb{R}, \tag{7.66}$$

where $P(z) := Q(z, e_1)$, $z \in \mathbb{C}$. This means that

$$Q(t,x) = \sum_{j=0}^{d} a_j \tau^j |x|^{\ell-2j}$$

where $d := \deg P$ and where $a_j := P^{(j)}(0)/j!$. Since $x \mapsto |x|^{\ell-2d}$ is a polynomial function if and only if $\ell - 2d$ is a non-negative integer the proof is complete. ■

Next we are going to determine the polynomial solutions of (7.64).

Proposition 7.74. *Let $k \in \mathbb{N}_0$, and let*

$$Q_k(\tau,\xi) = \sum_{j=0}^{k} b_j \tau^j |\xi|^{2k-2j}$$

where

$$b_j := (w/2)^{k-j} \binom{k}{j} \prod_{i=0}^{k-j-1} (2i+n)^{-1}.$$

Moreover, let $Q: \mathbb{R} \times \mathbb{R}^n \longrightarrow \mathbb{C}$ be a polynomial function which is invariant of degree $2k$. Then $q(D)Q = 0$ if and only if $Q = zQ_k$ for some $z \in \mathbb{C}$.

Proof. By Proposition 7.73 Q is of the form (7.65) with $\ell := 2k$. We may assume that $Q \neq 0$ so that the largest index $d \in \mathbb{N}_0$ satisfying $a_d \neq 0$ is well-defined. Then $\ell \geq 2d$. Now, for any $a = 2b \in \mathbb{N}_0$ we have

$$\partial_{\xi_j} |\xi|^a = b\,|\xi|^{2(b-1)}\, 2\xi_j = a\,\xi_j\, |\xi|^{2(b-1)}$$

and

$$\partial_{\xi_j}^2 |\xi|^a = a\,|\xi|^{a-2} + a\,\xi_j\,(b-1)\,|\xi|^{2(b-2)}\, 2\xi_j = a\,|\xi|^{a-2} + a\,(a-2)\,\xi_j^2\,|\xi|^{a-4}$$

and hence $\Delta_\xi |\xi|^a = a(n+a-2)|\xi|^{a-2}$. Consequently, we deduce

$$q(D)Q = \sum_{j=0}^{d} w\,j\,a_j\,\tau^{j-1}|x|^{\ell-2j} - \sum_{j=0}^{d} a_j(\ell-2j)(n+\ell-2j-2)\,\tau^j|x|^{\ell-2j-2} =$$

$$= \sum_{j=0}^{d-1} [w(j+1)a_{j+1} - a_j(\ell-2j)(n+\ell-2j-2)]\,\tau^j|x|^{\ell-2j-2} -$$

$$- a_d(\ell-2d)(n+\ell-2d-2)]\,\tau^d|x|^{\ell-2d-2}.$$

Since $a_d \neq 0$ this vanishes identically if and only if

(7.67.A)) $(\ell-2d)(n+\ell-2d-2)=0$

and

(7.67.B) $w\,(j+1)\,a_{j+1}-a_j\,(\ell-2j)(n+\ell-2j-2)=0\,,$ $0\leq j\leq d-1.$

Since $\ell-2d$ is even and non-negative the equation $n+\ell-2d-2=0$ can only be valid if $\ell=2d$ and $n=2$. Hence (7.67.A) is equivalent to $\ell=2d$, i.e. $d=k$. This, in particular, implies that $(\ell-2j)(n+\ell-2j-2)\geq 2n>0$ if $0\leq j\leq d-1$. Hence, if (7.67.A) is assumed to be valid then (7.67.B) amounts to

$$a_j=\frac{w}{2}\,\frac{j+1}{(d-j)(n+\ell-2j-2)}\,a_{j+1}\,,\qquad 0\leq j\leq d-1.$$

By induction this is seen to be equivalent to

$$a_j=\left(\frac{w}{2}\right)^{d-j}\prod_{i=j}^{d-1}\frac{i+1}{(d-i)(n+2(d-i-1))}\,a_d\qquad 0\leq j\leq d-1.$$

Since

$$\prod_{i=j}^{d-1}(i+1)=\frac{d!}{j!}\,,\quad \prod_{i=j}^{d-1}(d-i)=(d-j)!\,,\quad\text{and}\quad \prod_{i=j}^{d-1}(n+2(d-i-1))=\prod_{i=0}^{d-j-1}(n+2i)$$

the proof is complete. ■

In view of the fact that by Proposition 2.19 and by (2.8) a temperate distribution is invariant of degree m if and only if its Fourier transform $\hat{T}$ is invariant of degree $m^*=-m-n-2$ one obtains the following version of Proposition 7.74.

Proposition 7.74'. *The equation*

(7.64)' $qS=0$

has a non-trivial solution $S\in\mathcal{E}'(\{0\})$ which is invariant of degree m if and only if $k:=-(m+n+2)/2$ is a non-negative integer. In this case every such solution is a complex multiple of $Q_k(-D)\,\delta_0$ where Q_k is defined in Proposition 7.74. ■

It is now easy to determine all almost invariant solutions of (7.64) if $v\neq 0$:

Theorem 7.75. *Suppose that $v\neq 0$. Then the equation $q(D)\,T=0$ has a non-trivial solution $T\in\mathcal{D}'(\mathbb{R}\times\mathbb{R}^n)$ which is almost invariant of degree m if and only if m is a non-negative even integer. All such solutions constitute a 1-dimensional space consisting of the complex multiples of the polynomial functions $Q_{m/2}$ (defined in Proposition 7.74).*

Proof. By Theorem 6.45 it suffices to study temperate solutions of (7.64). Since then the equation (7.64) is equivalent to $q\hat{T} = 0$ and since $v \neq 0$ implies that $q^{-1}(0) \cap \mathbb{R} \times \mathbb{R}^n = \{0\}$ we have to consider only those temperate solutions T such that $\operatorname{supp} \hat{T} \subset \{0\}$, i.e. T is induced by a polynomial function Q. Consequently, since all almost invariant polynomial functions are invariant the assertion follows by Proposition 7.74. ■

We now come to the case $v = 0$. Here we first of all determine a sequence of distinguished almost invariant solutions of (7.64)'. To this end we observe that by

$$(7.68) \qquad \langle S, \varphi \rangle := \int_{S^{n-1}} \varphi(-1/u, \vartheta)\, d\vartheta, \qquad \varphi \in C^\infty(\mathbb{R} \times (\mathbb{R}^n \backslash \{0\})),$$

a distribution $S \in \mathcal{E}'(\mathbb{R} \times (\mathbb{R}^n \backslash \{0\}))$ is well-defined solving the equation (7.64)' and being invariant under the action of $O(n)$ on the space variables ξ. It follows by the assertions (ii) and (iv) of Proposition 6.16 and by Proposition 6.35 that for every $k \in \mathbb{N}_0$ the distribution

$$(7.69) \qquad S_{m,k} := S_{m,\omega_k}$$

on $\mathbb{R} \times \mathbb{R}^n$ is a solution of (7.64)' which is almost invariant such that the support of $(\partial_m - m)^{k+1} S_{m,k}$ is contained in $\{0\}$. Moreover, (6.30) implies that

$$(7.70) \qquad (\partial_M - m) S_{m,k} = -S_{m,k-1}, \qquad k \in \mathbb{N}.$$

Since in case $0 \notin \operatorname{supp} \varphi$ or $\operatorname{Re} m < 0$ we have

$$\varphi_{m,\omega_k}(\zeta) = \int_0^{+\infty} s^{-m} \varphi(M_s \zeta)\, \omega_k(s) \frac{ds}{s}, \qquad \zeta \in (\mathbb{R} \times \mathbb{R}^n) \backslash \{0\},$$

we obtain by introducing polar coordinates that in case $0 \notin \operatorname{supp} \varphi$ or $\operatorname{Re} m > -n-2$ we have

$$(7.69)' \qquad \langle S_{m,k}, \varphi \rangle = (-1)^k \int_{\mathbb{R}^n} |\xi|^{m+2} \varphi(-|\xi|^2/u, \xi)\, \omega_k(|\xi|)\, d\xi .$$

Next we compute $(\partial_M - m) S_{m,0}$:

Proposition 7.76. *(i)* $S_{m,0}$ *is invariant if and only if* $-m-n-2 \notin 2\mathbb{N}_0$.

(ii) If $\ell := -(m+n+2)/2$ *is a non-negative integer then*

(7.71) $\quad (\partial_M - m) S_{m,o} = d_\ell \, Q_\ell(-D)\,\delta_o$

where Q_ℓ *is defined in Proposition 7.74 and where* $d_\ell := |S^{n-1}|\,(-w)^{-\ell}/\ell!$.

Proof. By Proposition 6.35 we have $R := (\partial_M - m) S_{m,o} = Q_m'(S)$, i.e. $R = Q(-D)\,\delta_o$ where

$$Q := \sum_{\alpha \in \mathfrak{A}_m} \frac{i^{|\alpha|}}{\alpha!} a_\alpha Z^\alpha$$

with

$$\mathfrak{A}_m := \{ (j,\beta) \in \mathbb{N}_o \times \mathbb{N}_o^n ;\ 2j + |\beta| = -m-n-2 \}$$

and

$$a_\alpha := \langle S_\zeta , \zeta^\alpha \rangle = (-u)^{-j} \int_{S^{n-1}} \vartheta^\beta \, d\vartheta \,, \qquad \alpha = (j,\beta) \in \mathfrak{A}_m .$$

Since the last integral vanishes if β_i is odd for some $i \in \mathbb{N}_n$ it follows that $Q = 0$ in case $-m-n-2$ is odd. So in this case $S_{m,o}$ is quasihomogeneous. Suppose now that the assumption of (ii) is satisfied. It follows from Corollary 2.36.(ii) that R is a solution of (7.64)', as well. Moreover, by Propositions 6.24.(i) and 6.26.(iv) we conclude that R is invariant of degree m. Hence, by Proposition 7.74' there is a constant $d_\ell \in \mathbb{C}$ such that $Q = d_\ell Q_\ell$. To compute d_ℓ we fix $\chi \in C_o^\infty(\mathbb{R} \times \mathbb{R}^n)$ such that $\chi \equiv 1$ on a neighbourhood of 0 and define $\varphi \in C_o^\infty(\mathbb{R} \times \mathbb{R}^n)$ by $\varphi(\tau,\xi) := \tau^\ell \chi(\tau,\xi)$. Then

$$\langle Q(-D)\delta_o , \varphi \rangle = (Q(D)\varphi)(0) = \frac{1}{\ell!}\, i^\ell (-u)^{-\ell}\, |S^{n-1}|\, i^{-\ell} \ell!$$

and

$$\langle Q_\ell(-D)\delta_o , \varphi \rangle = (Q_\ell(D)\varphi)(0) = i^{-\ell} \ell! \,.$$

This implies the assertion on d_ℓ, and the proof is complete. ∎

Now we can formulate

Theorem 7.77. *Suppose that* $v = 0$. *Then for every* $N \in \mathbb{N}_o$ *the space of solutions* $T \in \mathcal{D}'(\mathbb{R} \times \mathbb{R}^n)$ *of (7.64)' which are invariant of degree m and of order* $\leq N$ *is (N+1)-dimensional, a basis being given by the distributions* $S_{m,j}$, $0 \leq j \leq N$, *in case* $\ell := -(m+n+2)/2$ *does not belong to* $2\mathbb{N}_o$ *and by the distributions* $Q_\ell(-D)\,\delta_o$ *and* $S_{m,j}$, $0 \leq j \leq N-1$, *in case* ℓ *is a non-negative integer.*

For the proof we require the following lemma describing the invariant distributions on the set $X := \mathbb{R}\times(\mathbb{R}^n\setminus\{0\})$. It is also valid in case $v\neq 0$. Here we make use of the C^∞ diffeomorphism $\Psi: X\longrightarrow X$ defined by

$$\Psi(t,x) := (t|x|^2,x), \quad \text{i.e.} \quad \Psi^{-1}(s,x) = (s|x|^{-2},x). \tag{7.72}$$

Note that if $f: X\longrightarrow\mathbb{C}$ is a function which is invariant of degree m then - compare the proof of (7.66) - we have

$$|x|^{-m}(f\circ\Psi)(t,x) = f(t,e_1), \qquad (t,x)\in X,$$

where e_1 is any unit vector in $\mathbb{R}^n$. That this remains valid for arbitrary distributions on X is the content of

Lemma 7.78. *If $X = \mathbb{R}\times(\mathbb{R}^n\setminus\{0\})$ then for every $T\in\mathcal{D}'(X)$ the following conditions are equivalent:*

(a) T is invariant of degree m;

(b) $\partial_{x_j}(|x|^{-m}T\circ\Psi) = 0$ for every $j\in\mathbb{N}_n$, and if $n=1$ then T is invariant under the transformation $(t,x)\mapsto(t,-x)$;

(c) there is a distribution $R\in\mathcal{D}'(\mathbb{R})$ such that

$$T = (R\otimes|x|^m)\circ\Psi^{-1}. \tag{7.73}$$

The distribution R satisfying (7.73) is unique and will be denoted by T^1.

Proof. (a)⇒(b). First we are going to show that

$$x_j\,\partial_{x_k}T = x_k\,\partial_{x_j}T, \qquad j,k\in\mathbb{N}_n. \tag{7.74}$$

We may suppose that $j<k$. Then for $\vartheta\in\mathbb{R}$ we define $S_\vartheta\in SO(1+n)$ by

$$(t,x)\mapsto(t,x_1,\dots,x_{j-1},x_j\cos\vartheta - x_k\sin\vartheta, x_{j+1},\dots,x_{k-1},x_j\sin\vartheta + x_k\cos\vartheta, x_{k+1},\dots,x_n).$$

Since $T\circ(S_\vartheta)^{-1} = T$ we obtain

$$0 = \frac{d}{d\vartheta}\langle T,\varphi\circ S_\vartheta\rangle\Big|_{\vartheta=0} = \langle T, -\partial_{x_j}\varphi\, x_k + \partial_{x_k}\varphi\, x_j\rangle, \qquad \varphi\in C_0^\infty(X),$$

which immediately implies (7.74).

Now, by the chain rule we have

$$\partial_{x_j}(|x|^{-m}T\circ\Psi) =$$

$$= \partial_{x_j}(|x|^2)^{-m/2}\,T\circ\Psi + |x|^{-m}\,2tx_j\,(\partial_t T)\circ\Psi + |x|^{-m}(\partial_{x_j}T)\circ\Psi =$$

$$= |x|^{-m-2}[-mx_j\, T\circ\Psi + 2t|x|^2 x_j(\partial_t T)\circ\Psi + \sum_{k=1}^{n} x_k x_k(\partial_{x_j} T)\circ\Psi].$$

Since $t|x|^2$ is equal to the first component of $\Psi(t,x)$ inserting (7.74) leads to

$$|x|^{-m-2} x_j [2t\partial_t T + \sum_{k=1}^{n} x_k \partial_{x_k} T - mT]\circ\Psi .$$

Since by Euler's equation the term in brackets vanishes the derivation of (b) is complete.

<u>*(b)⇒(c)*</u>. We first suppose that $n\geq 2$. By the theory of distributions the condition (b) implies that there is a unique distribution $R\in\mathcal{D}'(\mathbb{R})$ such that

$$(7.73)' \qquad |x|^{-m}\, T\circ\Psi = R\otimes I$$

where I denotes the distribution on $\mathbb{R}^n\setminus\{0\}$ induced by the constant function 1. Obviously, this is equivalent to (7.73).

We suppose now that $n=1$. Then the theory of distributions yields the existence of unique distributions $R_+, R_- \in\mathcal{D}'(\mathbb{R})$ satisfying

$$|x|^{-m}\, T\circ\Psi\big|_{\mathbb{R}\times]0,\pm\infty[} = R_\pm\otimes(I\big|_{]0,\pm\infty[}) .$$

Since T and hence $|x|^{-m}\, T\circ\Psi\big|_{\mathbb{R}\times]0,\pm\infty[}$ are invariant under the transformation $(t,x)\mapsto(t,-x)$ it follows that $R_+ = R_-$.

<u>*(c)⇒(a)*</u>. That T is invariant under the action of O(n) on the x-variables is obvious. For arbitrary $s\in]0,+\infty[$ and $\varphi\in C_0^\infty(X)$ we obtain

$$\langle T\circ M_s,\varphi\rangle = s^{-n-2}\langle T,\varphi\circ M_{1/s}\rangle =$$

$$= s^{-n-2}\langle R\otimes|x|^m,\varphi\circ M_{1/s}\circ\Psi\,\det\Psi'\rangle =$$

$$= s^m\langle R_t,\int_{\mathbb{R}^n}(|x|/s)^{m+2}\varphi(t(|x|/s)^2,x/s)\,d(x/s)\rangle =$$

$$= s^m\langle R_t,\int_{\mathbb{R}^n}|y|^{m+2}\varphi(t|y|^2,y)\,dy\rangle = s^m\langle T,\varphi\rangle . \quad\blacksquare$$

As a consequence of the preceding lemma we are going to derive

Proposition 7.79. *We set $X:=\mathbb{R}\times(\mathbb{R}^n\setminus\{0\})$. Suppose that $v=0$. Let $N\in\mathbb{N}_0$. Then the space of solutions $T\in\mathcal{D}'(X)$ of (7.64)' which are almost invariant of degree m and of order $\leq N$ is (N+1)-dimensional, a basis being given by the distributions*

$$S_{m,j}\big|_X , \qquad 0\leq j\leq N.$$

Proof: by induction on N. Let $T \in \mathcal{D}'(X)$ be an invariant solution of (7.64)'. By Ψ we denote the diffeomorphism defined by (7.72). Since $(q \circ \Psi)(\tau,\xi) = |\xi|^2 p_w(\tau)$ where $p_w(\tau) := u\tau + 1$ we deduce - taking Lemma 7.78 into account - that

$$(P_w T) \circ \Psi = (p_w T^1) \otimes |\xi|^{m+2}.$$

Consequently, $P_w T = 0$ if and only if $p_w T^1 = 0$. Since $(p_w)^{-1}(0) = \{-1/u\}$ and since $-1/u$ is a zero of p_w of order 1 the equation $p_w T^1 = 0$ is satisfied if and only if $T^1 = c\delta_{-1/u}$ for some $c \in \mathbb{C}$. Since from (7.69)' one deduces that

$$(S_{m,o}|_X)^1 = \delta_{-1/u}$$

the proof of the case $N = 0$ is complete.

In order to prove the induction step we fix $N \in \mathbb{N}$ and a solution $T \in \mathcal{D}'(X)$ of (7.64)' which is almost invariant of degree m and of order $\leq N$. Since then $(\partial_M - m)^N T$ is a solution of (7.64)' which is invariant of degree m it is equal to $c\, S_{m,o}|_X$ for some $c \in \mathbb{C}$. Since it follows from (7.70) that $(\partial_M - m)^N S_{m,N} = (-1)^N S_{m,o}$ we conclude that $R := T - (-1)^N c\, S_{m,N}|_X$ is almost invariant of degree m and of order $\leq N-1$. Since R is a solution of (7.64)', as well, the induction hypothesis implies that it is a linear combination of the distributions $S_{m,j}|_X$, $0 \leq j \leq N-1$, i.e. T is a linear combination of the distributions $S_{m,j}|_X$, $0 \leq j \leq N$. That the latter distributions are linearly independent follows from (7.70) and the fact that $S_{m,o}|_X$ is quasihomogeneous. ■

Proof of Theorem 7.77. In view of what was said in the text preceding Proposition 7.76 we already know that the distributions mentioned in the assertion are almost invariant solutions of (7.64)'. Conversely, let $T \in \mathcal{D}'(\mathbb{R} \times \mathbb{R}^n)$ be a solution of (7.64)' which is almost invariant of degree m and of order $\leq N$. Then by Proposition 7.79 we find a sequence of complex numbers c_i, $0 \leq i \leq N$, such that the support of the distribution

$$R := T - \sum_{j=0}^{N} c_j S_{m,j}$$

is contained in $\mathbb{R} \times \{0\}$. Since the supports of T and $S_{m,j}$ are contained in $q^{-1}(0)$ this implies that $\operatorname{supp} R \subset \{0\}$. Since R is almost invariant and solves (7.64)' we deduce from Proposition 7.74' that R vanishes in case $\ell := -(m+n+2)/2$ is not a non-negative integer and is quasihomogeneous of degree m, i.e. $(\partial_M - m) R = 0$ in case $\ell \in \mathbb{N}_0$. Now, in the latter case, in view of

$$(\partial_M - m)^{N+1} S_{m,j} = (-1)^j (\partial_M - m)^{N+1-j} S_{m,o} = 0, \qquad 0 \le j < N,$$

(see (7.70)) we deduce by Proposition 7.76.(ii) and (7.70) that

$$c_N \, d_\ell \, Q_\ell(-D)\, \delta_o = (-1)^N (\partial_M - m)^{N+1} T = 0$$

so that $c_N = 0$ in this case, as desired. ■

In view of Proposition 2.19 and (2.8) Theorem 7.77 can be reformulated as

Theorem 7.77'. *Suppose that* $v=0$. *Then for every* $N \in \mathbb{N}_o$ *the space of solutions* $T \in \mathcal{D}'(\mathbb{R}\times\mathbb{R}^n)$ *of (7.64) which are almost invariant of degree* m *and of order* $\le N$ *is (N+1)-dimensional, a basis being given by the distributions*

$$T_{m,j} := \mathcal{F}^{-1}(S_{-m-n-2,j}), \qquad 0 \le j \le N,$$

in case $\ell := m/2$ *does not belong to* $\mathbb{N}_o$ *and by the distributions* T_{Q_ℓ} *(see Proposition 7.74) and* $T_{m,j}$, $0 \le j \le N-1$, *in case* ℓ *is a non-negative integer.* ■

From Theorems 7.77' amd 7.72 an explicit formula for $T_{-n,o}$ is obtained:

(7.75) $\qquad T_{-n,o} = \text{const}\,(E^+ - E^-).$

This shows, in particular, that the singular and the analytic singular support of $T_{-n,o}$ coincide with $\{0\}\times\mathbb{R}^n$. This is valid in general:

Theorem 7.80. *Let* $T \in \mathcal{S}'(\mathbb{R}\times\mathbb{R}^n)$ *be an almost invariant solution of the equation (7.64) which is not induced by a polynomial function. Then its singular and its analytic singular support coincide and are equal to* $\{0\}\times\mathbb{R}^n$.

Proof. "*a-sing supp* $T \subset \{0\}\times\mathbb{R}^n$". In view of Theorem 7.77' it suffices to fix $k \in \mathbb{N}_o$ and to assume that $T = T_{m,k}$. First we deal with *the case "*$\mathrm{Re}\, m < 0$*"*. We fix $\sigma \in \{1,-1\}$, $a,b,R \in\,]0,+\infty[$ such that $a<b$, and $\varphi \in C_o^\infty([\sigma a, \sigma b]\times K(0,R))$. We observe that by the Fourier inversion formula, by (7.68), by Fubini's Theorem and by Lebesgue's Dominated Convergence Theorem we have

$$(2\pi)^{n+1}(-1)^k \langle T_{m,\omega_k}, \varphi\rangle = (-1)^k \langle S_{m^*,\omega_k}, \check{\hat{\varphi}}\rangle =$$

$$= \int_{S^{n-1}} (\check{\hat{\varphi}})_{m,\omega_k}(-1/u, \vartheta)\, d\vartheta = \lim_{\varepsilon\to 0} \int_\varepsilon^{1/\varepsilon} s^{-m}\, \omega_k(s) \int_{S^{n-1}} \hat{\varphi}(s^2/u, -s\vartheta)\, d\vartheta\, \frac{ds}{s}.$$

Next we are going to change the contour of integration. To this end we have to identify the integrand as a holomorphic function on a suitable open subset of $\mathbb{C}$. Employing that branch of the logarithm function which is holomorphic on the right half plane $\mathfrak{H} := \{\zeta \in \mathbb{C};\ \mathrm{Re}\,\zeta > 0\}$ and real-valued on $]0,+\infty[$ we extend the function $t \mapsto t^{-m-1}\omega_k(t)$ to a holomorphic function $h_m : \mathfrak{H} \longrightarrow \mathbb{C}$ satisfying the estimate

$$(7.76) \qquad |h_m(\zeta)| \le C\,|\zeta|^{\ell}\,(1+|\log|\zeta||)^k, \qquad \zeta \in \mathfrak{H},$$

where $C := \pi^k \exp(\pi\,|\mathrm{Im}\,m|)$ and $\ell := -\mathrm{Re}\,m - 1$.

Moreover, denoting by H the supporting function of the set $[\sigma a, \sigma b] \times \overline{K}(0,R)$ we see that $H(\tau, -\upsilon) = -a|\tau| + R|\upsilon|$ for every $\upsilon \in \mathbb{R}^n$ and for every $\tau \in \mathbb{R}$ satisfying $\sigma\tau \le 0$. Since for every $\zeta \in \Omega := \{z \in \mathbb{C};\ \mathrm{Re}\,z > 0,\ \sigma\,\mathrm{sign}(u)\,\mathrm{Im}\,z \le 0\}$ we have

$$\sigma\,\mathrm{Im}\,\zeta^2/u = 2\frac{\sigma}{u}\,\mathrm{Re}\,\zeta\,\mathrm{Im}\,\zeta \le 0$$

it follows that with the abbreviation $A := 2a/|u|$ we have

$$H(\mathrm{Im}(\zeta^2/u, -\zeta\vartheta)) = -(A|\mathrm{Re}\,\zeta| - R)\,|\mathrm{Im}\,\zeta|, \qquad \zeta \in \overline{\Omega},\ \vartheta \in S^{n-1}.$$

Consequently, in view of the Paley-Wiener-Schwartz theorem by

$$F(\zeta) := \int_{S^{n-1}} \hat{\varphi}(\zeta^2/u, -\zeta\vartheta)\, d\vartheta, \qquad \zeta \in \mathbb{C},$$

an entire holomorphic function is well-defined such that for every $N \in \mathbb{N}$ there is a constant C_N such that

$$(7.77) \qquad |F(\zeta)| \le C_N\,(1+|\zeta|)^{-N} \exp(-(A|\mathrm{Re}\,\zeta| - R)\,|\mathrm{Im}\,\zeta|), \qquad \zeta \in \overline{\Omega}.$$

Now, we define $\gamma_\varepsilon : [0, \pi/4] \longrightarrow \Omega$ by $s \mapsto \varepsilon \exp(-is\sigma\,\mathrm{sign}\,u))$, set $c := (1 - i\sigma\,\mathrm{sign}\,u)/\sqrt{2}$, and - noting that $\gamma_\varepsilon(\pi/4) = \varepsilon c$ - obtain by Cauchy's theorem that

$$\int_{\varepsilon}^{1/\varepsilon} s^{-m}\,\omega_k(s) \int_{S^{n-1}} \hat{\varphi}(s^2/u, -s\vartheta)\, d\vartheta\, \frac{ds}{s} =$$

$$= c\int_{\varepsilon}^{1/\varepsilon} h_m(cs)\,F(cs)\,ds + \int_{\gamma_\varepsilon} h_m(\zeta)\,F(\zeta)\,d\zeta - \int_{\gamma_{1/\varepsilon}} h_m(\zeta)\,F(\zeta)\,d\zeta, \qquad \varepsilon \in]0,1[.$$

In view of (7.76) and by the fact that $\ell > -1$ it follows that

$$\lim_{\varepsilon \to 0} \varepsilon \sup\{|h_m(\zeta)|;\ \zeta \in \mathfrak{H},\ |\zeta| = \varepsilon\} = 0.$$

Since F is bounded on K(0,1) this implies that

$$\lim_{\varepsilon \to 0} \int_{\gamma_\varepsilon} h_m(\zeta)\,F(\zeta)\,d\zeta = 0.$$

Combining (7.76) and (7.77) we likewise obtain that

$$\lim_{\varepsilon\to 0} \int_{\gamma_{1/\varepsilon}} h_m(\zeta)\, F(\zeta)\, d\zeta = 0\,.$$

Moreover, the estimates (7.76) and (7.77) show that the function $]0,+\infty[\longrightarrow\Omega$, $s\mapsto h_m(cs)\,F(cs)$, is absolutely integrable. Consequently, by Lebesgue's theorem, again, we obtain that

$$\langle T_{m,k}, \varphi\rangle = (-1)^k\,(2\pi)^{-n-1}\, c\int_0^{+\infty} h_m(cs)\, F(cs)\, ds\,.$$

Now we can show that near the support of φ $T_{m,k}$ is induced by a real analytic function. Since for arbitrary $s\in\,]0,+\infty[$, $\vartheta\in S^{n-1}$, $x\in\mathbb{R}^n$, and $t\in\sigma]0,+\infty[$ we have

$$\mathrm{Re}\,\langle -i(c^2s^2/u, -cs\vartheta), (t,x)\rangle \le -s^2|t|/|u| + s|x|/\sqrt{2})$$

it follows that on the set $X_\sigma := (\sigma]0,+\infty[)\times\mathbb{R}^n$ by

$$f_\sigma(t,x) := (-1)^k\,(2\pi)^{-n-1}\, c\int_0^{+\infty}\int_{S^{n-1}} h_m(cs)\exp(-s^2|t|/|u| + ics\langle x,\vartheta\rangle)\, d\vartheta\, ds$$

a continuous function $f_\sigma: X_\sigma \longrightarrow \mathbb{C}$ is well-defined satisfying (by the Fubini and the Fubini-Tonelli theorem):

$$\langle T_{m,k}, \varphi\rangle = \int_{X_\sigma} \varphi(z)\, f(z)\, dz\,.$$

It remains to be proved that F is real analytic. To do this we fix $\varepsilon>0$, set $\rho := \sqrt{2}\,R|u|/\varepsilon$ and observe that for $t\in\mathbb{R}\setminus\,]-\varepsilon,\varepsilon[$, $x\in K(0,R)$, and $s\in[\rho,+\infty[$ we have

$$\mathrm{Re}(-s^2|t|/|u| + ics\langle x,\vartheta\rangle) \le -\eta s^2 \quad\text{where } \eta := \frac{\varepsilon}{2|u|}\,.$$

It follows that in the double integral

$$g_\rho(t,x) := \int_\rho^{+\infty}\int_{S^{n-1}} h_m(cs)\exp(-s^2|t|/|u| + ics\langle x,\vartheta\rangle)\, d\vartheta\, ds$$

we may interchange differentiation and integration so that

$$(\partial_t^k\partial_x^\beta g_\rho)(t,x) =$$

$$= (-\sigma/|u|)^k\,(ic)^{|\beta|}\int_\rho^{+\infty}\int_{S^{n-1}} s^{2k+|\beta|}\, h_m(cs)\,\vartheta^\beta\exp(-s^2|t|/|u| + ics\langle x,\vartheta\rangle)\, d\vartheta\, ds$$

for arbitrary $k\in\mathbb{N}_0$ and $\beta\in\mathbb{N}_0^n$. By (7.76) and since we may assume without loss of generality that $\rho\ge 1$ we obtain a constant C' such that

$$|(\partial_t^k\partial_x^\beta g_\rho)(t,x)| \le C'\,|u|^{-k}\,|S^{n-1}|\,B(k+|\beta|+L)$$

where L is the smallest natural number strictly larger than $\frac{\ell-1}{2}$ and where

$$B(j) := \int_1^{+\infty} s^{2j} \exp(-\eta s^2)\, 2s\, ds = \eta^{-j-1} \int_\eta^{+\infty} r^j \exp(-r)\, dr \le \eta^{-j-1} C(j)$$

with

$$C(j) := \int_0^{+\infty} r^j \exp(-r)\, dr = j!\,, \qquad j \in \mathbb{N}_0\,.$$

Since $(k+|\beta|+L)! \le 2^{k+|\beta|+L} k!\,(|\beta|+L)!$, since $(|\beta|+L)! \le 2^{|\beta|+L} L!\,|\beta|!$, and since $|\beta|! \le 2^{n|\beta|}\beta!$ this shows that g_ρ is real analytic on $(\mathbb{R}\setminus]-\varepsilon,\varepsilon[) \times K(0,R)$. Since by inserting the Taylor expansion of the exponential function one sees that the funktion $\dot{\mathbb{R}} \times \mathbb{R}^n \longrightarrow \mathbb{C}$ defined by

$$(t,x) \mapsto \int_0^{\rho} \int_{S^{n-1}} h_m(cs) \exp(-s^2|t|/|u| + ics\langle x,\vartheta\rangle)\, d\vartheta\, ds$$

is real analytic, as well, we conclude that f_σ is real analytic, indeed.

<u>*The general case.*</u> It suffices to show that for arbitrary $m \in \mathbb{C}$ and $k \in \mathbb{N}_0$ we have

$$\text{(7.78)} \qquad WF_A(T_{m,k}) = WF_A(T_{m+2,k})\,.$$

For then the general case follows from the special case proved above by induction. Now, from (7.69)' we deduce that $|\xi|^2 S_{m^*-2,k}$ is equal to $S_{m^*,k}$ on $\mathbb{R} \times (\mathbb{R}^n\setminus\{0\})$. Since the support of both distributions is contained in $q^{-1}(0)$ and since

$$\mathbb{R} \times \{0\} \cap q^{-1}(0) = \{(0,0)\}$$

we conclude that the distributions differ by a distribution with support contained in $\{(0,0)\}$. This implies that the distributions $-\Delta_\xi T_{m+2,k}$ and $T_{m,k}$ differ by a polynomial function. Consequently,

$$\text{(7.79.a)} \qquad WF_A(\Delta_\xi T_{m+2,k}) = WF_A(T_{m,k})\,.$$

Moreover, the equation (7.64) means that $w\,\partial_t T_{m+2,k} = \Delta_\xi T_{m+2,k}$ so that

$$\text{(7.79.b)} \qquad WF_A(\partial_t T_{m+2,k}) = WF_A(T_{m,k})\,.$$

Since $WF_A(\Delta_\xi T_{m+2,k}) \subset WF_A(T_{m+2,k})$ it follows by (7.79.a) that

$$WF_A(T_{m,k}) \subset WF_A(T_{m+2,k})\,.$$

Conversely, by Theorem 8.6.1 in Hörmander [11] it follows from (7.79) that

$$WF_A(T_{m+2,k}) \subset WF_A(T_{m,k}) \cup [\mathrm{Char}(\Delta_\xi) \cap \mathrm{Char}(\partial_t)]$$

where *Char(P)* denotes *the characteristic set of P*. Since

$$\mathrm{Char}(\Delta_\xi) = \mathbb{R}\times\mathbb{R}^n\times\{(\tau,0);\ \tau\in\mathbb{R}\setminus\{0\}\}$$

and

$$\mathrm{Char}(\partial_t) = \mathbb{R}\times\mathbb{R}^n\times\{(0,\xi);\ \xi\in\mathbb{R}^n\setminus\{0\}\}$$

the intersection of both sets is empty so that $WF_A(T_{m+2,k}) \subset WF_A(T_{m,k})$. So the proof of (7.78) is complete.

"*$\{0\}\times\mathbb{R}^n\subset$ sing supp T*". If sing supp $T\subset\{0\}$ it then follows by Proposition 2.23 that sing supp $\hat{T}\subset\{0\}$ which in view of supp $\hat{T}\subset q^{-1}(0)$ implies that supp $\hat{T}\subset\{0\}$ contradicting the assumption on T. Consequently, in view of the inclusion already proven above we find $\xi\in\mathbb{R}^n\setminus\{0\}$ such that $(0,\xi)$ belongs to sing supp T. Since T is invariant it follows that $\{(0,\xi);\ \xi\in\mathbb{R}^n\setminus\{0\}\}$ lies in sing supp T. Since the latter is a closed subset of $\mathbb{R}\times\mathbb{R}^n$ the desired inclusion follows. ∎

Chapter VIII

Extending (Almost) Quasihomogeneous Distributions on X_+ to the Whole of X

As in Chapter 6 we assume that (1.14) holds and that $X \neq X_+$. In section (a) the theory of section 4.(e) is carried over to the following setting: for any fixed C^∞ function $\varkappa : X_+ \longrightarrow]0,+\infty[$ which is quasihomogeneous of degree 1 every distribution on the hypersurface $S^\varkappa$ is extended to an (almost) quasihomogeneous distribution on X. In particular, this leads to an alternative description of almost quasihomogeneous distributions on X (see Theorem 8.8 below).

In section (b) it is shown how the method of taking quasihomogeneous averages as well as the construction of section (a) can be used to extend (almost) quasihomogeneous distributions on X_+ to (almost) quasihomogeneous distributions on the whole of X. Employing these (almost) quasihomogeneous extensions, for quasihomogeneous C^∞ functions $q : X \longrightarrow \mathbb{C}$ without zeros in X_+ one can easily solve the equation $qS = T$ in the set of almost quasihomogeneous distributions (see Theorem 8.24 below).

While the extensions constructed in section (b) are not always unique, in section (c) for every $f \in \mathcal{S}(V)$ a distinguished extension of T_{f_m} to the whole of V is constructed. Its Fourier transform is computed in section (d). This leads to a formula for the Fourier transform of u_m if u is any temperate distribution with M-temperate support.

It should be pointed out that the present chapter is completely independent of Chapter 7 and can be read subsequent to Chapter 6.

(a) Pulling Back Distributions on $S^{\varkappa}$ to (Almost) Quasihomogeneous Distributions on X

We fix a C^{∞} function $\varkappa : X_+ \longrightarrow]0,+\infty[$ which is quasihomogeneous of degree 1 (this is possible by Corollary 3.36 and Theorem 3.39). Recall from sections 1.(f) and 4.(e) that $S^{\varkappa}$ is a C^{∞} submanifold of X and that $\mathscr{D}'(S^{\varkappa})$ is identified with the dual of $C_0^{\infty}(S^{\varkappa})$ via the density $\omega_M|_{S^{\varkappa}}$ - see (4.32). In analogy with Lemma 6.23 we have

Lemma 8.1. *If K is a compact subset of X then the set $S^{\varkappa} \cap M_0^{-1}(K)$ is compact.*

Proof. By Lemma 6.23 we find a compact subset L of X such that the set L_M contains $M_0^{-1}(K)$. By Remark 6.8 the set $S^{\varkappa} \cap \overline{L_M}$ is compact. Since by Lemma 5.46 $S^{\varkappa} \cap M_0^{-1}(K)$ is a closed subset of $S^{\varkappa} \cap \overline{L_M}$ the assertion follows. ∎

The following proposition corresponds to Proposition 6.24.

Proposition 8.2. *Let $\nu \in \mathscr{D}'(S^{\varkappa})$. By $\langle Q'_m \nu, \varphi \rangle := \langle \nu, (Q_{m^*}\varphi)|_{S^{\varkappa}} \rangle$ a distribution $Q'_m \nu \in \mathscr{D}'(X)$ is well-defined having the following properties:*

(i) $Q'_m \nu$ is almost quasihomogeneous of degree m and of order $\leq N(m^*)$;

(ii) the support of $Q'_m \nu$ is contained in $M_0(\operatorname{supp} \nu) \cap X$, the latter being a closed subset of $X \setminus X_+$ and hence closed in X;

(iii)
$$Q'_m \nu = \sum_{\alpha \in \mathfrak{A}^+_{m^*}} \frac{1}{\alpha!} (-\partial)^{\alpha} \, \widehat{\vartheta^{\alpha} \nu}$$
where $\widehat{\nu}$ is defined by

$$\widehat{\nu} := Q'_{-\mu} \nu . \tag{8.1}$$

Proof. Let K be a compact subset of X. By Lemma 8.1 $L := S^{\varkappa} \cap M_0^{-1}(K)$ is a compact subset of $S^{\varkappa}$. In view of Corollary 5.47 the map $C_0^{\infty}(K) \longrightarrow \mathscr{D}(L)$, $\varphi \mapsto (Q_{m^*}\varphi)|_{S^{\varkappa}}$, is well-defined and, of course, linear and continuous. Consequently, $Q'_m \nu$ is a well-defined distribution on X.

(i): By Proposition 5.48.(iv) and Proposition 5.45 we deduce for every $t \in]0,+\infty[$ that for $N := N(m^*)$ we have

$$Q_{m^*}(\varphi \circ M_{1/t}) = (Q_{m^*}\varphi) \circ M_{1/t} = t^{-m^*} \sum_{j=0}^{N} \omega_j(1/t)\,(\partial_M - m^*)^j\, Q_{m^*}\varphi\,.$$

This implies the assertion.

_(ii)_: Since $S^{\varkappa}$ is a closed subset of X $\operatorname{supp}\nu$ is a closed subset of X, as well. By Lemmata 6.7 and 6.9.(ii) $M_0(\operatorname{supp}\nu) \cap X$ is a closed subset of X. Now, if K is a compact subset of X such that $K \cap M_0(\operatorname{supp}\nu) = \emptyset$ then $\operatorname{supp}\nu$ does not intersect the set $M_0^{-1}(K)$ so that in view of Corollary 5.47 we conclude that $\langle Q'_m\nu, \varphi\rangle = 0$ for every $\varphi \in C_0^{\infty}(K)$.

(iii): this is clear from the definitions. ■

Next we state the analogue of Propositions 6.15 (for $I =]0,+\infty[$) and 6.35. To this end we fix a $(1+k)$-tuple $\nu = (\nu_0, \dots, \nu_k)$ of distributions $\nu_i \in \mathcal{D}'(S^{\varkappa})$, $0 \le i \le k$.

Proposition 8.3. *By (4.33) a distribution $\nu_m \in \mathcal{D}'(X)$ is well-defined having the following properties:*

(i) $\operatorname{supp}\nu_m \subset \bigcup_{i=0}^{k} \overline{(\operatorname{supp}\nu_i)_M}$;

(ii) *ν_m is almost quasihomogeneous of degree m and of order $\le k + N(m^*) + 1$ such that*

$$(\partial_M - m)^j \nu_m = (\nu_j, \dots, \nu_k)_m + \sum_{i=0}^{j-1} (\partial_M - m)^{j-i-1} Q'_m \nu_i\,, \qquad j \in \mathbb{N}_k\,, \tag{8.2.a}$$

and

$$(\partial_M - m)^j \nu_m = \sum_{i=0}^{k} (\partial_M - m)^{j-i-1} Q'_m \nu_i\,, \qquad j \ge k+1\,. \tag{8.2.b}$$

(iii) by $\mathfrak{M}_\nu(m) := \nu_m$, $m \in \mathbb{C} \setminus (-\mathfrak{A}(M) - \mu)$, a meromorphic function $\mathfrak{M}_\nu : \mathbb{C} \longrightarrow \mathcal{D}'(X)$ is defined, its poles lying in $(-\mathfrak{A}(M) - \mu)$; for every $m \in (-\mathfrak{A}(M) - \mu)$ $a_0(m; \mathfrak{M}_\nu)$ equals ν_m.

Note that ν_m extends the distribution defined in Proposition 4.24.(i) with X replaced by X_+. That both distributions are denoted by one and the same symbol should therefore not lead to confusion.

Proof. Let K be a compact subset of X. Then by Remark 6.8 $L := S^{\varkappa} \cap \overline{K_M}$ is a compact subset of $S^{\varkappa}$. Since in view of Theorem 5.37.(ii) the map $C_0^{\infty}(K) \longrightarrow \mathcal{D}(L)$ defined by $\varphi \mapsto \varphi_{m^*}|_{S^{\varkappa}}$ is well-defined, linear and continuous it follows that ν_m is a well-defined distribution on $S^{\varkappa}$.

(i): Since $\operatorname{supp} \nu_i$ is a closed subset of X Lemma 6.9.(iii) shows that $\overline{K_M} \cap \operatorname{supp} \nu_i$ is empty for any compact subset K of X such that $K \cap \overline{(\operatorname{supp} \nu_i)_M}$ is empty. In view of Theorem 5.37.(ii) the assertion follows.

(ii): from (6.31) we deduce that

$$t^{-m} \langle \nu_m \circ M_t, \varphi \rangle = t^{m^*} \sum_{i=0}^{k} \langle \nu_i, (\varphi \circ M_{1/t})_{m^*,\omega_i}|_{S^\varkappa} \rangle = A + B$$

where

$$A := \sum_{i=0}^{k} \sum_{j=0}^{i} \omega_j(t) \langle \nu_i, \varphi_{m^*,\omega_{i-j}}|_{S^\varkappa} \rangle$$

and

$$B := \sum_{i=0}^{k} \sum_{j=0}^{\infty} (-1)^j \omega_{i+1+j}(t) \langle \nu_i, (\partial_M - m^*)^j Q_{m^*} \varphi|_{S^\varkappa} \rangle .$$

Changing the order of summation and substituting $I = i-j$ we obtain

$$A = \sum_{j=0}^{k} \sum_{i=j}^{k} \omega_j(t) \langle \nu_i, \varphi_{m^*,\omega_{i-j}}|_{S^\varkappa} \rangle =$$

$$= \sum_{j=0}^{k} \omega_j(t) \Big[\sum_{I=0}^{k-j} \langle \nu_{I+j}, \varphi_{m^*,\omega_I}|_{S^\varkappa} \rangle \Big] = \sum_{j=0}^{k} \omega_j(t) (\nu_j, \ldots, \nu_k)_m .$$

Moreover, substituting $J = i+1+j$ and changing the order of summation leads to

$$B = \sum_{i=0}^{k} \sum_{J=i+1}^{\infty} \omega_J(t) \langle \nu_i, (-\partial_M + m^*)^{J-i-1} Q_{m^*} \varphi|_{S^\varkappa} \rangle =$$

$$= \sum_{J=0}^{\infty} \omega_J(t) \sum_{i=0}^{\min\{J-1,k\}} \langle \nu_i, (-\partial_M + m^*)^{J-i-1} Q_{m^*} \varphi|_{S^\varkappa} \rangle .$$

Consequently, in view of $(\partial_M - m^*)^j (Q_{m^*}\varphi) = Q_{m^*}((\partial_M - m^*)^j \varphi)$ and (2.5) the assertion follows.

(iii): this follows from the corresponding assertions about $\mathfrak{M}_{\varphi,\omega_i}$ - see Theorem 5.37.(iv) - because (compare (6.14))

$$\langle a_j(m;\mathfrak{M}_\nu), \varphi \rangle = (-1)^j \sum_{i=0}^{k} \langle \nu_i, a_j(m^*;\mathfrak{M}_{\varphi,\omega_i}) \rangle, \qquad \varphi \in C_0^\infty(X),\ j \in \mathbb{Z} . \tag{8.3}$$ ∎

If the distributions ν_i are induced by functions and if $\operatorname{Re} m > -\mu$ then ν_m is easily computed:

Remark 8.4. *Let $g_0, \ldots, g_k \in \mathcal{L}^1_{loc}(S^\varkappa; \tilde{\varkappa})$. If $\operatorname{Re} m > -\mu$ then*

$$(T_{g_0}, \ldots, T_{g_k})_m = T_f$$

where $f \in \mathcal{L}^1_{loc}(X)$ is defined by

$$f/_{X_+} := \varkappa^m \sum_{i=0}^{k} \omega_i \circ \varkappa \, g_i \circ \rho_{\varkappa} \quad \textit{(see Notation 1.71)} \quad \textit{and} \quad f/_{X \setminus X_+} :\equiv 0 .$$

Proof. First of all we observe that by Remark 1.73'.(i) f is almost quasihomogeneous of degree m and of order $\leq$ N and that for every $k \in \{0\} \cup \mathbb{N}_N$ the restriction of its kth order deficiency to $S^{\varkappa}$ is equal to g_k. Hence, by Lemma 1.90 f is locally integrable. By (4.34) $(T_{g_0}, \dots, T_{g_k})_m$ and T_f coincide on X_+. Since they are both almost quasihomogeneous of degree m and since by the assumption on m we have $m \notin (-\mathfrak{A}(M) - \mu)$ they coincide everywhere by Corollary 6.28. ∎

Likewise, if from the outset the components of ν are given as restrictions to $S^{\varkappa}$ of the deficiencies of a locally integrable function on X then we have

Remark 8.5. *Let $q_0 \in \mathcal{L}^1_{loc}(X)$ be almost quasihomogeneous of degree m and of order $\leq k$. Then $d(T_{q_0})_m = T_{q_0}$.*

Proof. Note that by Proposition 1.51.(i) the deficiencies $q_1, \dots, q_k$ of q_0 are locally integrable, as well. In view of Theorem 4.25.(i) the support of the distribution $c := d(T_{q_0})_m - T_{q_0}$ is contained in $X \setminus X_+$. By Proposition 8.3.(i) and by Remark 2.30.(i) and Proposition 2.37.(i) the support of c is also contained in the union of the sets $\operatorname{supp} q_i$, $0 \leq i \leq k$. Hence, if $\operatorname{supp} q_i \subset X_+$ for every $i \in \{0\} \cup \mathbb{N}_k$ then $c = 0$. On the other hand, if $\operatorname{supp} q_i \setminus X_+ \neq \emptyset$ for some $i \in \{0\} \cup \mathbb{N}_k$ then by Proposition 1.91 we have $\operatorname{Re} m > -\mu$, and c is equal to zero by Corollary 6.28. ∎

The preceding remark is a special case of Theorem 8.8 below. This theorem completes the description of almost quasihomogeneous distributions on X with a $(k+1)^{th}$ order deficiency the support of which is contained in $X \setminus X_+$. It requires special cut-off functions the existence of which is established in

Lemma 8.6. *Let $h \in \mathbb{N}_0$. One can choose functions $\psi_0, \dots, \psi_k \in C^{\infty}_M(X)$ satisfying (4.18) for $\mathfrak{J} = \mathbb{N}_0$, (4.19), and, in addition,*

$$(8.4) \qquad (\psi_0)_{0, \omega_{k+j}} = (-1)^j \, \omega_j \circ \varkappa , \qquad j \in \mathbb{N}_h .$$

Proof. Since the function $\omega_h \circ \varkappa$ is almost quasihomogeneous of degree 0 and of order $\leq$ h it follows from Remark 1.74 and Theorem 3.48 that we can find a function $\psi \in C^{\infty}_{M}(X_+)$ satisfying $\psi_{0,\omega_j} = (-1)^j \omega_j \circ x$ for $j \in \{0\} \cup \mathbb{N}_h$. In particular, we have $\psi_0 \equiv 1$. Hence, the assertion follows by Lemma 4.12.(ii) when we take into account that $(\psi_0)_{0,\omega_{k+j}} = (-\partial_M)^k \psi_{0,\omega_{k+j}} = \psi_{0,\omega_j}$. ■

Choosing $\psi_0, \dots, \psi_k$ as in Lemma 8.6 with h replaced by k+1+h and applying Lemma 6.47 we obtain

Lemma 8.7. *Let $\chi : X \longrightarrow \mathbb{C}$ be any C^{∞} function such that $\operatorname{supp} \chi$ is a weakly $(M, [1, +\infty[)$-bounded subset of X, and let $h \in \mathbb{N}_0$. Then we have:*

(i) The following conditions are equivalent:

(a) there exists a function $\chi_x \in C^{\infty}(X)$ satisfying

$$(\chi_x|_{X^o})_{0,\omega_j} = (-1)^{j+1} \omega_{j+1} \circ x|_{(X^o)_+}, \qquad j \in \mathfrak{J}, \tag{8.5}$$

for $\mathfrak{J} = \{0\} \cup \mathbb{N}_{k+h}$ such that the support of $\chi - \chi_x$ is a weakly $(M, [1, +\infty[)$-bounded subset of X and, at the same time, an M-bounded subset of $(X^o)_+$;

(b) $\chi|_{M_0(X) \cap X} \equiv 1$.

(ii) The following conditions are equivalent:

(a) there exists a function $\chi_x \in C^{\infty}(X)$ satisfying (8.5) for $\mathfrak{J} = \{0\} \cup \mathbb{N}_{k+h}$ such that the support of $\chi - \chi_x$ is a weakly M-bounded subset of X;

(b) $(\chi|_{X^o})_{0,\omega_{k+h}}$ extends to a C^{∞} function $g : X_+ \longrightarrow \mathbb{C}$ such that $(-\partial_M)^{k+h+1} g \equiv 1$. ■

Note that the existence of functions χ satisfying the condition (b) in (ii) is established by the assertion (i). Of course, if (5.17) holds, i.e. if $X = X^o$ then (b) is trivially satisfied, and the formulation of Lemma 8.7 becomes much simpler.

For the following theorem we fix a function χ_x having the properties in condition (a) of Lemma 8.7.(i) for $h = N(m^*)$. Moreover, by assuming that χ be identically equal to 1 on a neighbourhood of $X \cap M_0(X)$ we achieve that χ_x has the same property.

Theorem 8.8. *For every $T \in \mathcal{D}'(X)$ each of the (equivalent) conditions (a) - (c) of Theorem 6.37.(i) is equivalent to*

(d) the (1+k)-tuple d(T) *of distributions $((\partial_M - m)^i T)|_{S^x}$, $0 \le i \le k$, is well-defined, and $T = d(T)_m + Q'_m(\chi_x T)$.*

Proof. (a)⇒(d): Since the restriction of $(\partial_M - m)^i T$ to X_+ is almost quasihomogeneous of degree m and of order $\le k$ it follows from Theorem 2.42 that the distributions $\nu_i := (\partial_M - m)^i T|_{S^\varkappa}$, $0 \le i \le k$, are well-defined and from Theorem 4.25 that ν_m and T coincide on X_+. Since by Proposition 8.3.(ii) $T - \nu_m$ is almost quasihomogeneous of degree m and since $\chi_\varkappa$ is equal to 1 on a neighbourhood of $X \setminus X_+$ $T - \nu_m$ equals $\chi_\varkappa(T - \nu_m)$ so that by Proposition 6.27 we have

$$T - \nu_m = Q'_m(\chi_\varkappa(T - \nu_m)).$$

By Theorem 5.37.(vi) (applied to $Q_{m*}\varphi$ instead of q_0) and by (8.5) we deduce for arbitrary $\varphi \in C_0^\infty(X)$ and $i \in \mathbb{N}_0$ that

$$(\chi_\varkappa Q_{m*}\varphi)_{m*,\omega_i}|_{S^\varkappa \cap X^0} =$$

$$= \sum_{j=0}^{N} \binom{i+j}{i} (\chi_\varkappa|_{X^0})_{0,\omega_{i+j}}|_{S^\varkappa \cap X^0} ((\partial_M - m^*)^j Q_{m*}\varphi)|_{S^\varkappa \cap X^0} \equiv 0 .$$

Since by Corollary 5.47 and Remark 5.60 the support of $(\chi_\varkappa Q_{m*}\varphi)_{m*,\omega_i}$ is contained in X^0 we conclude that $Q'_m(\chi_\varkappa \nu_m) = 0$ so that (d) is proved.

(d)⇒(a): this follows from Propositions 8.3.(ii) and 6.24. ∎

Corollary 8.9. *Let $\psi_0, \dots, \psi_k \in C_M^\infty(X)$ be as in Lemma 8.6. If $T \in \mathcal{D}'(X)$ satisfies one (and hence each) of the conditions (a) - (d) of Theorems 6.37.(i) and 8.8 then*

$$\sum_{i=0}^{k} (\psi_i (\partial_M - m)^i T)_{m,\omega_k} = d(T)_m \tag{8.6}$$

where d(T) is defined in Theorem 8.8.

Proof. Since in view of (8.4) and (8.5) we can take χ_ψ to be equal to $\chi_\varkappa$ in Theorem 6.42, we obtain the assertion by combining the equation (6.43) and the one in condition (d) of Theorem 8.8. ∎

Corollary 8.10. *Suppose that (6.52) holds. Then every distribution $d \in \mathcal{D}'(X)$ with support contained in $X \setminus X_+$ which is almost quasihomogeneous of degree m is of the form*

$$d = \sum_{i=0}^{N} (\partial_M - m)^i Q'_m \nu_i \tag{8.7}$$

for some choice of distributions $\nu_0, \dots, \nu_N \in \mathcal{D}'(S^\varkappa)$ where $N := N(m^*)$.

Proof. By Corollary 6.51 we find a distribution $T \in \mathcal{D}'(X)$ such that $(\partial_M - m)^{N+1} T = d$.

Since T is almost quasihomogeneous of degree m Theorem 8.8, (8.2.b), and Corollary 6.28 show that the assertion is valid for $\nu_i = (\partial_M - m)^{N-i}\, T|_{S^\varkappa}$. ■

If $m \in (-\mathfrak{A}(M)-\mu)$ then not every distribution on X satisfying the condition (a) of Theorem 6.37.(i) is of the form ν_m for some $\nu \in \mathfrak{D}'(S^\varkappa)^{1+k}$. This is a consequence of the following remark which is proved by an application of Theorem 4.25.(ii) to $T := \nu_m|_{X_+}$.

Remark 8.11. *If* $\mathrm{supp}\, \nu_m \subset X \setminus X_+$ *then* $\nu = 0$ *and hence* $\nu_m = 0$. ■

If $\ker M = \{0\}$ then Theorem 6.45 shows that ν_m is temperate. We are going to approximate its Fourier transform. To this end we set

$$(8.8) \qquad w_{j,\varepsilon}(t) := \omega_j(t)\, e^{-\varepsilon t}, \qquad t, \varepsilon \in]0,+\infty[,\ j \in \mathbb{N}_0,$$

and state a preparatory lemma.

Lemma 8.12. *If* $\ker M = \{0\}$ *then for arbitrary* $\varepsilon \in]0,+\infty[$ *and* $j \in \mathbb{N}_0$ *by*

$$K^\varepsilon_{m,j}(\xi,x) := (e^{-i\langle\xi,\,\cdot\,\rangle})_{m,w_{j,\varepsilon}}(x) \qquad (\xi,x) \in V^* \times V,$$

a function $K^\varepsilon_{m,j} \in \mathcal{S}^\infty_{M^*\times M}(V^*\times V)$ *(see Notation 5.34) is well-defined such that*

$$(\hat{\varphi})_{m,w_{j,\varepsilon}}(\xi) = \int_V K^\varepsilon_{m,j}(\xi,x)\,\varphi(x)\,dx, \qquad \xi \in V^*,\ \varphi \in \mathcal{S}(V),$$

and

$$(\hat{\varphi})_{m,w_{j,\varepsilon}}(x) = \int_{V^*} K^\varepsilon_{m,j}(\xi,x)\,\varphi(\xi)\,d\xi, \qquad x \in V,\ \varphi \in \mathcal{S}(V^*).$$

Proof. We first observe that the function $f := e^{-i\langle\cdot,\,\cdot\cdot\rangle}$ belongs to $\mathcal{O}_{\mathfrak{M}}(V^*\times V)$ so that by Theorem 5.42 - applied to $(V^*\times V, M^*\times M)$ instead of (V,M) - $f_{m,v_{j,\varepsilon}}$ is a well-defined C^∞ function belonging to $\mathcal{S}^\infty_{M^*\times M}(V^*\times V)$ where the weight function $v_{j,\varepsilon} :]0,+\infty[\longrightarrow \mathbb{R}$ is defined by $v_{j,\varepsilon}(t) := \omega_j(2t)\, e^{-2\varepsilon t}$. Note that by (1.64) $v_{j,\varepsilon}$ is a linear combination of the functions $w_{i,2\varepsilon}$, $0 \le i \le j$. If $\mathrm{Re}\, m < 0$ we deduce by substituting $t = 2s$ that

$$(8.9) \quad K^\varepsilon_{m,j}(\xi,x) = \int_0^{+\infty} t^{-m} \exp(-i\langle M^*_{t/2}\xi, M_{t/2}x\rangle)\, \omega_j(t)\, e^{-\varepsilon t}\, \frac{dt}{t} = 2^{-m} f_{m,v_{j,\varepsilon}}(\xi,x).$$

By Theorem 5.42 and by the principle of analytic continuation this remains valid for every $m \in \mathbb{C} \setminus \mathfrak{A}_\varepsilon(M)$ (see (5.31)); in fact, this establishes an equality of meromorphic functions of $m \in \mathbb{C}$. Comparing the Laurent coefficients at the points $m \in \mathfrak{A}_{\varepsilon}(M)$ we obtain that

$$K^{\varepsilon}_{m,j}(\xi,\cdot) = 2^{-m} \sum_{j=0}^{\infty} \frac{1}{j!} (-\log 2)^j \, a_{-j}(m;\mathfrak{M}_{f,v_{j,\varepsilon}})(\xi,\cdot).$$

Since the Laurent coefficients $a_{-j}(m;\mathfrak{M}_{f,v_{j,\varepsilon}})$ belong to $\mathcal{S}^{\infty}_{M^*\times M}(V^* \times V)$ this completes the proof of the first part of the assertion.

As for the second part, if $\operatorname{Re} m < 0$ then for every $\varphi \in C^{\infty}_{0}(V^*)$ Fubini's theorem gives

$$(\hat{\varphi})_{m,w_{j,\varepsilon}}(x) = \int_0^{+\infty} t^{-m} \hat{\varphi}(M_t x)\, \omega_j(t)\, e^{-\varepsilon t} \frac{dt}{t} =$$

$$= \int_{V^*} \int_0^{+\infty} t^{-m} \exp(-i\langle \xi, M_t x\rangle)\, \omega_j(t)\, e^{-\varepsilon t} \frac{dt}{t}\, \varphi(\xi)\, d\xi = \int_{V^*} K^{\varepsilon}_{m,j}(\xi,x)\, \varphi(\xi)\, d\xi .$$

Again by the principle of analytic continuation this remains valid for every $m \in \mathbb{C} \setminus \mathfrak{A}_{\varepsilon}(M)$. This means that $\mathfrak{M}_{\hat{\varphi},w_{j,\varepsilon}}$ is equal to the meromorphic function $h : \mathbb{C} \longrightarrow \mathcal{S}_M(V)$ defined by

$$h(m)(x) := \mathfrak{S}(K^{\varepsilon}_{m,j}(\cdot,x)), \qquad x \in V,\ m \in \mathbb{C} \setminus \mathfrak{A}_{\varepsilon}(M),$$

where $\mathfrak{S} : \mathcal{S}_{M^*}(V^*) \longrightarrow \mathbb{C}$ denotes the continuous linear operator defined by

$$\mathfrak{S}(f) := \int_{V^*} (f\varphi)(\xi)\, d\xi .$$

Since for any fixed $x \in V$ the function $g_x : \mathbb{C} \longrightarrow \mathcal{S}_{M^*}(V^*)$ defined by $g_x(m) := K^{\varepsilon}_{m,j}(\cdot,x)$, $m \in \mathbb{C} \setminus \mathfrak{A}_{\varepsilon}(M)$, is meromorphic it follows by the integral formula for the Laurent coefficients that $a_j(m;h)(x) = \mathfrak{S}(a_j(m;g_x))$ for every $j \in \mathbb{Z}$. Hence the second formula in the assertion follows for $m \in \mathfrak{A}_{\varepsilon}(M)$, as well.

The first formula in the assertion is obtained by interchanging the roles of (V,M) and (V^*,M^*) and taking (8.9) into account. ■

Proposition 8.13. *Suppose that* $\ker M = \{0\}$ *and* $X = V$. *Then for every* $\varepsilon > 0$ *by*

$$\nu_{m,\varepsilon}(\xi) := \sum_{i=0}^{k} \langle \nu_i, K^{\varepsilon}_{m^*,i}(\xi,\cdot)/_{S^{\varkappa}} \rangle$$

a function $\nu_{m,\varepsilon} \in \mathcal{S}^{\infty}_{M^*}(V^*)$ *is well-defined such that*

$$\lim_{\varepsilon \to 0} T_{\nu_{m,\varepsilon}} = \mathcal{F}(\nu_m) \quad \text{in the topology of } \mathcal{S}'(V^*).$$

For the proof we require the following consequence of the mean value theorem:

Lemma 8.14. *Let $\mathfrak{B}$ be a bounded subset of $\mathcal{S}(V)$. Then*

(i) for every $\eta>0$ $\displaystyle\sum_{z\in\mathbb{Z}^n,\,|z|\le r}\eta^n\psi(\eta z)$ *converges to* $\displaystyle R_\eta(\psi):=\sum_{z\in\mathbb{Z}^n}\eta^n\psi(\eta z)$

uniformly for $\psi\in\mathfrak{B}$ as $r\to+\infty$;

(ii) $R_\eta(\psi)$ converges to $\int_V\psi(x)\,dx$ uniformly for $\psi\in\mathfrak{B}$ as $\eta\to 0$. ∎

Proof of Proposition 8.13. The assumptions on M and X imply that $(S^{n-1})_M=\dot{V}$ so that $S^\varkappa=S^\varkappa\cap(S^{n-1})_M$ is compact by Remark 6.8. Hence $\nu_{m,\varepsilon}$ is well-defined and belongs to $\mathcal{S}^\infty_{M^*}(V^*)$ by the first part of Lemma 8.12.

To prove the second part of the assertion we observe that by Proposition 5.43 and Remark 5.40 for every $\varphi\in\mathcal{S}(V^*)$ we have

$$\langle\mathcal{F}(\nu_m),\varphi\rangle=\sum_{i=0}^{k}\langle\nu_i,(\hat{\varphi})_{m^*,\omega_i}\big|_{S^\varkappa}\rangle=\lim_{\varepsilon\to 0}\sum_{i=0}^{k}\langle\nu_i,(\hat{\varphi})_{m^*,w_{i,\varepsilon}}\big|_{S^\varkappa}\rangle.$$

Now, let h be the maximum of the orders of the distributions ν_i on the (here compact!) manifold $S^\varkappa$. Then by Theorem 5.42 the set

$$\mathfrak{B}:=\{\partial_x^\alpha K^\varepsilon_{m^*,i}(\cdot,x)\,\varphi\ ;\ x\in S^\varkappa,\ |\alpha|\le h,\ 0\le i\le k\}$$

is a bounded subset of $\mathcal{S}(V^*)$. Hence by Lemma 8.14 we have

$$\lim_{\eta\to 0}\sup\Big\{\Big|\partial_x^\alpha\Big(\int_{V^*}K^\varepsilon_{m^*,i}(\xi,x)\,\varphi(\xi)\,d\xi-\eta^n\sum_{z\in\mathbb{Z}^n}K^\varepsilon_{m^*,i}(\eta z,x)\,\varphi(\eta z)\Big)\Big|\ ;\ x\in S^\varkappa,\ |\alpha|\le h\Big\}=0$$

for every $0\le i\le k$. Combining this with Lemma 8.12 we obtain

$$\langle\nu_i,(\hat{\varphi})_{m^*,w_{i,\varepsilon}}\big|_{S^\varkappa}\rangle=\lim_{\eta\to 0}\eta^n\sum_{z\in\mathbb{Z}^n}\langle\nu_i,K^\varepsilon_{m^*,i}(\eta z,\cdot)\big|_{S^\varkappa}\rangle\,\varphi(\eta z).$$

By Lemma 8.14 the right-hand side is equal to

$$\int_{V^*}\langle\nu_i,K^\varepsilon_{m^*,i}(\xi,\cdot)\big|_{S^\varkappa}\rangle\,\varphi(\xi)\,d\xi.$$

Summing over $i\in\{0\}\cup\mathbb{N}_k$ we obtain the second part of the assertion. ∎

We close this section by giving a duality interpretation of the formula in Theorem 8.8. It is analogous to the assertion of Remark 4.30. Below we shall denote the duality bracket $\mathfrak{A}'_{m,k}(\mathfrak{D}(X))\times\mathfrak{A}^\infty_{m^*,k}(\mathfrak{D}(X))\longrightarrow\mathbb{C}$ defined in Proposition 6.59 for the case $G=\{Id_V\}$ by $\langle\cdot,\cdot\cdot\rangle_M$. Moreover, we set $N:=N(m^*)$.

By applying Proposition 2.45 with (X,ν) replaced by $(X_+,\log\varkappa)$ we decompose every $f\in\mathfrak{A}^\infty_{m^*,k}(\mathfrak{D}(X))$ according to

$$(8.10)\qquad f = P_{\mathfrak{A}}(f) + P_D(f)$$

where

$$P_{\mathfrak{A}}(f) := \sum_{\ell=0}^{k} \omega_\ell \circ \varkappa \; P_\ell(f) \qquad \text{and} \qquad P_D(f) := \sum_{\ell=0}^{N} \omega_{k+1+\ell} \circ \varkappa \; P_{k+1+\ell}(f)$$

with

$$P_\ell(f) := \sum_{i \in \mathbb{N}_0} (-1)^i \omega_i \circ \varkappa \; (\partial_M - m^*)^{i+\ell} f$$

being quasihomogeneous of degree m^* for every $\ell \in \mathbb{N}_0$. By Proposition 2.45 $P_{\mathfrak{A}}$ is a projection onto the subspace $\mathfrak{A}\mathfrak{Q}^\infty_{m^*,k}(X)$. Hence P_D is a projection onto the subspace $\operatorname{im} P_D$. Note that

$$P_{k+1+\ell}(f) = P_\ell((\partial_M - m^*)^{k+1} f).$$

Consequently, in view of Proposition 5.59 and Notation 5.55 one deduces that

$$(8.11)\qquad \operatorname{im} P_D = \Big\{ \sum_{\ell=0}^{N} \omega_{k+1+\ell} \circ \varkappa \; P_\ell(R|_{X_+}) \, ; \, R \in Q_{m^*}(\mathfrak{D}(X)) \Big\}.$$

Moreover, choosing $\varphi \in \mathfrak{D}(X)$ such that $f = \varphi_{m^*,\omega_k}$ and taking (5.63) into account we observe that

$$\langle d(T)_m , f \rangle_M = \langle d(T)_m , \varphi \rangle =$$

$$= \sum_{i=0}^{k} \langle (\partial_M - m)^i T\big|_{S^\varkappa} , \varphi_{m^*,\omega_i}\big|_{S^\varkappa} \rangle =$$

$$= \sum_{i=0}^{k} (-1)^{k-i} \langle (\partial_M - m)^i T\big|_{S^\varkappa} , (\partial_M - m^*)^{k-i} f\big|_{S^\varkappa} \rangle .$$

Since by Lemma 2.34, Remark 1.74, and Proposition 2.31 we have that

$$(\partial_M - m^*)^i P_D(f)\big|_{S^\varkappa} = \sum_{\ell=0}^{N} [\omega_{k+1+\ell-i} \circ \varkappa \; P_{k+\ell+1}(f)]\big|_{S^\varkappa} \equiv 0, \qquad 0 \le i \le k,$$

we conclude that $\langle d(T)_m , P_D(f) \rangle_M = 0$.

Moreover, since by Proposition 3.51 we find $\psi \in C_0^\infty(X_+)$ such that $P_{\mathfrak{A}}(f) = \psi_{m^*,\omega_k}$ and since $\operatorname{supp} Q'_m(\chi_\varkappa T) \subset X \setminus X_+$ we see that

$$\langle Q'_m(\chi_\varkappa T), P_{\mathfrak{A}}(f) \rangle_M = \langle Q'_m(\chi_\varkappa T), \psi \rangle = 0.$$

Consequently, we derive from condition (d) of Theorem 8.8 that

$$(8.12)\qquad \langle T, f \rangle_M = \Big[\sum_{i=0}^{k} (-1)^{k-i} \langle (\partial_M - m)^i T\big|_{S^\varkappa} , (\partial_M - m^*)^{k-i} P_{\mathfrak{A}}(f)\big|_{S^\varkappa} \rangle \Big] +$$

$$+ \langle Q'_m(\chi_\varkappa T), P_D(f) \rangle_M .$$

In view of Theorem 4.29 and Remark 4.30 this equation exhibits $(-1)^k \langle \cdot , \cdot\cdot \rangle_M$

as the direct sum of the canonical duality bracket $\mathfrak{D}'(S^{\varkappa})^{1+k} \times \mathfrak{D}(S^{\varkappa})^{1+k} \longrightarrow \mathbb{C}$ and the duality bracket between the spaces $Q'_m(\mathfrak{D}(X))$ and $\operatorname{im} P_D$ induced by $(-1)^k \langle \cdot, \cdot\cdot \rangle_M$. In view of Remark 4.30 the first of these duality brackets can also be looked upon as the bracket between the space $\mathfrak{A}\mathfrak{Q}'_{m,k}(X_+)$ (see Notation 4.28) and its pre-dual $\mathfrak{A}\mathfrak{Q}^{\infty}_{m,k}(X_+)$.

(b) Extending (Almost) Quasihomogeneous Distributions on X_+ to the Whole of X

The basic result of this section is

Theorem 8.15. *Every almost quasihomogeneous distribution on X_+ extends to an almost quasihomogeneous distribution on X.*

In the proof, actually, we are going to be more precise by keeping track of the almost quasihomogeneous orders of the extensions. Indeed, let $k \in \mathbb{N}_0$, and let $T \in \mathfrak{D}'(X_+)$ be almost quasihomogeneous of degree m and of order $\leq k$. Then there are essentially two different ways of extending T to X. The first one is based on section (a): by Theorem 2.42 we know that $(\partial_M - m)^i T|_{S^{\varkappa}}$ is a well-defined distribution on $S^{\varkappa}$ for every $i \in \{0\} \cup \mathbb{N}_k$, and from Theorem 4.25.(i) we deduce that the distribution $d(T)_m$ defined in Proposition 8.3 extends T where *d(T)* denotes the (1+k)-tuple of distributions $(\partial_M - m)^i T|_{S^{\varkappa}}$, $0 \leq i \leq k$. The second way makes use of the functions $\psi_0, \ldots, \psi_k \in C^{\infty}_M(X_+)$ satisfying the assumptions of Theorem 6.42: by Remark 6.14 for every $i \in \{0\} \cup \mathbb{N}_k$ the distribution $\psi_i (\partial_M - m)^i T$ - which a priori belongs to $\mathfrak{D}'(X_+)$ - can be identified with a distribution in $\mathfrak{D}'_M(X)$ so that by Proposition 6.15 $(\psi_i (\partial_M - m)^i T)_{m,\omega_k}$ is a well-defined distribution on X, and by Proposition 4.13 - applied to (X_+, k) instead of (X, N) - the distribution

$$(8.13) \qquad \dot{T} := \sum_{i=0}^{k} (\psi_i (\partial_M - m)^i T)_{m,\omega_k}$$

extends T. (Since $\operatorname{ord}_M(T)$ may be strictly smaller than k it would be more precise to write $\dot{T}_k$ instead of $\dot{T}$; however, this ambiguity should cause no confusion).

These extensions are almost quasihomogeneous but in case $m \in (-\mathfrak{A}(M)-\mu)$ not necessarily of the same order as T (see Propositions 6.35 and 6.24). Moreover, they depend on the choices of $\varkappa$ and ψ_i. Different extensions differ from each other by almost quasihomogeneous distributions with support contained in $X \setminus X_+$. Hence, we conclude from Corollary 6.28 that the extensions are unique except when $m \in (-\mathfrak{A}(M)-\mu)$. From Corollary 8.9 we deduce

Remark 8.16. *Suppose that $m \notin (-\mathfrak{A}(M)-\mu)$ or that (8.4) is satisfied. Then the distribution $\dot{T}$ defined by (8.13) is equal to $d(T)_m$.* ■

Summarizing the discussion above we obtain

Theorem 8.15'. *Under the preceding assumptions the distribution $\dot{T} \in \mathcal{D}'(X)$ defined by (8.13) extends T to the whole of X.*

$\dot{T}$ is almost quasihomogeneous of degree m of order $\leq k+N+1$ where $N := N(m^*)$. More precisely, if $m \notin (-\mathfrak{A}(M)-\mu)$ then $\dot{T}$ is almost quasihomogeneous of order $\leq k$, being the unique almost quasihomogeneous extension of T; if $m \in (-\mathfrak{A}(M)-\mu)$ then as an almost quasihomogeneous extension of T $\dot{T}$ is unique up to an additive term of the form described in Proposition 6.27.(i). ■

If $\operatorname{Re} m > -\mu$ and if T is induced by a locally integrable function on X_+ then its extension $\dot{T}$ has, in fact, already been computed in Proposition 1.91:

Example 8.17. *Let $q_0 \in \mathcal{L}^1_{loc}(X_+)$ be almost quasihomogeneous of degree m and of order $\leq k$. If $\operatorname{Re} m > -\mu$ then for every $i \in \{0\} \cup \mathbb{N}_k$ the extension f_i of the ith order deficiency q_i of q_0 to X defined by $f_i|_{X \setminus X_+} :\equiv 0$ is locally integrable on X, and $(\partial_M - m)^i \dot{T}_{q_0} = T_{f_i}$.*

<u>*Proof.*</u> The first assertion follows by Proposition 1.91. Since T_{f_i} is almost quasihomogeneous of degree m the second assertion follows in view of Proposition 2.31 by the uniqueness part of Theorem 8.15'. ■

The following proposition contains rules of computation for the extensions $\dot{T}$.

Proposition 8.18. *(i)* $q\dot{T} = (qT)^{\bullet}$ *for every quasihomogeneous* C^{∞} *function* $q : X \to \mathbb{C}$;

(ii) $\dot{T} \circ A = (T \circ A)^{\bullet}$ *for every* $A \in GL(V,V)$ *commuting with* M *provided that* $m \notin (-\mathfrak{A}(M)-\mu)$ *or* $\psi_i \circ A = \psi_i$ *(resp.* $\varkappa \circ A = \varkappa$ *)*;

(iii) $\dot{T}_{\mathfrak{G}} = (T_{\mathfrak{G}})^{\bullet}$ *if* $\mathfrak{G}$ *satisfies the assumptions of Remark 2.67.(ii) and if* $m \notin (-A(M)-\mu)$ *or each of the functions* ψ_i *(resp.* $\varkappa$*) is* G*-invariant.*

Proof. *(i)*: this is an immediate consequence of (4.8) (see Proposition 6.16) and Corollary 2.36.(ii).

(ii): Since $A^{-1}(X_+) = (A^{-1}(X))_+$ one observes that $\dot{T} \circ A$ extends $T \circ A$. Since by Remark 2.67.(i) $\dot{T} \circ A$ is almost quasihomogeneous of degree m and of the same order as $\dot{T}$ the uniqueness part of Theorem 8.15' implies the desired equality provided that m does not belong to $(-\mathfrak{A}(M)-\mu)$. To deal with the exceptional case, by Proposition 6.16.(iv) and (2.38) one deduces for every $m \in \mathbb{C}$ from (8.13) that

$$\dot{T} \circ A = \sum_{i=0}^{k} (\psi_i \circ A\, (\partial_M - m)^i (T \circ A))_{m,\omega_k} .$$

If $m \in (-\mathfrak{A}(M)-\mu)$ then the right-hand side equals $(T \circ A)^{\bullet}$ by the assumption on the functions ψ_i.

(iii): Since $A(X_+) = X_+$ one observes in view of (2.35) that $(\dot{T})_{\mathfrak{G}}$ extends $T_{\mathfrak{G}}$. Since by Remark 2.67.(ii) $(\dot{T})_{\mathfrak{G}}$ is almost quasihomogeneous of degree m and of order not larger than that of $\dot{T}$ the uniqueness part of Theorem 8.15' implies the desired equality in case m does not belong to $(-\mathfrak{A}(M)-\mu)$. To deal with the exceptional case "$m \in (-\mathfrak{A}(M)-\mu)$", by Proposition 6.16.(v), by (2.35) and the assumption on the functions ψ_i, and by (2.39) one deduces for every $m \in \mathbb{C}$ that

$$(\dot{T})_{\mathfrak{G}} = \sum_{i=0}^{k} [(\psi_i (\partial_M - m)^i T)_{\mathfrak{G}}]_{m,\omega_k} = \sum_{i=0}^{k} [\psi_i (\partial_M - m)^i T_{\mathfrak{G}}]_{m,\omega_k} = (T_{\mathfrak{G}})^{\bullet} . \quad ■$$

Note that in case $A \in GL(V,V)$ commutes with M we have

$$(8.14) \qquad (\psi_i \circ A)_{0,\omega_j} = (\psi_i)_{0,\omega_j} \circ A , \qquad j \in \mathbb{N}_0 ,$$

so that the functions $\psi_i \circ A$, $0 \le i \le k$, satisfy the conditions (4.18) and (4.19), as well. Moreover, in view of

$$(8.14)' \qquad ((\psi_i)_G)_{0,\omega_j} = ((\psi_i)_{0,\omega_j})_G , \qquad j \in \mathbb{N}_0 ,$$

the G-invariant functions $(\psi_i)_G$, $0 \le i \le k$, continue to satisfy the conditions (4.18) and (4.19).

The rules for the derivatives are more complicated; for simplicity we compute the exceptional cases only under the special assumption that M be semi-simple.

Proposition 8.19. *(i)* $\partial^\alpha \dot{T} = (\partial^\alpha T)^\cdot$ *for every* $\alpha \in \mathfrak{A}$ *such that* $m - \alpha_M \notin (-\mathfrak{A}(M) - \mu)$.

(ii) Suppose that M is semi-simple, and let $\alpha \in \mathfrak{A}$ *such that* $|\alpha| = 1$; *if* $\alpha_M + m^*$ *belongs to* $\mathfrak{A}(M)$ *then*

$$\partial^\alpha \dot{T} - (\partial^\alpha T)^\cdot = Q'_{m-\alpha_M}\Big[\sum_{i=0}^{k} \partial^\alpha(\omega_{k+1} \circ \varkappa)\, \psi_i\, (\partial_M - m)^i T\Big].$$

<u>*Proof.*</u> <u>*(i)*</u>: Since $\partial^\alpha \dot{T}$ and $(\partial^\alpha T)^\cdot$ are both almost quasihomogeneous extensions of $\partial^\alpha T$ the assertion follows by the uniqueness part of Theorem 8.15'.

<u>*(ii)*</u>: By χ we denote the function χ_ψ satisfying (6.41) fixed in section 6.(d). In view of Lemma 6.47 we may assume that χ is equal to 1 on a neighbourhood of $X \setminus X_+$ and that $\operatorname{supp} \chi \subset X^O$. We set $h := (\chi|_{X^O})_{O,\omega_k}$. From (3.8)', (3.11), (6.41), and (1.38) we deduce that

$$\sum_{i=0}^{k} [\psi_i|_{X^O} h]_{O,\omega_{k-i}} = \sum_{i=0}^{k} \sum_{j=0}^{2k+1} (-1)^j \binom{k-i+j}{j} (\psi_i|_{X^O})_{O,\omega_{k-i+j}} (\psi_O)_{O,\omega_{2k+1-j}}\Big|_{(X^O)_+}.$$

Since it follows from (3.11) and (4.19) that

$$(\psi_O)_{O,\omega_\ell} = (-\partial_M)^{k-\ell} (\psi_O)_{O,\omega_k} = 0, \qquad 0 \le \ell \le k-1,$$

we conclude, making use of (4.18), that the left-hand side is equal to the restriction of

$$\sum_{j=0}^{k+1} (-1)^j \Big[\sum_{i=0}^{k} \binom{k-i+j}{j} c_{k,i}\Big] (\psi_O|_{X^O})_{O,\omega_{k+j}} (\psi_O|_{X^O})_{O,\omega_{2k+1-j}}$$

to $(X^O)_+$. Since by Lemma 6.40 the sum in square brackets equals $(-1)^k \delta_{O,j} + \delta_{j,k+1}$ the whole expression vanishes. Hence it follows by the Leibniz rule and (3.6) and by (8.5) that

$$\sum_{i=0}^{k} (\partial^\alpha \psi_i|_{X^O} h)_{-\alpha_M,\omega_{k-i}} = \tag{8.15}$$

$$= \partial^\alpha \Big[\sum_{i=0}^{k} (\psi_i|_{X^O} h)_{O,\omega_{k-i}}\Big] - \sum_{i=0}^{k} (\psi_i|_{X^O} \partial^\alpha h)_{-\alpha_M,\omega_{k-i}} =$$

$$= (-1)^k \sum_{i=0}^{k} [\psi_i\, \partial^\alpha(\omega_{k+1} \circ \varkappa)]_{-\alpha_M,\omega_{k-i}}\Big|_{(X^O)_+}.$$

Now, since $\partial^\alpha \dot{T}$ and $(\partial^\alpha T)^\cdot$ both extend $\partial^\alpha T$ the support of $d := \partial^\alpha \dot{T} - (\partial^\alpha T)^\cdot$ is contained in $X \setminus X_+$. In particular, we have $d = \chi d$. Since d is almost quasihomogeneous of degree $m-\alpha_M$ Proposition 6.27.(i) shows that $d = Q'_{m-\alpha_M}(\chi d)$. Now, by Proposition 6.16.(i), by Lemma 2.34, and by the Leibniz rule we have

$$d = \sum_{i=0}^{k} [\partial^\alpha (\psi_i (\partial_M - m)^i T)_{m,\omega_k} - (\psi_i (\partial_M - m + \alpha_M)^i \partial^\alpha T)_{m-\alpha_M,\omega_k}] =$$

$$= \sum_{i=0}^{k} [\partial^\alpha(\psi_i (\partial_M - m)^i T) - \psi_i \partial^\alpha (\partial_M - m)^i T]_{m-\alpha_M,\omega_k} =$$

$$= \sum_{i=0}^{k} [(\partial^\alpha \psi_i)(\partial_M - m)^i T]_{m-\alpha_M,\omega_k}.$$

Inserting this into the right-hand side of the equation $d = Q'_{m-\alpha_M}(\chi d)$ we see that for every $\varphi \in C_0^\infty(X)$ we have $\langle d, \varphi \rangle = \langle T, \phi \rangle$ where

$$\phi := (-1)^k \sum_{i=0}^{k} (-1)^i (\partial_M - m^*)^i [(\partial^\alpha \psi_i)(\chi Q_{m^*+\alpha_M} \varphi)_{m^*+\alpha_M,\omega_k}].$$

Note that by (3.8)', Remark 5.60.(ii), and Corollary 5.47 we have $\phi|_{X \setminus X^o} \equiv 0$ and

$$\phi|_{X^o} = (-1)^k \sum_{i=0}^{k} (-1)^i (\partial_M - m^*)^i [(\partial^\alpha \psi_i)\, h\, Q_{m^*+\alpha_M} \varphi|_{X^o})].$$

Using (3.10), (5.63), (3.9), and (8.15) we deduce that on the set $(X^O)_+$ we have

$$\phi_{m^*,\omega_k} = (-1)^k \sum_{i=0}^{k} [(\partial^\alpha \psi_i)\, h\, Q_{m^*+\alpha_M} \varphi|_{X^o})]_{m^*,\omega_{k-i}} =$$

$$= (-1)^k \sum_{i=0}^{k} [(\partial^\alpha \psi_i|_{X^o})\, h]_{-\alpha_M,\omega_{k-i}}\, Q_{m^*+\alpha_M} \varphi =$$

$$= \sum_{i=0}^{k} [\psi_i\, \partial^\alpha (\omega_{k+1} \circ \varkappa)]_{-\alpha_M,\omega_{k-i}}\, Q_{m^*+\alpha_M} \varphi =$$

$$= \sum_{i=0}^{k} [\psi_i\, \partial^\alpha (\omega_{k+1} \circ \varkappa)\, Q_{m^*+\alpha_M} \varphi]_{m^*,\omega_{k-i}} =$$

$$= \Big[\sum_{i=0}^{k} (-1)^i (\partial_M - m^*)^i [\psi_i\, \partial^\alpha (\omega_{k+1} \circ \varkappa)\, Q_{m^*+\alpha_M} \varphi]\Big]_{m^*,\omega_k}.$$

Since by Remark 5.60 the support of the function on the left and of the one on the right is contained in $(X^O)_+$ these functions are equal, and the condition (b) of Theorem 6.37.(i) implies that

$$\langle T, \phi \rangle = \sum_{i=0}^{k} (-1)^i \langle T, (\partial_M - m^*)^i [\psi_i\, \partial^\alpha (\omega_{k+1} \circ \varkappa)\, Q_{m^*+\alpha_M} \varphi] \rangle =$$

$$= \sum_{i=0}^{k} \langle (\partial_M - m)^i T, \psi_i\, \partial^\alpha (\omega_{k+1} \circ \varkappa)\, Q_{m^*+\alpha_M} \varphi \rangle.$$

Consequently, the desired formula follows. ■

Example 8.20. *Let $y \in X_+$, and set $T := (\delta_y)_{m,\omega_k}/X_+$. Then*

$$\dot{T} - (\delta_y)_{m,\omega_k} = \sum_{j=0}^{\infty} \binom{k+j}{k} (\omega_{k+j+1} \circ \varkappa)(y)\, (\partial_M - m)^j Q'_m(\delta_y),$$

and in case M is semi-simple

$$\partial^\alpha \dot{T} - (\partial^\alpha T)^{\cdot} = \partial^\alpha (\omega_{k+1} \circ \varkappa)(y)\, Q'_{m-\alpha_m}(\delta_y) \qquad \textit{if } |\alpha| = 1.$$

Proof. Since by (8.4) and (2.5) we have

$$(-1)^{k+1} \langle (\partial_M - m)^j Q'_m((\psi_O)_{O,\omega_{2k+j+1}} \delta_y), \varphi \rangle =$$

$$= \langle \delta_y, \omega_{k+j+1} \circ \varkappa\, Q_{m*}[(\partial_M - m^*)^j \varphi] \rangle =$$

$$= (-1)^j (\omega_{k+j+1} \circ \varkappa)(y) \langle (\partial_M - m)^j Q'_m(\delta_y), \varphi \rangle, \qquad \varphi \in C_O^\infty(X),$$

the first equation follows by Remark 6.43.

For the proof of the second equation we first observe that by (6.30) and (3.8)'

$$\langle \partial^\alpha (\omega_{k+1} \circ \varkappa)\, \psi_i\, (\partial_M - m)^i (\delta_y)_{m,\omega_k}, Q_{m^*+\alpha_M} \varphi \rangle =$$

$$= (-1)^i \langle (\delta_y)_{m,\omega_{k-i}}, \partial^\alpha (\omega_{k+1} \circ \varkappa)\, \psi_i\, Q_{m^*+\alpha_M} \varphi \rangle =$$

$$= (-1)^k (\partial^\alpha (\omega_{k+1} \circ \varkappa)\, \psi_i\, Q_{m^*+\alpha_M} \varphi)_{m^*,\omega_{k-i}}(y) =$$

$$= (-1)^k \sum_{j=0}^{\infty} [(\partial_M + \alpha_M)^j \partial^\alpha (\omega_{k+1} \circ \varkappa)\, (\psi_i)_{O,\omega_{k-i}\omega_j}\, Q_{m^*+\alpha_M} \varphi](y).$$

Since by Lemma 2.34 and by Remark 1.74

$$(\partial_M + \alpha_M)^j \partial^\alpha (\omega_{k+1} \circ \varkappa) = \partial^\alpha \partial_M{}^j (\omega_{k+1} \circ \varkappa) = \begin{cases} \partial^\alpha (\omega_{k+1-j} \circ \varkappa) & \text{if } j \le k+1 \\ 0 & \text{if } j > k+1 \end{cases}$$

the right-hand side is equal to

$$(-1)^k \sum_{j=0}^{k+1} [\partial^\alpha (\omega_{k+1-j} \circ \varkappa)\, (\psi_i)_{O,\omega_{k-i}\omega_j}\, Q_{m^*+\alpha_M} \varphi](y).$$

Combining this with (1.38) and (4.18) we obtain

$$\langle Q'_{m-\alpha_M} \Big[\sum_{i=0}^{k} (\partial^\alpha (\omega_{k+1} \circ \varkappa)\, \psi_i\, (\partial_M - m)^i (\delta_y)_{m,\omega_k} \Big], \varphi \rangle =$$

$$= (-1)^k \sum_{j=0}^{k+1} \Big[\sum_{i=0}^{k} \binom{k-i+j}{j} c_{k,i} \Big] [\partial^\alpha (\omega_{k+1-j} \circ \varkappa)\, (\psi_O)_{O,\omega_{k+j}}\, Q_{m^*+\alpha_M} \varphi](y).$$

Since by Lemma 6.40 the sum in square brackets is equal to $(-1)^k \delta_{O,j} + \delta_{j,k+1}$ and since $\partial^\alpha (\omega_{k+1-j} \circ \varkappa)$ vanishes for $j = k+1$ we conclude by (4.19) that this is equal to

$$\partial^{\alpha}(\omega_{k+1}\circ x)(y)\ [Q_{m^*+\alpha_M}\varphi](y) = \partial^{\alpha}(\omega_{k+1}\circ x)(y)\ \langle Q'_{m-\alpha_m}(\delta_y),\varphi\rangle .$$

Hence the assertion follows from Proposition 8.19.(ii). ∎

Is every almost quasihomogeneous extension of T equal to $\dot{T}$ for a suitable choice of $\psi_0,\dots,\psi_k$? In general the answer is in the negative as the example $T=0$ shows. A reformulation of Proposition 6.48 gives a complete description of the extensions of T obtained by the method of Theorem 8.15':

Proposition 8.21. *For every $d\in\mathcal{D}'(X)$ which is almost quasihomogeneous of degree m with support contained in $X\setminus X_+$ the following equivalence is valid:*

There exists another (1+k)-tuple of functions $\psi'_0,\dots,\psi'_k\in C^{\infty}_M(X)$ satisfying the assumptions of Theorem 6.42 such that

$$\dot{T}+d=\sum_{i=0}^{k}(\psi'_i(\partial_M-m)^i T)_{m,\omega_k}$$

if and only if

$$d=Q'_m(\tilde{\psi}T)\ \text{ for some }\ \tilde{\psi}\in C^{\infty}_M(X). \tag{8.16}$$

Proof. Note first that

$$\sum_{i=0}^{k}(\psi_i(\partial_M-m)^i\dot{T})_{m,\omega_k}=\sum_{i=0}^{k}(\psi_i(\partial_M-m)^i T)_{m,\omega_k}=\dot{T}. \tag{8.17}$$

Consequently, applying Theorem 6.37 to $\dot{T}$ instead of T we deduce that $Q'_m(\chi_\psi\dot{T})$ vanishes. Hence, by Proposition 6.27.(i) it follows that $d=Q'_m(\chi_\psi d)=Q'_m(\chi_\psi S)$ where $S:=\dot{T}+d$. By Proposition 6.48 we have

$$S=\sum_{i=0}^{k}(\psi'_i(\partial_M-m)^i S)_{m,\omega_k} \tag{8.18}$$

for some choice of functions $\psi'_0,\dots,\psi'_k$ if and only if there exists $\tilde{\psi}\in C^{\infty}_M(X)$ such that $Q'_m(\chi_\psi S)=Q'_m(\tilde{\psi}T)$. Since by (8.17) the right-hand side of (8.18) is equal to

$$\sum_{i=0}^{k}(\psi'_i(\partial_M-m)^i T)_{m,\omega_k}$$

the proof is complete. ∎

Corollary 8.22. *The following conditions are equivalent:*

(a) for every extension $T'\in\mathcal{D}'(X)$ of T which is almost quasihomogeneous of de-

gree m one can choose the functions $\psi_0, \dots, \psi_k$ *in such a way that* T' *is equal to the distribution* $\dot{T}$ *defined in (8.13);*

(b) every distribution $d \in \mathcal{D}'(X)$ *with support contained in* $X \setminus X_+$ *which is almost quasihomogeneous of degree m satisfies (8.16).* ∎

Next we note an observation concerning the extension of more general distributions on X_+ to the whole of X. The proof reduces the general extension problem to the extension problem for quasihomogeneous distributions.

Theorem 8.23. *Suppose that (6.52) holds. Let R be a distribution on* X_+ *such that* $(\partial_M - m)R$ *extends to a distribution* $d \in \mathcal{D}'(X)$. *Then R extends to a distribution* $\dot{R} \in \mathcal{D}'(X)$ *such that* $(\partial_M - m)\dot{R} - d$ *is almost quasihomogeneous of degree m with support contained in* $X \setminus X_+$. $\dot{R}$ *is unique up to an additive distribution of the form described in Proposition 6.27.(i).*

<u>*Proof.*</u> By Theorem 6.49 we choose $S \in \mathcal{D}'(X)$ such that $(\partial_M - m)S = d$. Then the distribution $T := R - S|_{X_+}$ is quasihomogeneous of degree m. Consequently, $\dot{R} := S + \dot{T}$ is the desired extension. For the uniqueness assertion one has to look at Proposition 6.27. ∎

We close this section by pointing out a slightly different way for solving the equation (7.1) if (6.28) holds (compare Theorems 3.2.4 and 7.1.20 in Hörmander [11]). It turns out that in this case the assumptions (7.3) and (7.4) are not required.

Theorem 8.24. *Let* $q \in C^\infty(X)$ *be quasihomogeneous of degree* $\ell \in \mathbb{C}$ *having no zeros in* X_+. *Let* $T \in \mathcal{D}'(X)$ *be almost quasihomogeneous of degree m such that the support of its* $(1+k)^{th}$ *order deficiency is contained in* $X \setminus X_+$. *Then the equation (7.1) has a solution* $S \in \mathcal{D}'(X)$ *which is almost quasihomogeneous of degree* $m - \ell$.

<u>*Proof.*</u> Since $T|_{X_+}$ is almost quasihomogeneous of degree m and of order $\leq k$ the distribution $R := \frac{1}{q}(T|_{X_+})$ is almost quasihomogeneous of degree $m - \ell$ and of order $\leq k$. By Proposition 8.18.(i) the support of $d := T - q\dot{R}$ is contained in $X \setminus X_+$. Since d is almost quasihomogeneous of degree m it follows by Proposition 6.33 that the equation $qc = d$ has a solution $c \in \mathcal{D}'(X)$ which is almost quasihomogeneous of degree $m - \ell$. Hence $S := \dot{R} + c$ is the desired solution of (7.1). ∎

(c) Extending $T_{f_{m,\omega_k}}$ to the Whole of X

In this section we suppose that X satisfies (5.17). Let $f \in C^{\infty,0}(X)$ (see Definition 1.18.(i) with $V_1 = G_M(\sigma_+)$ and $V_2 = \ker M$) satisfy the assumptions of Theorem 5.37. Then f_{m,ω_k} is a well-defined function belonging to $C^{\infty,0}(X_+)$. Moreover, f_{m,ω_k} is almost quasihomogeneous of degree m and of order $\leq k+N(m)+1$ with $(k+1)^{th}$ order deficiency $(-1)^{k+1} Q_m(f)|_{X_+}$. In particular, it is almost quasihomogeneous of degree m and of order $\leq k$ provided that $m \notin \mathfrak{A}(M)$. If, in addition, m does not belong to $(-\mathfrak{A}(M)-\mu)$ then by Theorem 8.15' $T_{f_{m,\omega_k}}$ possesses a unique almost quasihomogeneous extension of order $\leq k$ to the whole of X, namely $\dot{T}_{f_{m,\omega_k}}$. If $m \in \mathfrak{A}(M)$ then by Theorem 8.15' $T_{f_{m,\omega_k}}$ uniquely extends to an almost quasihomogeneous distribution with $(k+1)^{th}$ order deficiency $(-1)^{k+1} T_{Q_m(f)}$. Finally, if $m \in (-\mathfrak{A}(M)-\mu)$ then by Theorem 8.15' $T_{f_{m,\omega_k}}$ has an almost quasihomogeneous extension which, however, is not unique. In the present section we show that there is a canonical choice for the extension which does not depend on $\varkappa$ or ψ. To find it we employ meromorphic functions again. We require

Lemma 8.25. *Let w be as in Chapter 3. Suppose that $m \in (-\mathfrak{A}(M)-\mu)$. Let $g \in C^0(X_+)$ be such that $|g|_{m,|w|}$ is well-defined by (3.1)' and continuous. Then by*

$$\langle Q'_m(T_{g\, w\circ\varkappa}), \varphi\rangle := \int_{X_+} g(x)\, w(\varkappa(x))\, (Q_{m*}\varphi)(x)\, dx\,, \qquad \varphi \in C_0^\infty(X), \tag{8.19}$$

a distribution $Q'_m(T_{g\,w\circ\varkappa}) \in \mathcal{D}'(X)$ is well-defined; it is equal to

$$\sum_{i=0}^{N} (-1)^i (\partial_M - m)^i\, Q'_m(T_{h_i})$$

where $h_i := g_{m,w\omega_i}|_{S^\varkappa}$ and $N := N(m^*)$. Moreover, we have

$$Q'_m(T_{h_0}) = \sum_{i=0}^{N} (\partial_M - m)^i\, Q'_m(T_{g\,(w\omega_i)\circ\varkappa})\,.$$

<u>Proof</u>. Let $\varphi \in C_0^\infty(X)$, and set $K := \operatorname{supp}\varphi$ and $q_i := (\partial_M - m^*)^i Q_{m*}\varphi|_{X_+}$. Then for every $i \in \mathbb{N}_0$ by Corollary 5.47 the support of q_i is contained in $M_0^{-1}(K)$ so that by Lemma 8.1 the set $\operatorname{supp} q_i \cap S^\varkappa$ is compact, i.e. (4.30) is satisfied for g instead of f. Consequently, the assertions follow by Lemma 4.23 and Proposition 5.48.(iii). ■

Note that by Propositions 5.11 and 5.13 the function $g=f$ satisfies the assumptions of Lemma 8.25 for $w=\omega_j$, $j\in\mathbb{N}_0$.

Theorem 8.26. *By $\mathfrak{N}_f(m):=\dot{T}_{f_{m,\omega_k}}$, $m\in\mathfrak{A}(M)\cup(-\mathfrak{A}(M)-\mu)$, a meromorphic function $\mathfrak{N}_f:\mathbb{C}\longrightarrow\mathcal{D}'(X)$ is defined, its poles lying in the set $\mathfrak{A}(M)\cup(-\mathfrak{A}(M)-\mu)$. Moreover,*

(i) if $m\in\mathfrak{A}(M)$ then with $N:=N(m)$ we have

$$a_0(m;\mathfrak{N}_f)=d(T_{f_m})_m-\sum_{S=0}^{N}(-1)^{S+k}\binom{S+k}{k}(\partial_M-m)^S(Q_mf)\,\dot{T}_{\omega_{S+k+1}\circ\varkappa} \tag{8.20}$$

and

$$a_{-j}(m;\mathfrak{N}_f)=\begin{cases}0 & \text{if } j\in\mathbb{N}_k\\ -\binom{j-1}{k}(\partial_M-m)^{j-k-1}T_{Q_m(f)} & \text{if } j\in\mathbb{N}\setminus\mathbb{N}_k\end{cases}; \tag{8.21}$$

(ii) if $m\in(-\mathfrak{A}(M)-\mu)$ then with $N:=N(m^*)$ we have

$$a_0(m;\mathfrak{N}_f)=d(T_{f_m})_m-\sum_{S=0}^{N}\binom{S+k}{k}(\partial_M-m)^SQ'_m(T_{f\omega_{S+k+1}\circ\varkappa}) \tag{8.22}$$

and

$$a_{-j}(m;\mathfrak{N}_f)=\begin{cases}0 & \text{if } j\in\mathbb{N}_k\\ \binom{j-1}{k}(\partial_M-m)^{j-k-1}Q'_m(T_f) & \text{if } j\in\mathbb{N}\setminus\mathbb{N}_k\end{cases}. \tag{8.23}$$

<u>*Proof*</u>. Let $\varphi\in C_0^\infty(X)$. Then in view of (5.63) for every $m\in\mathbb{C}$ we have

$$\langle d(T_{f_{m,\omega_k}})_m,\varphi\rangle=\sum_{i=0}^{k}\langle(\partial_M-m)^iT_{f_{m,\omega_k}}|_{S^\varkappa},\varphi_{m^*,\omega_i}|_{S^\varkappa}\rangle= \tag{8.24}$$
$$=\sum_{i=0}^{k}(-1)^i\int_{S^\varkappa}f_{m,\omega_{k-i}}(\vartheta)\,\varphi_{m^*,\omega_i}(\vartheta)\,d\tilde{\varkappa}(\vartheta)\,.$$

Since by Theorem 5.37.(iv) for every $i\in\{0\}\cup\mathbb{N}_k$ by $\mathbb{C}\setminus\mathfrak{A}(M)\ni m\mapsto f_{m,\omega_{k-i}}$ (resp. $\mathbb{C}\setminus(-\mathfrak{A}(M)-\mu)\ni m\mapsto\varphi_{m^*,\omega_i}$) a meromorphic function on $\mathbb{C}$ is defined with poles lying in the set $\mathfrak{A}(M)$ (resp. $(-\mathfrak{A}(M)-\mu)$) the first assertion is easily deduced from (8.24). Note that by Theorem 5.37.(ii) and since $S^\varkappa$ is a weakly M-bounded subset of X the support of $\varphi_{m^*,\omega_i}|_{S^\varkappa}$ is a compact subset of $S^\varkappa$. Moreover, for every $j\in\mathbb{Z}$ one obtains that

$$\langle a_j(m;\mathfrak{N}_f),\varphi\rangle= \tag{8.25}$$
$$=\sum_{i=0}^{k}(-1)^i\sum_{\ell\in\mathbb{Z}}\int_{S^\varkappa}a_\ell(m;\mathfrak{M}_{f,\omega_{k-i}})(\vartheta)\,(-1)^{j-\ell}\,a_{j-\ell}(m^*;\mathfrak{M}_{\varphi,\omega_i})(\vartheta)\,d\tilde{\varkappa}(\vartheta)\,.$$

(i): In this case $\mathfrak{M}_{\varphi,\omega_i}$ is holomorphic at m^* so that (8.25) and Theorem 5.37.(iv) imply that for every $j \in \mathbb{N}_0$ we have

$$\langle a_{-j}(m;\mathfrak{N}_f), \varphi \rangle = \delta_{j0} H - \sum_{i=0}^{k} \sum_{\ell=\max\{k-i+1,j\}}^{\infty} (-1)^{k+1+j} \binom{\ell-1}{k-i} I_{\ell-k+i-1,\ell-j,i}$$

where by H we abbreviate the right-hand side of (8.24) and where

$$I_{\ell,j,i} := \int_{S^\varkappa} q_\ell(\vartheta)\, a_j(m^*;\mathfrak{M}_{\varphi,\omega_i})(\vartheta)\, d\tilde{\varkappa}(\vartheta)$$

with $q_\ell := (-1)^\ell (\partial_M - m)^\ell Q_m f$. Since by Propositions 5.11 and 5.13 and by (1.38) we have

$$a_j(m^*;\mathfrak{M}_{\varphi,\omega_i}) = (-1)^j \varphi_{m^*,\omega_i\omega_j} = (-1)^j \binom{i+j}{j} \varphi_{m^*,\omega_{i+j}}, \qquad j \in \mathbb{N}_0,$$

we deduce by Lemma 4.23 and (1.38) that

$$I_{\ell,j,i} = (-1)^j \binom{i+j}{i} \sum_{s=0}^{\infty} \binom{i+j+s}{s} \int_{X_+} [q_{\ell+s}\, \varphi\, \omega_{i+j+s} \circ \varkappa](x)\, dx.$$

Applying this to $I_{\ell-k+i-1,\ell-j,i}$, substituting $S = s+\ell+i-k-1$ and inserting the result into the equation at the beginning of the proof we conclude in view of (8.24) that

$$(8.26)\qquad a_{-j}(m;\mathfrak{N}_f) - \delta_{j0}\, d(T_{f_{m,\omega_k}})_m =$$
$$= \sum_{i=0}^{k} \sum_{\ell=\max\{k-i+1,j\}}^{\infty} \sum_{S=\ell+i-k-1}^{N} (-1)^{\ell+k} \binom{\ell-1}{k-i} \binom{i+\ell-j}{i} \binom{S+k+1-j}{i+\ell-j} R_{j,S}$$

where the distributions $R_{j,S} \in \mathcal{D}'(X)$ are defined by

$$\langle R_{j,S}, \varphi \rangle := \int_{X_+} [q_S\, \omega_{S+k+1-j} \circ \varkappa](x)\, \varphi(x)\, dx.$$

Changing the order of summation according to

$$\sum_{\ell=\max\{k-i+1,j\}}^{\infty} \sum_{S=\ell+i-k-1}^{N} \ldots = \sum_{\substack{S=0\\ S\ge j+i-k-1}}^{N} \sum_{\ell=\max\{k-i+1,j\}}^{S+k+1-i} \ldots$$

and

$$\sum_{i=0}^{k} \sum_{\substack{S=0\\ S\ge j+i-k-1}}^{N} \ldots = \sum_{S=\max\{0,j-k-1\}}^{N} \sum_{i=0}^{\min\{k,S+k+1-j\}} \ldots$$

and substituting $L = \ell+i-k-1$ we obtain

$$(8.26)'\qquad a_{-j}(m;\mathfrak{N}_f) - \delta_{j0}\, d(T_{f_{m,\omega_k}})_m = \sum_{S=\max\{0,j-k-1\}}^{N} b_{j,S}\, R_{j,S}$$

where

$$b_{j,S} := \sum_{i=0}^{\min\{k,S+k+1-j\}} \sum_{L=\max\{0,j+i-k-1\}}^{S} (-1)^{L+i+1} \binom{L+k-i}{L} \binom{L+k+1-j}{i} \binom{S+k+1-j}{L+k+1-j}.$$

Changing the order of summation we see that

$$b_{j,S} = \sum_{L=\max\{0,j-k-1\}}^{S} (-1)^{L+1} \Big[\sum_{i=0}^{\min\{k,L+k+1-j\}} (-1)^i \binom{L+k-i}{L} \binom{L+k+1-j}{i} \Big] \binom{S+k+1-j}{L+k+1-j}.$$

Applying Lemma 1.76 to $(L+k+1-j, L, j-1)$ instead of (ℓ, k, j) we deduce that the sum in square brackets vanishes in case $j \in \mathbb{N}_k$ and equals $\binom{j-1}{j-k-1} = \binom{j-1}{k}$ in case $j \geq k+1$. Since in the latter case we have

$$\sum_{L=j-k-1}^{S} (-1)^{L+1} \binom{S+k+1-j}{L+k+1-j} = (-1)^{j+k} \sum_{\ell=0}^{S+k+1-j} (-1)^{\ell} \binom{S+k+1-j}{\ell} =$$

$$= (-1)^{j+k} (1-1)^{S+k+1-j} = (-1)^{j+k} \delta_{0,S+k+1-j}$$

it follows that

$$b_{j,S} = \begin{cases} 0 & \text{if } j \in \mathbb{N}_k \\ (-1)^{j+k} \binom{j-1}{k} \delta_{S,j-k-1} & \text{if } j \geq k+1 \end{cases}. \tag{8.27}$$

In order to compute $b_{0,S}$ we observe that for $j=0$ the term in square brackets equals

$$\sum_{i=0}^{k} (-1)^i \binom{L+k-i}{L} \binom{L+k+1}{i}. \tag{8.28}$$

To compute (8.28) we derive a variant of (1.86). Differentiating the equation

$$(t-1)^{L+k+1}/t = \sum_{i=0}^{L+k+1} (-1)^i \binom{L+k+1}{i} t^{L+k-i}$$

L times with respect to t and evaluating the result at $t=1$ we obtain

$$0 = \sum_{i=0}^{k} (-1)^i \binom{L+k+1}{i} \frac{(L+k-i)!}{(k-i)!} + (-1)^{L+k+1} (-1)^L L!$$

so that (8.28) is equal to $(-1)^k$. Hence we conclude that

$$b_{0,S} = \sum_{L=0}^{S} (-1)^{L+1} (-1)^k \binom{S+k+1}{L+k+1} = \sum_{\ell=k+1}^{S+k+1} (-1)^{\ell} \binom{S+k+1}{\ell} = \tag{8.29}$$

$$= (1-1)^{S+k+1} - \sum_{\ell=0}^{k} (-1)^{\ell} \binom{S+k+1}{\ell} = -(-1)^k \binom{S+k}{k},$$

the last equality being valid by formula (4.20). Combining this with the other equations we see that the proof of (i) is complete.

(ii): The proof is analogous to that of (i). Under the present assumptions the function $\mathfrak{M}_{f,\omega_{k-i}}$ is holomorphic at m so that it follows by (8.25) and by (5.64) (applied to φ instead of f) that

$$\langle a_{-j}(m;\mathfrak{R}_f),\varphi\rangle = \delta_{j0}H - \sum_{I=0}^{k}\ \sum_{\substack{L=0\\ j+L\geq I+1}}^{\infty} (-1)^{I+j+L}\binom{j+L-1}{I} J_{L,j+L-I-1,k-I}$$

where, again, by H we denote the right-hand side of (8.24) and where

$$J_{L,j,i} := \int_{S^{\varkappa}} a_L(m;\mathfrak{M}_{f,\omega_i})(\vartheta)\,[(\partial_M - m^*)^j Q_{m*}\varphi](\vartheta)\,d\tilde{\varkappa}(\vartheta).$$

Since by Proposition 5.11 and 5.13 and by (1.38) we have

$$a_L(m;\mathfrak{M}_{f,\omega_i}) = (-1)^L f_{m,\omega_i\omega_L} = (-1)^L\binom{i+L}{i} f_{m,\omega_{i+L}}$$

we deduce by Proposition 5.48.(iii) (applied to $P(x,\partial) = \partial_M - m^*$), by Lemma 8.25, and by (1.38) and (2.5) that

$$J_{L,j,i} = (-1)^{L+j}\binom{i+L}{i}\sum_{s=0}^{\infty}\binom{i+L+s}{s}\langle(\partial_M - m)^{j+s} Q'_m(T_{f\omega_{i+L+s}}\circ\varkappa),\varphi\rangle.$$

Substituting $i = k-I$ and $S = s+j+L-I-1$ we obtain

$$J_{L,j+L-I-1,i} = (-1)^{k+I+1}\binom{i+L}{i}\sum_{S=j+L+i-k-1}^{N}\binom{S+k+1-j}{i+L}\langle R_{j,S},\varphi\rangle$$

where this time we set $N := N(m^*)$ and

$$R_{j,S} := (-1)^{k+j}(\partial_M - m)^S Q'_m(T_{f\omega_{S+k+1-j}}\circ\varkappa).$$

Putting everything together and substituting $\ell = L+j$ we conclude in view of (8.24) that (8.26) holds for the present choice of $R_{j,S}$, as well. It follows that (8.26)' is valid. Hence, in view of (8.27) and (8.29) the assertion is proved. ■

Notation 8.27. $\dot{f}_{m,\omega_k} := a_0(m;\mathfrak{R}_f)$; and, in particular, $\dot{f}_m := \dot{f}_{m,\omega_0}$.

Theorem 8.28. *$\dot{f}_{m,\omega_k}$ is an extension of $T_{f_{m,\omega_k}}$ to the whole of X having the following properties:*

(i) $(\partial_M - m)^i \dot{f}_{m,\omega_k} = (-1)^i \dot{f}_{m,\omega_{k-i}}$ for every $i \in \mathbb{N}_k$;

(ii) if $m \notin \mathfrak{A}(M)\cup(-\mathfrak{A}(M)-\mu)$ then $\dot{f}_{m,\omega_k}$ is almost quasihomogeneous of degree m and of order $\leq k$;

(iii) if $m \in \mathfrak{A}(M)$ then $\dot{f}_{m,\omega_k}$ is almost quasihomogeneous of degree m with $(1+k)^{th}$ order deficiency $(-1)^{1+k}T_{Q_m f}$;

(iv) if $m \in (-\mathfrak{A}(M)-\mu)$ then $\dot{f}_{m,\omega_k}$ is almost quasihomogeneous of degree m with $(1+k)^{th}$ order deficiency $(-1)^k Q'_m(T_f)$ (see (8.19)).

Proof. By Theorems 8.15' and 8.26 we know that $\mathfrak{N}_f(m)$ extends $T_{f_{m,\omega_k}}$ for every $m \in \Omega := \mathbb{C} \setminus (\mathfrak{A}(M) \cup (-\mathfrak{A}(M)-\mu))$. By the integral formula for the Laurent coefficients of the meromorphic functions $\mathfrak{N}_f$ and $\mathfrak{M}_{f,\omega_k}$ (see Theorem 5.37.(iv)) this implies the first part of the assertion.

(i): Let $i \in \{0\} \cup \mathbb{N}_k$. By $h_i : \mathbb{C} \longrightarrow \mathfrak{D}'(X)$ we denote the meromorphic function defined by $\Omega \ni m \mapsto (\partial_M - m)^i \mathfrak{N}_f(m)$. By (5.63) and by the assertion already proved above we have $h_i(m)|_{X_+} = (-1)^i \mathfrak{N}_{f,k-i}(m)|_{X_+}$, $m \in \Omega$, where by $\mathfrak{N}_{f,k-i}$ we denote the meromorphic function defining $\dot{f}_{m,\omega_{k-i}}$. By the uniqueness part of Theorem 8.15' we conclude that h_i and $(-1)^i \mathfrak{N}_{f,k-i}$ coincide on Ω and, hence, are equal as meromorphic functions. Consequently, in view of (8.21) and (8.23) for $j \in \mathbb{N}_k$, the assertion follows by Remark 2.54 (applied to $j=0$).

(ii): this follows from Theorem 8.15'.

(iii): In view of (8.21) for $j \in \mathbb{N}_k$ and Proposition 2.53 this follows from (ii) and (8.21) for $j = 1+k$. An alternative argument is based on the uniqueness part of Theorem 8.15': by (5.63) the distribution $(\partial_M - m)^{k+1} \dot{f}_{m,\omega_k}$ and the distribution induced by $(-1)^{1+k} Q_m f$ both extend $(\partial_M - m)^{k+1} f_{m,\omega_k}$ and are almost quasihomogeneous and consequently coincide on the whole of X.

(iv): In view of (8.23) for $j \in \mathbb{N}_k$ and Proposition 2.53 this follows from (ii) and (8.23) for $j = 1+k$. ∎

If $\mathrm{Re}\, m > -\mu$ then the distribution $\dot{f}_{m,\omega_k}$ is computed as in Example 8.17:

Remark 8.29. *If $\mathrm{Re}\, m > -\mu$ then $\dot{f}_{m,\omega_k} = T_q$ where $q : X \longrightarrow \mathbb{C}$ is the extension of f_{m,ω_k} defined by $q/_{X \setminus X_+} :\equiv 0$. Note that q is locally integrable, indeed, and in case $\mathrm{Re}\, m > 0$ even continuous (in Proposition 1.91 q was denoted by $\dot{f}_{m,\omega_k}$).*

Proof. In view of Theorem 5.37.(iii) we deduce from Proposition 1.91 that T_q is an extension of $T_{f_{m,\omega_k}}$ which is almost quasihomogeneous of degree m. Since $\mathrm{Re}\, m > -\mu$ Theorem 8.15' shows that T_q is the unique almost quasihomogeneous extension of $T_{f_{m,\omega_k}}$. Hence by Theorem 8.28 it coincides with $\dot{f}_{m,\omega_k}$. ∎

We close this section by verifying that the standard rules of computation which are valid for f_{m,ω_k} carry over to their extensions $\dot{f}_{m,\omega_k}$. Note that in a sense this is in contrast to Proposition 8.19.

Proposition 8.30. *(i) Let* $N \in \mathbb{N}_0$ *and* $\ell \in \mathbb{C}$, *and let* $P_0 : X \times V^* \longrightarrow \mathbb{C}$ *be a* C^∞ *copolynomial function which is almost quasihomogeneous of degree* ℓ, *of type* $M \times (-M)^*$, *and of order* $\leq N$. *Then*

$$(8.30) \qquad P_0(x,\partial)\,\dot{f}_{m,\omega_k} = \sum_{j=0}^{N} (-1)^j \binom{k+j}{k} (P_j(x,\partial)f)^{\bullet}_{m+\ell,\omega_{k+j}}$$

and

$$(8.30)' \qquad (P_0(x,\partial)f)^{\bullet}_{m+\ell,\omega_k} = \sum_{j=0}^{N} \binom{k+j}{k} P_j(x,\partial)\,\dot{f}_{m,\omega_{k+j}}$$

where the copolynomial functions P_j *are defined by (4.7).*

(ii) For every $\ell \in \mathbb{C}$ *and every* C^∞ *function* $q : X \longrightarrow \mathbb{C}$ *which is almost quasihomogeneous of degree* ℓ *we have*

$$(8.31) \qquad q\,\dot{f}_{m,\omega_k} = \sum_{j=0}^{N} (-1)^j \binom{k+j}{k} (q_j f)^{\bullet}_{m+\ell,\omega_k}$$

and

$$(8.31)' \qquad (qf)^{\bullet}_{m+\ell,\omega_k} = \sum_{j=0}^{N} \binom{k+j}{k} q_j\,\dot{f}_{m,\omega_{k+j}}$$

where $q_j := (\partial_M - \ell)^j q$.

(iii) $\dot{f}_{m,\omega_k} \circ A = (f \circ A)^{\bullet}_{m,\omega_k}$ *for every* $A \in GL(V,V)$ *commuting with* M;

(iv) $(\dot{f}_{m,\omega_k})_{\mathfrak{G}} = (f_{\mathfrak{G}})^{\bullet}_{m,\omega_k}$ *if* $\mathfrak{G}$ *satisfies the assumptions of Remark 2.67.(ii).*

Proof. We first observe that as functions of $m \in \Omega := \mathbb{C} \setminus (\mathfrak{A}(M) \cup (-\mathfrak{A}(M) - \mu))$ all the terms in the equations of the proposition are holomorphic and define meromorphic functions on the whole of $\mathbb{C}$. Hence, by making use of the integral formula (2.10) for the Laurent coefficients one easily reduces the proof to the case "$m \in \Omega$".

(iii): By Theorem 5.37.(vii) we have $T_{f_{m,\omega_k}} \circ A = T_{(f \circ A)_{m,\omega_k}}$. Consequently, for every $m \in \Omega$ we deduce the desired equality from Proposition 8.18.(ii).

(iv): By Theorem 5.37.(viii) we have $(T_{f_{m,\omega_k}})_{\mathfrak{G}} = T_{(f_{\mathfrak{G}})_{m,\omega_k}}$. Hence, for every $m \in \Omega$ the desired equation is obtained by Proposition 8.18.(iii).

(i): Here a proof by direct computation does not seem to work. However, in view of (5.34) and (5.34)' (see Theorem 5.37.(v)) the equations (8.30) and (8.30)' are valid for the restrictions to X_+. Since the distributions on both sides of these equations areare almost quasihomogeneous of degree $m + \ell$ the uniqueness part

of Theorem 8.15' implies that (8.30) and (8.30)' are valid for every $m \in \Omega$.

(ii): This is a special case of (i). ∎

(d) The Fourier Transform of $\dot{\varphi}_{m,\omega_k}$ if φ Belongs to $\mathcal{S}(V)$

In this section we suppose that $X = V$, and we fix a function $\varphi \in \mathcal{S}(V)$.

Theorem 8.31. *$\dot{\varphi}_{m,\omega_k}$ is a temperate distribution satisfying*

$$\mathcal{F}(\dot{\varphi}_{m,\omega_k}) = (-1)^k (\dot{\hat{\varphi}})_{m^*,\omega_k}.$$

A consequence is that the Fourier transform has quasihomogeneous eigenvectors:

Corollary 8.32. *Let $\lambda \in \mathbb{C}$ be such that $\hat{\varphi} = \lambda\varphi$, and set $m_0 := -\frac{\mu}{2}$. Then $S := \varphi_{m_0,\omega_k}$ is almost quasihomogeneous of degree m_0 and of order $\leq k$ satisfying*

$$\hat{S} = (-1)^k \lambda S. \quad ∎$$

For the proof of Theorem 8.31 the following lemma is required.

Lemma 8.33. *Suppose that*

(8.32) $$-\mu < \operatorname{Re} m < 0.$$

Then there is a constant C such that for arbitrary $\varphi, \chi \in \mathcal{L}^1(V) \cap \mathcal{L}^\infty(V)$ the function $f:]0,+\infty[\times V \longrightarrow \mathbb{C}$, $(t,x) \mapsto t^{-m-1} \varphi(M_t x)\chi(x)$, is absolutely integrable satisfying the estimate

$$\|f\|_{\mathcal{L}^1} \leq C(\|\varphi\|_{\mathcal{L}^1} \|\chi\|_{\mathcal{L}^\infty} + \|\varphi\|_{\mathcal{L}^\infty} \|\chi\|_{\mathcal{L}^1}).$$

Proof. Let $\ell \in \mathbb{C}$ be such that $\operatorname{Re} \ell > 0$. Since $\chi \in \mathcal{L}^1(V)$ and since φ is essentially bounded it follows by the Fubini-Tonelli theorem that the function

$$f_\ell :]0,1] \times V \longrightarrow \mathbb{C}, \quad (t,x) \mapsto t^{\ell-1} \varphi(M_t x)\chi(x),$$

belongs to $\mathcal{L}^1(]0,1] \times V)$ satisfying

$$\|f_\ell\|_{\mathcal{L}^1} \le \int_0^1 t^{\operatorname{Re}\ell-1}\,dt\ \ \|\varphi\|_{\mathcal{L}^\infty}\ \|\chi\|_{\mathcal{L}^1}.$$

Since the function $F:]1,+\infty[\times V \longrightarrow]0,1]\times V$, $(s,y)\mapsto(1/s, M_s y)$, is a diffeomorphism satisfying $\det DF(s,y) = -s^{-2}\det M_s = -s^{\mu-2}$ the change of variables formula shows that the function $f_\ell\circ F\,|\det DF|$, mapping (s,y) to $s^{\mu-\ell-1}\varphi(y)\,\chi(M_s y)$, belongs to $\mathcal{L}^1(]1,+\infty[\times V)$ satisfying

$$\|f_\ell\circ F\det DF\|_{\mathcal{L}^1} = \|f_\ell\|_{\mathcal{L}^1}.$$

Now, if $\ell=-m$ then f_ℓ is equal to the restriction of f to $]0,1]\times V$; if $\ell=m+\mu$ and if in the definition of f_ℓ the roles of φ and χ are interchanged then the function $f_\ell\circ F\,|\det DF|$ coincides with the restriction of f to $]1,+\infty[\times V$. Combining this with the preceding information one obtains the assertion with

$$C := \max\left\{\int_0^1 t^{-\operatorname{Re}m-1}\,dt,\ \int_0^1 t^{\operatorname{Re}m+\mu-1}\,dt\right\}.\quad\blacksquare$$

Proof of Theorem 8.31. By Proposition 5.41 the functions $\varphi_{m,\omega_{k-i}}$, $0\le i\le k$, belong to $\mathfrak{T}_M^\infty$ (see Notation 5.39). We choose $\psi_0,\dots,\psi_k$ as in Corollary 6.39. By Remark 5.40 this implies that the function

$$\psi := \sum_{i=0}^{k}(-1)^i\,\psi_i\,\varphi_{m,\omega_{k-i}}$$

belongs to $\mathcal{S}(V)$, its support being M-temperate. Consequently, by Proposition 6.22 $(T_\psi)_{m,\omega_k}$ is temperate for every $m\in\mathbb{C}$. Since in view of (4.10) we have

$$\mathfrak{N}_\varphi(m) = \sum_{i=0}^{k}(\psi_i\,(\partial_M-m)^i\,T_{\varphi_{m,\omega_k}})_{m,\omega_k} = (T_\psi)_{m,\omega_k}$$

for arbitrary $m\in\Omega:=\mathbb{C}\setminus(\mathfrak{A}(M)\cup(-\mathfrak{A}(M)-\mu))$ it follows that $\mathfrak{N}_\varphi(m)$ is temperate for every $m\in\Omega$. Moreover, since there is a compact subinterval J of $]0,+\infty[$ such that for arbitrary $m\in\Omega$ and $\chi\in\mathcal{S}(V)$ we have

$$\langle\mathfrak{N}_\varphi(m),\chi\rangle = (-1)^k\langle T_\psi,\chi_{m^*,\omega_k}\rangle =$$
$$= \sum_{i=0}^{k}(-1)^{k-i}\int_{\varkappa^{-1}(J)}\psi_i(x)\,\varphi_{m,\omega_{k-i}}(x)\,\chi_{m^*,\omega_k}(x)\,dx$$

we conclude by Proposition 6.22 that $\mathfrak{N}_\varphi$ is meromorphic with values in $\mathcal{S}'(V)$. Hence the first part of the assertion follows.

In order to deduce the formula for the Fourier transform we first observe that it suffices to prove it if (8.32) holds. For by the principle of analytic continuation it is then valid for every $m\in\Omega$, and to derive it for $m\in\mathfrak{A}(M)\cup(-\mathfrak{A}(M)-\mu)$ we choose $\varepsilon>0$ so small that

$$\overline{K}(m,\varepsilon)\cap(\mathfrak{A}(M)\cup(-\mathfrak{A}(M)-\mu))=\{m\},$$

define $\gamma:[0,2\pi]\longrightarrow\mathbb{C}$ by $t\mapsto\varepsilon e^{it}$, and obtain for arbitrary $j\in\mathbb{Z}$ and $\chi\in\mathcal{S}(V^*)$ that

$$\langle\mathcal{F}(a_j(m;\mathfrak{N}_\varphi)),\chi\rangle=\frac{1}{2\pi i}\int_\gamma\langle\mathfrak{N}_\varphi(z+m),\hat{\chi}\rangle z^{-j-1}\,dz=$$

$$=(-1)^k\frac{1}{2\pi i}\int_\gamma\langle\mathfrak{N}_{\hat{\varphi}}(-z-m-\mu),\chi\rangle z^{-j-1}\,dz=$$

$$=(-1)^{k+j}\frac{1}{2\pi i}\int_{-\gamma}\langle\mathfrak{N}_{\hat{\varphi}}(z-m-\mu),\chi\rangle z^{-j-1}\,dz\,;$$

and since $\int_{-\gamma}\dots=\int_\gamma\dots$ this is equal to $(-1)^{k+j}\langle a_j(m^*;\mathfrak{N}_{\hat{\varphi}}),\chi\rangle$, indeed.

Now suppose that (8.32) holds. By Remark 8.29 and by the definition of φ_{m,ω_k} we then conclude that

$$\langle\dot{\varphi}_{m,\omega_k},\chi\rangle=\int_{V_+}\varphi_{m,\omega_k}(x)\,\chi(x)\,dx=\int_{V_+}\int_0^{+\infty}t^{-m}\varphi(M_tx)\,\chi(x)\,\omega_k(t)\,\frac{dt}{t}\,dx$$

for every $\chi\in C_0^\infty(V)$. Since $\dot{\varphi}_{m,\omega_k}$ is temperate it follows by Lemma 8.33 and by the Fubini theorem that

$$(8.33)\qquad \langle\dot{\varphi}_{m,\omega_k},\chi\rangle=\int_0^{+\infty}t^{-m}\int_V\varphi(M_tx)\,\chi(x)\,dx\;\omega_k(t)\,\frac{dt}{t}\,,\qquad \chi\in\mathcal{S}(V).$$

Now we fix $\chi\in\mathcal{S}(V^*)$. Applying (8.33) to $\hat{\chi}$ instead of χ, using Parseval's equation, taking the equality $\mathcal{F}(\varphi\circ M_t)=t^{-\mu}\,\hat{\varphi}\circ M^*_{1/t}$ (see (2.8)') into account and substituting $s=1/t$ we obtain

$$\langle\dot{\varphi}_{m,\omega_k},\hat{\chi}\rangle=\int_0^{+\infty}t^{-m}\int_V\varphi(M_tx)\,\hat{\chi}(x)\,dx\;\omega_k(t)\,\frac{dt}{t}=$$

$$=\int_0^{+\infty}t^{-m-\mu}\int_{V^*}\hat{\varphi}(M^*_{1/t}\xi)\,\chi(\xi)\,d\xi\;\omega_k(t)\,\frac{dt}{t}=$$

$$=(-1)^k\int_0^{+\infty}s^{-m^*}\int_{V^*}\hat{\varphi}(M^*_s\xi)\,\chi(\xi)\,d\xi\;\omega_k(s)\,\frac{ds}{s}\,.$$

The last double integral is equal to $\langle(\dot{\hat{\varphi}})_{m^*,\omega_k},\chi\rangle$ as one sees by applying (8.33) to $(V^*,M^*,m^*,\hat{\varphi})$ instead of (V,M,m,φ). ■

Example 8.34. *$\varphi_0\equiv 1\iff(\dot{\hat{\varphi}})_{-\mu}=(2\pi)^n\delta_0$.*

Proof. Since $\dot{\varphi}_0$ is the unique almost quasihomogeneous extension of T_{φ_0} we conclude that $\varphi_0 \equiv 1 \Longleftrightarrow \dot{\varphi}_0 = T_1 \Longleftrightarrow \mathcal{F}(\dot{\varphi}_0) = (2\pi)^n \delta_0$. By Theorem 8.31 we have $\mathcal{F}(\dot{\varphi}_0) = (\dot{\hat{\varphi}})_{-\mu}$. Hence the assertion follows. ∎

Closing this section we fix a C^∞ function $\psi \in \mathcal{O}_{\mathfrak{M}}(V)$ with M-temperate support. Recall that its Fourier transforms $\hat{\psi}$ lies in $\mathcal{O}'_C(V^*)$, the space of convolution operators on $\mathcal{S}(V^*)$.

Theorem 8.35. *Let $\chi \in \mathcal{S}(V^*)$. Then the convolution product $\check{\hat{\psi}} * \dot{\chi}_{m,\omega_k}$ is induced by a function in $\mathcal{S}(V^*)$, and for every $u \in \mathcal{S}'(V)$ we have*

$$\langle \mathcal{F}((\psi u)_{m,\omega_k}), \chi \rangle = (2\pi)^{-n} \langle \hat{u}, \check{\hat{\psi}} * \dot{\chi}_{m,\omega_k} \rangle .$$

Note that this is a formula for the Fourier transform of u_{m,ω_k} in terms of $\hat{u}$ provided that ψ is equal to 1 on the support of u. Recall that by Lemma 5.53 one can choose ψ so as to have this property provided that supp u is M-temperate.

Proof of Theorem 8.35. By Remark 5.40 the function $\psi \hat{\chi}_{m^*,\omega_k}$ belongs to $\mathcal{S}(V)$. Hence it follows from Theorem 8.31 and from the Fourier inversion formula that

$$\mathcal{F}(\psi \hat{\chi}_{m^*,\omega_k}) = \mathcal{F}(\psi \dot{\hat{\chi}}_{m^*,\omega_k}) = (2\pi)^{-n} \hat{\psi} * \left(\mathcal{F}(\dot{\hat{\chi}}_{m^*,\omega_k})\right) =$$

$$= (2\pi)^{-n} (-1)^k \hat{\psi} * \dot{\hat{\hat{\chi}}}_{m,\omega_k} = (-1)^k \hat{\psi} * \dot{\check{\chi}}_{m,\omega_k} .$$

In particular, $\hat{\psi} * \dot{\check{\chi}}_{m,\omega_k}$ belongs to $\mathcal{S}(V^*)$. Consequently, we conclude by the Fourier inversion formula that

$$\langle \mathcal{F}((\psi u)_{m,\omega_k}), \chi \rangle = (-1)^k \langle \psi u, (\hat{\chi})_{m^*,\omega_k} \rangle =$$

$$= (-1)^k (2\pi)^{-n} \langle \check{\hat{u}}, \mathcal{F}(\psi (\hat{\chi})_{m^*,\omega_k}) \rangle = (2\pi)^{-n} \langle \hat{u}, \check{\hat{\psi}} * \dot{\chi}_{m,\omega_k} \rangle . \quad ∎$$

Chapter IX

Quasihomogeneous Wave Front Sets

The present chapter contains the basic theory of quasihomogeneous wave front sets of type M where this time M is a linear endomorphism of V^* such that all of its eigenvalues have positive real part. These types of wave front sets generalize the classical notion of (homogeneous) wave front set (see Hörmander [11]) which appears as the special case $M = \mathrm{Id}_{V^*}$. As in the classical case, they lead to a refined description of the singularities of distributions. In fact, the main result of the present chapter (Theorem 9.34 in section (c) below) shows that the singularities of distributions on V which are quasihomogeneous of type M^* are best described in terms of quasihomogeneous wave front sets of type M.

With the quasihomogeneous wave front sets the basic idea is as follows: for every $v \in \mathcal{E}'(V)$ one keeps track of the behaviour of $\hat{v}$ along the quasihomogeneous rays $\{M_\tau \xi ;\ \tau \in [1,+\infty[\}$, $\xi \in S_{V^*}$, where by S_{V^*} we denote the unit sphere of V^* with respect to a scalar product satisfying (1.79) with V replaced by V^*. In case M is a real diagonal matrix quasihomogeneous wave front sets have been introduced by R. Lascar in [12] in the context of quasihomogeneous pseudodifferential operators.

In section (a) we treat the basic properties of the quasihomogeneous wave front sets keeping as close as possible to Hörmander's way of presenting the theory of (homogeneous) wave front sets in [11] (compare also § 1.6 in Hörmander [10]). In a similar spirit in section (b) we introduce Gevrey type versions of quasihomogeneous wave front sets which include those by Rodino [16] (see also Liess-Rodino [13] and the literature cited there).

Section (c) is devoted to the main theorem of this chapter alluded to above. In addition, two further propositions on wave front set inclusions are given. Again, all these results generalize corresponding results in [11].

Concrete examples are treated in section (d). Assuming that M is a real diagonal matrix with entries of the form $p=(r,1,\dots,1)$ for some $r\in[1,+\infty[$, we compute the quasihomogeneous wave front sets of the invariant fundamental solutions of the heat and of the Schrödinger operator studied in section 7.(f).

(a) The Wave Front Set $WF_M(T)$

As already indicated above, for the whole chapter we fix $M\in L(V^*)$ such that $\operatorname{Re}\lambda>0$ for every $\lambda\in\sigma(M)$. We assume the Conventions 1.24 and 1.24* to be valid with M replaced by M^*, considering M^* as an element of L(V) via the canonical identification of V^{**} with V. Note that, in particular, Notations 1.25, 1.29, and 1.33 as well as Remark 1.43 are to be understood in this sense.

In this and the following section X is not required to be quasihomogeneous of any type. Let $T\in\mathcal{D}'(X)$, and let $x_0\in X$. Then by the Paley-Wiener theorem x_0 does not belong to sing supp T if and only if there is a test function $\varphi\in C_0^\infty(X)$ satisfying $\varphi(x_0)\neq 0$ such that

$$(9.1)\qquad \sup\{\,|\mathcal{F}(\varphi T)(\xi)|\,|\xi|^N;\ \xi\in V^*\setminus K(0,1)\} < +\infty \quad \text{for every } N\in\mathbb{N}.$$

Since $V^*\setminus K(0,1)=\bigcup_{\tau\in[1,+\infty[}M_\tau(S_{V^*})$ where S_{V^*} denotes the unit sphere in V^* and since by Lemma 1.7 for every $\varepsilon>0$ there exist constants $c_\varepsilon, d_\varepsilon>0$ such that

$$(9.2)\qquad c_\varepsilon\tau^{\lambda_{min}-\varepsilon}\,|\xi|\le|M_\tau\xi|\le d_\varepsilon\tau^{\lambda_{max}+\varepsilon}\,|\xi|,\qquad \tau\in[1,+\infty[,\ \xi\in V^*,$$

where $\lambda_{min}:=\min\{\operatorname{Re}\lambda;\ \lambda\in\sigma\}$ and $\lambda_{max}:=\max\{\operatorname{Re}\lambda;\ \lambda\in\sigma\}$ the condition (9.1) is equivalent to

$$(9.1)'\qquad \sup\{\,\tau^N|\mathcal{F}(\varphi T)(M_\tau\xi)|;\ \xi\in S_{V^*},\ \tau\in[1,+\infty[\} < +\infty \qquad \text{for every } N\in\mathbb{N}.$$

Definition 9.1. *$WF_M(T)$* is by definition the complement in $X\times S_{V^*}$ of the set of all $(x_0,\xi_0)\in X\times S_{V^*}$ such that for some $\varphi\in C_0^\infty(X)$ satisfying $\varphi(x_0)\neq0$ there is a neighbourhood W of ξ_0 in $V^*\setminus\{0\}$ such that for every $N\in\mathbb{N}$

$$\sup\{\,|\mathcal{F}(\varphi T)(M_\tau\xi)|;\ \xi\in W\}=O(\tau^{-N}) \text{ as } \tau\to\infty.$$

In order to show that in Definition 9.1 one may take any test function φ whose support is concentrated near x_0 we prove a few standard lemmata in a precise form suitable for later applications, as well. For this we have to touch upon a few technical points. First of all we recall that the Fourier transform of any $\nu \in \mathcal{E}'(V)$ extends to a holomorphic function (again denoted by $\hat{\nu}$) on the complexification $V^* \times V^*$ of V^*; namely, we have

$$\hat{\nu}(\xi,\upsilon) := \langle \nu, \exp(-i\langle \xi,\cdot\rangle + \langle \upsilon,\cdot\rangle)\rangle, \qquad (\xi,\upsilon) \in V^* \times V^*.$$

Denoting the real-complex coordinates of ξ (resp. υ) $\in V^*$ by (y,z) (resp. (v,w)) we shall write (compare Notation 1.25.A)

$$(9.3) \qquad (\xi,\upsilon)^\alpha := \prod_{j=1}^{d} (y_j + i v_j)^{\beta_j} \prod_{j=1}^{d} (z_j + i w_j)^{\gamma_j} (\overline{z_j} + i \overline{w_j})^{\delta_j}, \qquad \alpha = (\beta,\gamma,\delta) \in \mathfrak{A}.$$

Moreover, we recall that if K is a compact subset of V its *supporting function* $H_K : V^* \longrightarrow \mathbb{R}$ is defined by

$$H_K(\upsilon) := \max\{\langle \upsilon, x\rangle;\ x \in K\};$$

for $\nu \in \mathcal{E}'(V)$ we also write $H_\nu := H_{\operatorname{supp}\nu}$. Finally, in section (b) we have to distinguish two cases, namely, whether or not M is semi-simple. The reason appears in the following slightly more precise form of (9.2) we have to work with; in order to be able to handle both cases simultaneously we introduce the notation

$$(9.4) \qquad \mathcal{E}_M := \begin{cases} \{0\} & \text{if M is semi-simple} \\ ]0, \min\{1, \lambda_{min}\}[& \text{if M is not semi-simple} \end{cases}$$

and notice that by Lemma 1.7 we can choose a family of constants $c_\varepsilon, d_\varepsilon > 0$, $\varepsilon \in \mathcal{E}_M$, in such a way that

$$(9.2)' \quad c_\varepsilon^2 \sum_{\lambda \in \sigma} \tau^{2\mathrm{Re}\lambda - 2\varepsilon} |\xi_\lambda|^2 \le |M_\tau \xi|^2 \le d_\varepsilon^2 \sum_{\lambda \in \sigma} \tau^{2\mathrm{Re}\lambda + 2\varepsilon} |\xi_\lambda|^2, \qquad \xi \in V^*,\ \tau \in [1,+\infty[,$$

for every $\varepsilon \in \mathcal{E}_M$ where by $\xi \mapsto \xi_\lambda$ we denote the canonical projection of V^* onto $G_M(\lambda)$.

Lemma 9.2. *Let K be a compact subset of V, let $\eta \in \mathcal{E}_M$, $\varepsilon > 0$, and $k, \ell \in \mathbb{N}_0$. Then there are constants $a, b, C \in]0,+\infty[$ such that for every $\chi \in C_0^\infty(K)$ and for arbitrary constants $A, N \in [0,+\infty[$ and $B \in [1,+\infty[$ the following implication holds:*

if for every $\beta \in \mathfrak{A}$ satisfying $|\beta| \le k+\ell+1$ the estimates

$$(9.5) \qquad |\partial^{\alpha+\beta}\chi| \le A\, B^{\mathrm{Re}\,\alpha_M}, \qquad \alpha \in \mathfrak{A},\ \mathrm{Re}\,\alpha_M \le N,$$

are valid then for arbitrary $\gamma \in \mathfrak{A}$, $\tau \in [1,+\infty[$, $\xi \in V^* \setminus K(0,\varepsilon)$, *and* $\upsilon \in V^*$ *we have*

$$(9.6) \qquad |(\partial^{\gamma}\hat{\chi})(M_\tau \xi, \upsilon)| \le$$
$$\le a A \gamma! C^{|\gamma|} (1+|M_\tau \xi|)^{-k} \exp(H_\chi(\upsilon))\, \tau^{-\ell(\lambda_{min}-\eta)} (b B^{1+\eta}/\tau)^N.$$

Proof. First of all we fix $\rho \in \mathfrak{E}_M \cap [0,\eta]$ such that

$$(9.7) \qquad \operatorname{Re}\lambda/(\operatorname{Re}\lambda - \rho) \le 1+\eta\,, \qquad \lambda \in \sigma.$$

Next we fix $\xi \in V^* \setminus K(0,\varepsilon)$ and $\tau \in [1,+\infty[$. Then we first choose $\lambda \in \sigma$ such that $|\xi_\lambda| \ge |\xi|/\sqrt{n}$ and then fix $j \in \mathbb{N}_n$ such that $|(M_\tau \xi_\lambda)_j| \ge |M_\tau \xi_\lambda|/\sqrt{n}$ (here, for a moment, we work with the real coordinates with respect to the canonical $\mathbb{R}$-basis associated with the real-complex basis $\mathfrak{E}^*$ of V^* defined in Convention 1.24*; in this way we achieve that the estimates below are valid for the case $\upsilon \ne 0$, as well). Finally, we fix $m \in \mathbb{N}_0$ such that $N - \operatorname{Re}\lambda + \rho < m(\operatorname{Re}\lambda - \rho) \le N$. Since we may assume that $\varepsilon \le \sqrt{n}$ we deduce from (9.2)' - applied to ρ instead of ε) of ε - that

$$\tau^{\ell(\lambda_{min}-\eta)+N} \le \tau^{(\ell+m+1)(\operatorname{Re}\lambda-\rho)} \le (c_\rho |\xi_\lambda|)^{-\ell-m-1} |M_\tau \xi_\lambda|^{\ell+m+1} \le$$
$$\le (\sqrt{n}\,(c_\rho \varepsilon))^{\ell+1+m} (\sqrt{n}\,|(M_\tau \xi_\lambda)_j|)^{\ell+m+1} \le$$
$$\le (n/(c_\rho \varepsilon))^{\ell+1+N/(\lambda_{min}-\rho)} |(M_\tau \xi)^{(\ell+m+1)e_j}|.$$

Now we fix $\vartheta \in \mathbb{N}_0^n$ satisfying $|\vartheta| \le k$ and set $\Theta := \vartheta + (\ell+m+1)e_j$. It then follows that

$$|(M_\tau\xi)^{\Theta} (\partial^{\gamma}\hat{\chi})(M_\tau\xi,\upsilon)| \le |(M_\tau\xi,\upsilon)^{\Theta} (\partial^{\gamma}\hat{\chi})(M_\tau\xi,\upsilon)| =$$
$$= |\mathcal{F}(\partial^{\Theta}(x^{\gamma}\chi))(M_\tau\xi,\upsilon)| \le \int_K dx\, \|\partial^{\Theta}(x^{\gamma}\chi)\|_{L_\infty} \exp(H_\chi(\upsilon))\,, \qquad \upsilon \in V^*.$$

We set $\mathfrak{A}_\Theta := \{(\beta,\gamma,\delta) \in \mathfrak{A};\ \beta = (\Theta_1,\dots,\Theta_d),\ \forall i \in \mathbb{N}_c: \gamma_i + \delta_i = \Theta_{d+i} + \Theta_{d+c+i}\}$. In view of Notation 1.25.B.(i) we find a family of integers $a_{\Theta,\omega}$, $\omega \in \mathfrak{A}_\Theta$, such that

$$\partial^{\Theta} = \sum_{\omega \in \mathfrak{A}_\Theta} a_{\Theta,\omega} \partial^{\omega}, \qquad \text{and} \qquad \sum_{\omega \in \mathfrak{A}_\Theta} |a_{\Theta,\omega}| \le 2^{\Theta_{d+1}+\dots+\Theta_n}.$$

We fix $\omega \in \mathfrak{A}_\Theta$ and $\gamma \in \mathfrak{A}$. By the Leibniz rule we have

$$\partial^{\omega}(x^{\gamma}\chi) = \sum_{\beta \le \gamma,\omega} \binom{\omega}{\beta} \frac{\gamma!}{(\gamma-\beta)!} x^{\gamma-\beta} \partial^{\omega-\beta}\chi\,.$$

Now, for every $\beta \in \mathfrak{A}$ satisfying $\beta \le \omega$ we fix $\alpha, \tilde{\beta} \in \mathfrak{A}$ such that $\omega - \beta = \alpha + \tilde{\beta}$, $|\alpha| \le m$, $|\tilde{\beta}| \le k + \ell + 1$, and $\operatorname{Re}\alpha_M = |\alpha| \operatorname{Re}\lambda$. Applying (9.5) to $\tilde{\beta}$ instead of β and taking the estimates $m \le N/(\operatorname{Re}\lambda - \rho)$ and (9.7) into account we deduce

$$|\partial^{\omega-\beta}\chi| \le A B^{|\alpha|\operatorname{Re}\lambda} \le A B^{(1+\eta)N}.$$

Moreover, setting $C := \max\{1+|x|;\ x \in K\}$ we estimate $|x^{\gamma-\beta}|$ on supp χ by $C^{|\gamma|}$. Finally,

$$\sum_{\beta \le \omega} \binom{\omega}{\beta} = 2^{|\omega|} \le 2^{k+\ell+m+1} \le 2^{k+\ell+1}\, 2^{N/(\lambda_{min}-\rho)}.$$

Choosing a constant $d \in]0,+\infty[$ such that

$$(1+|\zeta|)^k \le d \max\{|\zeta^\vartheta|;\ \vartheta \in \mathbb{N}_o^n,\ |\vartheta| \le k\}, \qquad \zeta \in V^*,$$

and combining all the estimates we obtain the assertion with

$$b := (4n/(c_\rho \varepsilon))^{1/(\lambda_{min}-\rho)} \quad \text{and} \quad a := d\,(4n/(c_\rho \varepsilon))^{\ell+1}\, 4^k \int_K dx. \quad \blacksquare$$

Lemma 9.3. *Let K be a compact subset of V, and let $\eta \in \mathscr{E}_M$ and $\varepsilon, r, s \in]0,+\infty[$. Then there exist constants $b, c \in]0,+\infty[$ such that for arbitrary $A, N, W \in]0,+\infty[$ and $B \in [1,+\infty[$ the following implication is valid: if $v \in \mathscr{E}'(V)$ satisfies*

$$(9.8) \qquad |\hat{v}(\xi)| \le W(1+|\xi|)^s, \qquad \xi \in V^*,$$

and if $\chi \in C_o^\infty(K)$ fulfills (9.5) for every $\beta \in \mathfrak{A}$ satisfying the estimate $|\beta| \le k+\ell+1$ for $k := s+n+1$ and some $\ell > (s\lambda_{max}+r)/\lambda_{min}$ then

$$(9.9) \qquad |\mathscr{F}(\chi v)(M_\tau \xi, \upsilon)| \exp(-H_\chi(\upsilon)) \le$$

$$\le c A W \tau^{-r} (b B^{1+\eta}/\tau)^N + c A \sup\{|\hat{v}(M_\tau \zeta)|;\ \zeta \in K(\xi_o, 2\varepsilon)\}$$

for arbitrary $\xi_o \in S_{V^*}$, $\xi \in K(\xi_o, \varepsilon)$, $\upsilon \in V^*$, and $\tau \in [1,+\infty[$.

<u>*Proof.*</u> Since $\mathscr{F}(\chi v) = (2\pi)^{-n} \hat{\chi} * \hat{v}$ the substitution $\zeta \mapsto M_\tau \zeta$ leads to

$$\mathscr{F}(\chi v)(M_\tau \xi, \upsilon) = (2\pi)^{-n} \tau^\mu \int_{V^*} \hat{\chi}(M_\tau(\xi-\zeta), \upsilon)\, \hat{v}(M_\tau \zeta)\, d\zeta.$$

If $\zeta \in Z := V^* \setminus K(\xi_o, 2\varepsilon)$ and $\xi \in K(\xi_o, \varepsilon)$ then $|\xi-\zeta| \ge |\zeta-\xi_o| - |\xi-\xi_o| > \varepsilon$ so that Lemma 9.2 gives

$$|\hat{\chi}(M_\tau(\xi-\zeta), \upsilon)| \le a A \exp(H_\chi(\upsilon))\, (1+|M_\tau(\xi-\zeta)|)^{-s-n-1}\, \tau^{-\ell(\lambda_{min}-\eta)}\, (bB^{1+\eta}/\tau)^N.$$

Since $(1+|M_\tau \zeta|) \le (1+|M_\tau(\xi-\zeta)|)(1+|M_\tau \xi|)$ and – by (9.2)' –

$$1+|M_\tau \xi| \le 1 + d_\eta \tau^{(\lambda_{max}+\eta)}(1+\varepsilon) \le C\tau^{\lambda_{max}+\eta}$$

where $C := 1+(1+\varepsilon)d_\eta$ it follows by (9.8) that

$$\tau^\mu \left| \int_Z \hat{\chi}(M_\tau(\xi-\zeta), \upsilon)\, \hat{v}(M_\tau \zeta)\, d\zeta \right| \le$$

$$\le a A \exp(H_\chi(\upsilon))\, W \tilde{c}\, C^s\, \tau^{s(\lambda_{max}+\eta)-\ell(\lambda_{min}-\eta)}\, (bB^{1+\eta}/\tau)^N$$

where $\tilde{c} := \int_{V^*} (1+|M_\tau(\xi-\zeta)|)^{-n-1} d(M_\tau\zeta) = \|(1+|\cdot|)^{-n-1}\|_{L_1(V^*)} < +\infty$.

By the Paley-Wiener theorem one finds a constant $\tilde{d}$ only depending on K such that $|\hat{\chi}(\zeta,\upsilon)| \le \tilde{d}A(1+|\zeta|)^{-n-1}\exp(H_\chi(\upsilon))$, $\zeta,\upsilon \in V^*$. Consequently,

$$\tau^\mu \int_{K(\xi_0,2\varepsilon)} |\hat{\chi}(M_\tau(\xi-\zeta),\upsilon)|\, d\zeta \le \int_{V^*} |\hat{\chi}(\zeta,\upsilon)|\, d\zeta \le \tilde{d}A\tilde{c}\exp(H_\chi(\upsilon)).$$

Since we may assume that η is so small that $s(\lambda_{max}+\eta)-\ell(\lambda_{min}-\eta) \le -r$ the assertion follows with $c := \tilde{c}\max\{aC^s, \tilde{d}\}$. ∎

Lemma 9.4. *Let K be a compact subset of X, and let F be a closed subset of S_{V^*} such that $K\times F \cap WF_M(T) = \emptyset$. Then there exist open neighbourhoods $\tilde{K}$ of K in X and $\tilde{F}$ of F in $V^*\setminus\{0\}$, a sequence $(C_N)_{N\in\mathbb{N}}$ of positive constants and another constant k_0 such that*

$$sup\{\tau^N |\mathcal{F}(\varphi T)(M_\tau\xi,\upsilon)| \exp(-H_\chi(\upsilon));\ \tau\in[1,+\infty[,\ \xi\in\tilde{F},\ \upsilon\in V^*\} \le$$
$$\le C_N\ sup\{|\varphi^{(\alpha)}(x)|;\ x\in X,\ Re\,\alpha_M \le N+k_0\}$$

for arbitrary $N\in\mathbb{N}$ and $\varphi\in C_0^\infty(\tilde{K})$.

Proof. *Step 1:* We fix $(x_0,\xi_0)\in K\times F$. Then we can choose $\psi\in C_0^\infty(X)$ and $\varepsilon>0$ such that $\psi(x_0)=2$ and

$$\text{(9.10)}\qquad sup\{\tau^N |\mathcal{F}(\psi T)(M_\tau\zeta)|;\ \zeta\in K(\xi_0,2\varepsilon),\ \tau\in[1,+\infty[\} < +\infty, \qquad N\in\mathbb{N}.$$

We choose a compact neighbourhood U of x_0 such that $|\psi|\ge 1$ on U. Applying Lemma 9.3 (to $v=\psi T$ and $B=1$) and combining the result with (9.10) we find a constant $k_0\in\mathbb{N}$ and a sequence of constants $C_N\in]0,+\infty[$, $N\in\mathbb{N}$, such that

$$sup\{\tau^N |\mathcal{F}(\chi\psi T)(M_\tau\xi,\upsilon)| \exp(-H_\chi(\upsilon));\ \tau\in[1,+\infty[,\ \xi\in K(\xi_0,\varepsilon),\ \upsilon\in V^*\} \le$$
$$\le C_N\ sup\{\|\partial^\beta\chi\|_{L_\infty};\ Re\,\beta_M \le N+k_0\}$$

for arbitrary $\chi\in C_0^\infty(U)$ and $N\in\mathbb{N}$. Now, let $\varphi\in C_0^\infty(U)$. Since $|\psi|\ge 1$ on U there is $\chi\in C_0^\infty(U)$ such that $\chi\psi=\varphi$. Since by the Leibniz rule there are constants A_β not depending on φ such that

$$\|\partial^\beta\chi\|_{L_\infty} \le A_\beta\ sup\{\|\partial^\alpha\varphi\|_{L_\infty};\ \alpha\le\beta\}, \qquad \beta\in\mathfrak{A},$$

the conclusion follows for $(\tilde{K},\tilde{F}) = (U, K(\xi_0,\varepsilon))$.

Step 2: We fix $x_0\in X$. By Step 1 and by the compactness of F we find a finite open covering $\mathfrak{W}$ of F and a family $(U_W)_{W\in\mathfrak{W}}$ of neighbourhoods of x_0 such that for every $W \in \mathfrak{W}$ the conclusion holds with $(\tilde{K},\tilde{F})$ replaced by (U_W, W). But then the conclusion is also valid for $(\tilde{K},\tilde{F}) = (\bigcap_{W\in\mathfrak{W}} U_W, \bigcup\mathfrak{W})$.

Step 3: By Step 2 and by the compactness of K we find a finite open covering $\mathfrak{U}$ of K and a family $(F_U)_{U\in\mathfrak{U}}$ of open neighbourhoods of F in $V^*\setminus\{0\}$ such that for every $U\in\mathfrak{U}$ the conclusion holds for (U, F_U) instead of $(\tilde{K},\tilde{F})$. Now we set $\tilde{F} := \bigcap_{U\in\mathfrak{U}} F_U$ and fix a compact neighbourhood $\tilde{K}$ of K in $\bigcup\mathfrak{U}$. Moreover, we let $(\chi_U)_{U\in\mathfrak{U}}$ be a C^∞ partition of unity on $\tilde{K}$ subordinated to the covering $\mathfrak{U}$. Applying the assertion to $\chi_U\varphi$ and observing that by the Leibniz rule we have

$$|\partial^\beta(\chi_U\varphi)| \le B_\beta \sup\{|\varphi^{(\alpha)}(x)|;\ \alpha\le\beta,\ x\in X\}, \qquad \beta\in\mathfrak{A},$$

where the B_β are constants not depending on φ we deduce the assertion for any $\varphi\in C_0^\infty(\tilde{K})$. ■

We can now prove the following standard fact.

Proposition 9.5. *The projection of $WF_M(T)$ to the first variable is equal to sing supp T.*

Proof. If $x_0 \in X\setminus \text{sing supp } T$ then by the discussion preceding Definition 9.1 and by Proposition 1.72.(ii) the condition (9.1)' holds for some $\varphi\in C_0^\infty(X)$ satisfying $\varphi(x_0)\neq 0$ so that $WF_M(T)$ does not intersect the set $\{x_0\}\times S^{n-1}$. If, on the other hand, the last condition is valid then there is a compact neighbourhood U of x_0 such that $WF_M(T)\cap U\times S_{V^*} = \emptyset$. Hence by Lemma 9.4 the condition (9.1)' holds for every $\varphi\in C_0^\infty(U)$. Consequently, x_0 does not belong to sing supp T. ■

The following remark is concerned with elementary symmetry properties of $WF_M(T)$.

Remark 9.6. *(i) If T is real valued then $WF_M(T)$ is invariant under the transformation $(x,\xi)\mapsto(x,-\xi)$.*

(ii) If $\overline{T} = \check{T}$ then $WF_M(T)$ is invariant under the map $(x,\xi)\mapsto(-x,\xi)$.

(iii) Let W be another $\mathbb{C}$-vector space, and let $N\in L(W^*,W^*)$ and $A\in GL(W,V)$ be such that $N\circ A^* = A^*\circ M$. Then

$$WF_N(T\circ A) = \{(y, A^*\xi);\ (Ay,\xi)\in WF_M(T)\}.$$

In particular, if $W = V$, if A^* and M commute, and if T is A-invariant then $WF_M(T)$ is invariant under the transformation $(x,\xi)\mapsto(A^{-1}x, A^*\xi)$.

Proof. *(i).* This follows from the fact that

$$\mathcal{F}(\varphi T)(-\zeta) = \overline{\mathcal{F}(\overline{\varphi}\,\overline{T})(\zeta)} = \overline{\mathcal{F}(\overline{\varphi T})(\zeta)}, \qquad \zeta \in V^*.$$

(ii). This is a consequence of $\mathcal{F}(\check{\varphi}T) = \mathcal{F}(\check{\varphi}\check{\overline{T}}) = \overline{\mathcal{F}(\overline{\varphi}\,T)}$.

(iii). The assumption on A implies that $(A^*)^{-1} \circ N_\tau = M_\tau \circ (A^*)^{-1}$, $\tau \in]0,+\infty[$. Consequently, making use of (2.8) we deduce that

$$|\det A|\, \mathcal{F}(\varphi\, T \circ A) \circ N_\tau \circ A^* = \mathcal{F}(\varphi \circ A^{-1}\, T) \circ (A^*)^{-1} \circ N_\tau \circ A^* =$$

$$= \mathcal{F}(\varphi \circ A^{-1}\, T) \circ M_\tau, \qquad \tau \in]0,+\infty[.$$

This implies the assertion. ■

As for the behaviour of quasihomogeneous wave front sets under linear partial differential operators the usual results remain valid:

Proposition 9.7. *(i)* $WF_M(fT) \subset WF_M(T)$ *for every* $f \in C^\infty(X)$;

(ii) $WF_M(\partial_j T) \subset WF_M(T)$ *for every* $j \in \mathbb{N}_n$.

Proof. *(i)*: this follows immediately from Lemma 9.4.

(ii): Let $\alpha \in \mathfrak{A}$ be such that $|\alpha| = 1$. Then by the Leibniz-rule and in view of

$$(M_\tau \xi)^\alpha = \tau^{\alpha M} \sum_{i=0}^{N_\alpha} \omega_i(\tau)\, P_{\alpha,i}(\xi)$$

(see Remark 1.30) we have

$$\text{(9.11)} \qquad \mathcal{F}(\varphi\,(-i\partial)^\alpha T)(M_\tau \xi) =$$

$$= \tau^{\alpha M}\Big(\sum_{i=0}^{N_\alpha} \omega_i(\tau)\, P_{\alpha,i}(\xi)\, \mathcal{F}(\varphi T)(M_\tau \xi)\Big) + i\,\mathcal{F}(\varphi^{(\alpha)}T)(M_\tau \xi), \qquad \xi \in V^*.$$

Consequently, since $\operatorname{supp} \varphi^{(\alpha)} \subset \operatorname{supp} \varphi$ the assertion follows from Lemma 9.4. ■

The following result deals with the converse of Proposition 9.7 for differential operators with quasihomogeneous principal part. We fix $m \in \operatorname{Re}(\mathfrak{A}(M))$, and let

$$\text{(9.12)} \qquad P = P(x,D) = \sum_{\operatorname{Re}\alpha_M \le m} a_\alpha(x)\, \mathbf{D}^\alpha \qquad (\text{where } \boldsymbol{D}^\alpha := (-i\partial)^\alpha)$$

be a differential operator of (quasihomogeneous) order $m \in \operatorname{Re}(\mathfrak{A}(M))$ with coefficients $a_\alpha \in C^\infty(X)$. Its principal part $P_m : X \times V^* \longrightarrow \mathbb{C}$ is defined by

$$P_m(x,\xi) = \sum_{\operatorname{Re}\alpha_M = m} a_\alpha(x)\, \xi^\alpha, \qquad (x,\xi) \in X \times V^*.$$

Note that even if M is semi-simple $P_m(x,\cdot)$ need not be quasihomogeneous. However, if M is semi-simple and if, in addition, $\sigma(M) \subset \mathbb{R}$ then $P_m(x,\cdot)$ is quasi-homogeneous of degree m.

Theorem 9.8. *Under the preceding assumption we suppose that for some $\tilde{m} \in \mathfrak{A}(M)$ $P_m(x,\cdot)$ is quasihomogeneous of degree $\tilde{m}$ for every $x \in X$. Then for every $T \in \mathcal{D}'(X)$ we have*

$$(9.13) \qquad WF_M(T) \subset WF_M(PT) \cup P_m^{-1}(0).$$

<u>*Proof*</u>. The proof of Theorem 8.3.1 in Hörmander [11] is to be suitably modified. Let $(x_0, \xi_0) \in X \times S_{V^*}$ be such that it does not belong to the right-hand side of (9.13). Then we can choose compact neighbourhoods K of x_0 in X and W of ξ_0 in $V^* \setminus \{0\}$, a sequence of constants C_N, and a distribution $v \in \mathcal{E}'(X)$ such that

$$(9.14) \qquad v = PT \text{ near } K;$$

$$(9.15) \qquad |\hat{v}(M_\tau \xi)| \leq C_N \tau^{-N}, \qquad \xi \in W,\ N \in \mathbb{N},\ \tau \in [1,+\infty[;$$

$$(9.16) \qquad \inf\{|P_m(z)|;\ z \in K \times W\} > 0.$$

Note that (9.14) amounts to

$$(9.14)' \qquad \langle T, {}^tP\varphi \rangle = \langle v, \varphi \rangle, \qquad \varphi \in C_0^\infty(K).$$

We fix a function $\chi \in C_0^\infty(K)$ equal to 1 on a fixed neighbourhood U of x_0 and set $u := \chi T$. In order to prove that (x_0, ξ_0) does not belong to the left-hand side of (9.13) we have to verify that $\hat{u}(M_\tau \xi) = O(\tau^{-N})$ uniformly for ξ near ξ_0 as $\tau \to \infty$. To this end we fix $N \geq m$, $\tau \geq N$ and $\xi \in W$. The Leibniz rule shows that for every $w \in C_0^\infty(K)$ we have

$$(9.17) \qquad {}^tP\Big(w \exp(-i\langle M_\tau\xi, \cdot\rangle)/P_m(\cdot, M_\tau\xi)\Big) = \Big(w - \sum_{\rho \in \mathfrak{B}_m} R_\rho w\Big) \exp(-i\langle M_\tau \xi, \cdot\rangle)$$

where $\mathfrak{B}_m := \{\rho \in \mathrm{Re}(\mathfrak{A}(M));\ 0 < \rho \leq m\}$ and

$$(9.18) \qquad R_\rho := -\sum_{(\alpha,\beta,\gamma,\delta,\varepsilon) \in \mathfrak{C}_{m,\rho}} \frac{\alpha!}{\beta!\gamma!\delta!\varepsilon!} (i\partial)^\gamma a_\alpha (i\partial)^\delta \Big(\frac{1}{P_m(\cdot, M_\tau\xi)}\Big) (M_\tau\xi)^\varepsilon (i\partial)^\beta$$

with $\mathfrak{C}_{m,\rho} := \{(\alpha,\beta,\gamma,\delta,\varepsilon) \in \mathfrak{A}^5;\ \mathrm{Re}\,\alpha_M \leq m,\ \beta+\gamma+\delta+\varepsilon = \alpha,\ \mathrm{Re}\,\varepsilon_M = m-\rho\}$. Note that R_ρ is a differential operator of quasihomogeneous order $\leq \rho$ (for $\mathrm{Re}\,\varepsilon_M = m-\rho$ implies $\mathrm{Re}\,\beta_M = \mathrm{Re}\,\alpha_M - \mathrm{Re}(\gamma+\delta)_M + \rho - m \leq \rho$). For any $\nu \in]0,+\infty[$ defining differential operators

$$S_\nu := 1 + \sum_{k\in\mathbb{N}} \sum_{\rho_1+\ldots+\rho_k\le\nu} R_{\rho_1}\ldots R_{\rho_k}$$

and

$$E_\nu := \sum_{k\in\mathbb{N}} \sum_{\rho_1+\ldots+\rho_k>\nu\ge\rho_2+\ldots+\rho_k} R_{\rho_1}\ldots R_{\rho_k}$$

we derive the following identity of operators

$$1 = S_\nu - \sum_{\rho\in\mathfrak{B}_m} R_\rho \circ S_\nu + E_\nu .$$

Combining this with (9.17) - applied to $w = S_\nu(\chi)$ - and taking (9.14)' into account we conclude that

$$\hat{u}(M_\tau\xi) = \tau^{-\tilde{m}} \mathcal{F}(\psi_\nu v)(M_\tau\xi) + \mathcal{F}(e_\nu T)(M_\tau\xi) \tag{9.19}$$

where $\psi_\nu := S_\nu(\chi)/P_m(\cdot,\xi)$ and $e_\nu := E_\nu(\chi)$ and where $\tilde{m}\in\mathfrak{A}(M)$ is chosen in such a way that P_m is quasihomogeneous of degree $\tilde{m}$. Since $\operatorname{Re}\tilde{m} = m$ we deduce that

$$|P_m(\cdot, M_\tau\xi)| = \tau^m |P_m(\cdot,\xi)| . \tag{9.20}$$

Moreover, making use of (9.2)' for every $\eta\in\mathfrak{E}_M$ we find a constant C_η such that

$$|(M_\tau\xi)^\alpha| \le (C_\eta)^{|\alpha|} |\xi|^{|\alpha|} \tau^{\operatorname{Re}\alpha_M+\eta|\alpha|}, \qquad \xi\in V^*,\ \tau\in[1,+\infty[. \tag{9.21}$$

Employing the last two conditions we deduce that for arbitrary $\rho_1,\ldots,\rho_k\in\mathfrak{B}_m$, $\gamma\in\mathfrak{A}$, and $\eta < \min\mathfrak{B}_m$ there is a constant C such that

$$|\partial^\gamma R_{\rho_1}\ldots R_{\rho_k}\chi| \le C\tau^{-\rho_1-\ldots-\rho_k+\eta} \le C, \qquad \tau\in[1,+\infty[,$$

so that by Lemma 9.3 and (9.15) we conclude that $\mathcal{F}(\psi_N v)(M_\tau\xi) = O(\tau^{-N})$ uniformly for $\xi\in W$ as $\tau\to\infty$. Moreover, we deduce that for every $\gamma\in\mathfrak{A}$ there is a constant C_γ such that

$$|\partial^\gamma e_N| \le C_\gamma \tau^{-N+\eta}, \qquad \tau\in[1,+\infty[.$$

Now let q be the (quasihomogeneous) order of T on K. Then for some constant C_K (which is independent of e_N and τ) we have

$$|\mathcal{F}(e_N T)(M_\tau\xi)| \le C_K \sup\{\|\partial^\beta(e_N\exp(-i\langle M_\tau\xi,\cdot\rangle))\|_{L_\infty};\ \operatorname{Re}\beta_M\le q\}. \tag{9.22}$$

By the Leibniz rule we obtain

$$\partial^\beta(e_N\exp(-i\langle M_\tau\xi,\cdot\rangle)) = \exp(-i\langle M_\tau\xi,\cdot\rangle)\sum_{\gamma\le\beta}\binom{\beta}{\gamma}(M_\tau\xi)^{\beta-\gamma}\partial^\gamma e_N .$$

Combining this with (9.21) we obtain a constant C' such that

$$|\partial^\beta(e_N\exp(-i\langle M_\tau\xi,\cdot\rangle))| \le C'\tau^{q+1-\eta}\sup\{|\partial^\gamma e_N|;\ \gamma\le\beta\},\quad \operatorname{Re}\beta_M\le q . \tag{9.23}$$

Consequently, $\mathcal{F}(e_N T)(M_\tau \xi) = O(\tau^{-N+1+q})$ uniformly for $\xi \in W$ as $\tau \to \infty$. Combining this with the estimate for $|\mathcal{F}(\psi_N v)(M_\tau \xi)|$ and with (9.19) we conclude that (x_0, ξ_0) does not belong to $WF_M(T)$, indeed. ∎

We now come to the generalization of Theorem 4.20 alluded to in section 4.(d). Here we require

Notation 9.9. (i) By $\varkappa_M$ we denote the function $\varkappa_+ : V^* \setminus \{0\} \longrightarrow]0, +\infty[$ defined in Proposition 1.70 for $X = V^*$.

(ii) By $\rho_M : V^* \setminus \{0\} \longrightarrow S_{V^*}$ we denote the C^∞ retraction defined in Notation 1.71.(i) if $X = V^* \setminus \{0\}$ and $\varkappa = \varkappa_M$.

Remark 9.10. *Let $N: V \longrightarrow V$ be a non-zero linear map, and suppose that X is quasihomogeneous of type N. Moreover, let I be a closed subset of $]0, +\infty[$, and let Z be a closed subset of $X \times S_{V^*}$ which is contained in $L \times S_{V^*}$ for some (N, I)-bounded subset L of X. Then the set*

$$Z^M_{N,I} := \{(N_{1/t} x, \rho_M(N_t^* \xi)) \,;\, t \in I, \ (x, \xi) \in Z\}$$

is a closed subset of $X \times S_{V^*}$.

Proof. Let $(y_\infty, \upsilon_\infty) \in \overline{Z^M_{N,I}} \cap X \times S_{V^*}$. We choose sequences $(t_j)_{j \in \mathbb{N}}$ in I and $(x_j, \xi_j)_{j \in \mathbb{N}}$ in Z such that

$$(y_\infty, \upsilon_\infty) = \lim_{j \to \infty} (N_{1/t_j} x_j, \rho_M(N_{t_j}^* \xi_j)).$$

By choosing a subsequence if necessary we achieve that (ξ_j) converges to a point $\xi_\infty \in S_{V^*}$ as $j \to \infty$. Since $K := \{y_\infty\} \cup \{N_{1/t_j} x_j \,;\, j \in \mathbb{N}\}$ is a compact subset of X it follows by the assumption on L that the t_j stay in a compact subinterval of $]0, +\infty[$. Hence we may assume that (t_j) converges to some $t_\infty \in I$ as $j \to \infty$. By continuity we conclude that

$$\lim_{j \to \infty} x_j = \lim_{j \to \infty} N_{t_j}(N_{1/t_j} x_j) = N_{t_\infty} y_\infty =: x_\infty \in X$$

and

$$\lim_{j \to \infty} \rho_M(N_{t_j}^* \xi_j) = \rho_M(N_{t_\infty}^* \xi_\infty).$$

Consequently,

$$(y_\infty, \upsilon_\infty) = (N_{1/t_\infty} x_\infty, \rho_M(N_{t_\infty}^* \xi_\infty)).$$

Since Z is closed in $X \times S_{V^*}$ the pair (x_∞, ξ_∞) belongs to Z so that $(y_\infty, \upsilon_\infty)$ lies in $Z^M_{N,I}$. ■

Theorem 9.11. *Let X, N, and I be as in Remark 9.10, and suppose that N^* and M commute. Moreover, suppose that X is locally N-bounded. Let $u \in \mathcal{D}'(X)$ be such that $\operatorname{supp} u$ is an (N,I)-bounded subset of X. And let $w:]0,+\infty[\longrightarrow \mathbb{C}$ be a locally integrable function with support contained in I. If we denote by $u_{m,w}$ the distribution on X defined by (4.3) with M replaced by N then*

$$WF_M(u_{m,w}) \subset (WF_M(u))^M_{N,I}, \qquad m \in \mathbb{C}.$$

Proof. We set $Z := WF_M(u)$. Let $(x_0, \xi_0) \in X \times S_{V^*} \setminus Z^M_{N,I}$. By Remark 9.10 we can choose $\varepsilon > 0$ such that

$$(9.24) \qquad \overline{K}(x_0, 2\varepsilon) \times \overline{K}(\xi_0, 2\varepsilon) \cap Z^M_{N,I} = \emptyset .$$

In order to prove that this is equivalent to

$$(9.24)' \qquad N_{1/s}(\overline{K}(x_0, 2\varepsilon)) \times (\rho_M \circ N^*_s)(\overline{K}(\xi_0, 2\varepsilon)) \cap Z = \emptyset , \qquad s \in 1/I ,$$

we fix $(k,\ell) \in \overline{K}(x_0, 2\varepsilon) \times \overline{K}(\xi_0, 2\varepsilon)$, $(x,\xi) \in Z$ and $s \in 1/I$, and setting $t := \varkappa_M(N^*_{1/s}\xi)$ observe that $\ell = \rho_M(N^*_{1/s}\xi) \Rightarrow N^*_{1/s}\xi = M_t \ell \Rightarrow \xi = N^*_s(M_t \ell) = M_t(N^*_s \ell) \Rightarrow \xi = \rho_M(\xi) = = \rho_M(N^*_s \ell)$. Interchanging the roles of ℓ and ξ and replacing $1/s$ by s we see that the converse implication is valid, as well. Hence

$$(k,\ell) = (N_s x, \rho_M(N^*_{1/s}\xi)) \Longleftrightarrow (x,\xi) = (N_{1/s} k, \rho_M(N^*_s \ell))$$

so that by the definition of $Z^M_{N,I}$ the equivalence of (9.24) and (9.24)' follows.

By Proposition 3.42 we choose an open neighbourhood W of $\operatorname{supp} u$ such that W is an (N,I)-bounded subset of X. Since $K := \overline{K}(x_0, \varepsilon)$ is a compact subset of X it follows that $J := \{t \in 1/I;\ N_t(W) \cap K \neq \emptyset\}$ is a compact subset of $]0,+\infty[$. Let $\varphi \in C^\infty_0(K)$. Hence, using (4.4), substituting $s = \frac{1}{t}$, and setting $\nu := \operatorname{tr} N$ we deduce

$$\begin{aligned} \mathcal{F}(\varphi u_{m,w})(\xi) &= \langle u_{m,w}, \varphi \exp(-i\langle \xi, \cdot \rangle) \rangle = \\ &= \int_{1/J} s^{-m} \langle u \circ N_s, \varphi \exp(-i\langle \xi, \cdot \rangle) \rangle\, w(s) \frac{ds}{s} = \\ &= \int_J t^{m+\nu} \langle u, (\varphi \exp(-i\langle \xi, \cdot \rangle)) \circ N_t \rangle\, w(1/t) \frac{dt}{t} = \\ &= \int_J t^{m+\nu} \mathcal{F}(\varphi \circ N_t u)(N^*_t \xi)\, w(1/t) \frac{dt}{t}, \qquad \xi \in V^*. \end{aligned}$$

By continuity for every $s \in J$ we find an open neighbourhood I_s of s in J such that

$$(9.25)\qquad \text{(a)}\ N_t^*(\overline{K}(\xi_0,\varepsilon)) \subset N_s^*(K(\xi_0,2\varepsilon)), \quad \text{and}$$
$$\text{(b)}\ N_{1/t}(\overline{K}(x_0,\varepsilon)) \subset N_{1/s}(K(x_0,2\varepsilon)), \qquad t \in I_s.$$

Since J is compact we find a finite subset S of J such that $J = \bigcup_{s\in S} I_s$. In view of (9.24)' we find compact neighbourhoods K_s of $N_{1/s}(\overline{K}(x_0,2\varepsilon))$ in X and F_s of $(\rho_M \circ N_s^*)(\overline{K}(\xi_0,2\varepsilon))$ in S_{V^*} such that

$$(9.26)\qquad K_s \times F_s \cap WF_M(u) = \emptyset, \qquad s \in S.$$

Now we choose $c \in [1,+\infty[$ such that $\frac{1}{c} \le \varkappa_M(N_t^*\xi)$ for arbitrary $t \in J$ and $\xi \in \overline{K}(\xi_0,\varepsilon)$. Since $N_t^*(M_\tau\xi) = M_{\tau\varkappa_M(N_t^*\xi)}\,\rho_M(N_t^*\xi)$ it follows in view of (9.25.a) that for every $N \in \mathbb{N}$ we have

$$(9.27)\qquad \sup\{\tau^N |\mathcal{F}(\varphi u_{m,w})(M_\tau\xi)|;\ \tau \in [c,+\infty[,\ \xi \in \overline{K}(\xi_0,\varepsilon)\} \le$$
$$\le C_N \sup\{\tau^N |\mathcal{F}(\varphi \circ N_t u)(M_\tau\zeta)|;\ \zeta \in F_s,\ \tau \in [1,+\infty[,\ t \in I_s,\ s \in S\}$$

where $C_N := c^N \int_J t^{\mathrm{Re}\, m+\nu-1} |w(1/t)|\, dt$. Since in view of (9.25.b) we have

$$(9.28)\qquad \operatorname{supp} \varphi \circ N_t = N_{1/t}(\operatorname{supp}\varphi) \subset N_{1/s}(\overline{K}(x_0,2\varepsilon)) \subset K_s, \qquad t \in I_s,\ s \in S,$$

and since by (1.49) - applied to $P = P_\alpha$ (see Remark 1.43) and $P^*_{\alpha,i} := (\partial_{N^*} - \alpha_N)^i P$ - we have

$$(9.29)\qquad \partial^\alpha(\varphi \circ N_t) = t^{\alpha N} \sum_{i=0}^{\infty} \omega_i(t)\, (P^*_{\alpha,i}(\partial)\varphi) \circ N_t, \qquad \alpha \in \mathfrak{A},\ t \in J,$$

an application of Lemma 9.4 to the sets (K_s, F_s) - the assumption being satisfied by (9.26) - shows that the right-hand side and hence the left-hand side of (9.27) is bounded for every $N \in \mathbb{N}$. This means that (x_0,ξ_0) does not belong to $WF_M(u_{m,w})$. ∎

(b) The Wave Front Set with Respect to $C^{M,L}$

Let $L: [0,+\infty[\longrightarrow [1,+\infty[$ be an increasing continuous function such that $L(0) = 1$ and

(9.30) (a) $L(t) \geq 1+t$, and (b) $L(t+s) \leq \Lambda^{s+1} L(t)$, $t, s \in [0,+\infty[$,

for some constant Λ. Note that (9.30.b) implies

(9.31) $L(t) \leq \Lambda^{t+1}$, $t \in [0,+\infty[$.

Sometimes, in addition we postulate:

(9.32) $\forall d>1 \; \exists C_d>1 \; \forall t \in [0,+\infty[: \; L(dt) \leq C_d \, L(t)^d$.

Example 9.12. *The following choices for L satisfy the preceding assumptions:*

(i) $L=A$ *where* $A(t):=1+t$;

(ii) $L(t)=(1+t)^\rho$ *for some* $\rho \in]1,+\infty[$;

(iii) $L(t)=e^{at}$ *for some* $a \in]0,+\infty[$.

Moreover, if L satisfies the preceding assumptions then so does L^d *for every* $d>1$. ■

Notation 9.13. For any $N \in \mathbb{N}_0$ we write $L_N := L(N)$.

Definition 9.14. Let $h:\mathfrak{A} \longrightarrow]0,+\infty[$ be an additive function. By $C^{h,L}(X)$ we denote the space of C^∞ functions $f:X \longrightarrow \mathbb{C}$ such that for every compact subset K of X there is a constant C such that

(9.33) $|f^{(\alpha)}(x)| \leq C\,(C\,L_\alpha)^{h(\alpha)}$, $x \in K, \; \alpha \in \mathfrak{A}$,

where $L_\alpha := (L \circ h)(\alpha)$.

Since

(9.34) $\min\{h(\iota);\, \iota \in \mathfrak{A},\, |\iota|=1\}\,|\alpha| \leq h(\alpha) \leq \max\{h(\iota);\, \iota\in\mathfrak{A},\, |\iota|=1\}\,|\alpha|$, $\alpha \in \mathfrak{A}$,

we observe

Remark 9.15. *If* $\sup\{L(2t)/L(t);\, t \in [0,+\infty[\}$ *is finite then in (9.33) the term* L_α *can be replaced by* $L_{|\alpha|}$. ■

Remark 9.16. $C^{h,L}(X)$ *is a* $\mathbb{C}$*-algebra which is closed under differentiation.*

Proof. Let $f,g \in C^{h,L}(X)$, let K be a compact subset of X, and let C be a constant such that (9.33) is satisfied for f and g. Since L is increasing we obtain by the Leibniz rule that for every $x \in K$

$$|(fg)^{(\alpha)}(x)| \le \sum_{\beta\le\alpha} \binom{\alpha}{\beta} |f^{(\beta)}(x)\, g^{(\alpha-\beta)}(x)| \le$$

$$\le C^2 \sum_{\beta\le\alpha} \binom{\alpha}{\beta} (C L_\beta)^{h(\beta)} (C L_{\alpha-\beta})^{h(\alpha-\beta)} \le$$

$$\le C^2 (C L_\alpha)^{h(\alpha)} \sum_{\beta\le\alpha} \binom{\alpha}{\beta} = C^2 2^{|\alpha|} (C L_\alpha)^{h(\alpha)} .$$

In view of (9.34) this implies that fg belongs to $C^{h,L}(X)$. Since by (9.30.b) we have

$$(L_{\alpha+\iota})^{h(\alpha+\iota)} \le \Lambda^{(h(\iota)+1)[h(\alpha)+h(\iota)]} (L_\alpha)^{h(\iota)} (L_\alpha)^{h(\alpha)} \quad \text{and} \quad L_\alpha \le \Lambda^{h(\alpha)+1}$$

it follows that for every $\iota \in \mathfrak{A}$ the function $\partial^\iota f$ belongs to $C^{h,L}(X)$, as well. ■

We now introduce the spaces $C^{M,L}(X)$. As already indicated above, we have to distinguish the cases where M is semi-simple and is not. In order to be able to handle both cases simultaneously we introduce

Notation 9.17. From now on we set $h(\alpha) := \mathrm{Re}\,\alpha_M$ and define

$$C^{M,L}(X) := \bigcap_{\eta\in\mathfrak{E}_M} C^{h,L^{1+\eta}}(X)$$

where $\mathfrak{E}_M$ is defined in (9.4).

We suppose from now on that

(9.35) $\qquad \lambda_{min} \ge 1.$

In view of Lemma 1.3.6 in Hörmander [11] this assumption is necessary for the following lemma to hold.

Lemma 9.18. *Let K be a compact subset of V, and let $\mathcal{U}$ be a finite open covering of K. Then there exist constants C and A_β, $\beta \in \mathfrak{A}$, and functions $\chi_{N,U} \in C_0^\infty(U)$, $U \in \mathcal{U}$ and $N \in \mathbb{N}$, with values in [0,1] such that*

$$\sum_{U\in\mathcal{U}} \chi_{N,U} \equiv 1 \text{ on } K$$

and such that for arbitrary $\beta \in \mathfrak{A}$ $\chi_{N,U}$ satisfies (9.5) for $(A,B) = (A_\beta, CN)$.

Proof. Since (9.35) implies that

$$(9.35)' \qquad |\alpha| \le \operatorname{Re}\alpha_M , \qquad \alpha \in \mathfrak{A},$$

it suffices to prove the assertion for the special case $M = \mathrm{Id}_V$.

Let $\mathfrak{V} = (V_U)_{U \in \mathfrak{U}}$ be an open covering of K such that $\overline{V_U} \subset U$ for every $U \in \mathfrak{U}$. Then we choose $\varepsilon > 0$ such that

$$K + \overline{K}(0,2\varepsilon) \subset \bigcup_{U \in \mathfrak{U}} V_U, \quad \text{and} \quad V_U + \overline{K}(0,2\varepsilon) \subset U \quad \text{for every } U \in \mathfrak{U}.$$

By Theorem 1.4.5 in Hörmander [11] we choose a C^∞ partition of unity on the compact subset $K + \overline{K}(0,2\varepsilon)$ subordinated to the covering $\mathfrak{V}$, consisting of functions ψ_U with values in $[0,1]$. Applying Theorem 1.4.2 in Hörmander [11] to the sequence defined by

$$d_j := \begin{cases} \varepsilon/2N & \text{if } j \in \mathbb{N}_N \\ \varepsilon 2^{-j} & \text{otherwise} \end{cases}$$

we obtain functions $\tilde{\varphi}_N \in C_0^\infty(K(0,2\varepsilon))$ with values in $[0,1]$ equal to 1 on $\overline{K}(0,\varepsilon)$ such that $|\partial^\alpha \tilde{\varphi}_N| \le (BN)^{|\alpha|}$ if $|\alpha| \le N$ where B is a constant not depending on N. Then $C := 1/|K(0,\varepsilon)| \ge 1/\int \tilde{\varphi}_N \, dx$ so that $\varphi_N := \tilde{\varphi}_N / \int \tilde{\varphi}_N \, dx$ satisfies

$$|\partial^\alpha \varphi_N| \le C\,(BN)^{|\alpha|} \quad \text{if } |\alpha| \le N.$$

We set $\chi_{N,U} := \varphi_N * \psi_U$. Then $\operatorname{supp}\chi_{N,U} \subset K(0,2\varepsilon) + V_U \subset U$ and

$$\sum_{U \in \mathfrak{U}} \chi_{N,U} = \varphi_N * \Big(\sum_{U \in \mathfrak{U}} \psi_U \Big) \equiv 1 \text{ on } K.$$

Since $|\partial^{\alpha+\beta}\chi_{N,U}| = |(\partial^\alpha \varphi_N) * (\partial^\beta \psi_U)| \le \|\partial^\alpha \varphi_N\|_{L_\infty} \|\partial^\beta \psi_U\|_{L_1}$ the proof is complete. ∎

After these preparations we can prove

Proposition 9.19. *Suppose that (9.32) is satisfied in case M is not semi-simple. Let $u \in \mathcal{E}'(X)$ and $x_0 \in X$. Then $u_{/U}$ belongs to $C^{M,L}(U)$ for some open neighbourhood U of x_0 if and only if there are an open neighbourhood U of x_0 and a bounded sequence of distributions u_N, $N \in \mathbb{N}$, in $\mathcal{E}'(X)$ coinciding on U with u such that for a suitable family of constants C_η, $\eta \in \mathfrak{E}_M$, we have*

$$(9.36) \qquad |\hat{u}_N(M_\tau \xi)| \le C_\eta \,(C_\eta (L_N)^{1+\eta}/\tau)^N, \qquad \tau \in [1,+\infty[, \ \eta \in \mathfrak{E}_M,$$

for every $\xi \in S_{V^*}$.

Proof. "$\Rightarrow$": We fix $\varepsilon > 0$ such that $u \in C^{M,L}$ on $K(x_0, 3\varepsilon)$. By Lemma 9.18 we choose a constant $C > 1$ and a sequence of functions $\chi_N \in C_0^\infty(K(x_0, 2\varepsilon))$, $N \in \mathbb{N}$, equal to 1 on $K(x_0, \varepsilon)$ such that $|\partial^\alpha \chi_N| \le C\,(CN)^{h(\alpha)}$ if $|\alpha| \le N+1$ and such that $\sup\{\|\partial^\beta \chi_N\|_{L_\infty};\ N \in \mathbb{N}\}$ is finite for every $\beta \in \mathfrak{A}$. It follows from the last condition that the sequence of distributions $u_N := \chi_N u$ is bounded in $\mathcal{E}'(X)$. Moreover, by the Leibniz rule we obtain

$$|\partial^\alpha u_N| \le \sum_{\beta \le \alpha} \binom{\alpha}{\beta} |\partial^\beta \chi_N\, \partial^{\alpha-\beta} u| \le C^2 \sum_{\beta \le \alpha} \binom{\alpha}{\beta} (CN)^{h(\beta)} (C\, L_{\alpha-\beta})^{h(\alpha-\beta)}.$$

Since $N \le L_N$ (by (9.30.a)) and $L_{\alpha+e_j-\beta} \le L_{\alpha+e_j} \le L(N+\lambda_{max}) \le \Lambda^{1+\lambda_{max}} L_N$ (by (9.30.b)) this implies that

$$|\partial^{\alpha+\iota} u_N| \le C^2 (D\,L_N)^{h(\iota)} (D\,L_N)^{h(\alpha)} \qquad \text{if } \operatorname{Re}\alpha_M \le N \text{ and } \iota \in \mathfrak{A}_1$$

where $D := 2C\Lambda^{1+\lambda_{max}}$. Fixing $\eta \in \mathfrak{E}_M$ and applying Lemma 9.2 (to $k = 0 = \ell$) we conclude that

$$|\hat{u}_N(M_\tau \xi)| \le a\, C^2\, (D\,L_N)^{\lambda_{max}} (b\,(D\,L_N)^{1+\eta}/\tau)^N, \qquad \xi \in S_{V^*},\ \tau \in [1, +\infty[.$$

Since by (9.31) we have $L_N \le \Lambda^{N+1}$ the assertion follows.

"$\Leftarrow$": First of all we fix $\eta \in \mathfrak{E}_M$ and then $\rho \in \mathfrak{E}_M$ such that $(1+\rho)^3 \le 1+\eta$. Let $\alpha \in \mathfrak{A}$. We fix $k \in \mathbb{N}$ such that $k \ge \mu+1$ and choose $N \in \mathbb{N}$ such that

$$(9.37) \qquad N-1 \le (1+\rho)\,h(\alpha) < N.$$

We deduce from (9.2)' that there is a constant $B_\rho > 0$ not depending on α such that

$$(9.38) \qquad |(M_\tau \xi)^\alpha| \le B_\rho\, \tau^{(1+\rho)h(\alpha)}, \qquad \tau \in [1, +\infty[,\ \xi \in S_{V^*}.$$

Combining this with (9.36) and (9.37) we obtain for arbitrary $\tau \in [1, +\infty[$ and $\xi \in S_{V^*}$

$$\begin{aligned}
&|(M_\tau \xi)^\alpha \mathcal{F}(u_{N+k})(M_\tau \xi)| \le \\
&\le B_\rho\, \tau^{(1+\rho)h(\alpha)}\, C_\rho\, (C_\rho\, (L_{N+k})^{1+\rho}/\tau)^{N+k} \le \\
&\le \tau^{-\mu-1} B_\rho\, C_\rho\, (C_\rho\, (L_{N+k})^{1+\rho})^{N+k}.
\end{aligned}$$

Employing quasihomogeneous polar coordinates (see Proposition 1.86) and the Fubini-Tonelli theorem we conclude that the function defined by $\zeta \mapsto \zeta^\alpha \mathcal{F}(u_{N+k})(\zeta)$ is integrable on the set $Z := V^* \setminus K(0,1)$ such that

$$\int_Z |\zeta^\alpha \mathcal{F}(u_{N+k})(\zeta)|\, d\zeta \le D_\rho\, (C_\rho\, (L_{N+k})^{1+\rho})^{N+k}$$

for some constant D_ρ not depending on N and α. Since $(u_N)_{N\in\mathbb{N}}$ is a bounded sequence in $\mathcal{E}'(V)$ the functions $|\mathcal{F}(u_{N+k})|$ are bounded on K(0,1) by a constant not depending on N. Consequently, the function defined by $\zeta \mapsto \zeta^\alpha \mathcal{F}(u_{N+k})(\zeta)$ belongs to $\mathcal{L}^1(V^*)$, its L^1-norm being bounded by $E_\rho(E_\rho(L_{N+k})^{1+\rho})^{N+k}$ where E_ρ is a constant not depending on N and α. Hence, $\partial^\alpha u_{N+k} = (2\pi)^{-n}\mathcal{F}(\zeta^\alpha\mathcal{F}(u_{N+k}))^\vee$ is a continuous function on V. Since this is so for every N and every α it follows that u is a C^∞ function on U satisfying

$$|\partial^\alpha u| \le \|\mathcal{F}(\partial^\alpha u_{N+k})\|_{L_1} \le E_\rho\,(E_\rho\,(L_{N+k})^{1+\rho})^{N+k}.$$

Now, by (9.31) we have

$$(L_{N+k})^{(1+\rho)(k+1)} \le \Lambda^{(N+k+1)(1+\rho)(k+1)} =$$
$$= \Lambda^{(k+2)(1+\rho)(k+1)}\,\Lambda^{(N-1)(1+\rho)(k+1)}$$

and by (9.30.b)

$$(L_{N+k})^{(1+\rho)(N-1)} \le (\Lambda^{k+2}L_{N-1})^{(1+\rho)(N-1)}.$$

Moreover, by (9.37) and (9.32) we have

$$L_{N-1} \le L((1+\rho)h(\alpha)) \le C_{1+\rho}\,(L_\alpha)^{1+\rho}.$$

Finally, combining these estimates and making use of (9.37), again, we conclude

$$(E_\rho(L_{N+k})^{(1+\rho)})^{N+k} \le$$
$$\le E_\rho^{k+1}\,\Lambda^{(k+2)(1+\rho)(k+1)}\,E_\rho^{N-1}\,(C_{1+\rho}\Lambda^{2k+3}(L_\alpha)^{1+\rho})^{(1+\rho)(N-1)} \le$$
$$\le E_\rho^{k+1}\,\Lambda^{(k+2)(1+\rho)(k+1)}\,(F\,(L_\alpha)^{(1+\rho)^3})^{h(\alpha)}$$

where $F := [E_\rho(C_{1+\rho}\Lambda^{(2k+3)})^{(1+\rho)}]^{1+\rho}$. In view of our choice of ρ this shows that the restiction of u to U is induced by a $C^{h,L^{1+\eta}}$ function. ■

Definition 9.20. Let $T\in\mathcal{D}'(X)$. By $WF_{M,L}(T)$ we denote the complement in $X\times S_{V^*}$ of the set of all $(x_0,\xi_0)\in X\times S_{V^*}$ such that there exist a bounded sequence $(u_N)_{N\in\mathbb{N}}$ in $\mathcal{E}'(X)$, open neighbourhoods U of x_0 in X and W of ξ_0 in $V^*\setminus\{0\}$ and a family of positive constants C_η, $\eta\in\mathcal{C}_M$, such that for every $N\in\mathbb{N}$ u_N is equal to T on U and such that (9.36) is satisfied for every $\xi\in W$.

Remark 9.21. *In the definition of $WF_{M,L}$ it suffices to postulate the estimate in (9.36) for $\tau \ge (L_N)^{1+\eta}$.*

Proof. Since (u_N) is a bounded sequence in $\mathcal{E}'(X)$ we can find constants W and s such that $|\mathcal{F}(u_N)| \le W|\cdot|^s$ on $V^*\setminus K(0,1)$. Hence, choosing $k \in \mathbb{N}$ such that $k > s\lambda_{max}$ we deduce from (9.2)' that for any compact neighbourhood K of S_{V^*} we find a constant C such that $|M_\tau \xi|^s \le C\tau^k$, $\tau \in [1,+\infty[$ and $\xi \in K$, so that by (9.31) we conclude that

$$|\mathcal{F}(u_N)(M_\tau \xi)| \le WC\tau^k \le WC(L_N)^{(1+\eta)k} \le WC\Lambda^{(N+1)(1+\eta)k}$$

for arbitrary $\tau \in [1,(L_N)^{1+\eta}]$ and $\xi \in K$. ■

The analogue of Lemma 9.4 is valid:

Lemma 9.22. *Let K be a compact subset of X, and let F be a closed subset of S_{V^*} such that $WF_{M,L}(T) \cap K \times F = \emptyset$. Then there exist neighbourhoods $\widetilde{K}$ of K in X and $\widetilde{F}$ of F in $V^*\setminus\{0\}$ such that for any $\eta \in \mathcal{G}_M$, for any family of positive constants A_β, $\beta \in \mathfrak{A}$, and for any $r \in \mathbb{N}_0$ there is a constant a such that the following implication holds:*

If $(\chi_N)_{N\in\mathbb{N}}$ is any sequence in $C_0^\infty(\widetilde{K})$ satisfying

$$|\partial^{\alpha+\beta}\chi_N| \le A_\beta\,(A_\beta\,(L_N)^{1+\eta})^{h(\alpha)}, \qquad \alpha,\beta \in \mathfrak{A},\ h(\alpha) \le N, \tag{9.39}$$

then

$$|\mathcal{F}(\chi_N T)(M_\tau \xi, \upsilon)| \le a\,\exp(H_{\chi_N}(\upsilon))\,\tau^{-r}\,(a\,(L_N)^{1+\eta}/\tau)^N$$

for arbitrary $\tau \in [1,+\infty[$, $\xi \in \widetilde{F}$, $\upsilon \in V^*$, and $N \in \mathbb{N}$.

Proof. Of course, we may assume that (A_β) is *increasing* (i.e. $A_\beta \le A_\gamma$ if $\beta \le \gamma$). Before coming to the actual proof of the lemma we observe that if $(A'_\beta)_{\beta\in\mathfrak{A}}$ is another increasing family of positive constants and if (φ_N) is any sequence in $C^\infty(X)$ such that (9.39) is satisfied for $(\varphi_N|_{\widetilde{K}}, A'_\beta)$ instead of (χ_N, A_β) then the functions $\varphi_N\chi_N$ satisfy (9.39) with A_β replaced by $2^{|\beta|+1}\max\{A_\beta, A'_\beta, A_\beta A'_\beta\}$. Indeed, given $\alpha,\beta \in \mathfrak{A}$ such that $h(\alpha) \le N$ then for every $\gamma \le \alpha+\beta$ we write $\gamma = \gamma_\alpha + \gamma_\beta$ and $\alpha+\beta-\gamma = \Gamma_\alpha + \Gamma_\beta$ where $\gamma_\alpha, \gamma_\beta, \Gamma_\alpha, \Gamma_\beta$ belong to $\mathfrak{A}$ satisfying the equations $\gamma_\alpha + \Gamma_\alpha = \alpha$ and $\gamma_\beta + \Gamma_\beta = \beta$ and obtain by the Leibniz rule that

$$|\partial^{\alpha+\beta}(\chi_N\varphi_N)| \le \sum_{\gamma\le\alpha+\beta} \binom{\alpha}{\beta} |\partial^{\gamma_\alpha+\gamma_\beta}\chi_N|\,|\partial^{\Gamma_\alpha+\Gamma_\beta}\varphi_N| \le$$

$$\le \sum_{\gamma\le\alpha+\beta} \binom{\alpha}{\beta} A_{\gamma_\beta}(A_{\gamma_\beta}(L_N)^{1+\eta})^{h(\gamma_\alpha)}\,A'_{\Gamma_\beta}(A'_{\Gamma_\beta}(L_N)^{1+\eta})^{h(\Gamma_\alpha)} \le$$

$$\leq 2^{|\alpha+\beta|} A_\beta A'_\beta (\max\{A_\beta, A'_\beta\} (L_N)^{1+\eta})^{h(\alpha)} .$$

Step 1: We fix $x_0 \in K$ and $\xi_0 \in F$. Then there exist an open neighbourhood U of x_0 in X, a constant $\varepsilon > 0$, and a bounded sequence $(u_N)_{N\in\mathbb{N}}$ in $\mathcal{E}'(X)$ such that for every $N \in \mathbb{N}$ u_N is equal to T on U and (9.36) is satisfied for $\xi \in \overline{K}(\xi_0, 2\varepsilon)$. We first prove the assertion for $(U, \overline{K}(\xi_0, \varepsilon))$ instead of $(\tilde{K}, \tilde{F})$. Since (u_N) is bounded in $\mathcal{E}'(X)$ we find constants W and $s \in [0, +\infty[$ such that u_N satisfies (9.8) for every $N \in \mathbb{N}$. We set $A := \sup\{A_\beta ; |\beta| \leq k+\ell+1\}$ where k and ℓ are chosen so as to fulfill the requirements of Lemma 9.3. In view of (9.39) an application of Lemma 9.3 yields constants c, b, d such that

$$|\mathcal{F}(\chi_N T)(M_\tau \xi, \upsilon)| = |\mathcal{F}(\chi_N u_{N+r})(M_\tau \xi, \upsilon)| \leq$$

$$\leq \exp(H_{\chi_N}(\upsilon)) \left(c A \tau^{-r} (b (A L_N)^{1+\eta}/\tau)^N + d A C_\eta (C_\eta (L_{N+r})^{1+\eta}/\tau)^{N+r} \right)$$

for arbitrary $\xi \in K(\xi_0, \varepsilon)$, $\upsilon \in V^*$, and $\tau \in [1, +\infty[$. Since by (9.30.b) and (9.31) we have $L_{N+r} \leq \Lambda^{r+1} \min\{\Lambda^N, L_N\}$ we deduce the desired estimates.

Step 2: By Step 1 and by the compactness of F we find a finite open covering $\mathfrak{W}$ of F and a family $(U_W)_{W\in\mathfrak{W}}$ of open neighbourhoods of x_0 such that for every $W \in \mathfrak{W}$ the conclusion holds for (U_W, W) instead of $(\tilde{K}, \tilde{F})$. It then follows that the conclusion remains valid for $(\tilde{K}, \tilde{F}) = (\bigcap_{W\in\mathfrak{W}} U_W, \bigcup \mathfrak{W})$.

Step 3: By Step 2 and by the compactness of K we find a finite open covering $\mathfrak{U}$ of K and a family $(F_U)_{U\in\mathfrak{U}}$ of open neighbourhoods of F in $V^* \setminus \{0\}$ such that for every $U \in \mathfrak{U}$ the conclusion holds for (U, F_U) instead of $(\tilde{K}, \tilde{F})$. Now we set $\tilde{F} := \bigcap_{U\in\mathfrak{U}} F_U$ and fix a compact neighbourhood $\tilde{K}$ of K in $\bigcup \mathfrak{U}$. Moreover, by Lemma 9.18 for every $N \in \mathbb{N}$ we choose a C^∞ partition of unity $(\chi_{N,U})_{U\in\mathfrak{U}}$ on $\tilde{K}$ subordinated to the covering $\mathfrak{U}$ and satisfying the conditions of Lemma 9.18. Since by the argument at the beginning of the proof for every $U \in \mathfrak{U}$ the functions $\chi_N \chi_{N,U}$ satisfy (9.39) with possibly new constants it follows that the desired estimates are valid for $\xi \in \tilde{F}$ with χ_N replaced by $\chi_N \chi_{N,U}$. Since $\sum_{U\in\mathfrak{U}} \chi_N \chi_{N,U} = \chi_N$ the assertion follows by the triangle inequality. ■

Combining Proposition 9.19 with Lemma 9.22 one obtains the following analogue of Proposition 9.5.

Proposition 9.23. *Under the assumption of Proposition 9.19 the projection of the* *set $WF_{M,L}(T)$ on the first factor is equal to* $\operatorname{sing\,supp}_{M,L}(T)$ *which is by definition the smallest closed subset of X outside which T is induced by a $C^{M,L}$ function.* ∎

From the formulas computed in the proof of Remark 9.6 we immediately deduce

Remark 9.24. *The assertions of Remark 9.6 remain valid if WF_M is replaced by $WF_{M,L}$ throughout.* ∎

As another consequence of Lemma 9.22 we obtain

Proposition 9.25. *$WF_{M,L}(fT) \subset WF_{M,L}(T)$ for every $f \in C^{M,L}(X)$.*

Proof. Let $(x,\xi) \in X \times S_{V^*} \setminus WF_{M,L}(T)$, and let $\varepsilon > 0$ be such that $\overline{K}(x,2\varepsilon) \subset X$ and $\overline{K}(x,2\varepsilon) \times \overline{K}(\xi,\varepsilon) \cap WF_{M,L}(T) = \emptyset$. By Lemma 9.18 we find constants $C, A_\beta \geq 1$ and functions $\chi_N \in C_0^\infty(K(x,2\varepsilon))$ equal to 1 on $\overline{K}(x,\varepsilon)$ with values in $[0,1]$ such that for arbitrary $N \in \mathbb{N}$ and $\beta \in \mathfrak{A}$ the estimates (9.5) are satisfied for (χ_N, A_β, CN) instead of (χ, A, B).

Now, by Remark 9.16 for every $\beta \in \mathfrak{A}$ $\partial^\beta f$ belongs to $C^{M,L}(X)$. Hence, for any $\eta \in \mathfrak{G}_M$ we find a constant A'_β such that

$$|\partial^{\alpha+\beta} f(x)| \leq A'_\beta \, (A'_\beta \, (L_N)^{1+\eta})^{h(\alpha)}, \qquad x \in K(x,2\varepsilon),\ \alpha \in \mathfrak{A},\ h(\alpha) \leq N.$$

Applying the argument at the beginning of the proof of Lemma 9.22 to $\varphi_N = f$ we see that the functions $f\chi_N$ satisfy the estimates (9.39) with possibly new constants A_β. Consequently, Lemma 9.22 implies that (x,ξ) does not belong to $WF_{M,L}(T)$. ∎

In view of (9.11) another application of Lemma 9.22 yields

$$WF_{M,L}(D^\alpha T) \subset WF_{M,L}(T).$$

Combining this with Proposition 9.25 we obtain

Proposition 9.26. *$WF_{M,L}(P(x,\partial)T) \subset WF_{M,L}(T)$ for every linear differential operator $P(x,\partial)$ on X with coefficients in $C^{M,L}(X)$.* ∎

As for the converse inclusion, the expected assertion holds:

Theorem 9.27. *Let $m \in Re\mathfrak{A}(M)$, and let P be a differential operator as in Theorem 9.8. Suppose that its coefficients are real analytic functions $a_\alpha : X \longrightarrow \mathbb{C}$. Moreover, suppose that its quasihomogeneous principal part $P_m : X \times V^* \longrightarrow \mathbb{C}$ defined in the text preceding Theorem 9.8 is quasihomogeneous of degree $\tilde{m}$. Then*

$$(9.40) \qquad WF_{M,L}(T) \subset WF_{M,L}(P(x,\partial)T) \cup P_m^{-1}(0), \qquad T \in \mathcal{D}'(X).$$

Proof. The proof of Theorem 8.6.1 in Hörmander [11] is suitably modified. Let $(x_0, \xi_0) \in X \times S_{V^*}$ be such that it does not belong to the right-hand side of (9.40). Then we can choose compact neighbourhoods K of x_0 in X and W of ξ_0 in $V^* \backslash \{0\}$, a family of constants C_η, $\eta \in \mathfrak{E}_M$, and a bounded sequence $(v_N)_{N \in \mathbb{N}}$ in $\mathcal{E}'(X)$ such that for every $N \in \mathbb{N}$ (9.14) and hence (9.14)' are valid with v replaced by v_N, such that

$$(9.41) \qquad |\hat{v}_N(M_\tau \xi)| \le C_\eta \,(C_\eta L_N^{1+\eta}/\tau)^N, \qquad \xi \in W, \ \tau \in [1,+\infty[, \ \eta \in \mathfrak{E}_M,$$

and such that (9.16) holds.

Applying Lemma 9.18 to $M = Id_V$ we find a sequence of functions $\chi_N \in C_0^\infty(K)$ equal to 1 on a fixed neighbourhood U of x_0 and constants C_β such that

$$(9.42) \qquad |\partial^{\alpha+\beta}\chi_N| \le C_\beta \,(C_\beta N)^{|\alpha|}, \qquad \alpha, \beta \in \mathfrak{A}, \ |\alpha| \le 3N.$$

Then the distributions $u_N := \chi_N T$ form a bounded sequence in $\mathcal{E}'(X)$, and in order to prove that (x_0, ξ_0) does not belong to the left-hand side of (9.40) we have to verify (9.36) for ξ near ξ_0. To this end we fix $\xi \in W$, $N \ge m$, $\eta \in \mathfrak{E}_M$, and $\tau \ge N^{1+\eta}$. From the arguments in the proof of Theorem 9.8 we conclude that (9.19) is valid with χ replaced by χ_N and for $\nu := (1+\eta)N$, i.e.

$$(9.43) \qquad \mathcal{F}(u_N)(M_\tau \xi) = \tau^{-\tilde{m}} \mathcal{F}(\psi_N v_N)(M_\tau \xi) + \mathcal{F}(e_N T)(M_\tau \xi)$$

where this time we set $\psi_N := S_\nu(\chi_N)/P_m(\cdot, \xi)$ and $e_N := E_\nu(\chi_N)$. The proof will be complete once we have verified (9.36) for both summands on the right-hand side separately.

Estimating $\partial^\alpha R_{\rho_1} \dots R_{\rho_k}\chi_N$. In view of the meaning of $\tilde{m}$ we observe that $R_{\rho_1} \dots R_{\rho_k}\chi_N$ is a sum of terms of the form

$$(9.44) \qquad \tau^{-k\tilde{m}} (M_\tau \xi)^{\varepsilon_1 + \dots + \varepsilon_k} \, b_1 D^{\beta_1}(b_2 \dots D^{\beta_{k-1}}(b_k D^{\beta_k}\chi_N) \dots)$$

where the ε_j are such that $\mathrm{Re}(\varepsilon_j)_M = m-\rho_j$, where the b_j belong to a set $\mathfrak{H}$ of real analytic functions which can be extended to uniformly bounded holomorphic functions on a common complex neighbourhood of K, and where the $\beta_j \in \mathfrak{A}$ satisfy the estimate $\mathrm{Re}(\beta_j)_M \leq \rho_j$, i.e.

$$(9.45) \qquad |\beta_1+\ldots+\beta_k| \leq \mathrm{Re}(\beta_1+\ldots+\beta_k)_M \leq \rho_1+\ldots+\rho_k .$$

In order to estimate (9.44) we fix $\eta' \in \mathfrak{E}_M$ such that $(m-\rho)(1+\eta)\eta' \leq \rho\eta$ for every ρ belonging to the set $\mathfrak{B}_m$ defined in the text following (9.17), i.e.

$$-m+(m-\rho)(1+\eta') = -\rho-\rho\eta'+m\eta' \leq -\rho/(1+\eta), \qquad \rho \in \mathfrak{B}_m ,$$

so that we deduce from (9.21) - applied to η' instead of η - that

$$|\tau^{-k\tilde{m}}(M_\tau\xi)^{\varepsilon_1+\ldots+\varepsilon_k}| \leq$$

$$\leq c^{|\varepsilon_1+\ldots+\varepsilon_k|}(C_{\eta'})^{km-\rho_1-\ldots-\rho_k}\tau^{-km+(km-\rho_1-\ldots-\rho_k)(1+\eta')} \leq$$

$$\leq (c^m C_{\eta'}^m)^k \tau^{-(\rho_1+\ldots+\rho_k)/(1+\eta)}, \qquad \tau \in [1,+\infty[,$$

where $c := \max\{|\xi|;\ \xi \in W\}$. Next we observe that the conclusion of Lemma 8.6.3 in Hörmander [11] remains valid if the derivatives $(-i\partial)_j$, $j \in \mathbb{N}_n$, are replaced by $(-i\boldsymbol{\partial})^\alpha$, $\alpha \in \mathfrak{A}^1$. Hence, writing the derivatives of any function g appearing in (9.44) in the form

$$D^\beta g = D^{\iota_1}(1\, D^{\iota_2}(1\ldots D^{\iota_\ell} g)\ldots)) ,$$

where $\ell := |\beta|$ and $\iota_1,\ldots,\iota_\ell$ belong to $\mathfrak{A}^1$, we can apply Hörmander's Lemma and obtain a constant E not depending on N, β_ℓ, ρ_j and $b_j \in \mathfrak{H}$ such that

$$|D^{\beta_0}(b_1 D^{\beta_1}(b_2\ldots D^{\beta_{k-1}}(b_k D^{\beta_k}\chi_N)\ldots))| \leq E^{k+3N+1}(3N)^{|\beta_0+\ldots+\beta_k|}$$

provided that $|\beta_0+\ldots+\beta_k| \leq 3N$. Let ϑ be the maximum number of summands in R_ρ, $\rho \in \mathfrak{B}_m$. Then combining this with the estimates above and using (9.45) we obtain that

$$|D^\alpha R_{\rho_1}\ldots R_{\rho_k}\chi_N| \leq$$

$$\leq (\vartheta c^m C_{\eta'}^m)^k (3E)^{k+3N+1} N^{|\alpha|+\rho_1+\ldots+\rho_k}\tau^{-(\rho_1+\ldots+\rho_k)/(1+\eta)}$$

if $\tau \in [1,+\infty[$ and if

$$(9.46) \qquad |\alpha|+\rho_1+\ldots+\rho_k \leq 3N .$$

Since $\rho_j \geq \lambda_{min} \geq 1$ the estimate (9.46) implies $k \leq 3N$ so that we find another constant C not depending on N,τ,ξ and ρ_j such that

$$(9.47)\qquad |D^\alpha R_{\rho_1}\ldots R_{\rho_k}\chi_N| \le C^{N+1}\, N^{|\alpha|+\rho_1+\ldots+\rho_k}\, \tau^{-(\rho_1+\ldots+\rho_k)/(1+\eta)}$$

provided that (9.46) holds.

Estimating $\mathcal{F}(\psi_N v_N)(M_\tau\xi)$. Let Θ be the number of elements of $\mathfrak{B}_m$. Since $\rho_j \ge \lambda_{min} \ge 1$ the inequality $\rho_2+\ldots+\rho_k \le \nu$ implies that $k \le \nu+1$. Hence the number of summands in ψ_N is not larger than

$$(9.48)\qquad \sum_{k=0}^{[\nu]+1} \Theta^k \le (\nu+2)\,\Theta^{\nu+1} \le (e\,\Theta)^{\nu+1} = e\Theta\,((e\Theta)^{1+\eta})^N .$$

Employing the Leibniz rule, taking (9.16) into account, and combining (9.48) with (9.47) we obtain that

$$|D^\alpha \psi_N| \le C^{N+1} N^{|\alpha|} \quad \text{if } |\alpha| \le 2N$$

where C is another constant not depending on N,τ and ξ. Moreover, since the sequence (v_N) is bounded we find constants $W, s \in]0,+\infty[$ not depending on N such that every v_N satisfies (9.8). We set $k := s+n+1$ and choose numbers $\ell > s\lambda_{max}/\lambda_{min}$ and $\varepsilon > 0$ such that $\overline{K}(\xi_0, 2\varepsilon) \subset W$. Then by (9.41) and by Lemma 9.3 we conclude for $N \ge k+\ell+1$ that

$$|\mathcal{F}(\psi_N v_N)(M_\tau\xi)| \le$$

$$\le c\, C^{N+1} N^{k+\ell+1}\Big(W\,(bN^{1+\eta}/\tau)^N + C_\eta\,(C_\eta\,(L_N)^{1+\eta}/\tau)^N \Big) \le$$

$$\le C'_\eta\,(C'_\eta\,(L_N)^{1+\eta}/\tau)^N$$

for every $\xi \in K(\xi_0,\varepsilon)$ where C'_η is a constant not depending on N,τ and ξ. Here for the proof of the last inequality we made use of (9.30.a).

Estimating $\mathcal{F}(e_N T)(M_\tau\xi)$. Let q be the (quasihomogeneous) order of T on K. Since the number of summands in e_N is majorized by the left-hand side of (9.48) it follows by (9.48) and (9.47) that

$$|D^\gamma e_N| \le e\Theta\,((e\Theta)^{1+\eta})^N\, B^{N+1}\, N^{|\gamma|+\nu+m}\, \tau^{-N} \qquad \text{if } \mathrm{Re}\,\gamma_M \le q$$

where B is a constant not depending on ξ,γ,τ and N. Note that

$$N^{|\gamma|+\nu+m} \le (e^{q+m+(1+\eta)(q+1)})^N\,(N^{1+\eta})^{N-q-1}.$$

Consequently, taking (9.22) and (9.23) into account we conclude that

$$|\mathcal{F}(e_N T)(M_\tau\xi)| \le A\,(AN^{1+\eta}/\tau)^{N-q-1}$$

where A is still another constant not depending on ξ, τ and N.

Combining this with $L_N \geq N$ and (9.30.b) and with the estimate for $|\mathcal{F}(\psi_N v_N)(M_\tau \xi)|$ we deduce from (9.43) in view of Remark 9.21 that (x_0, ξ_0) does not belong to $WF_{M,L}(T)$, indeed. ■

Theorem 9.28. *The assertion of Theorem 9.11 remains valid if WF_M is replaced by $WF_{M,L}$.*

Proof. If $WF_M(u)$ is replaced by $WF_{M,L}(u)$ then the proof of Theorem 9.11 remains valid up to the estimates (9.27). Moreover, the conditions (9.28) and (9.29) hold. By Lemma 9.18 we find functions $\chi_N \in C_0^\infty(K(x_0, \varepsilon))$ equal to 1 on $K(x_0, \varepsilon/2)$ and constants A_β such that (9.39) is satisfied. By (9.29) we can change the constants A_β in such a way that $\chi_N \circ N_t$ satisfies (9.39) for every $t \in J$. Then in view of (9.28) an application of Lemma 9.22 to the sets (K_s, F_s), $s \in S$, shows that the right-hand side of (9.27) with φ replaced by χ_N is bounded by $C_\eta (C_\eta (L_N)^{1+\eta})^N$ where C_η is a constant not depending on N. This means that $(x_0, \xi_0) \notin WF_{M,L}(u_{m,w})$. ■

(c) Wave Front Sets of Almost Quasihomogeneous Distributions

In order to formulate a simple, but important inclusion for the quasihomogeneous wave front sets we require

Notation 9.29. If Y is a subset of V^* we define its *quasihomogeneous hull (of type M) at infinity* as the set $Y_{M,\infty}$ consisting of all $\xi \in S_{V^*}$ such that there exists a sequence $(\upsilon_j)_{j \in \mathbb{N}}$ in Y and a sequence (τ_j) in $[1, +\infty[$ converging to $+\infty$ such that

$$(9.49) \qquad \xi = \lim_{j \to \infty} M_{1/\tau_j} \upsilon_j .$$

Obviously, $Y_{M,\infty}$ is a closed subset of S_{V^*}. Moreover, if Y is closed and quasi-homogeneous of type M then $Y_{M,\infty} = Y \cap S_{V^*}$. For special Y a useful inclusion is given in

Remark 9.30. *Let $m \in \mathfrak{A}(M)$, let $P_m : V^* \longrightarrow \mathbb{C}$ be a polynomial function which is quasihomogeneous of degree m and of type M, and let $P : V^* \longrightarrow \mathbb{C}$ be a polynomial function such that $Q_\ell(P-P_m) = 0$ for every $\ell \in \mathfrak{A}(M)$ satisfying $\mathrm{Re}\,\ell \geq m$. Then $(P^{-1}(0))_{M,\infty} \subset (P_m)^{-1}(0)$.*

Proof. Let $\xi \in (P^{-1}(0))_{M,\infty}$, and choose a sequence $(\upsilon_j)_{j\in\mathbb{N}}$ in $P^{-1}(0)$ and a sequence (τ_j) in $[1,+\infty[$ converging to $+\infty$ such that (9.49) is satisfied. Then setting $R := P_m - P$ and $N := \max\{\mathrm{ord}_M(Q_\ell R)\,;\, \ell \in \Lambda(R)\}$ and making use of Proposition 1.34.(ii.a) and of (1.41) we deduce

$$P_m(M_{1/\tau_j}\upsilon_j) = \tau_j^{-m} P_m(\upsilon_j) = \tau_j^{-m} R(M_{\tau_j} M_{1/\tau_j}\upsilon_j) =$$

$$= \sum_{\ell\in\Lambda(R)} \sum_{k=0}^{N} \tau_j^{\ell-m}\, \omega_k(\tau_j)\, R_{\ell,k}(M_{1/\tau_j}\upsilon_j)\,.$$

In view of (9.49) and since by the assumption we have $\mathrm{Re}(\ell-m) < 0$ for every $\ell \in \Lambda(R)$ the right-hand side converges to 0 as $j\to\infty$. As the left-hand side tends to $P_m(\xi)$ as $j\to\infty$ this shows that $P_m(\xi) = 0$. ∎

Proposition 9.31. *$WF_{M,A}(T) \subset V \times (\mathrm{supp}\,\hat{T})_{M,\infty}$ for every $T \in \mathcal{S}'(V)$.*

Proof. Let U and W be bounded open subsets of V satisfying $\overline{U} \subset W$. By Lemma 9.18 we choose constants D, A_β and a sequence of functions $\chi_N \in C_0^\infty(W)$ equal to 1 on U such that for arbitrary $N \in \mathbb{N}$ and $\beta \in \mathfrak{A}$ the condition (9.5) is satisfied with (χ, A, B) replaced by (χ_N, A_β, DN). Let $\xi \in S_{V^*} \setminus (\mathrm{supp}\,\hat{T})_{M,\infty}$. Then there are $R, \varepsilon > 0$ such that

$$(9.50) \qquad |\xi - M_{1/\tau}\upsilon| \geq 3\varepsilon \qquad \text{if } \tau \in [R,+\infty[\text{ and } \upsilon \in Y := K(0,1) + \mathrm{supp}\,\hat{T}\,.$$

Finally, we choose a function $\varphi \in C^\infty(V^*)$ equal to 1 on $V^* \setminus K(0,2\varepsilon)$ and vanishing on $K(0,\varepsilon)$ and for every $\tau \in [R,+\infty[$ set $\varphi_\tau := \varphi \circ M_{1/\tau}$. Then for every $\chi \in C_0^\infty(V)$ one deduces from (9.50): if $\upsilon \in Y$ and $\zeta \in K(\xi,\varepsilon)$ then

$$\hat{\chi}(M_\tau\zeta - \upsilon) = (\varphi_\tau\hat{\chi})(M_\tau\zeta - \upsilon)$$

so that by the Fourier inversion formula we conclude that

$$\mathcal{F}(\chi T)(M_\tau\zeta) = \langle T, \chi \exp(-i\langle M_\tau\zeta, \cdot\rangle)\rangle =$$

$$= (2\pi)^{-n} \langle \check{\hat{T}}, \mathcal{F}(\chi \exp(-i\langle M_\tau\zeta, \cdot\rangle))\rangle =$$

$$= (2\pi)^{-n} \langle \hat{T}, (\varphi_\tau\hat{\chi})(M_\tau\zeta - \cdot\,)\rangle, \qquad \zeta \in K(\xi,\varepsilon)\,.$$

Consequently, since $\hat{T}$ is temperate one finds constants E and $m \in \mathbb{N}$ such that

$$|\mathcal{F}(\chi T)(M_\tau \zeta)| \le E\, S(\chi;\tau,\zeta), \qquad \chi \in C_0^\infty(V),\ \tau \in [R,+\infty[,\ \zeta \in K(\xi,\varepsilon),$$

where $S(\chi;\tau,\zeta) := \sup\{(1+|M_\tau \upsilon|)^m\, |\partial^\alpha(\varphi_\tau \hat{\chi})(M_\tau(\zeta-\upsilon))|;\ |\alpha| \le m,\ \upsilon \in V^*\}$. Note that $\varphi_\tau(M_\tau(\zeta-\upsilon)) = \varphi(\zeta-\upsilon) = 0$ if $\upsilon \in K(\zeta,\varepsilon)$ so that it suffices to take the supremum over all $\upsilon \in V^* \setminus K(\zeta,\varepsilon)$, only. By the Leibniz rule and by (1.49)' (applied to $P(\partial) = \partial^\beta$) we have

$$\partial^\alpha(\varphi_\tau \hat{\chi})(M_\tau(\zeta-\upsilon)) =$$

$$= \sum_{\beta \le \alpha} \binom{\alpha}{\beta} \tau^{-\beta M} \sum_{j=0}^{N_\beta} (-1)^j \omega_j(\tau)\, (P_{\beta,j}(\partial)\varphi)(\zeta-\upsilon)\, (\partial^{\alpha-\beta}\hat{\chi})(M_\tau(\zeta-\upsilon))$$

(for the definition of the polynomial functions $P_{\beta,j}$ see Remark 1.30 with M replaced by M^*). Now we fix $\eta \in \mathfrak{C}_M$ so small that $m\eta < 1$. By Lemma 9.2 we find constants a and b such that

$$|\partial^\gamma \hat{\chi}_N(M_\tau(\zeta-\upsilon))| \le a\,(1+|M_\tau(\zeta-\upsilon)|)^{-m}\, \tau^{-m\lambda_{max}-1}\, (b\,N^{1+\eta}/\tau)^N$$

for arbitrary $N \in \mathbb{N}$, $\tau \in [1,+\infty[$, $\zeta \in K(\xi,\varepsilon)$, $\upsilon \in V^* \setminus K(\zeta,\varepsilon)$, and $\gamma \in \mathfrak{A}$ satisfying $|\gamma| \le m$. Since the derivatives of φ are bounded and since

$$(1+|M_\tau \upsilon|) \le (1+|M_\tau(\zeta-\upsilon)|)(1+|M_\tau \zeta|)$$

and - by (9.2)' - $(1+|M_\tau \zeta|) \le (1+(1+\varepsilon)d_\eta)\, \tau^{\lambda_{max}+\eta}$ it follows that

$$S(\chi_N;\tau,\zeta) \le C_\eta\, (b\,N^{1+\eta}/\tau)^N, \qquad N \in \mathbb{N},\ \tau \in [R,+\infty[,\ \zeta \in K(\xi,\varepsilon),$$

where C_η is a constant only depending on η. In view of Remark 9.21 we conclude that the sequence of distributions $u_N := \chi_N T$ has the properties which imply that $U \times \{\xi\}$ does not intersect $WF_{M,A}(T)$. ■

Next we prove a version of Proposition 9.31 for distributions (not necessarily temperate) given by a Fourier representation formula. Let ν be a complex Radon measure on $V^* \times V^*$ such that for every compact subset K of X there is a constant m_K such that

(9.51) $$\int_{V^* \times V^*} (1+|\mathrm{Re}\,\zeta|)^{-m_K} \exp(H_K(-\mathrm{Im}\,\zeta))\, d|\nu|(\zeta) \quad \text{is finite.}$$

Then by

(9.52) $$\langle T_\nu, \varphi \rangle := \int_{V^* \times V^*} \hat{\varphi}(-\zeta)\, d\nu(\zeta), \qquad \varphi \in C_0^\infty(X),$$

a distribution $T_\nu \in \mathcal{D}'(X)$ is defined as is immediately derived from the Paley-Wiener estimates for $\hat{\varphi}$ when $\varphi \in C_o^\infty(K)$.

Proposition 9.32. *Under the preceding hypotheses we have*

$$WF_{M,A}(T_\nu) \subset X \times (\pi_{\mathbb{R}}(\operatorname{supp} \nu))_{M,\infty}$$

where $\pi_{\mathbb{R}} : V^* \times V^* \longrightarrow V^*$ *denotes the projection defined by* $\zeta \mapsto \operatorname{Re} \zeta$.

Proof. Let U and W be relatively compact open subsets of X satisfying $\overline{U} \subset W$. We set $K := \overline{W}$. We choose A_β, D and χ_N as in the proof of Proposition 9.31. From

$$\mathcal{F}(\chi_N T_\nu)(M_\tau \xi) = \langle T_\nu, \chi_N \exp(-i\langle M_\tau \xi, \cdot \rangle)\rangle = \int_{V^* \times V^*} \hat{\chi}_N(M_\tau \xi - \zeta)\, d\nu(\zeta)$$

we deduce that

$$(9.53) \qquad |\mathcal{F}(\chi_N T_\nu)(M_\tau \xi)| \le \int_{V^* \times V} (1 + |\operatorname{Re} \zeta|)^{-m_K} \exp(H_K(-\operatorname{Im} \zeta))\, d|\nu|(\zeta)\; S_N(M_\tau \xi)$$

where $S_N(M_\tau \xi) := \sup\{ |\hat{\chi}_N(M_\tau \xi - \zeta)| (1 + |\operatorname{Re} \zeta|)^{m_K} \exp(-H_K(-\operatorname{Im} \zeta))\, ;\, \zeta \in \operatorname{supp} \nu\}$. Now, we fix $\xi_o \in S_{V^*} \setminus (\pi_{\mathbb{R}}(\operatorname{supp} \nu))_{M,\infty}$. Then we find $\varepsilon > 0$ and $R \ge 1$ such that

$$|\xi_o - M_{1/\tau}\rho| \ge 2\varepsilon \qquad \text{if } \tau \in [R, +\infty[\text{ and } \rho \in \pi_{\mathbb{R}}(\operatorname{supp} \nu).$$

Lemma 9.2 shows that for any $\eta \in \mathcal{C}_M$ there are constants $a_\eta, b_\eta > 0$ such that

$$|\hat{\chi}_N(M_\tau \xi - \rho, -\upsilon)| \le$$
$$\le a_\eta (1 + |M_\tau \xi - \rho|)^{-m_K} \exp(H_K(-\upsilon))\, (b_\eta N^{1+\eta}/\tau)^N\, \tau^{-m_K(\lambda_{max}+\eta)}$$

for arbitrary $\xi \in K(\xi_o, \varepsilon)$, $\rho \in \pi_{\mathbb{R}}(\operatorname{supp} \nu)$, $\upsilon \in V^*$, $N \in \mathbb{N}$, and $\tau \in [R, +\infty[$. Since $(1 + |\rho|) \le (1 + |M_\tau \xi - \rho|)(1 + |M_\tau \xi|)$ and - by (9.2)' - $(1 + |M_\tau \xi|) \le C\tau^{\lambda_{max}+\eta}$ where $C := (1 + (1+\varepsilon) d_\eta)$ we have $(1 + |\rho|) \le (1 + |M_\tau \xi - \rho|)\, C\tau^{\lambda_{max}+\eta}$. Combining this with the estimate above we obtain $S_N(M_\tau \xi) \le a_\eta C^{m_K} (b_\eta N^{1+\eta}/\tau)^N$ so that from (9.53) we deduce that

$$|\mathcal{F}(\chi_N T_\nu)(M_\tau \xi)| \le C_{K,\eta} (b_\eta N^{1+\eta}/\tau)^N, \qquad \tau \in [R, +\infty[,\ \xi \in K(\xi_o, \varepsilon),\ N \in \mathbb{N},$$

where $C_{K,\eta}$ is a constant only depending on K and η. This shows that $U \times \{\xi_o\}$ does not intersect $WF_{M,A}(T_\nu)$. ■

We now come to the behaviour of the quasihomogeneous wave front sets of almost quasihomogeneous distributions. A preparatory remark is in order:

Remark 9.33. *Suppose that X is quasihomogeneous of type M^*. Let $T\in\mathcal{D}'(X)$ be almost quasihomogeneous of type M^*. Then for arbitrary $x\in X$ and $\xi\in S_{V^*}$ we have:*

$$(x,\xi)\in WF_M(T) \Longleftrightarrow \{(M_t^*x,\xi);\ t\in]0,+\infty[\}\subset WF_M(T) \Longleftrightarrow (\rho_{M^*}(x),\xi)\in WF_M(T)$$

(for the definition of ρ_{M^*} see Notation 9.9).

The assertion remains valid if WF_M is replaced by $WF_{M,L}$.

Proof. Let $m\in\mathbb{C}$ be such that T is almost quasihomogeneous of degree m. Then for arbitrary $\tau, t\in]0,+\infty[$, $\varphi\in C_0^\infty(X)$, and $\eta\in S_{V^*}$ we have

$$\mathcal{F}(\varphi T)(M_\tau\eta) = t^{-\mu}\langle T\circ M_{1/t}^*, [\varphi\exp(-i\langle M_\tau\eta,\cdot\rangle)]\circ M_{1/t}^*\rangle =$$

$$= t^{-m-\mu}\sum_{i=0}^{\infty}\omega_i(1/t)\langle(\partial_{M^*}-m)^iT, \varphi\circ M_{1/t}^*\exp(-i\langle M_{\tau/t}\eta,\cdot\rangle)\rangle =$$

$$= t^{-m-\mu}\sum_{i=0}^{\infty}(-1)^i\omega_i(t)\mathcal{F}(\varphi\circ M_{1/t}^*(\partial_{M^*}-m)^iT)(M_{\tau/t}\eta).$$

Since by Proposition 9.7 we have $WF_M((\partial_{M^*}-m)^iT)\subset WF_M(T)$ and since by Proposition 9.26 the same is valid with WF_M replaced by $WF_{M,L}$ the assertion follows. ∎

The following theorem is a straight-forward generalization of the corresponding results for homogeneous distributions in Hörmander [11] (see Theorems 8.1.8 and 8.4.18).

Theorem 9.34. *Let $T\in\mathcal{S}'(V)$ be such that $T/_{V\setminus\{0\}}$ is almost quasihomogeneous of type M^* or $\hat{T}/_{V^*\setminus\{0\}}$ is almost quasihomogeneous of type M. And let $x\in V\setminus\{0\}$ and $\xi\in V^*\setminus\{0\}$. Then*

(i) $(x,\rho_M(\xi))\in WF_M(T) \Longleftrightarrow (\xi,-\rho_{M^*}(x))\in WF_{M^*}(\hat{T})$;

(ii) the assertion (i) remains valid if WF_M is replaced by $WF_{M,L}$;

(iii.a) $(0,\rho_M(\xi))\in WF_M(T) \Longleftrightarrow (0,\rho_M(\xi))\in WF_{M,L}(T) \Longleftrightarrow \xi\in \operatorname{supp}\hat{T}$;

(iii.b) $x\in\operatorname{supp}T \Longleftrightarrow (0,-\rho_{M^*}(x))\in WF_{M^*}(\hat{T}) \Longleftrightarrow (0,-\rho_{M^*}(x))\in WF_{M^*,L}(\hat{T})$.

Proof of (i). Step 1: We first suppose that T is almost quasihomogeneous of degree m. By Remark 9.33 and Proposition 2.40.(i) we may assume that x lies in S_V

and ξ in S_{V^*}. In view of Proposition 2.40.(i) and the Fourier inversion formula it suffices to prove one of the implications, only.

Suppose that $(x,\xi) \notin WF_M(T)$. Then we find compact neighbourhoods K of x in V and F of ξ in $V^*\setminus\{0\}$ such that $K\times F \cap WF_M(T) = \emptyset$. We fix $\varepsilon > 0$ such that $\overline{K}(x,2\varepsilon) \subset \overset{\circ}{K}$ and choose $\varphi \in C_0^\infty(K)$ equal to 1 on $\overline{K}(x,2\varepsilon)$. Moreover, let U be an open neighbourhood of ξ in V^* such that $\overline{U} \subset \overset{\circ}{F}$ and choose $\chi \in C_0^\infty(F)$ equal to 1 on U. Note that by the Fourier inversion formula for arbitrary $y \in V$ and $t \in]0,+\infty[$ we have

$$\mathcal{F}(\chi\hat{T})(-M_t^*y) = (2\pi)^{-n}(\hat{\chi} * \hat{\hat{T}})(-M_t^*y) =$$

$$= \langle \check{T}, \hat{\chi}(-M_t^*y - (\cdot))\rangle = t^\mu \langle T\circ M_t^*, \hat{\chi}(M_t^*(\cdot - y))\rangle .$$

Since T is almost quasihomogeneous of degree m we deduce

$$\mathcal{F}(\chi\hat{T})(-M_t^*y) = t^{m+\mu}\sum_{i=0}^{\infty}\omega_i(t)\langle T_i, \tau_y(\hat{\chi}\circ M_t^*)\rangle, \qquad y\in V,\ t\in]0,+\infty[, \tag{9.54}$$

where $T_i := (\partial_{M^*} - m)^i T$. Now we decompose each T_i according to $T_i = \varphi T_i + (1-\varphi)T_i$ and first compute - making use of (2.8)' -

$$\langle \varphi T_i, \tau_y(\hat{\chi}\circ M_t^*)\rangle = \langle \varphi T_i, t^{-\mu}\mathcal{F}(e^{i\langle\cdot,y\rangle}\chi\circ M_{1/t})\rangle =$$

$$= \int_{V^*}\mathcal{F}(\varphi T_i)(\upsilon)\, e^{i\langle\upsilon,y\rangle}\chi(M_{1/t}\upsilon)\, d(M_{1/t}\upsilon) =$$

$$= \int_{V^*}\mathcal{F}(\varphi T_i)(M_t\upsilon)\exp(i\langle M_t\upsilon,y\rangle)\,\chi(\upsilon)\,d\upsilon .$$

Since $\operatorname{supp}\chi \subset F$ it follows that

$$|\langle\varphi T_i, \tau_y(\hat{\chi}\circ M_t^*)\rangle| \le \int_F |\chi(\upsilon)|\,d\upsilon\ \sup\{|\mathcal{F}(\varphi T_i)(M_t\upsilon)|;\ \upsilon\in F\}, \qquad t>0. \tag{9.55}$$

Since by Proposition 9.7 we have $WF_M(T_i) \subset WF_M(T)$ we deduce by means of Lemma 9.4 that this is equal to $O(t^{-N})$ as $t\to\infty$ for every N. Consequently, in view of (9.54) we can conclude that $(\xi,-x) \notin WF_{M^*}(\hat{T})$ once we have proved that

$$\sup\{|\langle(1-\varphi)T_i, \tau_y(\hat{\chi}\circ M_t^*)\rangle|;\ y\in\overline{K}(x,\varepsilon)\} = O(t^{-N}) \text{ as } t\to\infty$$

for every $N\in\mathbb{N}$. Since T and hence the T_i are temperate it suffices to show that for every N the set

$$\mathfrak{B}_N := \{t^N(1-\varphi)\,\tau_y(\hat{\chi}\circ M_t^*);\ y\in\overline{K}(x,\varepsilon),\ t\in[1,+\infty[\}$$

is a bounded subset of $\mathcal{S}(V)$. Now, since $1-\varphi$ vanishes on $K(x,2\varepsilon)$ we have

$$\operatorname{supp}(1-\varphi)\tau_y(\hat{\chi}\circ M_t^*) \subset V\setminus K(y,\varepsilon), \qquad y\in\overline{K}(x,\varepsilon). \tag{9.56}$$

Now, by (1.49) we have

$$(9.57) \qquad \partial^\beta(\hat{\chi}\circ M_t^*) = t^{\beta M}\sum_{j=0}^{N_\beta}\omega_j(t)\,(P_{\beta,j}(\partial)\,\hat{\chi}\,)\circ M_t^*, \qquad \beta\in\mathfrak{A},$$

(for the definition of the $P_{\beta,j}$ see Remark 1.30). Hence, by Lemma 9.2 for arbitrary $\beta\in\mathfrak{A}$ and $s\in\mathbb{N}$ there is a constant $C(\beta,s)$ such that

$$|\partial^\beta(\hat{\chi}\circ M_t^*)(z-y)| \le C(\beta,s)\,(1+|M_t^*(z-y)|)^{-s}\,t^{-N}$$

for arbitrary $y\in\overline{K}(x,\varepsilon)$, $z\in V\setminus K(y,\varepsilon)$ and $t\in[1,+\infty[$. Since $t\mapsto|M_t^*(z)|$ is increasing we have

$$(9.58) \qquad 1+|M_t^*(z-y)| \ge 1+|z-y| \ge (1+|z|)/(1+|y|), \qquad t\in[1,+\infty[,$$

so that the desired assertion about $\mathfrak{B}_N$ follows by the Leibniz rule.

Step 2: Now we suppose that T is not almost quasihomogeneous. Again by the Fourier inversion formula we may assume that T is almost quasihomogeneous on $\dot{V}=V\setminus\{0\}$. Then by Theorem 8.15 we find an almost quasihomogeneous distribution $S\in\mathcal{D}'(V)$ such that the support of $d:=T-S$ is contained in $\{0\}$. In particular, S is temperate, as well. It follows that $\hat{d}$ is induced by a polynomial function. Hence

$$(9.59) \qquad \text{(a)}\ \ WF_M(T)\cap\dot{V}\times S_{V^*} = WF_M(S)\cap\dot{V}\times S_{V^*}, \qquad \text{and}$$

$$\text{(b)}\ \ WF_{M^*}(\hat{T}) = WF_{M^*}(\hat{S}).$$

Applying Step 1 to S instead of T we deduce the desired assertion for T, as well.

Proof of (ii). Since the second step of the proof of the assertion (i) remains valid if WF_M is replaced by $WF_{M,L}$ we may assume that T is almost quasihomogeneous of degree m. As before, we may also assume that $x\in S_V$ and $\xi\in S_{V^*}$, and it suffices to prove one of the two implications, only.

We suppose that $(x,\xi)\notin WF_{M,L}(T)$ and choose K, F, ε, and W as in Step 1 of the proof of (i) after having replaced WF_M by $WF_{M,L}$. By Lemma 9.18 we choose constants C and A_β and sequences $(\varphi_N)_{N\in\mathbb{N}}$ in $C_0^\infty(K)$ and $(\chi_N)_{N\in\mathbb{N}}$ in $C_0^\infty(F)$ such that for every $N\in\mathbb{N}$ $\varphi_N\equiv 1$ on $K(x,2\varepsilon)$ and $\chi_N\equiv 1$ on W and for every $\beta\in\mathfrak{A}$ both φ_N and χ_N satisfy (9.5) with (A,B) replaced by (A_β, CN). We fix $i\in\mathbb{N}_0$. Then by Lemma 9.22 for every $\eta\in\mathfrak{E}_M$ we obtain a constant a_η such that

$$|\mathcal{F}(\varphi_N T_i)(M_t\xi)| \le a_\eta\, t^{-\mu-\mathrm{Re}\,m-1}\,(a_\eta\,(L_N)^{1+\eta}/t)^N, \qquad t\in[1,+\infty[,\ \xi\in F,\ N\in\mathbb{N}.$$

By (9.55) - applied to (φ_N, χ_N) instead of (φ, χ) - this implies

(9.60) $\quad |t^{m+\mu+1}\langle \varphi_N T_i, \tau_y(\hat{\chi}_N \circ M_t^*)\rangle| \le a'_\eta (a_\eta (L_N)^{1+\eta}/t)^N, \qquad t \ge 1,\ y \in V,\ N \in \mathbb{N},$

where a'_η is another constant depending only on η. Moreover, fixing $k \in \mathbb{N}$, making use of (9.57) and applying Lemma 9.2 to some $\ell > (1+r+\mathrm{Re}\, m+\mu+k\lambda_{max})/\lambda_{min}$ where $r := \max\{\mathrm{Re}\,\beta_M;\ \beta \in \mathfrak{A},\ |\beta| \le k\}$ we obtain a constant $C_{\eta,k}$ such that

$$|\partial^\beta(\hat{\chi}_N \circ M_t^*)(z-y)| \le C_{\eta,k}\,(1+|M_t^*(z-y)|)^{-k}\, t^{-\mu-\mathrm{Re}\, m-1}\,(C_{\eta,k} N^{1+\eta}/t)^N$$

for arbitrary $N \in \mathbb{N}$, $t \in [1,+\infty[$, $y \in K(x,\varepsilon)$, $z \in V\setminus K(y,\varepsilon)$, and $\beta \in \mathfrak{A}$ satisfying $|\beta| \le k$. In view of (9.56) and (9.58) and by the Leibniz formula one obtains new constants $C'_{\eta,k}$ such that

$$\sup\{(1+|z|)^k\,|\partial^\alpha[(1-\varphi_N)\,\tau_y(\hat{\chi}_N \circ M_t^*)](z)|;\ |\alpha| \le k,\ z \in V,\ y \in K(x,\varepsilon)\} \le$$
$$\le C'_{\eta,k}\, t^{-\mu-\mathrm{Re}\, m-1}\,(C_{\eta,k} N^{1+\eta}/t)^N, \qquad N \in \mathbb{N},\ t \in [1,+\infty[.$$

Since T_i is temperate it follows that for some $k \in \mathbb{N}$ there is a constant C''_η only depending on η such that

$$|t^{m+\mu+1}\langle (1-\varphi_N)\,T_i,\ \tau_y(\hat{\chi}_N \circ M_t)\rangle| \le C''_\eta\,(C_{\eta,k} N^{1+\eta}/t)^N$$

for arbitrary $N \in \mathbb{N}$, $t \in [1,+\infty[$, and $y \in K(x,\varepsilon)$. Combining this with (9.60) and (9.54) we conclude that $(\xi,-x) \notin WF_{M^*,L}(\hat{T})$, as desired.

<u>*Proof of (iii)*</u>. By the Fourier inversion formula, again, the assertion (iii.b) follows from (iii.a).

In order to prove (iii.a) we first suppose that for some $k \in \mathbb{N}$ the support of the distribution $T_k := (\partial_{M^*} - m)^k T$ is equal to $\{0\}$. In this case we have

$$\{0\} \times S_{V^*} = WF_M(T_k) \subset WF_M(T) \subset WF_{M,L}(T),$$

and since by Lemma 2.21 the distribution $(\partial_{M^*} - m^*)^k\,\hat{T}$ coincides with $(-1)^k\,\hat{T}_k$ and hence is induced by a non-trivial polynomial function on V^* the support of $\hat{T}$ equals the whole of V^*, as desired.

Next we suppose that for some $k \in \mathbb{N}_0$ T_k is induced by a polynomial function, i.e. the restriction of $\hat{T}$ to $V^*\setminus\{0\}$ is almost quasihomogeneous so that by Proposition 2.37 the support of $\hat{T}$ is quasihomogeneous which, in turn, implies that

$$(\mathrm{supp}\,\hat{T})_{M,\infty} = S_{V^*} \cap \mathrm{supp}\,\hat{T}.$$

Hence Proposition 9.31 shows that $WF_{M,L}(T) \subset V \times \mathrm{supp}\,\hat{T}$. Consequently, to complete the proof of (iii.a) we have to verify that the condition " $\xi \in S_{V^*} \cap \operatorname{supp} \hat{T}$ " implies " $(0,\xi) \in WF_{\mathbf{M}}(T)$ ".

To do this, we choose a polynomial functions P on V such that $u := T - P$ is almost quasihomogeneous of degree m and of type M^*. In fact, by Theorem 8.15 there is an extension $S \in \mathcal{S}'(V^*)$ of $\hat{T}|_{V^*\setminus\{0\}}$ which is almost quasihomogeneous of degree m^*; and since $\operatorname{supp}(\hat{T}-S) \subset \{0\}$ the distribution $T - \mathcal{F}^{-1}S$ is induced by a polynomial function P, and $u = \mathcal{F}^{-1}S$ is almost quasihomogeneous of type M^*, indeed, by Proposition 2.40.(i).

Next we observe that $WF_{\mathbf{M}}(T) = WF_{\mathbf{M}}(u)$ and $\hat{T} = \hat{u} + (2\pi)^n P(-D)\delta_o$ so that

$$(\operatorname{supp} \hat{T}) \setminus \{0\} = (\operatorname{supp} \hat{u}) \setminus \{0\}.$$

Therefore it suffices to prove the assertion for u instead of T.

Since $\hat{u}$, being equal to S, is almost quasihomogeneous of degree m^* and since by Proposition 2.40.(ii) we have $(\partial_{\mathbf{M}} - m^*)^j \hat{u} = (-1)^j \hat{u}_j$ where $u_j := (\partial_{\mathbf{M}^*} - m)^j u$ it follows that

$$\hat{u} = \tau^{-m^*} \sum_{j=0}^{k} \omega_j(\tau)\, \hat{u}_j \circ M_\tau \qquad \tau \in]0,+\infty[, \tag{9.61}$$

where $k := \operatorname{ord}_{\mathbf{M}}(\hat{u}) = \operatorname{ord}_{\mathbf{M}^*}(u)$.

Now, let $\xi \in S_{V^*}$, and suppose that $(0,\xi) \notin WF_{\mathbf{M}}(u)$. Then $(0,\xi) \notin WF_{\mathbf{M}}(u_j)$ for every $j \in \mathbb{N}_0$, and we can choose $\chi \in C_0^\infty(V)$ and $\varepsilon \in]0,1]$ such that $\chi \equiv 1$ on a neighbourhood of 0 and

$$\sup\{\tau^N |\mathcal{F}(\chi u_j)(M_\tau \rho)|;\ \rho \in K(\xi,\varepsilon),\ \tau \in [1,+\infty[\} < +\infty, \qquad N \in \mathbb{N},\ j \in \mathbb{N}_0. \tag{9.62}$$

Since

$$\mathcal{F}(\chi u_j)(M_\tau \rho) = (2\pi)^{-n} (\hat{\chi} * \hat{u}_j)(M_\tau \rho) = (2\pi)^{-n} \langle \hat{u}_j, \hat{\chi}(M_\tau \rho - (\cdot)) \rangle =$$

$$= (2\pi)^{-n} \tau^{\mu} \langle \hat{u}_j \circ M_\tau, (\hat{\chi} \circ M_\tau)(\rho - (\cdot)) \rangle = (\varphi_\tau * (\hat{u}_j \circ M_\tau))(\rho)$$

where $\varphi_\tau := (2\pi)^{-n} \tau^{\mu} \hat{\chi} \circ M_\tau$ we deduce from (9.61) that

$$(\varphi_\tau * \hat{u})(\rho) = \tau^{-m^*} \sum_{j=0}^{k} \omega_j(\tau)\, \mathcal{F}(\chi u_j)(M_\tau \rho).$$

It follows from (9.62) that $\varphi_\tau * \hat{u}$ converges to 0 uniformly on $K(\xi,\varepsilon)$ as $\tau \to +\infty$. On the other hand, making use of

$$\int_{V^*} \varphi_\tau(\upsilon)\,d\upsilon = (2\pi)^{-n} \int_{V^*} \hat{\chi}(M_\tau \upsilon)\, d(M_\tau \upsilon) = (2\pi)^{-n} \hat{\hat{\chi}}(0) = \chi(-0) = 1 \tag{9.63}$$

we are going to show that

(9.64) $\lim_{\tau\to\infty} \varphi_\tau * \psi = \psi$ in the topology of $\mathcal{S}(V^*)$ for every $\psi \in \mathcal{S}(V^*)$.

In view of $\langle \varphi_\tau * \hat{u}, \psi \rangle = \langle \check{\hat{u}}, \varphi_\tau * \check{\psi} \rangle$ it follows from (9.64) that $\varphi_\tau * \hat{u}$ converges weakly to $\hat{u}$ as $\tau \to +\infty$. By what was proved above this implies that $\hat{u}$ vanishes on $K(\xi, \varepsilon)$, i.e. ξ does not belong to supp $\hat{u}$, as was to be shown.

Now, for the proof of (9.64) we first observe that in view of $(\varphi_\tau * \psi)^{(\alpha)} = \varphi_\tau * \psi^{(\alpha)}$, $\alpha \in \mathbb{N}_o^n$, it suffices to show that for arbitrary $\psi \in \mathcal{S}(V^*)$ and $k \in \mathbb{N}$ we have

(9.64)' $\lim_{\tau\to\infty} \sup\{ (1+|\xi|)^k \, |(\varphi_\tau * \psi - \psi)(\xi)| \, ; \, \xi \in V^* \} = 0 .$

Now, using (9.63) and the mean value theorem we obtain

$$
\begin{aligned}
(\varphi_\tau * \psi - \psi)(\xi) &= \int_{V^*} \varphi_\tau(\upsilon) \, [\psi(\xi-\upsilon) - \psi(\xi)] \, d\upsilon = \\
&= (2\pi)^{-n} \int_{V^*} \hat{\chi}(M_\tau \upsilon) \, [\psi(\xi-\upsilon) - \psi(\xi)] \, d(M_\tau \upsilon) = \\
&= (2\pi)^{-n} \int_{V^*} \hat{\chi}(\zeta) \, [\psi(\xi - M_{1/\tau}\zeta) - \psi(\xi)] \, d\zeta = \\
&= -(2\pi)^{-n} \sum_{j=1}^{n} \int_{V^*} \hat{\chi}(\zeta) \, (M_{1/\tau}\zeta)_j \int_0^1 (\partial_j \psi)(\xi - s\, M_{1/\tau}\zeta) \, ds \, d\zeta .
\end{aligned}
$$

Moreover, by the triangle inequality and since by (1.79) $\tau \mapsto |M_\tau \zeta|$ is increasing we have

$$1 + |\xi| \le (1 + |s\, M_{1/\tau}\zeta|)(1 + |\xi - s\, M_{1/\tau}\zeta|) \le (1+|\zeta|)(1 + |\xi - s\, M_{1/\tau}\zeta|)$$

for arbitrary $\xi, \zeta \in V^*$, $s \in [0,1]$, and $\tau \in [1, +\infty[$. Putting everything together we arrive at

$$
\begin{aligned}
&(1+|\xi|)^k \, |(\varphi_\tau * \psi - \psi)(\xi)| \le \\
&\le \| M_{1/\tau} \| \sum_{j=1}^{n} \int_{V^*} (1+|\zeta|)^k \, |\hat{\chi}(\zeta)| \, |\zeta| \, d\zeta \, \sup\{ (1+|\upsilon|)^k \, |(\partial_j \psi)(\upsilon)| \, ; \, \upsilon \in V^* \} .
\end{aligned}
$$

Since $\hat{\chi}$ and $\partial_j \psi$ belong to $\mathcal{S}(V^*)$ and since by Remark 1.8.(i) $\lim_{\tau\to\infty} \| M_{1/\tau} \| = 0$, the condition (9.64)' follows. ■

(d) Quasihomogeneous Wave Front Sets of the Standard Fundamental Solutions of the Heat and of the Schrödinger Equation

Let q be as in section (f) of Chapter 7. We fix $r \in [1,+\infty[$, set $p := (r,1,\dots,1)$, and assume that M is of the form (1.1.a). Instead of WF_M and $C^{M,L}$ we shall write WF_p and $C^{p,L}$. In the present section the C^∞ and $C^{p,L}$ quasihomogeneous wave front sets (with respect to p) of the almost invariant (see Definition 7.70) fundamental solutions of q(D) are determined. The standard homogeneous wave front sets, of course, correspond to $r=1$, so the treatment includes C^∞ and analytic wave front sets as well as those with respect to C^L. In the course of the proof we give some rather precise estimates which might be interesting for their own sake. We also refer to the discussion following the proof of Theorem 9.35.B. Most of the material is taken from [7].

We recall that for any open subset Ω of $\mathbb{R}^{1+n}$ the space $C^{p,L}(\Omega)$ consists of all C^∞ functions $f:\Omega \longrightarrow \mathbb{C}$ such that for every compact subset K of Ω there is a constant C satisfying

$$(9.65) \qquad |f^{(\alpha)}(z)| \le C^{|\alpha|+1} L_\alpha^{\langle\alpha,p\rangle}, \qquad z \in K,\ \alpha \in \mathbb{N}_o^{1+n},$$

where $L_\alpha := L(r\alpha_1+\alpha_2+\dots+\alpha_{n+1})$. Note that $C^{p,A}(\Omega)$ is the standard inhomogeneous Gevrey class $\Gamma^p(\Omega)$ of type p which for $r=1$ coincides with the space of all real analytic functions on Ω.

From Proposition 9.7 and Theorem 9.27 we see that for any $E \in \mathcal{D}'(\mathbb{R}^{1+n})$ the following inclusions hold:

$$WF_p(q(D)E) \subset WF_p(E) \subset WF_{p,L}(E) \subset WF_{p,A}(E) \subset WF_{p,A}(q(D)E) \cup \mathbb{R}^{1+n} \times q_{w,p}^{-1}(0)$$

where $q_{w,p}$ denotes the quasihomogeneous (of type p) principal part of q which is given by

$$(9.66) \qquad q_{w,p}(\tau,\xi) = \begin{cases} iw\tau & \text{if } r>2 \\ q(\tau,\xi) & \text{if } r=2\,. \\ |\xi|^2 & \text{if } r<2 \end{cases}$$

If E is an almost invariant fundamental solution of q(D) then in view of (7.61) and by Theorem 7.80 E is real analytic outside $\{0\} \times \mathbb{R}^n$ so that in view of the equalities $WF_p(\delta_o) = \{0\} \times S^n = WF_{p,A}(\delta_o)$ the preceding inclusions become

(9.67) $\{0\}\times S^n \subset WF_p(E) \subset WF_{p,L}(E) \subset WF_{p,A}(E) \subset \{0\}\times S^n \cup (\{0\}\times \mathbb{R}^n)\times q_{w,p}^{-1}(0)$.

The equations (9.66) suggest that when determining the wave front sets of E one has to deal with three different cases. The first two of them can be handled together, they are easier than the third. In order to avoid repetitions, from now on we denote by E one of the distributions defined in Notation 7.71, i.e. in case $v\neq 0$ E is the unique invariant fundamental solution of q(D) whereas in case $v=0$ E is one of the two invariant fundamental solutions E^+ or E^- of q(D).

Theorem 9.35.A. *If $r\geq 2$ then $WF_p(E) = WF_{p,L}(E) = \{0\}\times S^n \cup W$ where*

$$W := \begin{cases} \emptyset & \text{if } v\neq 0 \\ \{((0,\mathrm{sign}(\sigma u)s\xi),(\tau,\xi));\ (\tau,\xi)\in S^n\cap q^{-1}(0),\ s>0\} & \text{if } v=0 \text{ and } r=2. \\ \{((0,\mathrm{sign}(\sigma u)s\xi),(0,\xi));\ \xi\in S^{n-1},\ s>0\} & \text{if } v=0 \text{ and } r>2 \end{cases}$$

For the case $r=2$ we note that

(9.68) $$S^n\cap q^{-1}(0) = \{(\tau_u,\rho_u\xi);\ \xi\in S^{n-1}\} \quad \text{if } v=0$$

where $\tau_u := \frac{1}{2}(u-\mathrm{sign}(u)\sqrt{u^2+4})$ *and* $\rho_u := \sqrt{|u\tau_u|}$.

<u>*Proof in case $v\neq 0$*</u>. In this case we have $q^{-1}(0)\cap S^n = \emptyset$. Hence for $r=2$ the assertion follows from (9.66) and (9.67). This means that the restriction T of E to $\mathbb{R}^{1+n}\setminus\{0\}$ belongs to the Gevrey class $\Gamma^{(2,1,\dots,1)}(\mathbb{R}^{1+n}\setminus\{0\})$. Since the latter is contained in $\Gamma^p(\mathbb{R}^{1+n}\setminus\{0\})$ if $r>2$ this implies that T belongs to $\Gamma^p(\mathbb{R}^{1+n}\setminus\{0\})$, i.e. $WF_{p,A}(T)=\emptyset$ for $r>2$, as well. ■

When $v=0$ the main step in the proof of Theorem 9.35.A is the following lemma which we are going to prove in subsection (d.1) below by microlocally cutting off a suitable real analytic function.

Lemma 9.36. *Suppose that $v=0$. Let $\xi_0\in S^{n-1}$. Then $WF_{p,A}(E)\cap \Omega\times\Delta=\emptyset$ where $\Delta := \{(\tau,\xi)\in\mathbb{R}\times\mathbb{R}^n;\ \mathrm{sign}(\sigma u)\langle\xi,\xi_0\rangle < 0\}$ and where*

(9.69) $$\Omega := \mathbb{R}\times X \quad \text{with} \quad X := \{x\in\mathbb{R}^n;\ \langle\xi_0,x\rangle > 0\}.$$

<u>*Proof of Theorem 9.35.A if $v=0$*</u>. Let $x_0\in\mathbb{R}^n\setminus\{0\}$ and $\xi\in\mathbb{R}^n\setminus(\mathrm{sign}(\sigma u)\,[0,+\infty[\,x_0)$. Then the compact convex segment $K := [-\mathrm{sign}(\sigma u)\xi, x_0]$ in $\mathbb{R}^n$ does not contain

the origin. Hence by the Hahn-Banach Separation Theorem we find $\xi_0 \in S^{n-1}$ such that $\langle \xi_0, x \rangle > 0$ for every $x \in K$, in particular: $\langle \xi_0, x_0 \rangle > 0$ and $\operatorname{sign}(\sigma u) \langle \xi, \xi_0 \rangle < 0$. Hence Lemma 9.36 implies that $((0,x_0),(\tau,\xi)) \notin WF_{p,A}(E)$ for every $\tau \in \mathbb{R}$. Consequently, by (9.67) we have

$$(9.70) \qquad \{\zeta \in S^n ;\ ((0,x_0),\zeta) \in WF_{p,A}(E)\} \subset$$

$$\subset [\mathbb{R} \times (\operatorname{sign}(\sigma u)\, [0,+\infty[\, x_0)] \cap q_{w,p}^{-1}(0) \cap S^n .$$

Since the analytic singular support of E equals $\{0\} \times \mathbb{R}^n$ the left-hand side of (9.70) is non-empty, and since in view of (9.66) and (9.68) for $r \geq 2$ the set on the right-hand side consists of one element only we conclude that equality holds. ∎

We now come to the more difficult case "$r < 2$". In this case it follows from (9.67) and (9.66) - or from (9.70) - that for every $x_0 \in \mathbb{R}^n \setminus \{0\}$ we have

$$(9.71) \qquad \{\zeta \in S^n ;\ ((0,x_0),\zeta) \in WF_{p,A}(E)\} \subset \{(1,0),(-1,0)\}.$$

As here the right-hand side consists of two elements, equality does not automatically follow from the non-emptiness of the left-hand side. In fact, it depends on L whether equality holds or not:

Theorem 9.35.B. *Suppose that $r<2$, and set $S := \sup\{k^2 L_k^{-r};\ k \in \mathbb{N}\}$.*

(i) *If $S < +\infty$ then $WF_p(E) = WF_{p,L}(E) = \{0\} \times S^n \cup W$ where*

$$W = \begin{cases} \emptyset & \text{if } v \neq 0 \\ (\{0\} \times (\mathbb{R}^n \setminus \{0\}) \times \{(-\operatorname{sign} u, 0)\} & \text{if } v = 0 \end{cases} .$$

(ii) *If $S = +\infty$ then $WF_{p,L}(E) = WF_{p,A}(E) = \{0\} \times S^n \cup (\{0\} \times \mathbb{R}^n) \times \{(1,0),(-1,0)\}$.*

The proof is based on the following more precise result establishing a microlocal decomposition of E. It will be derived in subsections (d.2) and (d.3) below. Note that the inclusion $(\{0\} \times \mathbb{R}^n) \times \{(1,0),(-1,0)\} \subset WF_A(E)$ is also a consequence of Theorem 8.5.6 in Hörmander [11].

Theorem 9.37. *Suppose that $r<2$. Let $\nu \in \{+1,-1\}$ such that $\nu = \operatorname{sign} u$ if $v = 0$. Let $\xi_0 \in S^{n-1}$, and define Ω and X as in (9.69). Then $E|_\Omega$ has a decomposition $E|_\Omega = E_+ + E_-$ where E_+ and E_- are distributions on Ω having the following properties:*

(9.72) $\quad WF_{p,A}(E_{\pm}) = (\{0\}\times X)\times\{(\pm\nu,0)\}$,

and E_+ is induced by a function $f\in\Gamma^{(2,1,\dots,1)}(\Omega)$ such that there are positive constants c and B satisfying

(9.73) $\quad |(\partial_t^j f)(0,s\xi_0)| \geq c^{j+1} j^{2j} s^{-2j-n} \qquad$ *if $s>0$ and $j\geq B(1+s)$.*

Proof of Theorem 9.35.B.(i). Note that the condition "$S<+\infty$" implies that the Gevrey class $\Gamma^{(2,1,\dots,1)}(\Omega)$ is contained in $C^{p,L}(\Omega)$. Consequently, the function f in Theorem 9.37 satisfies: $WF_{p,L}(T_f)=\emptyset$, so that $WF_{p,L}(E|_\Omega) = WF_{p,L}(E_-)$, and $WF_{p,L}(E|_\Omega)$ is contained in $(\{0\}\times X)\times\{(-\nu,0)\}$. By the choice of ν this implies: (a) if $v\neq 0$ then ν can be replaced by $-\nu$, and for every $x_0\in X$ the left-hand side of (9.71) is empty; and (b) if $v=0$ then for every $x_0\in X$ the left-hand side of (9.71) is contained in $\{(-\operatorname{sign} u,0)\}$ and - since it is non-empty - equal to $\{(-\operatorname{sign} u,0)\}$. Since for every $x_0\in\mathbb{R}^n\setminus\{0\}$ there is $\xi_0\in S^{n-1}$ such that $(0,x_0)\in\Omega$ the proof is complete. ■

Proof of Theorem 9.35.B.(ii). Let $x_0\in\mathbb{R}^n\setminus\{0\}$, and set $\xi_0 := x_0/|x_0|$. The hypothesis "$S=+\infty$" implies the existence of a strictly increasing sequence $(j_i)_{i\in\mathbb{N}}$ in $\mathbb{N}$ such that

$$j_i^2 \geq i\, L_{2j_i}^r , \qquad i\in\mathbb{N}.$$

Combining this with (9.73) we see that f does not belong to $C^{p,L}$ at $(0,x_0)$. Hence (9.72) implies that $((0,x_0),(\nu,0))\in W_{p,L}(E_+)$. Since by (9.72), as well, $W_{p,L}(E_+)$ and $W_{p,L}(E_-)$ do not intersect this means that $((0,x_0),(\nu,0))$ belongs to $W_{p,L}(E)$. By the choice of ν this implies: (a) if $v\neq 0$ then replacing ν by $-\nu$ we deduce that equality holds in (9.71); and (b) if $v=0$ then $W_{p,L}(E)$ contains $((0,x_0),(\operatorname{sign} u,0))$. Since by assertion (i) the point $((0,x_0),(-\operatorname{sign} u,0))$ belongs to $WF_p(E)\subset W_{p,L}(E)$ equality holds in (9.71) in the case "$v=0$", as well. ■

Discussion. In case $v=0$ the results of Theorems 9.35.A and 9.35.B on the wave front sets $WF_p(E^{\pm})$, $r\geq 1$, may be interpreted as follows. The high frequencies causing the C^∞ singularities of $E^{\pm}$ at the points $(0,x_0)$ of $\{0\}\times(\mathbb{R}^n\setminus\{0\})$ are concentrated near the direction $\eta_0 := (-\operatorname{sign} u,0)$ but nevertheless keep a little away from it: they are contained in the sets $\{(-s^r \operatorname{sign} u, s\xi);\ s\geq 1,\ \xi\in K(0,\varepsilon)\}$,

$\varepsilon > 0$, (which are the smaller the larger r is) in case $r < 2$ but stay in their complement in case $r \geq 2$ and ε is sufficiently small. Moreover, in contrast to the case $r < 2$ the case $r \geq 2$ shows that it depends on the point x_0 how the ξ-components of the high frequencies causing singularities look like. As for the $C^{p,L}$ singularities of E the interpretation of the results of Theorems 9.35.A and 9.35.B is more complicated since when varying r one does not only change the shape of the frequency domains involved in the definition of $WF_{p,L}(E)$ but also the type of Gevrey regularity described by $WF_{p,L}(E)$.

Finally, in case $v = 0$ we compute the quasihomogeneous wave front sets of the other almost invariant fundamental solutions of q(D).

Theorem 9.38. *Suppose that $v = 0$. Let F be an almost invariant fundamental solution of q(D) which is different from E^+ and E^- (see Notation 7.71).*

(i) If $r \geq 2$ then $WF_p(F) = WF_{p,A}(F) = WF_{p,A}(E^+) \cup WF_{p,A}(E^-)$.

(ii) If $r < 2$ then $WF_p(F) = WF_p(E^{\pm})$ and $WF_{p,L}(F) = WF_{p,L}(E^{\pm})$.

For the proof we require information on the quasihomogeneous wave front sets of the almost invariant solutions of (7.64) determined in section 7.(f):

Theorem 9.39. *Suppose that $v = 0$. Let $T \in \mathcal{S}'(\mathbb{R} \times \mathbb{R}^n)$ be any almost invariant solution of the equation (7.64) which is not induced by a polynomial function. Then its wave front set $WF_p(T)$ coincides with $WF_{p,A}(T)$ and is equal to*

$$\begin{cases} \{((0,s\xi),(0,\xi));\ \xi \in S^{n-1},\ s \in \mathbb{R}\} & \text{if } r > 2 \\ \{((0,s\xi),(\tau,\xi));\ (\tau,\xi) \in S^n \cap q^{-1}(0),\ s \in \mathbb{R}\} & \text{if } r = 2 \\ (\{0\} \times \mathbb{R}^n) \times \{(-\operatorname{sign} u, 0)\} & \text{if } r < 2. \end{cases}$$

Proof. First of all we verify that in case $r < 2$ we have

$$(q^{-1}(0))_{p,\infty} = \{(-\operatorname{sign} u, 0)\}.$$

Indeed, if $\zeta = (\tau,\xi) \in S^n$ belongs to the left-hand side then we find a sequence of points (τ_j, ξ_j), $j \in \mathbb{N}$, in $q^{-1}(0)$ and a sequence (s_j) in $]0,+\infty[$ converging to zero as $j \to \infty$ such that

$$\zeta = \lim_{j\to\infty} (s_j^r \tau_j, s_j \xi_j).$$

Since the condition on (τ_j,ξ_j) amounts to $u\tau_j = -|\xi_j|^2$ this implies

$$|\xi| = \lim_{j\to\infty} |s_j\xi_j| = |u|^{1/2} \lim_{j\to\infty} s_j |\tau_j|^{1/2} =$$
$$= |u|^{1/2}\Big(\lim_{j\to\infty} |s_j^r\tau_j|\Big)^{1/2} \lim_{j\to\infty} s_j^{1-r/2} = |u\tau|^{1/2} \lim_{j\to\infty} s_j^{1-r/2}.$$

As $r<2$ this is equal to 0. In view of $\operatorname{sign}\tau_j = -\operatorname{sign} u$ we conclude that $\zeta = (-\operatorname{sign} u, 0)$, as claimed above.

Since $\operatorname{supp}\hat{T} \subset q^{-1}(0)$ we deduce from Proposition 9.31, Remark 9.30, and (9.66) that

(9.74) $$WF_A(T) \subset (\mathbb{R}\times\mathbb{R}^n)\times\Sigma_p$$

where

(9.75) $$\Sigma_p := \begin{cases} \{(0,\xi);\ \xi\in S^{n-1}\} & \text{if } r>2 \\ S^n \cap q^{-1}(0) & \text{if } r=2 \\ \{(-\operatorname{sign} u, 0)\} & \text{if } r<2. \end{cases}$$

Next we observe that it suffices to prove the equalities outside the origin of $\mathbb{R}\times\mathbb{R}^n$ because since the wave front sets are closed it then follows that they contain the set $\{0\}\times\Sigma_p$ so that in view of (9.74) their intersection with $\{0\}\times S^n$ coincides with $\{0\}\times\Sigma_p$.

Since by Theorem 7.80 the analytic singular support of T is equal to $\{0\}\times\mathbb{R}^n$ the assertion follows in the case $r<2$.

For the proof of the other cases we observe that by the O(n)-invariance it follows by (7.74) and Theorem 9.27 that $WF_{p,A}(T)$ is contained in the set of all $((t,x),(\tau,\xi))$ in $(\mathbb{R}\times\mathbb{R}^n)\times S^n$ such that $x_j\xi_k - x_k\xi_j = 0$ for arbitrary $j,k\in\mathbb{N}_n$. Since in case $\xi\neq 0$ this means that $x = s\xi$ for some $s\in\mathbb{R}$ we conclude that

$$WF_{p,A}(T) \subset \{((t,s\xi),(\tau,\xi));\ t,s\in\mathbb{R},\ (\tau,\xi)\in\Sigma_p\}.$$

Since the distribution $S_{m*,k}$ defined in (7.69) is real valued it follows that the distribution $T_{m,k}$ defined in Theorem 7.77' satisfies the assumption of Remark 9.6.(ii). Hence, if $((0,x),(\tau,\xi))\in WF_{p,A}(T_{m,k})$ then $((0,-x),(\tau,\xi))\in WF_{p,A}(T_{m,k})$, and - since T is invariant under the map $(t,x)\mapsto(t,-x)$ - it follows by Remark 9.6.(iii) that $((0,x),(\tau,-\xi))\in WF_{M,A}(T_{m,k})$. Since by Theorem 7.80 the analytic singular support of $T_{m,k}$ is equal to $\{0\}\times\mathbb{R}^n$ the assertion follows in case $T = T_{m,k}$ for some $k\in\mathbb{N}_0$. In order to remove this restriction on T we employ Theorem 7.77' to find $N\in\mathbb{N}_0$ and constants $a_0,\dots,a_N\in\mathbb{C}$ such that $a_N\neq 0$ and

$$T - \sum_{j=0}^{N} a_j T_{m,j}$$

is induced by a polynomial function Q. Since $(\partial_p - m) S_{m,j} = -S_{m,j-1}$ it follows by Proposition 2.40.(ii) that $(\partial_p - m) T_{m,j} = T_{m,j-1}$, $j \in \mathbb{N}$. Moreover, since by Proposition 7.76 the support of $(\partial_p - m) S_{m,o}$ is contained in $\{0\}$, the distribution $(\partial_p - m)^N T_{m,N-1} = (\partial_p - m) T_{m,o}$ is induced by a (quasihomogeneous) polynomial function R. Consequently,

$$(\partial_p - m)^N T = a_N T_{m,o} + a_{N-1} T_R + (\partial_p - m)^N T_Q .$$

Since $a_N \neq 0$ this implies that

$$WF_{p,A}(T_{m,o}) \subset WF_{p,A}((\partial_p - m)^N T) \subset WF_{p,A}(T)$$

so that the assertion is valid for general T, as well. ■

Proof of Theorem 9.38. Since F is different from $E^{\pm}$ the distribution $T_{\pm} := F - E^{\pm}$ is a non-trivial solution of (7.64) which is almost invariant of degree $-n$. We set $\nu := \operatorname{sign} u$.

(i). Hence, if $r \geq 2$ Theorems 9.35.A and 9.39 show that

$$\{((0, \mp \nu s \xi), (\tau, \xi)) ; (\tau, \xi) \in \Sigma_p , s \in \mathbb{R}\} \subset WF_p(T_{\pm}) \setminus WF_p(E_{\pm}) .$$

From this it follows that

$$\{((0, \mp \nu s \xi), (\tau, \xi)) ; (\tau, \xi) \in \Sigma_p , s \in \mathbb{R}\} \subset WF_p(F) .$$

In view of $WF_{p,A}(F) \subset WF_{p,A}(E^{\pm}) \cup WF_{p,A}(T_{\pm})$ the inverse inclusion is valid with WF_p replaced by $WF_{p,A}$ by Theorems 9.35.A and 9.39.

(ii). We first suppose that the number S in Theorem 9.35.B is finite. Then it follows from Theorems 9.35.B and 9.39 that

$$WF_p(F) \subset WF_{p,L}(F) \subset (\{0\} \times \mathbb{R}^n) \times \{(-\nu, 0)\} \cup \{0\} \times S^n . \tag{9.76}$$

Since q(D) is not hypoelliptic the singular support of F intersects $\{0\} \times (\mathbb{R}^n \setminus \{0\})$. Since F is invariant under the action of O(n) on the spaces variables x it follows that $\{0\} \times (\mathbb{R}^n \setminus \{0\})$ is contained in sing supp F. Consequently, equality holds everywhere in (9.76)

Now we suppose that $S = +\infty$. Since by Theorem 9.39 $WF_{p,L}(T_{\pm})$ does not intersect $(\mathbb{R} \times \mathbb{R}^n) \times \{(\nu, 0)\}$ it follows from Theorem 9.35.B.(ii) that

$$\mathrm{WF}_{p,L}(F) \cap (\mathbb{R}\times\mathbb{R}^n\setminus\{0\})\times\{(\nu,0)\} =$$

$$= \mathrm{WF}_{p,L}(E^{\pm}) \cap (\mathbb{R}\times\mathbb{R}^n\setminus\{0\})\times\{(\nu,0)\} = (\{0\}\times\mathbb{R}^n\setminus\{0\})\times\{(\nu,0)\}.$$

Since $\mathrm{WF}_p(F)$ is contained in $\mathrm{WF}_{p,L}(F)$ the first part of the proof shows that $\mathrm{WF}_{p,L}(F)$ contains $(\{0\}\times\mathbb{R}^n\setminus\{0\})\times\{(-\nu,0)\}$. Hence the assertion follows. ■

(d.1) Proof of Lemma 9.36

The first step relies on a suitable Fourier representation formula for E which is valid for the case $v \neq 0$, as well.

Proposition 9.40. *For every* $s \in]0,+\infty[$ *we have*

$$\langle E,\check{\varphi}\rangle = (2\pi)^{-n-1}\int_{\mathbb{R}^n}\int_{\mathbb{R}}\frac{\hat{\varphi}}{q}(\tau-i\sigma s,\xi)\,d\tau\,d\xi, \qquad \varphi\in C_0^\infty(\mathbb{R}\times\mathbb{R}^n).$$

The proof requires a deformation of the contour of integration:

Lemma 9.41. *Let Z be an open subset of $\mathbb{C}^n$, let $F: Z \longrightarrow \mathbb{C}$ be a holomorphic function, and let $g: \mathbb{R}^n\times[0,1]\longrightarrow\mathbb{R}^n$ be a C^1 function such that*

$$\{\xi+ig(\xi,t);\ \xi\in\mathbb{R}^n,\ t\in[0,1]\}\subset Z.$$

Suppose that there are constants C and L such that

$$(9.77)\qquad |(\partial_{\xi_j}g)(\xi,t)| \le C(1+|\xi|)^L, \qquad (\xi,t)\in\mathbb{R}^n\times[0,1],$$

and

$$(9.78)\qquad |F(\xi+ig(\xi,t))| \le C(1+|\xi|)^L, \qquad (\xi,t)\in\mathbb{R}^n\times[0,1].$$

Moreover, let Ω be an open subset of $\mathbb{R}^n$ such that for every compact subset K of Ω there are constants L_K and c_K satisfying

$$(9.79)\qquad H_K(-g(\cdot,t)) \le L_K\log(1+|\cdot|)+c_K, \qquad t\in[0,1].$$

Then the distributions $\mu_t\in\mathcal{D}'(\Omega)$, $t\in[0,1]$, defined by

$$\langle\mu_t,\varphi\rangle = \int_{\Phi_t}\hat{\varphi}(-\zeta)\,F(\zeta)\,d\zeta, \qquad \varphi\in C_0^\infty(\Omega),$$

all coincide where $\Phi_t : \mathbb{R}^n \longrightarrow \mathbb{C}^n$ is defined by $\xi \mapsto \xi + ig(\xi,t)$.

Proof. By P_r^n we denote the polydisc in $\mathbb{R}^n$ of radius $r>0$ with center 0. Then $\alpha_r : P_r^n \times [0,1] \longrightarrow \mathbb{C}^n$, $(\xi,t) \mapsto \Phi_t(\xi)$, is an $(n+1)$-chain, its boundary $\partial\alpha_r$ consisting of the two n-chains $\gamma_j : P_r^n \longrightarrow \mathbb{C}^n$, $\xi \mapsto \Phi_j(\xi)$, $j \in \{0,1\}$, and the 2^n n-chains

$$G_k^\sigma : P_r^{n-1} \times [0,1] \longrightarrow \mathbb{C}^n,\ y \mapsto \Phi_{y_n}(f_k^\sigma(y)), \qquad k \in \mathbb{N},\ \sigma \in \{+1,-1\},$$

where $f_k^\sigma(y) := (y_1,\ldots,y_{k-1},\sigma r, y_k,\ldots,y_{n-1})$. By (9.77) $(G_k^\sigma)^*(d\zeta_1 \wedge \ldots \wedge d\zeta_n)$ is of the form $h_k^\sigma dy_1 \wedge \ldots \wedge dy_n$ where $h_k^\sigma : P_r^{n-1} \times [0,1] \longrightarrow \mathbb{C}$ is a continuous function satisfying

$$|h_k^\sigma| \le C_1 (1 + |f_k^\sigma|)^P \tag{9.80}$$

for suitable constants C_1 and P which are independent of r.

Now, let K be a compact subset of Ω. We choose $\varepsilon > 0$ such that $K_\varepsilon := K + \overline{K}(0,\varepsilon)$ is contained in Ω. By the Paley-Wiener Theorem we find a constant C_2 such that for every $\varphi \in C_0^\infty(K)$ the following estimate holds:

$$|\hat{\varphi}(\xi + i\upsilon)| \le C_2 \exp[-(L+P+L_{K_\varepsilon}+n)\log(1+|\xi|) + H_{K_\varepsilon}(\upsilon)], \qquad \xi,\upsilon \in \mathbb{R}^n.$$

By (9.78) and (9.80) and by (9.79) it follows that

$$\Big|\int_{G_k^\sigma} \check{\hat{\varphi}}\, F\, d\zeta_1 \wedge \ldots \wedge d\zeta_n\Big| \le \int_{P_r^{n-1}\times[0,1]} |(\check{\hat{\varphi}}\, F) \circ G_k^\sigma\, h_k^\sigma|\, dy \le C_2 \exp(c_{K_\varepsilon})\, C\, C_1\, B(r)$$

where

$$B(r) := (2r)^{n-1} \sup\{(1+|f_k^\sigma(y)|)^{-n};\ y \in P_r^{n-1} \times [0,1]\} \le (2r)^{n-1} r^{-n} = 2^{n-1}/r$$

converges to 0 as $r \to +\infty$. Consequently, since by Stokes' theorem

$$\int_{\partial\alpha_r} \check{\hat{\varphi}}\, F\, d\zeta_1 \wedge \ldots \wedge d\zeta_n = 0,$$

taking the limit as $r \to +\infty$ shows that $\mu_0 = \mu_1$. ■

Proof of Proposition 9.40. By the choice of σ (see Notation 7.71) we have $\sigma v = |v|$. Hence

$$(u\tau + v\sigma s + |\xi|^2)^2 + (-u\sigma s + v\tau)^2 = (u\tau + |\xi|^2)^2 + v^2 s^2 + v^2\tau^2 + u^2 s^2 + 2|\xi|^2 |v| s, \text{ i.e.}$$

$$|q(\tau - i\sigma s,\xi)|^2 = |q(\tau,\xi)|^2 + |w|^2 s^2 + 2|\xi|^2 |v| s, \qquad (\tau,\xi) \in \mathbb{R}\times\mathbb{R}^n,\ s \in \mathbb{R}. \tag{9.81}$$

Denoting by $\mathcal{F}_1$ the partial Fourier transform with respect to the time variable t we see that $\hat{E} = \mathcal{F}_1 \tilde{E}$ where $\tilde{E}$ is the partial Fourier transform of E with respect to the space variables x. Since $\tilde{E}$ is induced by the function (7.60) it follows for every $\psi \in \mathcal{S}(\mathbb{R} \times \mathbb{R}^n)$ that

$$\langle \hat{E}, \psi \rangle = \langle \tilde{E}, \mathcal{F}_1 \psi \rangle =$$

$$= \frac{\sigma}{w} \int_{\mathbb{R}^n} \int_{-\infty}^{+\infty} H(\sigma t) \exp(-\tfrac{t}{w} |\xi|^2)\, (\mathcal{F}_1 \psi)(t,\xi)\, dt\, d\xi =$$

$$= \frac{\sigma}{w} \int_{\mathbb{R}^n} \int_{0}^{+\infty} \exp(-t \tfrac{\sigma}{w} |\xi|^2)\, (\mathcal{F}_1 \psi)(\sigma t,\xi)\, dt\, d\xi =$$

$$= \lim_{\varepsilon \to 0} \int_{\mathbb{R}^n} \int_{0}^{+\infty} \frac{\sigma}{w} \exp[-t(\varepsilon + \tfrac{\sigma}{w} |\xi|^2)]\, (\mathcal{F}_1 \psi)(\sigma t,\xi)\, dt\, d\xi \,.$$

By the definition of $\mathcal{F}_1$ and by Fubini's theorem the integral with respect to t is equal to

$$\int_{\mathbb{R}} \int_{0}^{+\infty} f(t,\tau)\, dt\, \psi(\tau,\xi)\, d\tau$$

where

$$f(t,\tau) := \frac{\sigma}{w} \exp[-t(\varepsilon + i\sigma\tau + \tfrac{\sigma}{w} |\xi|^2)] = \frac{\sigma}{w} \exp[-t \tfrac{\sigma}{w} q(\tau - i\varepsilon\sigma,\xi)] =$$

$$= -\frac{1}{q(\tau - i\varepsilon\sigma,\xi)}\, \partial_t \exp[-t \tfrac{\sigma}{w} q(\tau - i\varepsilon\sigma,\xi)] \,.$$

Note that (9.81) shows that $q(\tau - i\varepsilon\sigma,\xi) \neq 0$, indeed; more precisely, we have

$$|q(\tau - i\varepsilon\sigma,\xi)| \geq |w|\,\varepsilon > 0 \,.$$

The main theorem of calculus implies that $\int_0^\infty f(t,\tau)\, dt = 1/q(\tau - i\varepsilon\sigma,\xi)$ so that

$$\langle \hat{E}, \psi \rangle = \lim_{\varepsilon \to 0} \int_{\mathbb{R}^n} \int_{\mathbb{R}} \frac{\psi(\tau,\xi)}{q(\tau - i\varepsilon\sigma,\xi)}\, d\tau\, d\xi \,. \tag{9.82}$$

Now we assume that $\psi \in \mathcal{F} C_0^\infty(\mathbb{R} \times \mathbb{R}^n)$ so that ψ is well-defined on $\mathbb{C} \times \mathbb{C}^n$ as an entire holomorphic function. Since by (9.81), again, we have

$$|q(\tau - i\sigma(\varepsilon + s),\xi)| \geq |w|\,\varepsilon > 0 \,, \qquad s \in [0,+\infty[\,,$$

one can deform the contour of integration by applying Lemma 9.41 to $g(\tau,\xi,t) := (\tau - i\sigma(ts + (1-t)\varepsilon, 0)$ so that the double integral on the right-hand side of (9.82) is equal to

$$\int_{\mathbb{R}^n} \int_{\mathbb{R}} \frac{\psi(\tau - i\sigma s,\xi)}{q(\tau - i\sigma(\varepsilon + s),\xi)}\, d\tau\, d\xi$$

for every $s \in]0,+\infty[$. By (9.81) we also have $|q(\tau - i\sigma(\varepsilon + s),\xi)| \geq |w|\, s$ for arbitrary

$s,\varepsilon \in]0,+\infty[$. Hence, in view of the Paley-Wiener estimates for ψ one concludes by Lebesgue's Dominated Convergence Theorem that the limit as $\varepsilon \to 0$ can be taken under the integral sign. This gives

$$(9.83) \qquad \langle \hat{E}, \psi \rangle = \int_{\mathbb{R}^n} \int_{\mathbb{R}} \frac{\psi}{q}(\tau - i\sigma s, \xi)\, d\tau\, d\xi .$$

An application of the Fourier inversion formula leads to the desired formula. ∎

Proof of Lemma 9.36. Suppose that $v = 0$, and let ξ_0 and Δ be as in the statement of Lemma 9.36. The first step in the proof consists in deforming the contour of integration in the formula of Proposition 9.40. This is based on the following estimate

$$(9.84) \qquad |q(\tau - i\sigma s, \xi + i\rho\xi_0)| \geq |u|\, s\,, \qquad s,\rho \in [0,+\infty[,\ (\tau,\xi) \in \Delta,$$

which follows from $|\operatorname{Im} q(\tau - i\sigma s, \xi + i\rho\xi_0)| = |u|\, s + 2\rho(-\operatorname{sign}(\sigma u)\langle \xi, \xi_0 \rangle)$.

We now fix a function $\chi \in C^1(\mathbb{R} \times \mathbb{R}^n)$ with values in $[0,1]$ and with support contained in Δ satisfying

$$(9.85) \qquad \text{for some } N \in \mathbb{R} \text{ the functions } (1+|\cdot|)^{-N} \partial_j \chi,\ j \in \mathbb{N}_n, \text{ are bounded.}$$

Then in view of (9.84) the assumptions of Lemma 9.41 are satisfied for $F = 1/q$ and $g(\zeta, t) = (-\sigma, t\chi(\zeta)\, |\zeta|\, \xi_0)$. Hence, setting

$$\Phi(\zeta) := (\tau - i\sigma, \xi + i\chi(\zeta)\, |\zeta|\, \xi_0), \qquad \zeta = (\tau,\xi) \in \mathbb{R} \times \mathbb{R}^n,$$

one obtains

$$\langle E, \varphi \rangle = (2\pi)^{-n-1} \int_{\mathbb{R} \times \mathbb{R}^n} \frac{\check{\hat{\varphi}}}{q}(\Phi(\zeta))\, J(\zeta)\, d\zeta\,, \qquad \varphi \in C_0^\infty(\Omega),$$

where $J(\zeta) := \det \Phi'(\zeta)$.

Now we fix $C, \varepsilon > 0$ and set

$$\Delta_{C,\varepsilon} := \{ (\tau,\xi) \in \mathbb{R} \times \mathbb{R}^n;\ -\operatorname{sign}(\sigma u)\langle \xi, \xi_0 \rangle > \varepsilon |\xi|,\ |\tau| < C |\xi|^r \}$$

and $\Delta_1 := \mathbb{R} \times \mathbb{R}^n \setminus \Delta_2$ where $\Delta_2 := \Delta_{C,\varepsilon} \setminus \mathbb{R} \times K(0,1)$. We suppose that χ is chosen in such way that, in addition, $\chi \equiv 1$ on Δ_2 and that χ is quasihomogeneous of degree 0 and of type p on $\mathbb{R} \times \mathbb{R}^n \setminus K(0,1)$. Then the derivatives of χ are even bounded so that (9.85) holds for $N = 0$. For $j \in \{1,2\}$ we define $E_j \in \mathcal{D}'(\Omega)$ by

$$\langle E_j, \varphi \rangle := (2\pi)^{-n-1} \int_{\Delta_j} \frac{\check{\hat{\varphi}}}{q}(\Phi(\zeta))\, J(\zeta)\, d\zeta\,, \qquad \varphi \in C_0^\infty(\Omega).$$

Then $E|_\Omega = E_1 + E_2$. Since $\Delta_{C,\varepsilon}$ is quasihomogeneous and since the set

$$\Delta_1 \setminus (\mathbb{R}\times\mathbb{R}^n \setminus \Delta_{C,\varepsilon}) = \Delta_{C,\varepsilon}\setminus\Delta_2 = \Delta_{C,\varepsilon} \cap \mathbb{R}\times K(0,1)$$

is contained in the bounded set $[-C,C]\times K(0,1)$ it follows by Proposition 9.32 that

$$\mathrm{WF}_{p,A}(E_1) \cap \Omega\times\Delta_{C,\varepsilon} = \emptyset. \tag{9.86}$$

Next we are going to show that a function $g:\Omega \longrightarrow \mathbb{C}$ is well-defined by

$$g(z) := \int_{\Delta_2} I(z,\zeta)\,d\zeta$$

where

$$I(z,\zeta) := (2\pi)^{-n-1} \exp\langle i\Phi(\zeta),z\rangle \frac{J(\zeta)}{q(\Phi(\zeta))}. \tag{9.87}$$

To this end we have to estimate I. Note that for $\zeta=(\tau,\xi)\in\Delta_2$ we have $\chi(\zeta)=1$ so that $\Phi(\zeta)=(\tau-i\sigma, \xi+i|\zeta|\xi_0)$ and

$$|\exp\langle i\Phi(\zeta),z\rangle| = \exp(\sigma t - |\zeta|\langle\xi_0,x\rangle), \qquad z=(t,x)\in\Omega,$$

and

$$J(\zeta) = \det\Big(\delta_{jk} + i\partial_{\xi_j}|\zeta|(\xi_0)_k\Big)_{(j,k)\in\mathbb{N}_n\times\mathbb{N}_n} = \tag{9.88}$$

$$= 1 + i\langle\nabla_\xi|\zeta|,\xi_0\rangle = 1 + i\langle\xi,\xi_0\rangle/|\zeta|.$$

Hence $|J|\le\sqrt{2}$ on Δ_2, and - in view of (9.84) -

$$|I(z,\zeta)| \le \frac{1}{|u|} e^{\sigma t}\exp(-|\zeta|\langle\xi_0,x\rangle), \qquad \zeta\in\Delta_2,\ z=(t,x)\in\Omega. \tag{9.89}$$

Since I is measureable and since $\exp(-|\cdot|\langle\xi_0,x\rangle)$ is absolutely integrable on Δ_2 g is well-defined, indeed. Since for any compact subset K of Ω the number

$$\inf\{\langle\xi_0,x\rangle\ ;\ (t,x)\in K \text{ for some } t\in\mathbb{R}\} \tag{9.90}$$

is positive Fubini's theorem shows that $\langle E_2,\varphi\rangle = \int_\Omega g(z)\varphi(z)\,dz$, $\varphi\in C_0^\infty(\Omega)$, i.e. $E_2 = T_g$. In order to verify that g is differentiable we fix $\alpha=(j,\beta)\in\mathbb{N}_o\times\mathbb{N}_o^n$. Then we have

$$D_z^\alpha I(z,\zeta) = \Phi(\zeta)^\alpha I(z,\zeta). \tag{9.91}$$

Moreover, since $|\xi|\ge 1$ we have

$$|\Phi(z)^\alpha| = |\tau-i\sigma|^j\,|(\xi+i|\zeta|\xi_0)^\beta| \le |\zeta|^j(|\xi|+|\zeta|)^{|\beta|} \le 2^{|\alpha|}\,|\zeta|^{|\alpha|},\ \zeta\in\Delta_2.$$

If $c=\langle\xi_0,x\rangle/2$ then from (9.89) and (9.91) we deduce that

$$|D_z^\alpha I(z,\zeta)| \le \frac{1}{|u|} 2^{|\alpha|} C_\alpha \exp(-c|\zeta|)\, e^{\sigma t}$$

where

$$C_\alpha := \sup\{ |\zeta|^{|\alpha|} \exp(-c|\zeta|);\ \zeta \in \Delta_2 \}.$$

Since for any $a>0$ the derivative of the function $[0,+\infty[\ni t \mapsto t^a \exp(-ct)$ vanishes at $t = a/c$, only, we have

$$\sup\{ t^a \exp(-ct)\,;\ t \in [0,+\infty[\} = \left(\frac{a}{ce}\right)^a, \qquad a>0. \tag{9.92}$$

Hence $C_\alpha \le \left(\frac{|\alpha|}{ce}\right)^{|\alpha|}$. Since

$$\int_{\Delta_2} \exp(-c|\cdot|) \le \int_{\mathbb{R}\times\mathbb{R}^n} \exp(-c|\cdot|) = c^{-n-1} \int_{\mathbb{R}\times\mathbb{R}^n} \exp(-|\cdot|)$$

we conclude that g is a C^∞ function satisfying

$$|g^{(\alpha)}(z)| \le B^{|\alpha|+1} \langle \xi_0, x\rangle^{-|\alpha|-n-1} |\alpha|^{|\alpha|} e^{\sigma t}, \qquad z=(t,x)\in\Omega,$$

where B is a constant not depending on z and α. In particular, g is real analytic. Consequently, $WF({}_{p,A}(E|_\Omega) = WF_{p,A}(E_1)$, and since Δ is the union of all the sets $\Delta_{C,\varepsilon}$ the assertion of Lemma 9.36 follows from (9.86). ■

(d.2) Proof of Theorem 9.37. Part 1: Establishing the Microlocal Decomposition of E

We follow the pattern of the proof of Lemma 9.36. Again, the first step is a deformation of the integration contour which is based on the following lemma. We set

$$\Gamma_o := \{ \tau \in \mathbb{R}\setminus]-\tfrac{1}{2},\tfrac{1}{2}[\,;\ \operatorname{sign}\tau = \operatorname{sign} u \text{ if } v=0 \} \times \mathbb{R}^n \quad \text{and} \quad \rho(\tau,\xi) := (|\tau| + |\xi|^2)^{1/2}.$$

Lemma 9.42. *There are constants $a, c_o, s_o \in]0,1[$ only depending on w such that*

$$|q(\tau - i\sigma s, \xi + i\upsilon)| \ge c_o \tag{9.93}$$

for arbitrary $s \in [0,s_o]$, $(\tau,\xi)\in\Gamma_o$, and $\upsilon\in\mathbb{R}^n$ satisfying $|\upsilon| \le a\rho(\tau,\xi)$.

<u>*Proof.*</u> We have $z := q(\tau - i\sigma s, \xi + i\upsilon) = u\tau + |v|s + |\xi|^2 - |\upsilon|^2 + i(v\tau - u\sigma s + 2\langle\xi,\upsilon\rangle)$. We suppose first that $v=0$ and choose $a\in]0,1[$ such that $a^2 < |u|$. Then the conditions

$\operatorname{sign}\tau = \operatorname{sign}u$ and $|\tau| \geq 1/2$ imply that

$$\operatorname{Re} z = |u||\tau| + |\xi|^2 - |\upsilon|^2 \geq |\tau|(|u| - a^2) + |\xi|^2(1-a^2) \geq (|u|-a^2)/2 =: c_o .$$

We suppose now that $v \neq 0$, fix $b \in]4|u|, +\infty[$, and choose $a \in]0,1[$ such that $a^2(1+1/b) < 1/2$ and $2a(b^2+b)^{1/2} < |v|$. If $|\xi| \geq (b|\tau|)^{1/2}$ then $|\upsilon| \leq a(1+1/b)^{1/2}|\xi|$ and

$$\operatorname{Re} z \geq -|u||\tau| + |\xi|^2[1-a^2(1+1/b)] \geq |\xi|^2(-|u|/b + 1/2) \geq b/8 .$$

On the other hand, if $|\xi| \leq (b|\tau|)^{1/2}$ then $|\langle \xi, \upsilon\rangle| \leq |\xi||\upsilon| \leq a(b^2+b)^{1/2}|\tau|$ so that

$$|\operatorname{Im} z| \geq [|v| - 2a(b^2+b)^{1/2}]/2 - |u| s .$$

Choosing s_o sufficiently small one arrives at the assertion. ■

Now we fix constants a, c, s_o having the properties in the assertion of Lemma 9.42, choose a C^1 function $\eta: \mathbb{R} \times \mathbb{R}^n \longrightarrow \mathbb{R}^n$ with support contained in Γ_O such that $|\eta| \leq a\rho$, and choose another C^1 function $\varkappa: \mathbb{R} \times \mathbb{R}^n \longrightarrow \mathbb{R}$ with values in $[0, s_o]$ such that

(9.94) $$\varkappa \equiv s_o \quad \text{on } \mathbb{R} \times \mathbb{R}^n \setminus \Gamma_O .$$

Moreover, let U be an open subset of $\mathbb{R}^n$ such that

(9.95) $$\langle \eta(\zeta), x \rangle \geq 0 , \qquad \zeta \in \mathbb{R} \times \mathbb{R}^n,\ x \in U .$$

If the partial derivatives of η and $\varkappa$ of order 1 are bounded by $\operatorname{const}(1+|\cdot|)^N$ for some $N \in \mathbb{N}$ then Proposition 9.40 and Lemma 9.41 (applied to $F = 1/q$, $g(\zeta, t) = (-\sigma(1-t)s_o - \sigma t \varkappa(\zeta), t\eta(\zeta))$ and $\Omega = \mathbb{R} \times U$) yield

(9.96) $$\langle E, \varphi \rangle = (2\pi)^{-n-1} \int_{\mathbb{R} \times \mathbb{R}^n} \frac{\check{\hat{\varphi}}}{q}(\Phi(\zeta))\, J(\zeta)\, d\zeta , \qquad \varphi \in C_O^\infty(\mathbb{R} \times U),$$

where $\Phi(\zeta) := \zeta + i(-\sigma\varkappa(\zeta), \eta(\zeta))$ and $J(\zeta) := \det \Phi'(\zeta)$.

Next we fix a C^1 function $\chi: \mathbb{R} \times \mathbb{R}^n \longrightarrow [0,1]$ with support contained in Γ_O which is quasihomogeneous of degree 0 and of type p on the complement of $K(0,1)$. Since the partial derivatives of χ are bounded all the preceding hypotheses are satisfied if we set

$$\eta := a\chi\rho\xi_O , \quad \varkappa := s_o(1-\chi) , \text{ and } U := X \text{ where } X \text{ is defined in (9.69).}$$

In order to define the desired microlocal decomposition of E we choose a quasihomogeneous (of type p) open subset Γ of $\mathbb{R} \times \mathbb{R}^n$ such that $\Gamma_\varepsilon \subset \Gamma \subset \Gamma_C$ for suitable constants $\varepsilon, C > 0$ where

(9.97) $\qquad \Gamma_\varepsilon := \{(\tau,\xi) \in \mathbb{R}\times\mathbb{R}^n ;\ \tau\nu > 0,\ |\xi| < \varepsilon|\tau|^{1/r}\}.$

Moreover, we fix $R \in [1,+\infty[$ and set

$$\Gamma_+ := \Gamma \setminus (]-R,R[\times\mathbb{R}^n) \quad \text{and} \quad \Gamma_- := \mathbb{R}\times\mathbb{R}^n \setminus \Gamma_+ .$$

In addition, we suppose that

(9.98) $\qquad \chi \equiv 1 \ \text{ on } \Gamma \setminus [-1,1]\times\mathbb{R}^n$

and define $E_\pm \in \mathcal{D}'(\Omega)$ by

$$\langle E_\pm , \varphi \rangle = (2\pi)^{-n-1} \int_{\Gamma_\pm} \frac{\check{\hat{\varphi}}}{q}(\Phi(\zeta))\, J(\zeta)\, d\zeta , \qquad \varphi \in C_0^\infty(\Omega).$$

Since $\Gamma_- \setminus (\mathbb{R}\times\mathbb{R}^n\setminus\Gamma) = \Gamma \cap\,]-R,R[\times\mathbb{R}^n$ is contained in the bounded set $]-R,R[\times K(0,CR^{1/r})$ it follows by Proposition 9.32 that

(9.99) $\qquad WF_{p,A}(E_-) \subset \Omega\times(\mathbb{R}\times\mathbb{R}^n\setminus\Gamma).$

In order to study E_+ we are going to show that a C^∞ function $f:\Omega \longrightarrow \mathbb{C}$ is well-defined by

$$f(z) := \int_{\Gamma_+} I(z,\zeta)\, d\zeta$$

where I is defined as in (9.87). To this end we have to estimate I. For $\zeta = (\tau,\xi) \in \Gamma_+$ we have by (9.98): $\chi(\zeta) = 1$ and $\varkappa(\zeta) = 0$ so that $\Phi(\zeta) = (\tau, \xi + ia\rho(\zeta)\xi_0)$ and

$$|\exp\langle i\Phi(\zeta), z\rangle| = \exp(-a\rho(\zeta)\langle\xi_0,x\rangle), \qquad z = (t,x)\in\Omega,$$

as well as

(9.100) $$J(\zeta) = \det\Big(\delta_{jk} + ia(\partial_{\xi_j}\rho)(\zeta)\,(\xi_0)_k\Big)_{(j,k)\in\mathbb{N}_0\times\mathbb{N}_0} =$$
$$= 1 + ia\langle\nabla_\xi\rho, \xi_0\rangle = 1 + ia\langle\xi,\xi_0\rangle/\rho(\zeta).$$

It follows that $|J| \le (1+a^2)^{1/2} \le \sqrt{2}$ on Γ_+ and - in view of (9.93) - that

(9.101) $\qquad |I(z,\zeta)| \le c_0^{-1} \exp(-a\rho(\zeta)\langle\xi_0,x\rangle), \qquad z=(t,x)\in\Omega,\ \zeta\in\Gamma_+ .$

Since I is measurable and since $\exp(-a\langle\xi_0,x\rangle\rho)$ is absolutely integrable on Γ_+ we conclude that f is well-defined. Since for any compact subset K of Ω the number (9.90) is positive, Fubini's theorem shows that $E_+ = T_f$.

In order to see that f is differentiable we fix $\alpha = (j,\beta) \in \mathbb{N}_0\times\mathbb{N}_0^n$ and $\zeta = (\tau,\xi)\in\Gamma_+$. Since $|\tau| \le \rho(\zeta)^2$ and $|\xi| \le \rho(\zeta)$ we deduce that

$$|\Phi(\zeta)^\alpha| = |\tau|^j |(\xi + ia\rho(\zeta)\xi_0)^\beta| \le |\tau|^j (|\xi| + a\rho(\zeta))^{|\beta|} \le (1+a)^{|\beta|}\rho(\zeta)^{j+|\alpha|}.$$

Setting $c := a\langle \xi_0, x\rangle/2$ and making use of (9.91) and (9.101) we conclude that

$$|D_z^\alpha I(z,\zeta)| \le c_0^{-1}\, 2^{|\beta|} B_\alpha \exp(-c\rho(\zeta)), \qquad z\in\Omega,$$

where $B_\alpha := \sup\{\rho(\zeta)^{j+|\alpha|}\exp(-c\rho(\zeta))\,;\ \zeta\in\Gamma_+\}$. It follows by (9.92) that

$$B_\alpha \le \left(\frac{j+|\alpha|}{ce}\right)^{j+|\alpha|}.$$

Since

$$(9.102)\qquad \int_{\Gamma_+} e^{-c\rho} \le \int_{\mathbb{R}\times\mathbb{R}^n} \exp[-\rho(c^2\tau, c\xi)]\, d\tau\wedge d\xi = c^{-n-2}\int_{\mathbb{R}\times\mathbb{R}^n} e^{-\rho},$$

combining the preceding estimates we see that f is a C^∞ function such that

$$(9.103)\qquad D^\alpha f(z) = \int_{\Gamma_+} \tau^j\,(\xi + ia\rho(\zeta)\xi_0)^\beta I(z,\zeta)\, d\zeta, \qquad z\in\Omega,$$

and

$$(9.104)\qquad |f^{(\alpha)}(z)| \le C^{|\alpha|+1}\langle \xi_0, x\rangle^{-j-|\alpha|-n-2}(j+|\alpha|)^{j+|\alpha|}, \qquad z=(t,x)\in\Omega,$$

where C is a constant not depending on z and α. In particular, this shows that f belongs to $\Gamma^{(2,1,\dots,1)}(\Omega)$.

To obtain information on $WF_{p,A}(E_+)$ we let $\varepsilon>0$ be any constant such that Γ_ε (see (9.97)) is contained in Γ and define $h:\Omega \longrightarrow \mathbb{C}$ by

$$h(z) := \int_{\Gamma_+\setminus\Gamma_\varepsilon} I(z,\zeta)\, d\zeta.$$

By the same arguments as with f we see that h is a well-defined C^∞ function such that (9.103) holds with (f,Γ_+) replaced by $(h,\Gamma_+\setminus\Gamma_\varepsilon)$. To find more precise estimates for the derivatives of h we observe that if $\zeta=(\tau,\xi)\in\Gamma_+\setminus\Gamma_\varepsilon$ then $|\tau| \le (|\xi|/\varepsilon)^r \le (\rho(\zeta)/\varepsilon)^r$ so that

$$|\Phi(\zeta)^\alpha| = |\tau|^j |(\xi + ia\rho(\zeta)\xi_0)^\beta| \le \varepsilon^{-rj}(1+a)^{|\beta|}\rho(\zeta)^{rj+|\beta|}.$$

In combination with (9.91) and (9.101) this yields

$$|D_z^\alpha I(z,\zeta)| \le c_0^{-1}\varepsilon^{-rj}\, 2^{|\beta|} A_\alpha \exp(-c\rho(\zeta)), \qquad \zeta\in\Gamma_+\setminus\Gamma_\varepsilon,\ z\in\Omega,$$

where $c := a\langle\xi_0,x\rangle/2$ and $A_\alpha := \sup\{\rho(\zeta)^{rj+|\beta|}\exp(-c\rho(\zeta))\,;\ \zeta\in\Gamma_+\setminus\Gamma_\varepsilon\}$. By (9.92) we deduce that

$$A_\alpha \le \left(\frac{rj+|\beta|}{ce}\right)^{rj+|\beta|}.$$

Combining the estimates above with (9.102) we find a constant C not depending on z and α such that

$$|h^{(\alpha)}(z)| \leq C^{|\alpha|+1} \langle \xi_0, x\rangle^{-rj-|\beta|-n-2} (rj+|\beta|)^{rj+|\beta|}, \qquad z=(t,x)\in\Omega.$$

In particular, this shows that h belongs to $C^{p,A}(\Omega)$, i.e. $WF_{p,A}(T_h) = \emptyset$ and

$$WF_{p,A}(E_+) = WF_{p,A}(T_{f-h}).$$

Since

$$\langle T_{f-h}, \varphi\rangle = (2\pi)^{-n-1} \int_{\Gamma_+\cap\Gamma_\varepsilon} \frac{\check{\hat{\varphi}}}{q}(\Phi(\zeta))\, J(\zeta)\, d\zeta, \qquad \varphi\in C_0^\infty(\Omega),$$

since Γ_ε is quasihomogeneous, and since $\Gamma_\varepsilon\setminus(\Gamma_+\cap\Gamma_\varepsilon) = \Gamma_\varepsilon\cap\,]-R,R[\times\mathbb{R}^n$ is contained in the bounded set $]-R,R[\times K(0,\varepsilon R^{1/r})$ it follows from Proposition 9.32 that $WF_{p,A}(T_{f-h}) \subset \Omega\times\overline{\Gamma_\varepsilon}$. Hence $WF_{p,A}(E_+) \subset \Omega\times\overline{\Gamma_\varepsilon}$. Since this is valid for sufficiently small ε we conclude that

$$(9.105) \qquad WF_{p,A}(E_+) \subset \Omega\times\{(\nu,0)\}.$$

In view of (9.99) this implies that $WF_{p,A}(E_+)\cap WF_{p,A}(E_-)$ is empty. Hence in view of (9.67) and (9.66) we conclude that

$$(9.106) \qquad WF_{p,A}(E_\pm) = WF_{p,A}(E)\cap\Omega\times\{(\pm\nu,0)\}.$$

Now we suppose for a moment that (9.73) is proved. This implies that f does not belong to Γ^p at the points of $\{0\}\times]0,+\infty[\xi_0$ so that in view of (9.105) and (9.106) we have $(\{0\}\times]0,+\infty[\xi_0)\times\{(\nu,0)\} \subset WF_{p,A}(E_+) \subset WF_{p,A}(E)$. Since this is valid for any $\xi_0\in S^{n-1}$ it follows that $(\{0\}\times\mathbb{R}^n)\times\{(\nu,0)\}\subset WF_{p,A}(E)$. Using (9.106) and (9.67), again, we conclude that

$$WF_{p,A}(E_+) = (\{0\}\times X)\times\{(\nu,0)\}.$$

If $\nu\neq 0$ then replacing ν by $-\nu$ we deduce that $(\{0\}\times\mathbb{R}^n)\times\{(-\nu,0)\}\subset WF_{p,A}(E)$. If $\nu=0$ then this follows from the fact that the singular support of E is equal to $\{0\}\times\mathbb{R}^n$: indeed, since $f\in C^\infty(\Omega)$ we have $WF_p(E_+)=\emptyset$ so that by (9.106) we see that $WF_p(E|_\Omega) = WF_p(E_-)\subset\Omega\times\{(-\nu,0)\}$. Consequently, in both cases one derives from (9.106) that $WF_{p,A}(E_-) = (\{0\}\times X)\times\{(-\nu,0)\}$, and the proof of (9.72) is complete.

It remains to prove (9.73).

(d.3) Proof of Theorem 9.37. Part 2: Estimating the Derivatives of f from Below

Here, again, it is necessary to deform the contour of integration. However, this time the deformation is across poles so that additional terms arise. This procedure is based on the following lemma and the discussion following it.

Lemma 9.43. *There is a constant $A \in]0,+\infty[$ only depending on w such that for arbitrary $\tau \in \mathbb{R}$, $b \in]0,+\infty[$, and $\lambda \in \mathbb{R}$ the following holds: If*

(9.107) $\quad Q(\lambda) := \lambda^2 + iw\tau + b = 0$

then

(9.108) $\quad |Re\,\lambda| \le (2|w\tau|)^{1/2}$

and - provided that $u\tau \ge 0$ if $v = 0$ -

(9.109) $\quad |Im\,\lambda| \ge A\,(|\tau| + |Re\,\lambda|^2 + b)^{1/2}.$

<u>*Proof*</u>. We set $c := Re\,\lambda$ and $d := Im\,\lambda$. Then the equation (9.107) amounts to

(9.107)' $\quad$ (a) $d^2 - c^2 = u\tau + b$, and (b) $2cd = -v\tau$.

If $|c| \le 2|d|$ then by (9.107.b)': $c^2 \le 2|cd| = |v\tau|$, i.e. $|c| \le |v\tau|^{1/2}$. If $|c| \ge 2|d|$ then by (9.107.a)': $c^2 = d^2 - u\tau - b \le |u\tau| + c^2/4$ so that $|c| \le (\frac{4}{3}|u\tau|)^{1/2}$. Hence the proof of (9.108) is complete.

If $\tau = 0$ then by (9.107.a)' we have $d^2 = |\tau| + c^2 + b$, and (9.109) is trivially valid if A is chosen to be ≤ 1. So we may assume that $\tau \ne 0$.

We first suppose that $v = 0$. Since in this case by the assumption we have $u\tau \ge 0$ it follows from (9.107.a)' that $|d| \ge |c|$ so that $c = 0$ by (9.107.b)'. Hence (9.107.a)' yields: $d^2 = |u\tau| + b \ge A^2(|\tau| + c^2 + b)$ if $A := \min\{1, \sqrt{|u|}\}$.

Next we suppose that $u = 0$. Then $d^2 = c^2 + b$, in particular: $|c| \le |d|$. Hence by (9.107.b)' we have $d^2 \ge |cd| = |v\tau|/2$. Consequently, $d^2 \ge \frac{1}{2}(c^2 + b + |v\tau|/2)$, and (9.109) is valid if $A^2 \le \min\{1/2, |v|/4\}$.

Finally, we suppose that $v \ne 0 \ne u$ and choose $A > 0$ such that $8|w|(1+5|w|)A^2 < v^2$ and $A^2 \le \min\{|u|, 1/3\}$. If $b \ge 3|u\tau|$ then by (9.107.a)' we have

$$d^2 = c^2 + u\tau + b \geq c^2 + |u\tau| + \frac{b}{3} \geq A^2(|\tau| + c^2 + b).$$

If $b \leq 3|u\tau|$ then the assumption that (9.109) be false leads in combination with (9.108) and (9.107.b)' to

$$8|w|(1+5|w|)A^2\tau^2 = 4\cdot 2|w\tau|A^2(|\tau| + 2|w\tau| + 3|w\tau|) \geq$$

$$\geq 4c^2A^2(|\tau| + c^2 + b) \geq (2cd)^2 = v^2\tau^2$$

which in view of "$\tau \neq 0$" contradicts the choice of A. Hence (9.109) is valid. ■

We now formulate a few additional technical assumptions. First of all we fix a constant $A > 0$ such that the assertion of Lemma 9.43 holds and assume that $a < A$ where a is the constant in Lemma 9.42 appearing in the definition of η. Moreover, we fix $\varepsilon > 0$. In order to simplify computations we assume that

$$R \geq \left(\max\{8|w|, 1\}/\varepsilon^2\right)^{r/(2-r)}.$$

This implies that with the abbreviation $d := 1/r$ we have

$$(9.110) \qquad (2|w\tau|)^{1/2} \leq \frac{\varepsilon}{2}|\tau|^d, \qquad \tau \in \mathbb{R}\setminus]-R, R[.$$

Defining Q as in (9.107) we are going to derive

$$(9.111) \qquad |Q(\pm\varepsilon|\tau|^d + is)| \geq \frac{\varepsilon^2}{2}|\tau|^{2d}, \qquad s \in \mathbb{R},\ \tau \in \mathbb{R}\setminus]-R, R[.$$

In fact, if λ_0 is a zero of Q with non-negative real part then

$$|\varepsilon|\tau|^d + is + \lambda_0| \geq \varepsilon|\tau|^d + \operatorname{Re}\lambda_0 \geq \varepsilon|\tau|^d$$

and - by (9.108) and (9.110) -

$$|\varepsilon|\tau|^d + is - \lambda_0| \geq \varepsilon|\tau|^d - \operatorname{Re}\lambda_0 \geq \frac{\varepsilon}{2}|\tau|^d$$

so that the estimate (9.111) follows in view of $Q(\lambda) = (\lambda + \lambda_0)(\lambda - \lambda_0)$.

Now we choose orthonormal coordinates in $\mathbb{R}^n_\xi$ such that $\xi_0 = (1,0)$, write $\xi = (\xi_1, \xi') \in \mathbb{R}\times\mathbb{R}^{n-1}$ and suppose that

$$\Gamma = \{(\tau,\xi) \in \mathbb{R}\times\mathbb{R}^n;\ \nu\tau > 0,\ \max\{|\xi_1|, |\xi'|\} < \varepsilon|\tau|^d\}$$

(we prefer this choice of Γ to the set Γ_ε defined in (9.97) since it makes the application of Fubini's theorem easier). Let $x \in \mathbb{R}^n$ such that $x_1 > 0$. Moreover, let $\alpha = (j, \beta) \in \mathbb{N}_0 \times \mathbb{N}_0^n$. We write $\beta = (k, \gamma) \in \mathbb{N}_0 \times \mathbb{N}_0^{n-1}$. Combining (9.103), (9.87) and (9.100) one obtains

$$(2\pi)^{n+1}(D^{\alpha}f)(0,x) = \int_{\Gamma_+} \tau^j (\xi + ia\rho(\zeta)\xi_0)^{\beta} \exp\langle i\xi - a\rho(\zeta)\xi_0, x\rangle \frac{1+ia(\partial_{\xi_1}\rho)(\zeta)}{q(\Phi(\zeta))} d\zeta .$$

By our choice of Γ and Γ_+ and by Fubini's theorem this is equal to

$$(9.112) \qquad \int_R^{+\infty} (\nu\tau)^j \int_{K'(0,\varepsilon\tau^d)} I_0(\tau,\xi',x_1)\,(\xi')^{\gamma} \exp\langle i\xi',x'\rangle\, d\xi'\, d\tau$$

where

$$I_0(\tau,\xi',x_1) := \int_{-\varepsilon\tau^d}^{\varepsilon\tau^d} \frac{1+ia(\partial_{\xi_1}\rho)(\zeta)}{q_W(\Phi(\zeta))} (\xi_1 + ia\rho(\zeta))^k \exp[(i\xi_1 - a\rho(\zeta))x_1]\, d\xi_1$$

where we write $\zeta = (\tau,\xi_1,\xi')$ and where we denote by q_W the polynomial obtained out of q by replacing w with $W := \nu w$. Estimating (9.112) from below requires

Deforming the contour of integration in I_0. We fix $\tau \in [R,+\infty[$ and $\xi' \in K'(0,\varepsilon\tau^d)$. To identify $I_0(\tau,\xi',x_1)$ as a path integral we define a meromorphic function $H:\mathbb{C} \longrightarrow \mathbb{C}$ by

$$H(\lambda) := \frac{\lambda^k e^{ix_1\lambda}}{q_W(\tau,\lambda,\xi')}$$

and define a path $\gamma_0 : [-\varepsilon\tau^d, \varepsilon\tau^d] \longrightarrow \mathbb{C}$ by $s \mapsto s + ia\rho(\tau,s,\xi')$. Then

$$(9.113) \qquad I_0(\tau,\xi',x_1) = \int_{\gamma_0} H(\lambda)\, d\lambda .$$

To deform the contour of integration we set $b := a\rho(\tau,\varepsilon\tau^d,\xi')$, and for any $c > b$ define $\gamma_{\pm} : [b,c] \longrightarrow \mathbb{C}$ by $s \mapsto \pm\varepsilon\tau^d + is$ and $\delta : [-\varepsilon\tau^d, \varepsilon\tau^d] \longrightarrow \mathbb{C}$ by $s \mapsto s + ic$. Then $Z := \gamma_0 + \gamma_+ - \delta - \gamma_-$ is a simple cycle. If c is sufficiently large then in view of Lemma 9.43 and (9.110) precisely one of the two poles of H lies in the interior of Z, namely the one with positive imaginary part; we denote it by $i\lambda_0$ where

$$\lambda_0 := \sqrt{iW\tau + |\xi'|^2} ,$$

here $\sqrt{\cdot}$ denotes the holomorphic branch of the complex square root with positive real part and with $\mathbb{C} \setminus]-\infty,0]$ as its domain of definition (note that $iW\tau > 0$ if $v = 0$). Since the other pole of H lies in the exterior of Z an application of the residue theorem and taking the limit as $c \to +\infty$ yields in view of (9.113):

$$(9.114) \qquad I_0(\tau,\xi',x_1) = h_-(\tau,\xi',x_1) - h_+(\tau,\xi',x_1) + 2\pi i \operatorname{res}(i\lambda_0; H)$$

where

$$h_{\pm}(\tau,\xi',x_1) := i\int_b^{+\infty} H(\pm\varepsilon\tau^d + is)\, ds .$$

In view of $q_W(\tau,\lambda,\xi') = (\lambda - i\lambda_o)(\lambda + i\lambda_o)$ the residue of H at $i\lambda_o$ is equal to

$$\frac{1}{2}(i\lambda_o)^{k-1}\exp(-x_1\lambda_o).$$

By inserting (9.114) we decompose (9.112) into a sum $J_-(x) - J_+(x) + \pi J_1(x)$ and are going to estimate each term separately. The third one will give the main contribution. So in order to obtain an estimate of (9.112) from below we have to estimate $J_\pm(x)$ from above whereas $J_1(x)$ is to be estimated from below.

Estimating $J_\pm(x)$ from above. If $s \geq b$ then $\varepsilon\tau^d \leq s/a$ so that $|\pm\varepsilon\tau^d + is| \leq (a^{-2}+1)^{1/2}s$. In view of (9.111) this gives

$$|H(\pm\varepsilon\tau^d + is)| \leq 2\varepsilon^{-2}(a^{-2}+1)^{k/2}\tau^{-2d}s^k\exp(-x_1 s).$$

Hence $|J_\pm(x)|$ is not larger than

$$2|K'(0,1)|\varepsilon^{|\gamma|+n-3}(a^{-2}+1)^{k/2}\int_R \tau^{j+(n-3+|\gamma|)d}\int_{a\varepsilon\tau^d}^{+\infty} s^k\exp(-x_1 s)\,ds\,d\tau.$$

Substituting first $x_1 s = 3t$, i.e. $ds = 3dt/x_1$, and then $\frac{a}{3}\varepsilon x_1\tau^d = \sigma$, i.e.

$$\tau = \left(\frac{3}{a\varepsilon x_1}\sigma\right)^r \quad\text{and}\quad d\tau = \left(\frac{3}{a\varepsilon x_1}\right)^r r\sigma^{r-1}d\sigma$$

one sees that the double integral above is equal to

$$r\left(\frac{3}{a\varepsilon x_1}\right)^{rj+n-3+|\gamma|+r}\left(\frac{3}{x_1}\right)^{k+1}\int_{a\varepsilon x_1 R^d/3}^{+\infty}\sigma^{rj+n-3+|\gamma|+r-1}\int_\sigma^{+\infty}t^k e^{-3t}\,dt\,d\sigma.$$

Here, in turn, the double integral is majorized by

$$\int_0^{+\infty}e^{-\sigma}d\sigma\int_0^{+\infty}t^{rj+n-4+|\beta|+r}e^{-2t}dt$$

provided that $rj+|\gamma|+n+r \geq 4$ which is the case if $j \geq 3-n$. Combining (9.92) with $\int_0^{+\infty}e^{-t}dt = 1$ one obtains that

$$\int_0^{+\infty}t^\ell e^{-2t}dt \leq (\ell/e)^\ell, \qquad \ell\in]0,+\infty[. \tag{9.115}$$

Consequently, the second integral above is not larger than

$$(rj+|\beta|+n+r-4)^{rj+|\beta|+n+r-4}.$$

Putting everything together one finds a constant C_1 only depending on ε, n, and a such that

$$|J_\pm(x)| \leq C_1^{|\alpha|+1}x_1^{-rj-|\beta|-n-r+2}|\alpha|^{rj+|\beta|} \qquad \text{if } j\geq 3-n. \tag{9.116}$$

Estimating $J_1(x)$ from below: the case "n=1". Substituting $\sigma = \sqrt{\tau}$, i.e. $\tau = \sigma^2$ and $d\tau = 2\sigma d\sigma$ we see that

$$\nu^j J_1(x) = i^k \int_R^{+\infty} \tau^{j+(k-1)/2} \exp(-x_1\sqrt{\tau}\sqrt{iW})\, d\tau\, (\sqrt{iW})^{k-1} =$$

$$= 2i^k(\sqrt{iW})^{k-1} \int_{\sqrt{R}}^{+\infty} \sigma^{2j+k} \exp(-x_1\sigma\sqrt{iW})\, d\sigma .$$

From this one obtains

$$|J_1(x)| \geq 2|w|^{(k-1)/2}\left(\frac{(j+|\alpha|)!}{(x_1\sqrt{|w|})^{j+|\alpha|+1}} - R^{|\alpha|+1}\right) \text{ if } j+|\alpha| \geq x_1\sqrt{|w|R} \tag{9.117}$$

by estimating the integral above with the help of the following lemma the full statement of which is required for the case "$n \geq 2$".

Lemma 9.44. *Let $c \in [0,+\infty[$, $k \in \mathbb{N}_0$, and $\lambda \in \mathbb{C}$ such that $Re\lambda > 0$. Then*

$$\left|\int_c^{+\infty} t^k e^{-\lambda t}\, dt - \frac{k!}{\lambda^{k+1}}\right| \leq \begin{cases} c\left(\frac{k}{e Re\lambda}\right)^k & \text{if } c Re\lambda \geq k \\ c^{k+1}\exp(-c Re\lambda) & \text{if } c Re\lambda \leq k \end{cases} .$$

Note that $c\left(\frac{k}{e Re\lambda}\right)^k \leq c^{k+1} e^{-k}$ if $c Re\lambda \geq k$.

Proof. Partial integration leads to

$$\int_0^{+\infty} t^{k+1} e^{-\lambda t}\, dt = \frac{k+1}{\lambda}\int_0^{+\infty} t^k e^{-\lambda t}\, dt$$

so that by induction one obtains that

$$\int_0^{+\infty} t^k e^{-\lambda t}\, dt = \frac{k!}{\lambda^{k+1}} .$$

Hence the term to be estimated is equal to

$$\int_0^c t^k e^{-\lambda t}\, dt$$

which is not larger than cS where $S := \sup\{g(t)\,;\, t \in [0,c]\}$ with $g(t) := t^k e^{-Re\lambda t}$. Since $g'(t) = 0$ if and only if $t = t_0 := \frac{k}{Re\lambda}$ the function g has a unique maximum at t_0. If $t_0 \geq c$ then $g|_{[0,c]}$ is increasing and $S = g(c)$; if $t_0 \leq c$ then $S = g(t_0)$. ■

Estimating $J_1(x)$ from below: the case "$n \geq 2$". We set

$$M(\tau,x) := \int_{K'(0,\varepsilon\tau^d)} (\xi')^\gamma (iW\tau + |\xi'|^2)^{(k-1)/2} \exp[\langle i\xi',x'\rangle - x_1(iW\tau + |\xi'|^2)^{1/2}]\, d\xi'$$

Introducing polar coordinates we find

$$M(\tau,x) = \int_{S^{n-2}} \vartheta^{\gamma} L(\tau,\vartheta,x)\, d\vartheta$$

where

$$L(\tau,\vartheta,x) := \int_0^{\varepsilon\tau^d} \rho^{|\gamma|+n-2} (iW\tau+\rho^2)^{(k-1)/2} \exp[i\rho\langle\vartheta,x'\rangle - x_1 (iW\tau+\rho^2)^{1/2}]\, d\rho .$$

Substituting $\rho = \sqrt{\tau s}$, i.e. $\rho^2 = \tau s$ and $\rho\, d\rho = \frac{\tau}{2} ds$ we obtain

$$(iW\tau+\rho^2)^{1/2} = \sqrt{\tau}\, \lambda(s) \qquad \text{where } \lambda(s) := (iW+s)^{1/2},$$

$$i\rho\langle\vartheta,x'\rangle - x_1 (iW\tau+\rho^2)^{1/2} = -\sqrt{\tau}\, x_1 \Lambda(s,\langle\vartheta,y\rangle)$$

where $y := x'/x_1$ and $\Lambda(s,t) := \lambda(s) - i\sqrt{s}\, t$, and

$$L(\tau,\vartheta,x) = \frac{1}{2}\tau^{(|\beta|+n-2)/2} \int_0^{\varepsilon^2\tau^{2d-1}} s^m \lambda(s)^{k-1} \exp[-\sqrt{\tau}\, x_1 \Lambda(s,\langle\vartheta,y\rangle)]\, ds$$

where $m := (|\gamma|+n-3)/2$. Substituting $\tau = \sigma^2$, i.e. $\frac{1}{2} d\tau = \sigma\, d\sigma$, leads to

$$\int_R^{+\infty} \tau^j L(\tau,\vartheta,x)\, d\tau = \int_{\sqrt{R}}^{+\infty} \sigma^{2j+|\beta|+n-1} \int_0^{g(\sigma)} s^m \lambda(s)^{k-1} \exp[-\sigma x_1 \Lambda(s,\langle\vartheta,y\rangle)]\, ds\, d\sigma$$

where $g(\sigma) := \varepsilon^2 \sigma^{4d-2}$. By Fubini's theorem this is equal to

$$\int_0^{+\infty} s^m \lambda(s)^{k-1} N(s,x_1,\langle\vartheta,y\rangle)\, ds$$

where

$$N(s,x_1,t) := \int_{h(s)}^{+\infty} \sigma^{j+|\alpha|+n-1} \exp[-\sigma x_1 \Lambda(s,t)]\, d\sigma$$

with

$$h(s) := \max\{\sqrt{R}, g^{-1}(s)\} \qquad \text{where} \qquad g^{-1}(s) = (\varepsilon^{-2} s)^{1/(4d-2)}.$$

Note that

$$(9.118) \qquad \forall s \in]0,+\infty[: \quad h(s) = \sqrt{R} \iff s \le T := g(\sqrt{R}) = \varepsilon^2 R^{2d-1}.$$

Note, moreover, that by the choice of R we have $T \ge 1$.

We are going to apply Lemma 9.44, making use of both of its cases. To distinguish between them we set $h_1(s) := h(s)\,\mathrm{Re}\,\lambda(s)$ and fix $\hat{s} \ge T$ such that h_1 is strictly increasing on $[\hat{s},+\infty[$ and such that $h_1([0,\hat{s}[) \cap h_1([\hat{s},+\infty[)$ is empty. We suppose now that

$$(9.119) \qquad j + |\alpha| + n - 1 \ge x_1 h_1(\hat{s}).$$

Then we can choose $s_\alpha \in [\hat{s}, +\infty[$ satisfying $j+|\alpha|+n-1 = x_1 h_1(s_\alpha)$. It follows that

$$\forall s \in]0,+\infty[: \; x_1 h_1(s) \le j+|\alpha|+n-1 \iff s \le s_\alpha .$$

By Lemma 9.44 - applied to $c = h(s)$ and $\lambda = x_1 \Lambda(s, \langle \vartheta, y\rangle)$ - we conclude that

$$(9.120) \qquad |J_1(x)| = \Big| \int_R \tau^j M(\tau,x)\, d\tau \Big| = \Big| \int_{S^{n-2}} \vartheta^\gamma \int_0^{+\infty} s^m \lambda(s)^{k-1} N(s, x_1, \langle\vartheta, y\rangle)\, ds\, d\vartheta \Big| \ge$$

$$\ge (j+|\alpha|+n-1)!\; x_1^{-j-|\alpha|-n} \, |K_1(y)| - K_2(x_1) - K_3(x_1)$$

where

$$K_1(y) := \int_{S^{n-2}} \vartheta^\gamma \int_0^{+\infty} \phi(s, \langle\vartheta, y\rangle)\, ds\, d\vartheta \quad \text{with} \quad \phi(s,t) := s^m \lambda(s)^{k-1} \Lambda(s,t)^{-j-|\alpha|-n},$$

$$K_2(x_1) := |S^{n-2}| \Big(\frac{j+|\alpha|+n-1}{e x_1}\Big)^{j+|\alpha|+n-1} \int_{s_\alpha}^{+\infty} \frac{s^m |\lambda(s)|^{k-1} h(s)}{[\mathrm{Re}\,\lambda(s)]^{j+|\alpha|+n-1}}\, ds ,$$

and

$$K_3(x_1) := |S^{n-2}| \int_0^{s_\alpha} \psi(x_1, s)\, ds$$

with

$$\psi(x_1, s) := s^m |\lambda(s)|^{k-1} h(s)^{j+|\alpha|+n} \exp[-h(s) x_1 \mathrm{Re}\,\lambda(s)] .$$

Estimating $K_2(x_1)$ from above. With the abbreviation $B := (|w|+1)^{1/2}$ we have

$$(9.121) \qquad \mathrm{Re}\,\lambda(s) \le |\lambda(s)| \le B \max\{\sqrt{s}, 1\} , \qquad s \in]0,+\infty[.$$

By Lemma 9.43 we find a constant $A > 0$ such that

$$(9.122) \qquad \mathrm{Re}\,\lambda(s) \ge A \sqrt{s} , \qquad s \in]0,+\infty[.$$

Hence, making use of $\mathrm{Re}\,\lambda(s)/|\lambda(s)| \le 1$ and of (9.118) we deduce that

$$\int_{s_\alpha}^{+\infty} \frac{s^m |\lambda(s)|^{k-1} h(s)}{[\mathrm{Re}\,\lambda(s)]^{j+|\alpha|+n-1}}\, ds \le \varepsilon^{-1/(2d-1)} B^k A^{-j-|\alpha|-n} \int_{s_\alpha}^{+\infty} s^\varkappa\, ds$$

where

$$\varkappa := m + \frac{1}{2}(k-j-|\alpha|-n) + \frac{1}{4d-2} = \frac{1}{2}(|\gamma|+n-3+k-j-|\alpha|-n) + \frac{1}{4d-2} = -j + j_0 - 1$$

with

$$j_0 := \frac{1}{2}\Big(\frac{1}{2d-1} - 3\Big) + 1 = \frac{1-d}{2d-1} = \frac{r-1}{2-r} .$$

Consequently, the last integral is finite and equals $s_\alpha^{-j+j_0}/(j-j_0)$ if $j > j_0$. By the definition of s_α and h_1, by (9.118) and by (9.121) we have

$$j+|\alpha|+n-1 \le B\,x_1\,\varepsilon^{-1/(2d-1)}\,s_\alpha^{\ell}$$

where

(9.123) $$\ell := \frac{1}{4d-2}+\frac{1}{2} = \frac{1+2d-1}{4d-2} = \frac{d}{2d-1} = \frac{1}{2-r}\,.$$

Since $\frac{2-r}{2d-1} = \frac{2-r}{2-r}\,r = r$ this implies

$$1/s_\alpha \le B^{2-r}\,\varepsilon^{-r}\,x_1^{2-r}\,(j+|\alpha|+n-1)^{r-2}.$$

Putting everything together we find a constant C_2 not depending on α and x_1 such that

(9.124) $$K_2(x_1) \le C_2^{|\alpha|+1}\,x_1^{-rj-|\beta|-n-r+2}\,|\alpha|^{rj+|\beta|}\,, \qquad j>j_0\,.$$

Estimating $K_3(x_1)$ from above. By (9.122), by the definition of h and by (9.123) we have

$$\exp[-h(s)\,x_1\,\mathrm{Re}\,\lambda(s)] \le \exp(-Q\,x_1\,s^{1/(2-r)})\,, \qquad s>0\,,$$

where $Q := A\,\varepsilon^{r/(r-2)}$. By (9.121), (9.122), and (9.118) we conclude that

$$\int_0^1 \psi(x_1,s)\,ds \le \frac{B^k}{A}\,(\sqrt{R}\,)^{j+|\alpha|+n}\int_0^1 s^{m-1/2}\exp(-Q\,x_1\,s^{1/(2-r)})\,ds\,.$$

Here the integral on the left-hand side is not larger than $\int_0^1 ds/\sqrt{s} = 2$. Again by (9.121), (9.122), and (9.118) we deduce that

$$\int_1^T \psi(x_1,s)\,ds \le \frac{B^k}{A}\,(\sqrt{R}\,)^{j+|\alpha|+n}\int_0^T s^{m+(k-1)/2}\exp(-Q\,x_1\,s^{1/(2-r)})\,ds\,.$$

Here the integral on the right-hand side is majorized by

$$T^{m+k/2}\int_0^T ds/\sqrt{s} = 2\,T^{m+(k+1)/2}\,.$$

Finally, the conditions (9.121), (9.122), and (9.118) also imply

$$\int_T^{s_\alpha} \psi(x_1,s)\,ds \le \frac{B^k}{A}\,\varepsilon^{-(j+|\alpha|+n)/(2d-1)}\int_0^{+\infty} s^{\ell}\exp(-Q\,x_1\,s^{1/(2-r)})\,ds$$

where $\ell := m-\frac{1}{2}+\frac{k}{2}+\frac{j+|\alpha|+n}{4d-2}$. Substituting $Q\,x_1\,s^{1/(2-r)} = 2t$, i.e. $s = \left(\frac{2t}{Qx_1}\right)^{2-r}$ and $ds = (Qx_1/2)^{r-2}\,(2-r)\,t^{1-r}\,dt$, shows that the last integral is equal to

$$(2-r)\;(Qx_1/2)^{(r-2)(\ell+1)}\int_0^{+\infty} t^{(2-r)\ell+1-r}\exp(-2t)\,dt\,.$$

Since

$$(2-r)\ell+1-r = \frac{2-r}{2}(|\gamma|+n-3-1+k+\frac{r}{2-r}(j+|\alpha|+n)+2]-1 =$$

$$= |\beta|+n-3+\frac{r}{2}(j+|\alpha|+n-|\beta|-n+2) = rj+|\beta|+n+r-3$$

it follows by (9.115) that the integral above is not larger than

$$[(rj+|\beta|+n+r-3)/e]^{rj+|\beta|+n+r-3}.$$

Putting everything together one finds a constant C_3 not depending on α and x_1 such that

$$K_3(x_1) \le C_3^{|\alpha|+1}[x_1^{-rj-|\beta|-n-r+2}|\alpha|^{rj+|\beta|}+1]. \tag{9.125}$$

<u>*Estimating $|K_1(y)|$ from below.*</u> Since

$$m+\frac{k-1}{2}-\frac{j+|\alpha|+n}{2} = \frac{1}{2}(|\gamma|+n-3+k-1-j-|\alpha|-n) = -j-2$$

we obtain

$$\phi(s,b) = (iW)^{-j}s^{-2}G(iW/s,b)$$

where

$$G(\lambda,b) := \lambda^j \frac{(\sqrt{\lambda+1})^{k-1}}{(\sqrt{\lambda+1}-ib)^{j+|\alpha|+n}}, \qquad \lambda \in U := \mathbb{C}\setminus]-\infty,-1],\ b \in \mathbb{R}.$$

Substituting $t=|w|/s$, i.e. $s=|w|/t$ and $s^{-2}ds = dt/|w|$ we obtain

$$\int_0^{+\infty} \phi(s,b)\,ds = (iW)^{-j-1}\int_0^{+\infty} G(tiZ,b)\,iZ\,dt$$

where $Z := W/|w|$. Note that $G(\cdot,b)$ is holomorphic on U. Hence we may deform the contour of integration. To do this we fix $c>0$ and define $\gamma_0:[0,c]\longrightarrow U$ by $\gamma_0(t):=tiZ$, $\gamma_1:[0,c]\longrightarrow U$ by $\gamma_1(t):=t$, and $\delta_c:[0,\vartheta]\longrightarrow U$ by $\delta_c(t):=ce^{it}$ where $\vartheta\in]-\pi,+\pi[$ is defined by the condition $iZ=e^{i\vartheta}$. We set $\iota := \operatorname{sign}\vartheta$. Then $\gamma_0-\iota\delta_c-\gamma_1$ is a simple cycle in U so that by Cauchy's theorem it follows hat

$$\int_{\gamma_0} G(\lambda,b)\,d\lambda = \int_{\gamma_1} G(\lambda,b)\,d\lambda + \iota\int_{\delta_c} G(\lambda,b)\,d\lambda.$$

If c is sufficiently large we have $|(ce^{it}+1)^{1/2}-ib| \ge \sqrt{c}/2$ and $|(ce^{it}+1)^{1/2}| \le 2\sqrt{c}$, and hence

$$|G(\delta_c(t),b)| \le 2^{k+1+j+|\alpha|+n}c^{\ell}$$

where

$$\ell := j+\frac{k-1}{2}-\frac{j+|\alpha|+n}{2} = \frac{1}{2}(k-1-|\beta|-n) = -\frac{1}{2}(|\gamma|+n+1) \le -\frac{3}{2}.$$

Since $|\delta_c'| \equiv c$ this implies that

$$\lim_{c\to\infty} \int_{\delta_c} G(\lambda,b)\,d\lambda = 0 .$$

Consequently,

$$\int_0^{+\infty} G(tiZ,b)\,iZ\,dt = \int_0^{+\infty} G(t,b)\,dt .$$

Substituting $t = s^2-1$, i.e. $s = \sqrt{t+1}$ and $dt = 2s\,ds$ one obtains

$$\int_0^{+\infty} G(t,b)\,dt = 2\int_1^{+\infty} \frac{(s^2-1)^j s^k}{(s-ib)^{j+|\alpha|+n}}\,ds . \tag{9.126}$$

Estimating the right-hand side of (9.126) from below may be difficult in general, but it is easy when $b=0$:

$$\int_0^{+\infty} G(t,0)\,dt \ge 2\int_{\sqrt{2}}^{+\infty} (1-s^{-2})^j s^{-|\gamma|-n}\,ds \ge 2^{-j+1}\int_{\sqrt{2}}^{+\infty} s^{-|\gamma|-n}\,ds =$$

$$= 2^{-j+1}\frac{1}{|\gamma|+n-1}(\sqrt{2})^{-|\gamma|-n+1} .$$

Finally, we have to deal with the integration over S^{n-2}. Using polar coordinates and setting $c := \sqrt{n-1}$ we obtain

$$\int_{S^{n-2}} \vartheta^\gamma\,d\vartheta\,\frac{c^{|\gamma|+n-1}}{|\gamma|+n-1} = \int_0^c \int_{S^{n-2}} (t\vartheta)^\gamma t^{n-2}\,d\vartheta\,dt = \int_{K'(0,c)} (x')^\gamma\,dx' .$$

If $\gamma \in (2\mathbb{N}_0)^{n-1}$ then $(x')^\gamma$ is non-negative, and the integral on the left-hand side is not smaller than

$$\int_{[-1,1]^{n-1}} (x')^\gamma\,dx' = \prod_{i=1}^{n-1} \frac{2}{\gamma_i+1} .$$

Putting everything together one obtains a positive constant C_4 not depending on α and x_1 such that

$$|K_1(0)| \ge C_4^{|\alpha|+1} \qquad \text{if } \gamma\in(2\mathbb{N}_0)^{n-1} . \tag{9.127}$$

<u>*Combining all the estimates.*</u> Suppose that $x'=0$, i.e. $y=0$. Note that

$$(j+|\alpha|+n-1)! \ge (|\alpha|/e)^{j+|\alpha|} .$$

Combining this with (9.117) and (9.116) if $n=1$ and with (9.120), (9.127), (9.124), (9.125), and (9.116) if $n\ge 2$ we find positive constants C, D, and ω not depending on α and x_1 such that with the abbreviation

$$c(s,\alpha) := D^{|\alpha|}\left(|\alpha|^{(r-2)j} s^{(2-r)(j+1)} + |\alpha|^{-j-|\alpha|} s^{j+|\alpha|+n}\right)$$

we have

$$|f^{(\alpha)}(0,x)|\, C^{-|\alpha|-1}(x_1)^{j+|\alpha|+n}\, |\alpha|^{-j-|\alpha|} \geq 3 - c(x_1,\alpha) \tag{9.128}$$

provided that $j+|\alpha|+n-1 \geq 2\omega x_1$, $j > \max\{2-n, j_0\}$ and $\gamma \in (2\mathbb{N}_0)^{n-1}$. The right-hand side of (9.128) is not smaller than 1 if x_1 stays in a fixed finite interval and j is sufficiently large and if $|\alpha|/j$ does not grow too fast. More precisely, setting $b := 2-r$ and fixing $\rho \geq 1$ we have

$$c(s,\alpha) = D^{|\alpha|}\, |\alpha|^{-bj}\, s^{b(j+1)}\, [1 + (s/|\alpha|)^{rj+|\beta|}\, s^{n-b}] \leq$$

$$\leq 2\,[D^{|\alpha|/b}\, |\alpha|^{-j}\, s^{j+1}]^{b} \qquad \text{if} \quad 0 \leq s \leq S_\rho(\alpha)$$

where $S_\rho(\alpha) := \min\{|\alpha|/\rho, \rho^{|\alpha|/(n-b)}\}$. Note that the term in square brackets, being smaller than

$$B^{|\alpha|}\,(s/|\alpha|)^{j+1} \qquad \text{where} \qquad B := e\,D^{1/b},$$

is not larger than 1 if $\rho^{j+1} \geq B^{|\alpha|}$ and $|\alpha| \geq \rho s$. Hence, observing that $S_\rho(\alpha) = |\alpha|/\rho$ if $\rho \geq e^{n-b}$ and setting $B_0 := \max\{e^{n-b}, \omega\}$ we arrive at

Theorem 9.37'. *There are positive constants C, B, B_0, and j_1 such that*

$$|f^{(\alpha)}(0,(s,0))| \geq C^{|\alpha|}\, s^{-j-|\alpha|-n}\, |\alpha|^{j+|\alpha|} \tag{9.129}$$

for every $s \in\,]0,+\infty[$ and every $\alpha = (j,k,\gamma) \in \mathbb{N}_0 \times \mathbb{N}_0 \times (2\mathbb{N}_0)^{n-1}$ satisfying the following conditions:

$$j \geq j_1 \qquad \text{and} \qquad |\alpha|/s \geq \max\{B_0, B^{|\alpha|/j}\}. \quad \blacksquare$$

Note that the assertion of Theorem 9.37' is consistent with (9.104), indeed.

Since (9.73) immediately follows the *proof of Theorem 9.37* is complete. ■

References

[1] Arnold, V.I.: *Geometrical Methods in the Theory of Ordinary Differential Equations.* Grundlehren der Math. Wiss. 250. Springer-Verlag, Berlin, Heidelberg, New York, Tokio 1983.

[2] Bernstein, I.N.: *The Analytic Continuation of Generalized Functions with Respect to a Parameter.* Functional Anal. Appl. 6, 273-285 (1972).

[3] Björk, J.E.: *Rings of Differential Operators.* North-Holland Publ. Co. Math. Library Vol. 21, Amsterdam, London 1979.

[4] De Wilde, M.: *Closed Graph Theorems and Webbed Spaces.* Pitman Research Notes in Math. 19, London, San Francisco, Melbourne 1978.

[5] Floret, K. und J. Wloka: *Einführung in die Theorie der lokalkonvexen Räume.* Springer Lecture Notes in Math. 56, Berlin, Heidelberg, New York 1968.

[6] Gårding, L.: *Transformation de Fourier des distributions homogènes.* Bull. Soc. math. France, 89, 381-428 (1961).

[7] von Grudzinski, O.: *On the Standard Fundamental Solutions of the Schrödinger and of the Heat Operator.* Preprint Univ. Kiel 1986

[8] Hörmander, L.: *On the Division of Distributions by Polynomials.* Ark. Mat. 3, 555-568 (1958).

[9] Hörmander, L.: *An Introduction to Complex Analysis in Several Variables.* 2nd ed., North-Holland Publ. Co., Amsterdam, London 1973.

[10] Hörmander, L.: *On the Existence and the Regularity of Solutions of Linear Pseudodifferential Equations.* L' Ens. Math. 17, 99-163 (1971).

[11] Hörmander, L.: *The Analysis of Linear Partial Differential Operators. Vol. I.* Grundlehren der Math. Wiss. 256. Springer-Verlag, Berlin, Heidelberg, New York, Tokio 1983.

[12] Lascar, R.: *Propagation des singularités des solutions d' équations pseudo-différentielles quasi homogènes.* Ann. Inst. Fourier (Grenoble) 27, 79-123 (1977).

[13] Liess, L. and L. Rodino: *Inhomogeneous Gevrey Classes and Related Pseudo-differential Operators.* Boll. Un. Mat. Ital. C (6) 3, no. 1, 233-323 (1984).

[14] Lojasiewicz, S.: *Sur le problème de division.* Studia Math. 18, 87-136 (1959).

[15] Ortner, N.: *Regularisierte Faltung von Distributionen. Teil 2: Eine Tabelle von Fundamentallösungen.* ZAMP 31, 155-173 (1980).

[16] Rodino, L.: *On the Gevrey Wave Front Set of the Solutions of a Quasielliptic Degenerate Equation.* Conference on linear partial and pseudodifferential operators (Torino 1982). Rend. Sem. Mat. Univ. Politec. Torino 1983, special issue, 221-234 (1984).

[17] Tougeron, J.C.: *Idéaux de fonctions différentiables.* Ergebnisse der Math. 71. Springer-Verlag, Berlin, Heidelberg, New York 1972.

Index